A Bridge to Advanced Mathematics

A Bridge to Advanced Mathematics

DENNIS SENTILLES
University of Missouri

DOVER PUBLICATIONS, INC.
Mineola, New York

Bibliographical Note

This Dover edition, first published in 2011, is an unabridged republication of the work originally published by The Williams & Wilkins Company, Baltimore, in 1975. A new preface by the author has been specially prepared for this edition.

Library of Congress Cataloging-in-Publication Data

Sentilles, Dennis.
A bridge to advanced mathematics / Dennis Sentilles.
p. cm.
Originally published: Baltimore : Wilkins, 1975.
Summary: "This helpful "bridge" book offers students the foundations they need to understand advanced mathematics, spanning the gap between practically oriented and theoretically oriented courses. Part 1 provides the most basic tools, examples, and motivation for the manner, method, and material of higher mathematics. Part 2 covers sets, relations, functions, infinite sets, and mathematical proofs and reasoning. 1975 edition"— Provided by publisher.
Includes bibliographical references and index.
ISBN-13: 978-0-486-48219-4 (pbk.)
ISBN-10: 0-486-48219-7 (pbk.)
1. Proof theory. 2. Arithmetic—Foundations. I. Title.

QA9.S44 2011
511.3'6—dc22

2011002266

Manufactured in the United States by Courier Corporation
48219701
www.doverpublications.com

PREFACE TO THE SECOND (DOVER) EDITION

It is surprising that this little book has survived all these years to reappear now thanks to the interest of Dover Publications. It was first written at almost the very moment when the next generation of prospective mathematics students became the first generation to turn instead toward computer science, a fate that might well have appealed to its author, had he been born but a very few years later. There seems not that much difference in satisfaction to be found between devising a beautiful proof or devising a beautiful program, except for greater immediacy of reward in the latter. These endeavors both demand rigor of mind applied to abstract conception that is largely existent at first only in the imagination. Experience in one is surely valuable to the other. But that seems the extent of it.

A fundamental mathematical result endures forever, the Pythagorean Theorem being perhaps the oldest example. A good mathematical result is a hard, solid thing, established via a demanding methodology, even as it exists only in the mind, valid wherever and whenever its underlying assumptions occur. That is what this book is about and I hope still serves: basic foundations and an introduction to the rigor of the fundamental methodology. Such a study was the great pleasure of my undergraduate studies.

Paul Cohen's work on the Continuum Hypothesis made it just barely into the original pages of this book. Since their writing, the 4-Color Problem, Fermat's Last Theorem, and the Poincaré Conjecture have all fallen to the mathematician's mind. Other than these advances, almost all of its contents serve as currently now as it did then.

To pull together what so many superior minds had done, so that another mind might immerse in it too—it was a labor of love, a happy moment that I was privileged to have had all those years ago.

Dennis Sentilles
Professor Emeritus of Mathematics
The University of Missouri

PREFACE

This text is what its title claims. A bridge to be crossed by the student of mathematics in his (usually) third semester of the university curriculum. Its intent is to span the gap that exists between a practically oriented calculus sequence and the theoretically oriented courses in algebra, analysis and other areas which typically follow in his third and fourth years.

I came to this course and the writing of this text by way of two experiences. The first was that I had taken such a course in the "foundations of mathematics" as an undergraduate and found it to be the most interesting and one of the most valuable of my studies. The second was my experience in attempting to teach advanced calculus to mathematics students who had no real idea of what post-18th century mathematics was like, in neither its techniques nor interests. I thought this was a disservice to students who were about to commit their third or even fourth year to the subject, including both those who would go on and those who needed to realize that core mathematics was not really what interested them. Hence I began teaching a course intended to bridge the gap between the "procedural" courses of the first year and "conceptual" courses of the later years, to be taken immediately following or concurrently with the last semester of basic calculus.

My colleagues and I both felt that the students leaving such a course should have a basic acquaintance with proof in mathematics, an awareness of what modern mathematics was like, and a sure knowledge of basic logic, set theory, relations, functions, and the real and natural number systems. I also thought that the students might profit from and enjoy some discussion and study of how mathematics got to be the way it is, an appearance quite different from what they imagined. Hence one finds in this text a few sections on the history and development of mathematics and the reasons for and uses of axiom systems. These ideas framed the basic course; it has considerably evolved from this earliest conception.

I found it necessary to devote more time to logic than I wished, one week to basic symbolic logic and one week to the uses of quantifiers and the actual meaning of mathematical statements. In response to the suggestions of several anonymous and very helpful reviewers, Chapter 1 on logic has been considerably expanded and includes far more than I actually cover in class. I find that Sections 1.1 through 1.9, 1.12 and 1.13 are quite enough for the student at this initial stage. Section 1.14 can be referred to for further explanation of details of logic encountered *in use* in later chapters.

Chapter 2 is devoted to a bit of history and axiomatics (Sections 2.1 through 2.6 to which I devote a week or so of the course), followed by an introduction to the most basic entities, the real and natural number systems. Peano's axioms and an axiom system for the reals are given. Mathematical induction and the completeness axiom are thoroughly discussed. These sections are followed by a consideration of intuitive infinity (2.9), in part historically oriented. A follow-up section involves practice in "reasoning at infinity" centered around a discussion of infinite decimals, a mildly familiar concept, but one that clearly involves the key idea of Cauchy-Weierstrass analysis, that analysis must be as rigorous as geometry and ultimately founded upon arithmetical operations of finite extent. This material is later used in the development of the theory of infinite sets in Chapter 6. I caution the instructor that (1) the students find much of this material on infinite decimals rather difficult, but (2) the existence and uniqueness (modulo the usual restrictions) of infinite decimal expansions, in base 2 as well as base 10, are needed to do several problems in Chapter 6 which definitely serve to illuminate the discussion therein of infinite sets. The instructor should, I feel, minimize the difficulties in this section (2.9) and allow for greater discussion of these when they again arise in Chapter 6 and are likely to be more meaningful to the student.

Part 1 of the test, which as a whole can be described as providing the most basic tools, examples and motivation for the manner, method and concern of advanced mathematics, closes with Chapter 2. One then comes to Part 2 of the text. This format is no gimmick. Whereas Part 1 was essentially a listing and very brief study of various topics somewhat but only mildly related to each other, Part 2 is an attempt to exhibit mathematical development as it often exists while at the same time, and of equal importance, introducing the student to the basic tools: sets, relations, functions, infinite sets and mathematical proof and mathematical reasoning. The intent is to exhibit these concepts and approaches as part of an organic whole, not as generalizations that one must learn with no goal or use in mind. The approach chosen is to raise some questions intuitively and to attempt to answer these precisely. These questions are detailed in the introduction to Part 2. Question I asks (roughly): can one, in sequence, remove all the points of a line segment? Question II asks: if one has a "connected" "object" and "distorts" it "without tearing," is the resulting object a connected one? Originally, only Question II was raised, but Question I has proven to be far more interesting to the students. *More importantly, a very complete and satisfying course, which concludes with Section 6.5 and the answer to Question I, can be taught within one semester to students who, because of their background, must begin on page 1 and proceed most thoroughly.* The instructor who has a more advanced class might well wish to devote only a couple of weeks or so to Part 1 of the text and make Part 2 the essence of the course. In doing so, in arriving at a precise meaning for "without tearing" and in answering Question II, the student will be throoughly introduced to the basic concept of continuity in a context for all occasions (Section 8.2). He then will be well prepared, it is felt, for the study of advanced calculus, modern topology, and other courses in continuous mathematics. Concerning Question I, it is essential that Chapter 4 (through Section 4.6.3) be covered in order to answer it. Finally, I believe that I have left the instructor with a good deal of responsibility for making this format effective, if indeed he choses to follow it. The organization of Part 2 around Question I or II, or both, will, I believe, not be felt strongly enough to be effective toward any significant realization on the student's part unless the instructor constantly keeps the Question before the student and shows how each new topic relates in some close way to some detail of the Question or Questions posed and shows how in the end these topics all come together to provide a definitive answer.

This brings up a final key decision about the content of this text. I wished to teach the art of mathematical proof. I needed a subject which was clear enough to allow easy concentration on the details and techniques of proof itself, while at the same time was abstract enough to allow a valid introduction to modern mathematics. I also wished to be able to say to these sophomores that this mathematics was at least of their own century, albeit before their time. But I also needed a subject that could be well founded intuitively so that I could easily motivate concepts and definitions and could easily manifest in a clear manner the meaning of results that were stated and proven most abstractly. Many texts meet this goal by choosing some subject from modern algebra or linear algebra or choose to develop the real number system. I felt that these approaches have two faults. One is that the student already knows a little about the subject and persists in cluttering up the development with half-forgotten memories of high school or college algebra. Another is that these subjects provide minimal aid in the upcoming confrontation with modern analysis, differential equations, advanced applied mathematics and probability, with their bewildering requirements of "for every ϵ there is a δ" and their necessity to operate constantly with inequality, approximation, and the technique of making "uncertainty" certain. One subject that I felt would meet all these problems was basic point set topology. Topology is not the subject of Part 2. Topology is the glue that holds together the mix of all those topics and techniques and ways of mathematical thought that I wished to introduce. (My own experience with point set topology is that it is a good place to learn to think, write and speak in mathematics.) The students have found topology interesting, have been able to prove theorems in it and, as one student remarked, "come to see that mathematics exists and can be done without number, direction or measurement." Pedagogically, topology has the strong point of enabling the student to *actually see* the need for, and the use of, the "for every" and "there exist" quantifiers, a point that, from my experience, presents considerable difficulty.

The above is a sketch of how this text evolved and how its content was chosen. The remaining problems concerned the level at which the text should be written, the style of presentation and the manner in which it might be taught. I began with the assumption that the student taking the course would have little or no concept of proof and only a vague sense of logic, that he might not be aware of what the mathematician finds interesting in mathematics and that he often regards our beautiful abstractions as but useless generalities. I have found no reason to change or temper these assumptions. In consequence, a good deal of the early part of the text is devoted to talking about mathematics rather than doing mathematics. This material can be covered quite rapidly by the student on his own time.

Concerning the style of presentation, I wished to be as lively and informal in the nontechnical parts of the text as I felt I could be. This has caused me considerable difficulty with certain reviewers and resulted in two thorough revisions of the text, both, I believe, to good effect. I wanted a strong contrast between the rigor and formality of a mathematically framed development and its proofs, and the informality of any discussion which serves to motivate, to see through, but not *to prove*.

This brings up then the question or level of mathematical rigor to be employed. I have attempted to introduce the topics intuitively, develop them correctly and formally, but not to the point at which I guessed boredom or distaste might set in. I found the depth of rigor herein to be quite sufficient for the classes I taught, two of which were special "honors" sections. More formal or sophisticated treatments are noted throughout the text. I did not intend to be as abstract or rigorous or generalized as the standard "foundations course" (as I judge these to be from other textbooks).

In teaching the course, I minimized the use of the formal lecture and followed as much as I could a firm belief that the joy of mathematics is in the doing, not in the hearing or repeating of it. I have found that I need to note the important points of each upcoming section, explain a key definition or theorem, etc., and indicate what the problems are about, *before* assigning the section and the problems I wish to cover. I can then let the text do much of what a lecture might do for that section and devote the next class period to a discussion and question and answer session on that section, centered around the solutions of the problems and the proofs of the unproven theorems. I have tried to get the students to present these themselves and in the latter half of the semester have succeeded in getting good participation.

In closing, some very special thanks are due. One must go to the students who let me know that they enjoyed the course and who gave me their opinions on the text. Another must go to several anonymous reviewers who were generous in praise and in criticism, and to several colleagues who read a section or two on those topics with which I did not feel a sufficient familiarity. More personally, I want to express my thanks to two instructors of my undergraduate years, Mrs. Elise Hughey of the English Department who encouraged me to write and write and write, and . . . , and Mr. L. R. Tappen who encouraged me to major in mathematics and taught me one whole lot of it. More recently, thanks must go to the initial editor for this book, Alexander Kugeshev, who, on the basis of some rather inadequate notes, encouraged me to write this text and kept me at it when his reviewers were not the least enthralled by my efforts. During the same period a good friend and experienced teacher, Don McInnis, was encouraging me to write a book that was to be somehow not typical of mathematics books at this level. Don later reviewed the manuscript under trying and hurried circumstances and gave me a number of suggestions and some good criticism. A very special thanks must go, too, to Terry Davis, who typed the book—yes—but also read it as she typed, criticized the sentences, encouraged the informality of the text, exhibited infinite patience with my handwriting and unending revisions, and who, being a student, but not of mathematics, tried to be and was a real help in making the book more readable and, hopefully, more interesting to read. I also want to thank Professor George Wald for his permission to use the material in Section 2.6 and for his suggested improvement. Finally I want to thank Sara Finnegan and the people at Williams & Wilkins for being willing and interested in publishing a text aimed at an essentially small market.

That I should ever write any book at all, and this one in this way, must be traced to many influences, not the least of which must be credited to the earnest and unending efforts of so many in my family to educate me, and to them I owe more than I can ever express. A final and special thanks must also go to my wife and kids for all the hours spent on this rather than with them.

CONTENTS

PART 1

"The study of science teaches one to think, while study of the classics teaches one to express thoughts."—John Stuart Mill

STARTING POINTS

Two only slightly noted events of this past midsummer came together to form an interesting backdrop to larger affairs of the day. A movie was released concerning the world of insects, its message being that these marvelously adaptable populations may well have it within themselves to some day inherit this earth. At nearly the same time there was discovered in Califorinia a mosquito which had evolved an immunity to all known pesticides. Some 2, 000 years earlier the mathematician Pappus had proven that the honeybee's hexagonal honeycomb uses a *minimum* amount of wax to store a *maximum* amount of honey.

For eons of time, ninety-nine percent of his existence on this earth, man was as much a part of the natural world as the bee and the mosquito, his population checked in the same manner as theirs, his mind only a bare shade more imaginative. As the discovery of Pappus implies and the immune mosquito reiterates, it is not our ability to raise structures vast and efficient unto our species, nor our awesome technology that so distinguishes us from the natural world. If an ancient tool maker of the African savannah were to suddenly appear within our midst, he could quickly come to swat our mosquito, wield our bayonet, even drive our automobile within a short time. But our language would be useless to him, Beethoven at his best but a cacophony of noise, Picasso frightening, and Pappus' discovery void of any meaning at all. Language, art, music and mathematics, the four corners of the religious, scientific, social and philosophic edifices that we most readily perceive as especially our own are the essentials of our uniqueness within the natural world and the measure of our distance from our earliest enlightenment. Of these four the oldest is uncertain, the most pure debatable, but the one least comprehended is almost universally agreed upon. The aim of this text is to do something about that.

One would have to shut his eyes and ears not to experience art, music or language. They have a real existence independent of self. Indeed less developed forms of these have long existed in the natural world: the porpoise communicates with his kind, nature abounds in artistic perfection and a melodious tune was heard long before man got here. As far as is known, mathematics and abstract thinking, the making of reality from nothing but the ephemeral imagination, is unique to man. To do mathematics is to make the imagination so real as to be able to act upon it, to think further about it. Mathematics is indeed all in one's mind—if you don't think about it, it does not exist for you. Transcendental numbers are to the mathematician no more than whole numbers to most men: the sum and distribution of chemical reactions and electrical charges within his brain. But they are as real to him as a blade of spring grass, real enough to be communicated to the audio engineer who uses them to design your stereo receiver.

Like a lot of other things, mathematics is not what it used to be. Never shackled by any need to cling to thoughts past and methods that have run their course, this "perhaps most original creation of the human spirit" is today a vibrant, ever growing way of thinking, whose contemporary patterns of thought are quite different from those of all previous centuries.[1] There has perhaps never been an age in which more than a very small percentage of the human species was even aware of contemporary mathematical thought, and this is surely true today. Even one who has studied the calculus must date his knowledge of mathematics at some point early in the 19 th century, unless he has an acquaintance with set theory, a subject of more recent vintage. The intent of this book is to bring to within the reader's grasp the core concepts and methods of 20 th century mathematical thought—logic, axiomatic systems, set theory, functions, relations, infinite sets and cardinal numbers—all of these to be used and glued together within a matrix of introductory topology, a singularly important offspring of the 20 th century which has, among other things, unified the centuries old ***limit concept*** into a coherent whole.

As with other institutions that have undergone significant growth and change, the development of mathematics has had its rough moments, not lacking at all in surprises, sometimes to the chagrin of its adherents. Being so much a product of the imagination, some of its concepts have no realization in the physical world, despite the special and powerful aid these may give in comprehending that world. A few circuits in the mind combine to produce a new concept and one seeks to open others that are closely allied. Sometimes too few circuits operate initially, and when others are finally opened they damage or destroy the initial circuit and a concept is amended or lost. One's imagination is a tenuous thing, a flickering pinpoint of light to be carefully guarded.

One of the demands of learning mathematics is that of learning to be exact in what one says and means, ***trying never to be misunderstood.*** This

[1] A. N. Whitehead: ***Science in the Modern World***, New York: The Macmillan Co., 1925.

must be carried to an extent that would be extreme in any other area of thought. Mathematics is constructed so that such an extreme is attainable, indeed absolutely necessary, and a non-affinity for this manner of mathematical thought and expression is perhaps the most unfathomable obstacle confronting one who wishes to do and to learn mathematics. To some this manner comes naturally, to others never, and they go on to poetry or music or other equally worthwhile things. If one is of a mind that from true belief all good things will follow, then mathematics, outside of it formularizations, will lack a compelling attraction. Doing mathematics requires a habitual concern for precision of expression and a strong sense for the essential, crux, point at hand, with a will to question every detail, no matter how unwelcome, until it is settled in a coherent way with nothing left to chance. Such is hard work. So that the reader will have some reason to recognize these needs, we begin this text with a stunning example of what did go wrong with one of the most simple, and at the same time profound, products of one brilliant man's ephemeral imagination: the original concept of a *set* and the resulting "Russell paradox." This tale concerns the work of three men: Georg Cantor, possessing a brilliant but troubled mind: Sir Bertrand Russell, philosopher, logician, author, mathematician, correspondent of presidents, dictators and prime ministers and sometimes inmate of British jails because of his profound pacifism; and Gottlob Frege, determined mathematician, dedicating years of labor to the establishment of a foundation for fundamental mathematics only to see it crumble upon completion.

A mathematician expounding his mathematics can be at times downright smug in his self-assurance. He can afford it, for his conclusions have probably been worked out in exhaustive detail based on certain fixed premises. If you should wish to challenge him he leaves you nothing but his premises to question. But that's the last place to try, for he cares not one whit for the truth of his premises but only for what can be drawn from them. He does however, remain open: his conclusions would be worthless if his premises were inconsistent in themselves. There just might be a complex of brain cells whose chemistry would make inoperable the separate circuits that make up his premises. As things stand today not even arithmetic has been shown to be internally consistent. On the other hand, arithmetic has not in 4, 000 years led to any inconsistency.

The Russell paradox is a famed argument of inconsistency involving the very foundation of one of the simplest of all abstract ideas, the idea of a *set* or "collection of definite distinguishable concepts of one's intellect, *thought of as a whole.*" This last phrase was the original definition of a set, given by the founder of set theory, Georg Cantor, in the late 19th century. Russell showed that the initial premise that this was a proper definition of a set was unacceptable, for it would lead to the following contradiction.

Imagine all the definite distinguishable concepts of your intellect.

Among these you will find a *collection* of definite distinguishable concepts of your intellect which does not contain itself (the collection itself) as one of its definite distiguishable concepts; that is, you can imagine a set which does not contain itself as one of its elements. Call *any* set which does not contain itself as one of its elements an *ordinary set*.

It is possible to develop Russell's paradox without ever considering the question of whether or not there are any other kinds of sets. However, if you really thought hard for a long enough time you might think of the following. An abstract idea is a definite distinguishable concept of your intellect, such as the idea of *two* or *two-ness*, or the idea of *set* for that matter. Consider the set of all abstract ideas. It is an abstract idea and thus contains itself. If you don't like that, ignore it, but don't fail to notice that it goes along with the above definition of set, which, unwarned, one might reasonably think is a good enough one.

At any rate, consider now the set of all ordinary sets and call it S; the set S exists because each ordinary set is a definite distinguishable concept of one's intellect. Here is Russell's paradox.

(1) The set S is an ordinary set.

For suppose not. Then S is not ordinary and thus contains itself as one of its elements. But every set in S (this now includes S itself) is ordinary. Thus S is ordinary. This is contradictory. Hence our supposition that S is not ordinary must be false. Therefore S is ordinary.

(2) The set S is not an ordinary set.

For supporse not. Then S is ordinary. Since S contains all ordinary sets, then S contains itself. But this means S is not ordinary. This is contradictory. Hence our supposition that S is not ordinary must be false. Therefore, S is ordinary.

There is no doubt that (1) and (2) are contradictory. In searching for what has gone wrong, one could question the logic used, but note that it is the same for both (1) and (2) and, as we shall soon see, this kind of reasoning is an inescapable consequence of what one usually accepts as good reasoning. The fault lies in the loose definition of a set. It admits the existence of the set of all ordinary sets, which by (1) and (2) is a contradiction in itself.

The Russell paradox was of only a momentary embarrassment to mathematics, but a very real disaster for Gottlob Frege. Frege had labored for over 10 years in the production of his *Grundgesetze der Arithmetik* (*The Fundamental Laws of Arithmetic*) in which he had often used the idea of the set of all sets without being aware of the inherent (and similar) contradictions in this notion. Just before the work was to be publshed, Frege received a letter from Russell containing the above paradox. Frege closed his work with the comment that "a scientist can hardly encounter anything more undesirable than to have the foundation collapse just as the work was finished. I was put in this position by Mr. Bertrand Russell when the work was almost through the press."

This text is not a study and resolution of the Russell Paradox. It is a study of the basic methods and approaches to mathematics that have to a considerable extent resulted from the awareness that such paradoxes could occur. Although it may be true that the definition of a set given by Cantor was not the ultimate in precision, it seems fair to say that a definition of this order of precision is acceptable to most people in most circumstances. Russell showed it not to be good enough in theses circumstances, and by implication that possibly a whole herd of other basic premises might be in a similar state of potential disarray. After all, a contradiction may presumably exist without its being discovered! With Russell's paradox, and not a small number of other examples and discoveries as detailed in Chapter 2, mathematicians were compelled to investigate, rethink and straighten out the foundations of mathematics. In consequence, the material in subsequent chapters exists as a minimal prerequisite for understanding today's mathematics.

Understanding is what it's all about. Genetically endowed with the proper formula within its tiny brain, the honeybee builds his honeycomb quite well. We have learned to store our formulas within machines to be used with a rapidity and efficiency we can never match. Understanding is one thing uniquely our own. Nothing in the sequel need be taken as *a priori* true. Understanding what you're doing and being certain of the underlying justification for whatever conclusions you reach, no matter how trivial, is the whole story. A great deal of emphasis is on the *manner* of mathematical thought.

chapter 1

"Humble thyself, impotent reason"—Pascal

LOGIC, LANGUAGE AND MATHEMATICS

1. INTRODUCTION

Despite some evidence to the contrary, each member of the species *Homo sapiens* seems to fancy himself as a cold, analytical and logical thinker in matters of great personal importance, and for the most part he is intuitively logical, though perhaps not always reasonable and coldly analytical, may be with good reason. Mathematics of course is famous for its objective, analytical judgment of things and as might be expected the logic of mathematics is more precise (you might claim tortuous!) than the logic of everyday affairs. This chapter is a survey of the most common patterns of logic and logical form found to be of constant use in mathematics. This will not be an in depth study of logic, but its thrust will be truly mathematical: to study the patterns of logical thought as entities in themselves, independently of any particular use or instance, but of course accompanied by examples and problems to give meaning to these patterns.

In plain fact, very little logic is used or needed to do most mathematics, and much of this is naturally though often only vaguely acquired by anyone with a strong interest in the subject. The intent herein is to remove whatever uncertainty there may be and to give to logic the closer attention it requires for use in mathematics. To serve this end it will be convenient to introduce some symbolic logic, but because a detailed capability in symbolic logic is not needed for mathematics, a lengthy study of symbolic logic will not be attempted.[1] If the reader finds symbolic logic interesting in itself and would like to know more about it, there are some admirable and detailed investigations of it referenced at the close of this chapter. The outlook in this chapter is that logic is only one of a number of tools used to do mathematics, and that a little logic can take one a long way.

Having stated that little logic is needed to do mathematics and that one is likely to be acquainted with much of it by now anyway, what can justify

[1] Nonetheless, you should be aware that symbolic logic is an important study in itself, having applications in computer science, the design and simplification of circuitry, and other areas having little to do with pure or applied mathematics. On the other hand, symbolic logic does have an extremely close connection with the interesting and important mathematical topics of Boolean algebras and lattice theory.

any effort to learn about logic abstracted from any particular use? Would it perhaps suffice to simply begin with mathematics, believing that one will pick up logic as the need arises? Probably not. It should prove worthwhile to bring to your attention the common forms of exposition and reasoning in mathematics isolated from any particular application, so that you will obtain a grasp of logic and mathematical exposition as something in itself, independently of how it is used. This done, you should find it easier to follow the course of reasoning in abstract mathematics and be more certain of its validity. More importantly, you will be better able to focus on the mathematics itself, its content, and not be distracted by its logical form.

Now to turn this all around and assert that mathematics is the quintessence of logic! Because of the exhaustive work of others, this assertion has some validity. The *Principia Mathematica* was a partially successful attempt by the philosopher-mathematicians Alfred North Whitehead and Bertrand Russell to construct the basic elements of mathematics (such as the idea of "one," etc.) as logical extensions of a very few basic logical forms expressed symbolically. To paraphrase Whitehead himself, mathematics is the exhibition of unexpected relationships between two or more abstracted concepts. The mode of exhibition is called *proof* and the logic of mathematics furnishes a set of rules and guidelines governing the validity, the exposition and the development of proof in mathematics. Put another way, to do mathematics is to expose the logical dependence of one concept in mathematics upon another. Because of this, mathematics can be reduced to an exercise in purely logical form, but this is not the most common way of doing it.

There is more to this matter of *proof* and it bears on the intent of this entire text. *Proof* is a very special kind of thing in mathematics.[2] In the carefully constructed world of mathematical thought, proof can meet an uncommon standard, not expected in other disciplines. It is hoped that in using this text the reader will come to an appreciation of, and an ability for, proof in mathematics, and an acquaintance with its power and limits. The standards of proof demanded are greater than one commonly pursues and are likely to be only slowly acquired. But at the same time there is more, much more, to proof than logic and the care to meet certain standards. Logic is one part of mathematical proof that can be learned. An equally important part of proof springs from the same force that causes an artist to paint, a poet to write. It is a feeling for one's subject, intuition and those other non-rational means by which one knows something is right and cannot be any other way. Logic is a part of thought that retrieves substance from ideas conceived in this way. As the would-be artist who cannot hold

[2] Appendix A is devoted to a general discussion of proof, how to obtain a proof as the mathematican does, but written with a certain suspicion that such can't be told, that one has to learn this by his own good means.

a steady brush cannot paint, the would-be scientist or mathematician who cannot explain and argue precisely for what he knows accomplishes little. Logic helps to give the mathematician the steady mind and sure course of reason that he needs.

2. LOGIC AND SYMBOLIC LOGIC

The logic chosen to do mathematics is basically the logic of most affairs with a few special cases tacked on to accommodate the natural evolution of mathematical thought over the past four thousand years. Nevertheless, it is the logic *chosen* for mathematics. It is not the way we must reason, but is the way we find it convenient to reason because our reasoning, for the most part, then coincides with the reasoning we grow into as we come to think and to use a language. The logic that certain mathematicians and logicians use would seem strange to us. They work with such logics because they find it useful, which is a point worth making now and which will be emphasized again: mathematics is what mathematicians make it, independently of (though not in ignorance of) themselves, our history, or the world around us. So, too, with the logic of mathematics. The system of logic we will study is the one that has thus far been the most useful and the least subject to philosophical dispute. It is not ordained by any deity, dictator or the constitution as the way things must be but exists as it is because it serves well. In a later section we will consider some special forms of set theoretic logic not accepted by all mathematicians as valid.

What is logic? A definition is difficult and in the end essentially impossible, but, following the suggestion of Russell and others who have studied the matter, one of the best answers is that the logic *of an argument* is that which is left over when the meaning of the argument has been removed. To find meaning in that, consider the following examples.

> EXAMPLE 1.2.1. If the policies of a government are just, then the power of enforcement is derived from the consent of the governed and is not derived from law sinstituted to enforce these policies. Therefore, if these policies do not have the consent of the governed or must be enforced by law, then they are unjust.

> EXAMPLE 1.2.2. If $2 \neq 4$, then green trees grow on barren hills and children are not happy. Therefore, if green trees do not grow on barren hills or children are happy, then $2 = 4$.

If you will closely examine these two arguments, the first of which has some content, the second of which is frivolous, you will see that they have something in common, something that remains when all meaning is taken away. There is a form or structure which remains and it is the same in both. Both arguments have the abstracted form:

If p, then q and not r. Therefore, if not q or r, then not p

where p, q and r represent the meaningful statements in either argument in the order that they occur. This *form is the logic* of both arguments and indicates what was meant above about logic: logic is that which remains when all meaning has been removed. *Thus the logic of an argument is the abstracted form of that argument as a thing independent of the content of the argument.* The study of logic is the study of the possible form and structure of argument. Form is always a big thing in mathematics, in logic it is almost everything. This point has more use than you might expect at first; keep it in mind, it might save you some time!

The logic of mathematics certifies both arguments above as valid arguments. It must certify both (or neither) because both have the same form. This must bring into question then the place of truth in logic. If logic is concerned only with form, and the same form can occur in the juxtaposition of the most frivolous statements, or the most weighty ones, or the most absurd and the most believable, then one cannot in logical manipulations concern himself with meaning or truth. In practice this has two consequences. The first is that one must have a means of validating the logical form of an argument independently of the content of the argument, independently of the truth or falsity of its constituent parts. This means is provided by symbolic logic and the method of truth-tables, which furnish a precise and concise procedure for validating the logical form of an argument independently of its content. The second consequence is that logic cannot be used to make value judgments. For example, in Example 1.2.1, logic only certifies that the second sentence is a consequence of the first, and furnishes no basis for judgment of the validity of the first sentence. At the same time this hardly makes logic irrelevant. One can use logic to determine whether or not the first sentence in Example 1.2.1 is a true consequence of that admirably logical document, our Declaration of Independence, though one (to emphasize it again) cannot use logic to verify that document as a true statement about men and government.

There is yet more to be said about these examples. These contain *all* of the usual and essentail constituents of a logical argument! This need by no means be clear now but should soon become so. Each of the sentences in these examples consists of a collection of declarative phrases related and ordered by the words *and*, *not*, and *if . . . then . . . .* The second sentence is a collection of the "logical opposites" of each of the phrases in the first sentence, related and ordered by the words *or* and *if . . . then . . . .* The problem of determining the logical validity of each argument comes down to a verification that the first combination (the first sentence) of the separate declarative phrases must imply the second combination without regard to the meaning or truth of each declarative phrase. That is to say, the problem of establishing the logical validity of these is one of showing that if p, q and r represent "statements" and if the combination: if p, then q and not r, is given, then the combination: if not q or r, then not p, must follow without regard to the truth of p, q and r. That (as we will see) the second combination of p, q and r follows logically from the first is a consequence of the logic chosen to do mathematics and, again, has no further justification.

These last few paragraphs furnish a sort of overview of the task and intent of a study of logic. If you have picked up the main points and see some sense in them, use them as a guide for the sequel. If, for one reason or another, you feel this whole business is rather hazy ("what is left over when all meaning has been removed" does have an air of self-contradiction about it!), then your sensibilities are merely yet to be satisfied. In either case, the task before us is to examine the ways in which a given collection of simple statements, and often their "logical opposites," related and ordered by the relatively few words *and*, *or* and *if . . . then . . .*, can be manipulated and reordered to produce further combinations which in some precise and determinable sense (which we call logical) are related to the given statements without (for the last time!) paying any heed to the meaning of the statements themselves. To serve this last end it is both natural and useful to introduce some notation, and hence some symbolic logic, as was done earlier to point out the form of Examples 1.2.1 and 1.2.2.

3. STATEMENTS, PROPOSITIONS, DISJUNCTION AND CONJUNCTION

As will be seen in detail in Chapter 2, mathematical validity and logical validity, like any value judgment, is a relative thing and must always rest on some agreeable but ultimately unprovable basis. The matter of choosing a basis for any part of mathematics is a complicated, interesting and deceptive one: the Russell paradox is an outstanding example of this deceptiveness. In a later chapter we will consider the most common kind of basis for much of mathematics, the axiomatic method. Roughly speaking, the degree of intuition and imprecision that one is willing to live with, in the establishment of a basis for whatever system is to be developed, is directly proportional to the pace with which that development establishes "useful" results and, in return, the ease with which paradoxes can (possibly, not necessarily) creep in. Thus, if one begins with an extremely rigorous and non-intuitive basis for logic, governed by strict rules of as simple a nature as possible leaving no possibility of subjective interpretation, it will take some time to develop, within this setting, the principal useful results of mathematical logic. At the same time, the likelihood of misunderstanding and inconsistency would be virtually eliminated. We will establish the validity of the most useful forms of logical discourse on the basis of an intuitively motivated agreement on a precise meaning for the relatively few words which connect a sequence of statements into what is called a logical argument. Such a basis will suffice for the material in this book, although it is far from the epitomy of logical rigor.

In line then with this aim, we begin with the agreement that by the word *statement* we mean a declarative sentence. Thus

x is a number greater than 2 .

is a statement, as is,

All statements in this book are false .

which indicates that one must narrow down the class of statements that can be assimilated into any sensible system of logic, for by the barest notions of intuitive logic this last statement is self-contradictory.

Certain statements have a character referred to as truth. Thus,

Fish are agile swimmers .

has a universal, if innocuous, character of truth, while

All water contains fish .

has an equivalent character of untruth (falsehood) while yet

Water pollution must end .

has a character of truth that is subjective. Other statements, such as

There is life on some other planet in our galaxy .

has an undetermined and perhaps undeterminable character of truth while the statement

Individual freedom and individual security are inversely proportional .

has a ring of truth to the author, but not perhaps to the reader.

Admitting that truth, like progress, is in the eye of the beholder, mathematical logic avoids the issue by only claiming a capability to investigate the *consequences* of statements which *already* have a settled-upon character of truth or falsehood and claims no ability to determine this character. In a later chapter we will consider the nature of the truth of statements in mathematics. For the purposes of basic mathematical logic we need consider only statements to which a "truth-value" is assumed to be assigned. To indicate that we are thinking of a statement as true, we will say that the truth-value is 1, whereas, to indicate we are thinking of it as untrue, we will say that the truth-value is 0. To have a name to remind us of the nature of such statements we will use the term *proposition*; a proposition is a statement which has a truth-value of either 1 or 0 and not both.

Here are some examples of propositions and non-propositions. The statements

$$2 + 3 = 4 .$$

or

$$x^2 + 1 = 2 \text{ has a solution}.$$

are propositions with truth-values 0 and 1, respectively. Neither of the statements

$$x \text{ is an even number}.$$

or

This statement is false.

is a proposition, but for different reasons. The first is not, because we cannot determine its truth-value without knowing what x is. The second is not because assigning either truth-value leads to the other, and, for our purposes, nonsense. In a later section we will see how statements of the first type are assimilated into the system of mathematical logic. For now, simply keep in mind that a proposition is a statement with a determined truth-value.

It was indicated earlier that mathematical logic concerns itself with the relationship of various combinations of statements, combined through the use of words *and, or, not* and *if . . . then . . . .* Once this fact was recognized by those who are credited with the founding of modern mathematical logic, most notably George Boole, it was only in character with the mathematical approach to things that these essentials of logic should be precisely defined and analyzed in as simple a way as possible.[3] The purpose of this section is to state precisely the meaning given in mathematics to the words *and* and *or.*

It was noted earlier that the logic of mathematics is the logic of most affairs. With this in mind consider the following propositions.

EXAMPLE 1.3.1.

(a) The number 2 is even or the number 2 is twice 1.
(b) The number 2 is even or the number 2 is odd.
(c) The number 2 is odd or the number 2 is even.
(d) The number 2 is odd or the number 2 is less than 1.

Ordinarily, one thinks of (b) and (c) as being true and (d) as being false. Notice that in (b) and (c) at least one of the propositions connected by the word *or* is false, while in (d) both are false.

Sentence (a) brings up a point of difference between the use of *or* in mathematics and its ordinary usage, where *or* can have two meanings. Mathematics cannot abide ambiquity of language, there being plenty enough other difficulties to satisfy everyone. In mathematics, a statement such as (a), where the two statements connected by *or* are both true, is

[3] G. Boole: *An Investigation of the Laws of Thought,* New York: Dover Publications Inc., 1951.

taken to be true. In ordinary usage, the heavens can go dark and stars go out over such statements as

We will meet at seven or not take in the movie.

for she, meeting me at seven, is disappointed when I, being a mathematician, takes this to mean we could do both. One should simply be aware of the special use of *or* in mathematics, and forget it in the appropriate circumstance. From this point on, the word *or* will only be used in the sense given by

DEFINITION 1.3.2. Let p and q denote propositions. The statement p *or* q is assigned the truth-value 1 if at least one of the propositions p, or q, has truth-value 1, and 0 otherwise. The proposition p *or* q is called the *disjunction* of p and q and will be denoted by $p \vee q$.

All this definition does is settle upon a single invariant meaning for the word *or* when it is used to form a new proposition $p \vee q$ from given propositions p, q. But in addition it does provide a valuable computational tool for validating logical forms. This is known as a table of truth-values or truth table for the proposition $p \vee q$. Definition 1.3.2 is equivalent to the truth-table for a statement $p \vee q$, for each pair of possible truth-values for p and q, given as follows

p	q	$p < q$
1	1	1
1	0	1
0	1	1
0	0	0

Notice that with this meaning assigned to *or*, Example 1.3.1(a) is considered a true statement.

Consider now

EXAMPLE 1.3.3.

(a) The number 2 is even and the number 2 is twice 1.
(b) The number 2 is even and the number 2 is odd.
(c) The number 2 is odd and the number 2 is even.
(d) The number 2 is odd and the number 2 is less than 1.

We ordinarily take these sentences to be true, false, false, and false, and carrying the notions of truth-value indicated here to generality we make the

DEFINITION 1.3.4. Let p and q denote propositions. The statement p and q is assigned a truth-value of 1 when both p and q have value 1 and is assigned a value of 0 otherwise. The proposition p and q is called the *conjunction* of p and q and is denoted by $p \wedge q$.

The definition is equivalent to the truth-table for conjunction

p	q	$p \wedge q$
1	1	1
1	0	0
0	1	0
0	0	0

Problems.

I. Assign truth-values to the following statements. If the statements are not conjunctions or disjunctions, rewrite them as such.

(1) $2 + 2 = 4$ or image is not reality.
(2) 4 is even or is divisible by 3.
(3) $2 + 2 = 4$ and 2 is not less than 3.
(4) $1 + 1 = 2$ but 1 is greater than 2.
(5) 4 is a number greater than 2 such that $2 + 2 = 4$.

II. Using truth-tables, show that $p \vee (q \vee r)$ and $(p \vee q) \vee r$ have the same truth-value, no matter what the truth-value of p, q and r. What about $p \wedge (q \wedge r)$ and $(p \wedge q) \wedge r$? You will need a truth-table with eight rows to do this.

4. THE NEGATION OF A PROPOSITION

In the next section we will verify the most common logical relationships between propositions formed by disjunction and those formed by conjunction. The key ideas needed to do this are that of the negation of a statement and that of logically equivalent propositions; in this section we turn to negation. The negation of a statement turns out to be much more complicated than might first appear but is also one of the most useful of logical concepts. After seeing how naturally this concept arises it will be necessary and desirable to make a thorough study of how to form negations of typically mathematical statements; you will find this to be one of the most useful things you learn.

In the usual sense of things the negation of a statement such as,

Men are in control of events.

is simply

Men are not in control of events.

or,

It is false that men are in control of events.

or similar unsettling things. It is clear that by the negation of a statement in the sense of this example we simply mean a statement whose value is 0 if the given statement has value 1 and whose value is 1 if the given statement has value 0. Thus the general

DEFINITION 1.4.1. Let p denote a proposition. The negation of p is that proposition which has a truth-value of 0 when p has a truth-value of 1 and a truth-value of 1 when p has a value of 0. The negation of p is denoted by $-p$ and read "not p."

This definition, too, amounts to a truth-table, the truth-table of negation, which is given by

p	$-p$
1	0
0	1

One should exercise a little caution in forming negations even for the relatively simple statements we will consider in the next few sections. For example, if p is given by

p: That pompous politician is too self-righteous.

then

$-p$: That pompous politician is not too self-righteous.

is a correct way to write this negation, while neither

That non-pompous politician is too self-righteous.

nor

That non-pompous politician is not too self-righteous.

is a logically correct negation, aside from the unwieldly grammar. In short, for the time being your good sense must and will carry you through.

It will even carry you through when you don't know what you're talking about, in logic as in other things; in fact, even more so in logic, as you will shortly see! For now, and for example, your good sense tells you that a correct negation of

p: The compact space is normal.

is

$-p$: The compact space is not normal.

and a correct negation of

q: The group G is not cyclic.

is

$$-q\text{: The group } G \text{ is cyclic}.$$

and you probably don't know what these statements mean.

Certain parts of the next problem indicate that more than your good sense may be required if these things get too complicated. You may have some difficulty in this problem because of the ordinary ambiguity of the written spoken language, an ambiguity appropriate to many circumstances. By its very nature, in complement to its extensive abstraction, mathematics cannot usually operate in a setting in which essential points are not either black or white. Precision of language usage in mathematics is ordinarily possible to the extent practical. However, throughout this chapter there will arise mild confusion over the exact logical form of a given statement, over how such sentences can be viewed in the form $p \wedge q$ or $p \vee q$ or other basic logical forms. This will occur even when the sentences are about mathematical things. The reason this will happen is that we have set no rules for the initial *formation* of sentences and use only ordinary grammar which is not always logically precise. This can sometimes cause difficulty in applying logic and symbolic logic to a given statement or statements. But it would do no good to try to handle this problem by giving a set of strict rules for the formation of sentences outside of the uses of the logical connectives we specifically define. It would do no good because nowhere else would you find these rules used to form sentences, not even in most subsequent mathematics courses. But do not misunderstand, it is possible to give strict rules for the formation of the substantive statements of main-stream mathematics; in fact, it is possible to form these statements and do mathematics using only symbols and no words at all. When this is done, using formal logic in mathematics is straight forward, but with an attendant increase in abstraction.

Problems. State the negation of each of the following statements.

(1) The grass is green in the fall.

(2) The function $f(x) = |x|$ is not differentiable.

(3) Do not judge the product by its package.

(4) All of us are asleep.

(5) All of us are not asleep.

(6) Walk a mile in my shoes.

(7) Irresponsible leaders are more dangerous than irresponsible citizens.

(8) There is something wrong.

(9) Science is good, it is technology that leads to harm.

(10) Man's most natural traits were forged in some long distant past and will not be changed in a single generation.

(11) Freedom is a word for nothing left to lose.

(12) That course is obscure to the dull-witted, boring to the brilliant and unfair to the practical minded!

5. LOGICAL EQUIVALENCE

Have you ever argued with friend or foe for hours only to find out you actually agreed, that most of the discussion was the product of only misunderstanding? Our study and use of logical equivalence probably wouldn't help much if such arguments are heated, but it is concerned with the same sort of thing: when do two propositions have the same truth-value? To be more precise about it, we need a vehicle of discussion by which we can relate one proposition formed by more than one logical operation to another perhaps more useful proposition. For example, we might wish to know how to negate a conjunction $p \wedge q$ in terms of $-p$ and $-q$. Logical equivalence furnishes such a vehicle

DEFINITION 1.5.1. If u and v are propositions, we will say that u and v are *logically equivalent* if they have the same truth-value.

Consider the propositions,

u: The times are changing and people are confused.

v: The times are not changing or people are not confused.

and the propositions

u': 2 is even and $2 = 1 + 1$.

v': 2 is not even or $2 \neq 1 + 1$.

as well as

u'': S is compact and S is normal.

v'': S is not compact or S is not normal.

With the meaning given previously to the words *and* and *or*, as well as in the usual way of speaking, each of the statements v, v' and v'' is the negation of the corresponding statement u, u' and u''. That is, if (say) v is false then u is true, or if v is true then u is false, and similarly for the remaining statements. Thus v, v', v'' are, respectively, logically equivalent to $-u$, $-u'$ and $-u''$.

There is a reason for choosing such disparate statements. It is to make clear the point that the notion of logical equivalence is dependent on *form*, not content. As things develop, it should become more and more clear that formal symbolic logic enables one to compute or calculate various logically equivalent forms of statements without regard to the content or meaning of the statements themselves. Thus using logic, one may be able to decide, as in the case of u'' and v'', when one statement has the same truth-value as another without knowing what the statements are about. This is an obvious strength and an obvious weakness which we reiterate: logic can be of no use in judging the merit of a statement but is of great use in determining the merit of another statement relative to the that of a given statement.

At this point something of substance can be gotten from the concepts so far considered. The following theorem states the most fundamental logical relationships between the words *and* and *or*. These are the ways by which we most commonly manipulate, in logical fashion, the words *and* and *or*, particularly in mathematics. The verification of these theorems also indicates the usefulness of truth-tables.

THEOREM 1.5.2. Let p and q denote any two propositions.

(a) The propositions $-(p \wedge q)$ and $-p \vee -q$ are logically equivalent.

(b) The propositions $-(p \vee q)$ and $-p \wedge -q$ are logically equivalent.

To verify this claim, we have only to show, according to Definition 1.5.1, that (in (a)) the propositions $-(p \wedge q)$ and $-p \vee -q$ have the same truth-value, and of course this is to be done regardless of the truth-value of p and q. Truth-tables make it a simple matter:

p	q	$p \wedge q$	$-(p \wedge q)$	$-p$	$-q$	$-p \vee -q$
1	1	1	0	0	0	0
1	0	0	1	0	1	1
0	1	0	1	1	0	1
0	0	0	1	1	1	1

The third and seventh columns vindicate the claim in (a). The reader is left to verify (b).

Problems.

I. State the negation of each of the following as a disjunction or conjunction, if it is a conjunction or disjunction, respectively.

(1) Democrats are spendthrifts and Republicans are penny pinchers.
(2) We are not the missing link and evolution is not certain.
(3) 2 is an even number and 2 is not an even number.
(4) It is them or us.
(5) G is a group or G has no proper subgroup.
(6) Image and reality are not the same.

Carry the negations of the following to the furthest extent possible.

(7) The set S is open and is closed or the set S is compact.
(8) It is false that S is compact and S is normal.
(9) The function f is Riemann integrable or bounded and has integral zero.
(10) Environmental laws are not made by man but are slowly realized by man to be laws of nature and the universe and are beyond appeasement or appeal.

II. Symbolize each part of (I) and compute the negations using Theorem 1.5.2. Do your results conform to your answers in (I)? Does all of this make sense?

III. Using truth-tables, establish the logical equivalence of

(1) p and $-(-p)$ for a given proposition p

(2) $-(-p \wedge -q)$ and $p \vee q$ for given propositions of p and q.

IV. Form a conjunction logically equivalent to $-(-p \vee -q)$.

V. Negate Camus' famous statement:

I would like to love justice and my country too.

Perhaps you should first rewrite this statement in a form more amenable to logical manipulation, yet preserving its meaning.

6. THE KERNEL OF LOGICAL THOUGHT: IMPLICATION

Without doubt the most common logical form found in mathematics or in any kind of ordinary argumentative discourse is the implication or conditional statement. An example of such a statement, whose content you might well take note of as you try to settle your view of the place of logic in mathematics, is

If, as Hadamand remarked, "logic only sanctions the conquests of intuition," then one has no need for logic when one has no intuition.

Another statement, of similar form, is

If we must fight to maintain the *status guo,* then all history is our enemy.

Yet another statement of this form, but more relaxing to the mind, is

If the moon is blue, then $1 + 1$ is 2.

Any statement of the form

If p then q

where p and q are statements, is called an implication, or conditional statement. The statement p is called the *hypothesis* of the implication and the statement q is called the *conclusion*. We will interest ourselves, as before, only in implications where p and q are in fact propositions. At the start of this section we want to see how truth-values should be assigned to such statements.

Mathematics always strives for the broadest possible applicability of its theories while at the same time maintaining viability and consistency. In part because of this the meaning given to implication statements in mathematical logic may at first seem at least a little strange. Mathematicians have found it extremely *convenient* and non-restrictive to assign truth-values to the implication

If p then q

even when p and q have *no material connection,* such as in the third example given above. Moreover, just as for the conjunction and disjunction of propositions, mathematicians wish to assign truth-values to such hypothetical statements for all possible truth-values of p and q; in short, even when the hypothesis is false. Again, it's the same old story that logic is concerned with form and not content, with structure and not truth.

That all being so, there is yet the problem of deciding how to assign truth-values in all cases.[4] Here is one way of looking at it. Confronted with a statement

$$\text{If } p \text{ then } q$$

one might ask, "When, and only when, would this statement *not* hold?" And from that corner of his brain where his thought ordering cells do their thing he should get the answer: "When and only when, p holds and q does not hold." That is, when the hypothesis does hold and the conclusion does not follow. Intuitively then, this means that

$$p \wedge -q$$

is a logical form which should be the *negation* of

$$\text{If } p \text{ then } q$$

if truth-values are to be assigned in conformity with ordinary thought processes. That is, an implication

$$\text{If } p \text{ then } q$$

should be logically equivalent to

$$-(p \wedge -q)$$

and the definition is made accordingly.

DEFINITION 1.6.1. Let p and q denote propositions. The statement *if p then q* is assigned a truth-value 0 when p has value 1 and q has value 0 and a truth-value of 1 otherwise. This proposition is called the implication of q by p and is denoted by $p \to q$, which is read "p implies q."

This definition is equivalent to the truth-table

q	q	$p \to p$
1	1	1
1	0	0
0	1	1
0	0	1

[4] For another, different approach to this matter, see G. A. Kraus: Motivation for defining the conditional, *American Mathematical Monthly* 75 (1968), 1103–1104. The *Monthly* is a mathematics journal dedicated to undergraduate mathematics and mathematics teaching.

which one can quickly see is the same as the truth-table for $-(p \wedge -q)$.

That being all well and good, an example might make this seem more reasonable, particularly in the cases where p has value 0. Let

p: New Orleans is in Alaska and Alaska is in the United States

q: New Orleans is in the United States .

It is in accord with your usual way of reasoning that *if p* were true, then q follows. That is, you would think of yourself as making good sense if you were to state

> If New Orleans is in Alaska and Alaska is in the United States, then New Orleans is in the United States .

where by *if* you really do mean if, though in reality there is no *if* about the validity of the hypothesis, it is not true. Again, logic is concerned only with the *form* of thought.

The question of a logical form which is logically equivalent to the denial or negation of a proposition $p \to q$ was essentially settled above in motivating the definition of $p \to q$. Thus, the expected result, stated formally.

THEOREM 1.6.2. Let p and q be propositions. The propositions $-(p \to q)$ and $p \wedge -q$ are logically equivalent.

You may wish to verify this using truth-tables.

Problems.

I. Give false propositions p and q for which the statement "If p, then q" would be thought of as a correct, logical statement.

II. Under what conditions could you be called a liar for telling a friend

> If I get off work early, then we will make it to the party.

Under what conditions would you consider yourself truthful?

III. Symbolize each of the following statements, symbolize its negation and write out this negation in ordinary language.

(1) If the package were the product, then there would be no need for consumer legislation.

(2) There is no need for reason if the wish is father to the thought.

IV. State the negation of each of the following statements as a conjunction, carrying this negation out as far as you can. Notice that you can operate in a logical manner without knowing what you're talking about!

(1) If S is compact and Hausdorff, then S is normal.

(2) If $|a| < b$, then $a < b$ and $-b < a$.

(3) If f is continuous on $[a, b]$ or of bounded variation on $[a, b]$, then f is Riemann integrable on $[a, b]$.

(4) If S is not finite, then S contains a denumerable subset.

(5) If S is normal, then S is compact and Hausdorff.

(6) If they associate themselves with all that is good in the country,

then they seem to think they can reject, with no consideration, any attempt to make it better, if we let them.

(7) You would better know your place in the universe if you would take a long walk every so often.

(8) For people to realize the importance of a decent environment, it is necessary that they see themselves, biologically, as no more then a super complex organism.

(9) Even if your eyes can't see it, your cells must live with it, and if they don't, you don't.

(10) If you do me no favors, then I'll tell you no lies.

(11) As long as it is cheaper to pollute and more costly to have clean air and water, we know what to expect.

V. Show that none of the following are negations of $p \rightarrow q$. Do this by truth-tables and by specific examples:

$$p \rightarrow -q,\ -p \rightarrow q,\ -p \rightarrow -q,\ -q \rightarrow -p,\ -q \rightarrow p\,,\ \text{and}\ q \rightarrow -p\,.$$

7. A FINAL CONNECTIVE FOR SYMBOLIC LOGIC: EQUIVALENCE

The statements

$$\text{If } a = 1\,,\ \text{then } a^2 = 1$$

and

$$\text{If } a^2 = 1\,,\ \text{then } a = 1$$

by no means say the same thing. Moreover, the first is a true statement about numbers while the second is false. On the other hand, the two statements

$$\text{If } a^2 = 1\,,\ \text{then } a = 1 \text{ or } a = -1$$

and

$$\text{If } a = 1 \text{ or } a = -1\,,\ \text{then } a^2 = 1$$

while again not saying the same thing, are nevertheless both true statements. In this section we introduce a final connective of symbolic logic and relate its use to the earlier topic of logical equivalence. First of all

DEFINITION 1.7.1. Let p and q be propositions. The proposition $q \rightarrow p$ is called the *converse* of the proposition $p \rightarrow q$.

The first example above shows that an implication statement and its converse need not have *any* important connection in the framework of logic, where content is not important. That is, aside from the fact that $p \rightarrow q$ and $q \rightarrow p$ are statements which differ in appearance, the truth of either one has no bearing on the validity of the other. In those *many* cases where these do have a common validity, as in the second example above, the mathematician would like to have a brief and universal way of expressing this.

DEFINITION 1.7.2. Let p and q be propositions. The statement "p if and only if q" is assigned a truth-value of 1 when p and q have the same truth-value, and is assigned the value 0 otherwise. The proposition "p if and only if q" is called the *equivalence* of p and q, and is denoted by $p \leftrightarrow q$ or p iff q.

This definition yields the truth-table for equivalence

p	q	$p \leftrightarrow q$
1	1	1
1	0	0
0	1	0
0	0	1

and comparing this to the truth-table for a statement and its converse we have

p	q	$p \rightarrow q$	$q \rightarrow p$	$(p \rightarrow q) \wedge (q \rightarrow p)$
1	1	1	1	1
1	0	0	1	0
0	1	1	0	0
0	0	1	1	1

This means then that the equivalence of p and q is merely a quick way of stating that both a statement $p \rightarrow q$ and its converse $q \rightarrow p$ are both true or both false simultaneously. More formally,

THEOREM 1.7.3. Let p and q be propositions. The propositions $p \leftrightarrow q$ and $(p \rightarrow q) \wedge (q \rightarrow p)$ are logically equivalent.

Equivalence statements occur quite frequently in mathematics. Some common examples are

$x + 1 = 2$ iff $x = 1$.
$|a| < b$ iff $-b < a < b$.
Let a, b be numbers greater than 0. Then $a^2 = b^2$ iff $a = b$.

With the notion of equivalence defined, we have completed the introduction of all the logical connectives needed to express precisely any statement in mathematics. Here are some examples.

EXAMPLE 1.7.4.
(a) The first statement given just above can be symbolized as

$$p \longleftrightarrow q$$

where

$$p\colon\ x + 1 = 2$$

and

$$q\colon\ x = 1\,.$$

(b) The last statement given just above can be symbolized as

$$p \rightarrow (q \longleftrightarrow r)$$

where

p: a, b are numbers greater than 0

q: $a^2 = b^2$

r: $a = b$.

(c) The following statement (which is about as logically complicated as these things get)

> If every continuous function on S is bounded iff S is pseudo-compact, and S is not pseudo-compact, then there is a continuous function on S which is not bounded

can be symbolized

$$[(p \longleftrightarrow q) \wedge -q] \rightarrow r$$

where

p: every continuous function on S is bounded

q: S is pseudo-compact

r: there is a continuous function on S which is not bounded.

From Theorem 1.7.3, $p \leftrightarrow q$ is logically equivalent to $(p \rightarrow q) \wedge (q \rightarrow p)$. From Theorem 1.5.2, it then follows that $-(p \leftrightarrow q)$ must be logically equivalent to $-(p \rightarrow q) \vee -(q \rightarrow p)$. Thus

THEOREM 1.7.5. Let p and q be propositions. The proposition $-(p \leftrightarrow q)$ is logically equivalent to $-(p \rightarrow q) \vee -(q \rightarrow p)$.

Problems.

I. Symbolize each of the following statements and its negation, and then write out this negation.

(1) If S is subset of E^n, then S is compact iff every continuous function on S is bounded.

(2) Let $a, b > 0$. Then $a^2 = b^2$ iff $a = b$.

(3) If S is normal, then S is Hausdorff whenever S is T_1.

(4) If H is a subgroup of G, then G/H is a group if H is normal.

(5) Given that S is a closed subset of X, it follows that S is compact.

II. State the converse of each statement in Problem IV, Section 6.

III. To the best of your knowledge, for which of the following pairs, does $p \leftrightarrow q$ have a truth-value of 1?

(1) p: n is a even number
q: $n = 2k$ for some whole number k.

(2) p: $0 \leq a \leq 1$
q: $0 \leq a^2 \leq 1$.

(3) p: $a = 2$
q: $a = \sqrt{4}$.

(4) p: f is a continuous function
q: f is a differentiable function.

(5) p: $3 + 1 = 2$
q: $3 = 1$.

(6) p: $3 + 1 = 2$
q: $1 + 1 = 2$.

8. THE PROOF OF IMPLICATIONS: DIRECT METHOD

What does the mathematician mean when he claims to have proven the implication $p \rightarrow q$, for given propositions p and q? He most emphatically does not mean that either p or q is a true statement about anything! He does mean that he has shown by one means or another that $p \rightarrow q$ has a truth-value of 1. Recalling the truth-table for implication, this must mean that he has shown that under the assumption that p is true, q must also be true. For p is either true or false. If p is false, then $p \rightarrow q$ has a value of 1 no matter what value q has. If, assuming p is true, he has been able to conclude that q is true, then all possibilities are exhausted and $p \rightarrow q$ is simply a true statement. In this section and the next we want to state and verify the most common methods of showing that an implication $p \rightarrow q$ is a true statement.

But there is perhaps a better way to convey what this section is about. Consider the human skeleton, it is quite a thing. But without its joints, it would be not much of a thing at all. A proof of a particular implication $p \rightarrow q$ is a skeleton of reasoning, a skeleton of that body of thought which led somebody to believe that given p, q must follow. This proof only holds his thoughts together, it is the bare bones of his reasoning and nothing more. What are its joints? They are the methods of proof described and characterized herein, and, as a skeleton has many different joints, a single proof may employ all these methods. These methods hold a proof together, giving it direction and flexibility, allowing one's reasoning to twist and turn, to latch onto this or that essential element of thought, and incorporate all into a single frame that can then go rigid on some printed page. Never more to be disturbed? Not hardly, but that's so esoteric a point it must be left for the reader to complete at some later date. To close this apt analogy, and so that he will better realize how little and how much this section does, one final point: there is a lot more to a skeleton than its joints!

There are two principal kinds of proof: direct proof and indirect proof. The most important form of proof in general, and of direct proof in particular, is the ***syllogism***, which has been around as a principle of logic since the days of Aristotle.

Suppose it is known that

If the light is on, then all is well.

and

If he is there, then the light is on.

By the most primitive of logical notions you would conclude

If he is there, then all is well.

This kind of reasoning is called syllogistic reasoning, and having for several sections carefully considered the connectives of logical thought we should be (and are) able to show that this kind of reasoning is merely a consequence of the meaning assigned to the connectives of logic and nothing more. That is, that the validity of this argument resides only in its form. Here is why

THEOREM 1.8.1. Let p, q and r be any three propositions. The proposition $[(p \to q) \land (q \to r)] \to (p \to r)$ has a truth-value of 1, ***no matter what*** be the truth-values of p, q and r.

Here, too, the finite reasoning by truth-tables comes to fore.

p	q	r	$p \to q$	$q \to r$	$p \to r$	$(p \to q) \land (q \to r)$	$[(p \to q) \land (q \to r)] \to (p \to r)$
1	1	1	1	1	1	1	1
1	1	0	1	0	0	0	1
1	0	1	0	1	1	0	1
1	0	0	0	1	0	0	1
0	1	1	1	1	1	1	1
0	1	0	1	0	1	0	1
0	0	1	1	1	1	1	1
0	0	0	1	1	1	1	1

One conclusion to be drawn from Theorem 1.8.1 is that you need not know what you're talking about in order to make sense, to utter logically valid statements. Pick *any* three propositions p, q and r and remark

If p implies q and q implies r, then p implies r

and you will be making sense, though perhaps not be recognized as sensible! In spite of this property of the syllogism, indeed because of it, the syllogism is the most frequent logical form found in mathematics. To examine its meaning and recognize its utility, we record

NOTE 1.8.2. Let n be a proposition. To show that n is a true proposition it suffices to do two things

(a) Find a proposition m such that (you can show) $m \to n$ is true.
(b) Show that m is true.

This altogether reasonable note is the key to using the syllogism, and is itself one of the most important ideas of this section. But reasonable is not enough, the validity of this note must have a basis in the logic so far developed. Here it is:

Suppose one has obtained (a) and (b). The claim is that n must be a true statement. Becase of (b), the possibilities for the implication $m \to n$ are restricted to rows (1) and (2) of the truth-table for implication. Bebause of (a), only row (1) remains pertinent. But in row (1), n is true. That is all.

Now suppose you wish to prove an implication $s \to t$. Let n denote the proposition $s \to t$. You can show n is true by doing the two things in Note 1.8.2. And here is where the syllogism comes in, it provides a candidate for m. The syllogism suggests that one take m to be a statement of the form

$$(s \to x) \wedge (x \to t) ,$$

for, by the syllogism, $m \to n$ is true no matter what x is, and condition (a) is automatically satisfied. Obtaining (b) is the problem, and this might be called the problem of finding a proof. The problem is to make a judicious choice of a proposition x for which you can actually show m to be true. But the syllogism does leave you absolutely free to choose x, you only aim being to make m true.

For example, consider a proof of the assertion (where a and b are numbers)

$$\text{If } a = b \text{, then } a^2 = b^2 .$$

It would suffice to prove the two implications

$$\text{If } a = b \text{, then the moon is made of green cheese}$$

and

$$\text{If the moon is made of green cheese, then } a^2 = b^2$$

though this would be a poor choice for x! A better choice for x would be

$$a^2 = ab \text{ and } ab = b^2$$

for, using the usual rules of algebra, one can prove

$$\text{If } a = b \text{ then } a^2 = ab \text{ and } ab = b^2$$

(by multiplying the given equation alternatively by a and by b) and

$$\text{If } a^2 = ab \text{ and } ab = b^2, \text{ then } a^2 = b^2$$

(by setting equals to equals). By the syllogism, the truth of these two implications assures the truth of the original implication.

This rather detailed yet simple discussion indicates how logic can be used to derive and verify forms of argument. Hopefully, it also indicates that the logic of a proof is really a property of the form in which a sequence of statements occur in the proof and not what the statements actually are.

Problems.

1. Given are statements s, t and u. Supposing the truth of s and t, does u follow? If not, indicate a situation which is a counterexample.

(1) s: If we have might, then we are right.
t: We are right.
u: We have might.

(2) s: If our cause is just, then we will succeed.
t: If we succeed, then we shall be acclaimed.
u: If we are acclaimed, then our cause is just.

(3) s: If a mirror turns left to right, then it must turn up to down.
t: A mirror turns left to right.
u: A mirror turns up to down.

(4) s: If one makes paper from dead trees, then a microfilm library helps to preserve the environment.
t: If a microfilm library helps preserve the environment, then we must be willing to accept the inconvenience of microfilmed books.
u: If one must kill a tree to make a few sheets of paper, then we must accept the inconvenience of microfilmed books.

(5) s: If we are burning fossile fuels, then we are using energy long ago absorbed by this earth.
t: We are burning fossil fuels faster than the earth stores new energy from the sun.
u: We are going to run out of gas.

(6) s: A society based on private avarice cannot endure in a situation of declining resources.
t: Resources are declining.
u: That society cannot endure.

(7) s: If the enemy is losing the war, his defections will increase.
t: His defections are increasing.
u: He is losing the war.

(8) s: If social institutions kept pace with the technology at their disposal, then bombs would be in museums.
t: If bombs are in museums, then children have a future.

u: If social institutions keep pace with the technology at their disposal, then children have a future.

II. Let p, q and r be statements. Suppose that one wishes to prove the truth of $(p \leftrightarrow q) \wedge (q \leftrightarrow r) \wedge (p \leftrightarrow r)$. Does it suffice to merely prove the truth of $(p \rightarrow q) \wedge (q \rightarrow r) \wedge (r \rightarrow p)$?

III. Prove, using the common laws of arithmetic and the syllogism, the truth of the implication $p \rightarrow q$ for p and q given as follows.

(1) p: $95 = 93$
q: $2 = 0$

(2) p: $2 = 1$
q: $4 = 1$

(3) p: a^2 is divisible by 2 for any whole number a
q: $\sqrt{2}$ is rational

(4) p: $2 = 1$
q: $1 + 1 = 2$.

9. THE PROOF OF IMPLICATIONS: INDIRECT METHODS

The syllogism is really the only important form of direct proof. There are two other important forms of proof, both referred to as indirect because they make use of negations and contradictions to assert truths. But they are as valid as the syllogism, for as we shall see their validity is based on the same agreements that establish the syllogism. The first of these methods is known as *reductio ad absurdum*—reduction to absurdity. Underlying its validity is a slightly controversial law of logic that also dates back to the time of Aristotle. It is expressed in the following

> THEOREM 1.9.1. *The Law of the Excluded Middle.* Let r be a proposition. The proposition $r \vee -r$ has a truth-value of 1, no matter what the truth-value of r is.

One can quickly verify this theorem using truth-tables. It does not perhaps seem controversial. Nevertheless, some highly reputable mathematicians (of the intuitionist school of thought) feel that a logical basis for mathematics should not rely on the unrestricted use of the "law of the excluded middle," essentially because this is a principle of logic originating in antiquity where it was used *only* (as in everyday use) to apply to statements about *finitely* many things. Since many, many statements in mathematics (e.g., The Goldbach conjecture: every even integer is the sum of two primes) involve infinitely many objects (there are infinitely many even integers) there is some validity to this objection. The Goldbach conjecture just mentioned has of yet not been shown to be true nor to be false and the reader may consult the excellent text by Meschkowski[5] to find out why one might wish to hold that it may be neither. We are developing a

[5] H. Meschkowski: *Evolution of Mathematical Thought*, San Francisco: Holden-Day Inc., 1965.

system of logic acceptable for most mathematics and mathematicians and this system does include the law of the excluded middle applied to statements about infinitely many objects.

It follows from Theorem 1.9.1 that $r \wedge -r$, being by Theorem 1.5.2(b) logically equivalent to $-(r \vee -r)$, must always be false for any proposition r. The idea behind the proof of an implication $p \rightarrow q$ by the method of *reductio ad absurdum*, is as follows: show that to assume the negation of $p \rightarrow q$ leads to a statement $r \wedge -r$ for some statement r, and hence that this negation cannot be true and thus $p \rightarrow q$ must be true. More precisely,

THEOREM 1.9.2. (*Reductio ad absurdum*) Let p, q and r be propositions. The proposition

$$[-(p \rightarrow q) \rightarrow (r \wedge -r)] \rightarrow (p \rightarrow q)$$

is true no matter what be the truth of p, q and r.

This, too, can be verified using truth-tables in a manner similar to that for the syllogism.

As with the syllogism, the use of this method of proof goes back to Note 1.8.2. Let n: $p \rightarrow q$. Find a proposition x such that

$$m\text{:} \quad -(p \rightarrow q) \rightarrow (x \wedge -x)$$

can be shown to be true. Then, by Note 1.8.2 and Theorem 1.9.2, n must be true. The syllogism, of course, may be used to prove m itself.

There is a more convenient way of stating this method of proof which is almost invariably the form that occurs in practice. Recalling from Theorem 1.6.2 that $p \wedge -q$ is logically equivalent to $-(p \rightarrow q)$, one would expect, and can show using truth-tables,

THEOREM 1.9.3. For propositions p, q and r,

$$[(p \wedge -q) \rightarrow (r \wedge -r) \rightarrow (p \rightarrow q)$$

has a truth-value of 1.

Here is an example of proof by this method. Suppose one is again to prove that for a and b real numbers,

$$\text{If } a = b\text{, then } a^2 = b^2\text{.}$$

A proof by *reductio ad absurdum* might then proceed as follows

Suppose that $a = b$ and $a^2 \neq b^2$. Then $a = b$ implies $a - b = 0$. Also, $a^2 \neq b^2$ implies $(a - b)(a + b) \neq 0$. This implies $a - b \neq 0$. Hence, the supposition $a = b$ and $a^2 \neq b^2$ implies $a - b = 0$ and $a - b \neq 0$. Therefore, $a = b$ implies $a^2 = b^2$.

The second and last important method of proof we consider is referred to as proof by contraposition. The statement

If it rains, then the grass is wet

says nothing more nor less than the statement

If the grass is not wet, then it does not rain .

Contrast this with the statement,

If the grass is wet, then it rains .

You can easily imagine plots of grass for which this last statement is false, while the first is true. Try as you might you should find that you cannot find such examples for the first and second statements. To see why this assertion is correct let us begin with the

DEFINITION 1.9.4. Let p and q be propositions. The ***contrapositive*** of the implication $p \to q$ is the implication $-q \to -p$.

and follow up with the

THEOREM 1.9.5. An implication and its contrapositive have the same truth-value. That is, the propositions $p \to q$ and $-q \to -p$ are logically equivalent.

Writing out the truth-table for these implications, we obtain

p	q	$p \to q$	$-q$	$-p$	$-q \to -p$
1	1	1	0	0	1
1	0	0	1	0	0
0	1	1	0	1	1
0	0	1	1	1	1

The conclusion to be derived from Theorem 1.9.5 is that wishing to prove $p \to q$ one might just as well prove $-q \to -p$, if this is for one reason or another more feasible. Wishing to then prove $-q \to -p$, one has access to the syllogistic and *reductio ad absurdum* methods for the contrapositive is again only an implication. And note that the converse and the contrapositive are entirely different things. The converse cannot be used to prove $p \to q$, the contrapositive can.

For example, wishing to prove

$$\text{If } a = b\,, \text{ then } a^2 = b^2$$

one can proceed as follows.

Suppose $a^2 \neq b^2$. Then $a^2 - b^2 = 0$. Since $a^2 - b^2 \neq 0$ then (upon factoring) $(a-b)(a+b) \neq 0$. Since $(a-b)(a+b) \neq 0$ then $a - b \neq 0$. Therefore $a \neq b$. Hence the proof.

Notice that this proof uses a syllogistic argument to verify the contrapositive of the assertion—a skeleton has many kinds of joints.

These then are the principal methods of proof in mathematics. They provide much of the formal logic you are likely to need in doing mathematics. They are the joints that hold the skeleton of reasoning in a proof together.

Problems.

(1) State the contrapositive of each statement in Problem IV of Section 6.

(2) Show that Example 1.2.1 is a correct logical argument.

(3) Is the conclusion of the following argument a logical consequence of the supposition?

Suppose that a man is honest and cannot be elected and that if a man is honest then he can be a politician and that if a man can be a politician then he can be elected. Then, if a man is honest, then he can be elected.

What form of argument is this? In what sense does logic assert that the conclusion is a true statement?

(4) Do you distinguish between logic, as described herein, and reason or logic, as used ordinarily? Give examples.

(5) In each of the following, two statements are given. The first is a theorem which you wish to prove and the second is what you're able to prove. In which cases do you have a proof of the first assertion? Symbolize each to check your answers.

(a) If f is differentiable then f is continuous.

(a′) If f is not continuous then f is not differentiable.

(b) If S is countable then S is finite or denumerable.

(b′) If S is countable and not finite then S is denumerable.

(c) Let f be defined on $[a, b]$. If f is bounded and the set of discontinuities of f has measure zero, then f is integrable.

(c′) If f is defined on $[a, b]$ and is not integrable, then f is not bounded or the set of discontinuities of f does not have measure 0.

(d) The same as (c).

(d′) If f is defined on $[a, b]$ and f is not integrable and f is bounded, then the set of discontinuities of f does not have measure 0.

(The transformation from (c′) to (d′) is an important one for a very practical reason. The hypothesis in (d′) gives one *three* assumptions to work with to obtain only *one* conclusion. Observe that $p \rightarrow (q \vee r)$ is logically equivalent to $(-q \wedge -r) \rightarrow -p$.)

(e) An even integer is the sum of two primes.

(e′) If the even number k is not the sum of two primes, then $2 = 1$.

(6) Symbolize the following theorem and its proof. If all implications in the proof are true, is the proof an instance of valid reasoning? You need not know the meaning of the statements to answer this question.

Theorem: If S is a finite set and T is a proper subset of S, then T is not an equivalent subset of S.

Proof: Suppose that S is a finite set and T is a proper subset of S and T is an equivalent subset of S. Since S is a finite set, then S has n elements. Since T is an equivalent subset of S, then T has n elements. Since T is a proper subset of S, then T has less than n elements. Hence the theorem.

(7) Use a *reductio ad absurdum* argument to prove the contrapositive of assertion: If $a = b$, then $a^2 = b^2$.

(8) Go to Problem III in Section 8 and try to prove, first by the method of contraposition and then by the method of *reductio ad absurdum*, the truth of the implication $p \rightarrow q$ in (1), (2), (3) and (4).

10. TAUTOLOGY

A *tautology* is a combination of propositions and/or their negations using the connectives $\wedge$, $\vee$, $\rightarrow$ and $\leftrightarrow$ as defined in previous sections such that this combination has a truth-value of 1 no matter what be the truth-value of the separate propositions involved. We have seen two important examples of tautologies, the syllogism, and the implication which justifies the *reductio ad absurdum* method of proof. Both these tautologies yield important methods of proof, and one can think of tautologies, or tautological statements, as laws of logic. In other studies a tautology is what is usually meant when one speaks of a law of logic.

Tautologies arise quite easily, more so than might be indicated by Section 9. In particular, when the connective $\leftrightarrow$ is involved, tautologies result any time this connective joins two logically equivalent statements. For, if m and n are logically equivalent, then they have the same truth-value, either 0 or 1, and by the truth-table for "$\leftrightarrow$", $m \leftrightarrow n$ must have value 1. With this in mind we can restate the rules of negation developed earlier in the following way.

THEOREM 1.10.1. Let p, q and r be propositions. Each of the following is a tautology.

(a) $-(p \vee q) \leftrightarrow (-p \wedge -q)$
(b) $-(p \wedge q) \leftrightarrow (-p \vee -q)$
(c) $-(p \rightarrow q) \leftrightarrow (p \wedge -q)$
(d) $p \leftrightarrow -(-p)$
(e) $-(p \wedge -p)$

We can also use the concept of tautology to summarize the laws of logic (methods of proof) previously developed.

THEOREM 1.10.2. For p, q and r propositions, each of the following is a tautology.

(a) (Law of the excluded middle) $p \vee -p$
(b) (The syllogism) $[(p \rightarrow q) \wedge (q \rightarrow r)] \rightarrow (p \rightarrow r)$
(c) (*Reductio ad absurdum*) $[(p \wedge -q) \rightarrow (r \wedge -r)] \rightarrow (p \rightarrow q)$

(d) (Law of detachment) $[p \wedge (p \to q)] \to q$
(e) (Law of contraposition) $(p \to q) \leftrightarrow (-q \to -p)$.

Notice that (d) is no more nor less than Note 1.8.2.

In the problems at the close of this section various tautologies are listed, all of which one can verify using truth-tables. Do not try to remember these things as abstract formulas to be used for reasoning. Look at them for the logic they express. See how they express, in many cases, the way you ordinarily reason without ever being aware of formal or symbolic logic. All of these tautologies indicate valid forms of reasoning, all can be used to check out the reasoning you will use in subsequent chapters.

Problems.

I. Show that for propositions p, q and r the following are tautologies

(1) $[p \vee (q \vee r)] \leftrightarrow [(p \vee q) \vee r]$
(2) $[p \wedge (q \wedge r)] \leftrightarrow [(p \wedge q) \wedge r]$
(3) $[p \wedge (q \vee r)] \leftrightarrow [(p \wedge q) \vee (p \vee r)]$
(4) $[p \vee (q \wedge r)] \leftrightarrow [(p \vee q) \wedge (p \wedge r)]$.

The next few provide guides as to how to prove implications with compound hypotheses or conclusions, by providing logically equivalent combinations of "simpler" implications.

(5) $[(p \wedge q) \to r] \leftrightarrow [p \to (q \to r)]$
(6) $[(p \vee q) \to r] \leftrightarrow [(p \to r) \wedge (q \to r)]$
(7) $[p \to (q \wedge r)] \leftrightarrow [(p \to q) \wedge (p \to r)]$
(8) $[p \to (q \vee r)] \leftrightarrow [(p \to q) \vee (p \to r)]$
(9) $[p \to (q \vee r)] \leftrightarrow [-r \to (p \to q)]$
(10) $[p \to (q \vee r)] \leftrightarrow [(p \wedge -r) \to q]$.

The following express variants of the method of *reductio ad absurdum.*

(11) $[(p \wedge -q) \to -p] \to (p \to q)$
(12) $[(p \wedge -q) \to q] \to (p \to q)$.

These last few can be used to prove outright propositions.

(13) $[-p \to p] \to p$
(14) $[-p \to (r \wedge -r)] \to p$
(15) $[-q \wedge (-p \to q)] \to p$.

II. With p and q propositions, replace the symbol $\triangle$ in the expression $(p \triangle q) \leftrightarrow (q \triangle p)$ alternatively by $\vee$, $\wedge$, $\to$, and $\leftrightarrow$. In which cases do tautologies result?

III. Each of the next statements is a tentative restatement of those in Problem I, Section 7. Which of these are logically equivalent to those given statements? You might try to use the tautologies listed above to determine this in some cases.

(1) If S is compact iff every continuous function on S is bounded, then S is a subset of E^n.
(2) If $a, b > 0$, then $a = b$ iff $a^2 = b^2$.
(3) If S is normal and S is T_1, then S is Hausdorff.

(4) If H is a subgroup of G and G/H is not a group, then H is not normal.
(5) It is false that S is a closed subset of X and S is not compact.

IV. Which of the tautologies in (I) justifies the logic of the Russell contradiction? Despite the almost unreasonable, or at least strange appearance of this tautology, this author asserts that he finds it more absolutely believable than many things he encounters. What reasons might one have for that assertion?

V. Which of the above tautologies justifies the logic of the following proof?

Theorem: $\sqrt{2}$ is irrational.

Proof: Suppose $\sqrt{2}$ is rational. Then $\sqrt{2} = s/t$ where s and t are integers with no common factor (i.e., s/t is in lowest terms). Hence $s^2 = 2t^2$ and therefore s is an even integer. Let us write $s = 2k$ where k is some integer. Then $4k^2 = 2t^2$ or $2k^2 = t^2$. Hence t must be an even integer. But then s and t have a factor, namely 2, in common, contradicting the fact that s/t is in lowest terms. Hence $\sqrt{2}$ is not rational.

VI. The following diagrams represent electrical circuits with switches p, q, and r. If a switch is closed, current flows—correspond this to a truth-value of 1. If open, current does not flow—correspond this to value of 0. If p is a switch, $-p$ is a switch which is closed when p is open, open when p is closed.

(1) Show that the second circuit carries current whenever the first does, but not conversely.

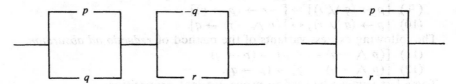

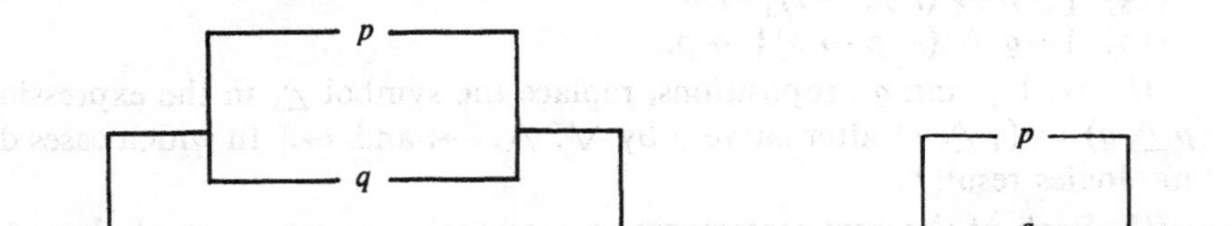

(2) Show that these circuits are "electrically equivalent."[6]

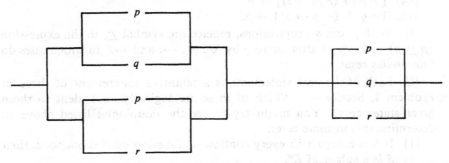

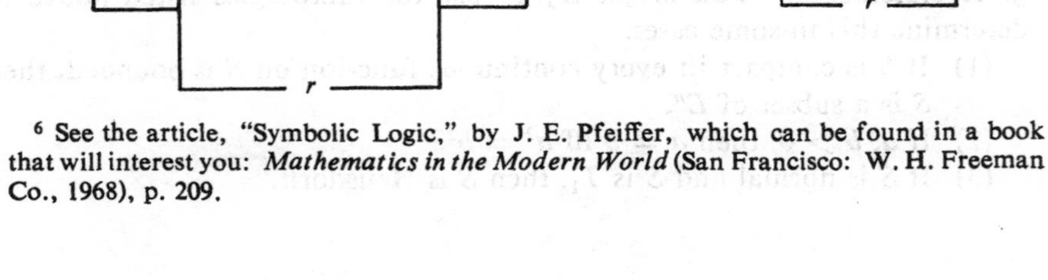

[6] See the article, "Symbolic Logic," by J. E. Pfeiffer, which can be found in a book that will interest you: *Mathematics in the Modern World* (San Francisco: W. H. Freeman Co., 1968), p. 209.

11. SOME ODDS AND ENDS

The previous sections contain enough information about the place and use of logic in mathematics, and about the meaning of the connectives *and*, *or*, *implies* and *if and only if*, to serve most purposes in the study of mathematics. The study of logic can easily fill a semester's work and there are many other things of basic importance to the study of higher mathematics. In subsequent chapters you will have to use fully the small amount of logic developed herein, and not much more. Through its use it will become more familiar and more meaningful. Because of what we have done in this chapter, you will know what justifies the form of your reasoning, nothing more. Hopefully, you will then be able to concentrate on content and not form because of this study of form.

The material of this section consists of bringing to your attention some common grammatical forms of expression found especially in mathematics. Mathematicians, because of the necessity for precision of expression, are limited in the ways they can state the mathematical results of their work. They do, however, find that expression and sometimes meaning flow more easily if words other than "if . . . then . . ." or "implies" are used to state implications. Some variants are listed below.

First of all, there are six common ways of writing an implication

(a) If p then p.

These are

(b) p implies q

(c) q if p

(d) p only if q

(e) p is a sufficient condition for q

(f) q is a necessary condition for p.

Thus, one might write any one of the following and yet mean the same thing.

(a′) If $a = b$, then $a^2 = b^2$

(b′) $a = b$ implies $a^2 = b^2$

(c′) $a^2 = b^2$ if $a = b$

(d′) $a = b$ only if $a^2 = b^2$

(e′) $a = b$ is a sufficient condition for $a^2 = b^2$

(f′) $a^2 = b^2$ is a necessary condition for $a = b$.

Secondly, an equivalence statement

(a″) p iff q

might instead be written as

(b″) p implies q and conversely

(c″) If p then q and conversely

(d″) p is a necessary and sufficient condition for q

(e″) q is a necessary and sufficient condition for p.

Problems.

(1) Restate each implication in Problem IV of Section 6 once using (in

the order given) the words *implies, sufficient, necessary, only if, if.* For example, restate (1) using *implies,* (2) using *sufficient,* etc.

(2) Is the following a tautology?

If p is a necessary condition for q and q necessary and sufficient for r, then r holds only if p holds.

12. NEGATION IN THE MATHEMATICAL IDIOM

One seemingly quite homogeneous trait of human nature is the tendency of one person to stereotype whole classes of other persons whom he may know little about. The idea seems to be one of placing each class of persons which shares a common property into a little mental box accompanied by some neat label by which all in that box can be quickly, if not fairly, dealt with. Examples abound. They usually begin with the word *all* or an equivalent: all students . . . , all professors . . . , all the poor . . . , all the rich . . . , all generals . . . , all politicians . . . .

One can speculate, though not prove, that the causes of such generalizations fit quite well with the causes for their appearance and extensive use in mathematics. An underlying cause for generalizations does often appear to involve the existence of complex issues involving the existence and actions of virtually unknown quantities combined with the need of he who generalizes to come to grips with the issues in ways that make sense to him. The need to generalize, albeit possibly incorrect, can serve a positive purpose. Generalization is indeed a recognized route to deeper understanding, the trick of it being to appropriately temper generalization as understanding increases.

This is to some mathematical point. Mathematics is complex. Mathematics is abstract. In the highly idealized world of mathematical thought one is always trying to understand the new and unknown and often in an unfamiliar setting. Not surprising then that the hypothesis of a mathematician is likely to be a generalization of the type

All x of this kind is a y of that kind .

For example, it may not be enough to know that a given statement is true about just a few numbers to draw any conclusion from it. For, if in the course of *abstract* reasoning begining with the given statement, one arrives at some number *abstractly defined,* how is one to know whether the given statement applies to this particular number or not? If, on the other hand, one knows that a given assumption is true of *all* numbers, then he is free to apply it to *any particular* number no matter how abstractly defined, no matter the way in which it arises. For such good reasons statements which assert that something is true *for all* or *for every* appear with maddening frequency in mathematics. As in everyday thought, the use of such assumptions is a technique for *simplification,* allowing one to obtain some definite conclusion.

For the idealized world of mathematical thought, the use of these words to idealize a situation is most appropriate. Yet one could conjecture that the use of such words in mathematics must limit the applicability of mathematics, as they might other subjects. This is not the case in part because of that one concept unique, thus far, to mathematics—infinity. In order to reach out to infinity and obtain something other than a paradox, the mathematician finds both power and a deft mental touch in the concept of *for all*. Consider the sequence of numbers

$$1\,,\ \frac{1}{2}\,,\ \frac{1}{3}\,,\ \frac{1}{4}\,,\ldots$$

which you would say has "limit" 0. What property does 0 have, with respect to this sequence, which distinguishes it from *every other* real number? That is, is there a statement about the numbers 1, 1/2, . . . and an arbitrary number L which is true only about the number $L = 0$? It took mathematicians several centuries to arrive at the best answer. Only the number $L = 0$ satisfies the statement

DEFINITION 1.12.6. For *every* positive number, a, there is a whole number, N, such that

$$\left|\frac{1}{n} - L\right| < a \quad \text{for } all \quad n \geqq N\,.$$

Through the use of the *for every* concept, the mathematically vague notion of "$1/n$ approaching 0 as n approaches infinity," which had existed in such a naive form for centuries and caused much misunderstanding of the calculus, can be replaced by quantitative statements subject to *arithmetical analysis*. This was a key step forward in the development of mathematics and it will be considered in detail in a subsequent chapter.

This is a brief case for the wide use of statements of generality, beginning with assumptions of the *for all* or *for every* type, in mathematics. In part because of statements of generality, another kind of statement occurs throughout most mathematical discussions. These are called existence statements and occur in two ways: in their own right and as negations of statements of generality.

Consider the statement, attributed to W. C. Braun,

No man can be a patriot on any empty stomach.

This is a statement of generality, referring to all men. Supposing it is true or false and not both, what is a statement which if true makes Braun's false and if false makes it true? An appropriate answer seems to be

There exists a man with an empty stomach and who is a patriot.

Notice that this negation involves more than adding a word such as *not* and in fact it is not readily apparent how this negation relates to the logic so far developed.

The negation of Braun's statement is called an existence statement. It asserts that something (a man) *exists* with a certain property. Such statements occur as negations of statements of generality and, nicely enough, vice versa! Consider the statement

There exists a man without a country.

What statement which, if true, makes this false; if false, makes this true? The answer appears to be

All men have a country.

Thus generality is the other side of the coin of existence.

Statements of generality and of existence and especially their negations are the subject matter of Sections 1.12, 1.13 and 1.14. Such statements predominate in mathematical discussions. Forgive the pedantic tone, but if the reader is to appreciate the finest thoughts that mathematics has to offer, then he must develop a natural feeling for the meaning and the use of such statements in an abstract setting. Unlike the logic of the past sections, which is more or less naturally acquired, the material in this section must really be learned. There are rules by which to learn it, and to a point the material is essentially technical and subject to mindless manipulation by the rules, but almost surely, one will not be able to use what he learns without constant direction unless he involves his mind a bit further. One can rule-book his way through this section, and will get through, but with only a few lasting thoughts of the journey. The rules to be given are only a step intermediate to understanding, as much as the rules for steering a canoe down a mountain stream are only intermediate to acquiring a feeling for canoeing that can free one to appreciate the journey itself.

In the following problems, the reader is asked to operate on statements of generality and existence intuitively, without specific rules. Following this we will make the whole matter more precise and derive general rules to encompass all of these examples.

Problems.

I. Give examples which show that each of the following statements is false

(1) All integers are even.

(2) All integers are not even.

II. What is a good way of expressing the negation of

All integers are even.

III. For each of the following statements, give a statement which, if true, makes the given statement false, and, if false, makes the given statement true.

(1) All of us are asleep.
(2) There is an x which is not a y.
(3) Everything is related to everything else. (This is, of course, the fundamental assumption of ecological studies.)
(4) There is an x such that $x^2 = -1$.
(5) If all of us are seated comfortably, then we are all asleep.

IV. Which of the latter two diagrams is the negation of the first in the sense of "Everything is related to everything else?"

V. Design a classroom with the following property: For every hour of the day past noon, there is a seat in the room such that every person seated to the right of this seat is bathed in sunlight.

VI. The problem is unlike any so far encountered in this chapter. Mathematics, like any trade or profession, has its special way of saying things suited to its purposes, and if one doesn't understand and have some feel for a trade language, he will miss a lot of whatever is going on. Try talking with an insurance agent or real estate agent for example. (And beware! "Ordinary Life" insurance is anything but ordinary simple life insurance!) This problem asks you to find *meaning* in the following typically mathematical statements. Each is a statement about a number, L—note that first. But each also involves other numbers: a (any real number) and n or N (whole numbers), upon which some qualifications are placed. The problem is to choose a particular value for L and show that the resulting statement is true about that value of L. Then choose a value for L for which the statement is false about L.

For example, consider Definition 1.12.1. Choose $L = 0$. Then the statement in Definition 1.12.1 is true. For if a is any positive number, then there does exist a whole number, N (take N to be greater than $1/a$) for which Definition 1.12.1 is true. For, if $N > 1/a$, then for all $n \geq N$ one has $|1/n - L| = |1/n - 0| = |1/n| = 1/n \leq 1/N < a$. Since this argument can be made for any number a, Definition 1.12.1 with $L = 0$ is a true statement.

For $L = 2$ the statement is false. For, *there is* (negation of generality) a number a, $a = 1$ such that no matter how one chooses N, then for an $n \geq N$, one has $|1/n - L| \geq 1 = a$. Now you try it. Find a number L for which the following are true and find a number L for which each is false. This problem is but an abstracted version of Problem V above.

(1) There is a positive number $a > 0$ and a number n such that $|1/n - L| \geq a$.
(2) There is a positive number $a > 0$ such that for any number N there

is an $n \geq N$ such that $|1/n - L| \geq a$.

(3) For every integer n there is a positive number $a > 0$ such that $|1/n - L| < a$.

(4) For every integer n there is a positive number $a > 0$ such that $|1/n - L| \geq a$.

(5) Is any one of the statements in (VI) a negation of Definition 1.12.1?

(6) Why is $L = 0$ the only number which satisfies Definition 1.12.1?

13. QUANTIFIERS AND PROPOSITIONAL FUNCTIONS

The previous discussions should leave one with several impressions. One is that the problem of negating statements can become more involved than one might have thought. Another is that it appears that if something is *not* true for *all*, then this *only* means *there is* at least *one* thing (number, person, etc.) for which it is not true, while if it is not true that *there is* a certain kind of thing (number, person, etc.) then *all* things (numbers, persons, etc.) are *not* of that kind. Finally, there ought to be a way to handle such an apparent rule in a more precise and universal way. There is, and this is what we will now take up. The key concept is that of a *propositional function.*

Consider the statement

$$x \text{ is an even integer} .$$

This is a statement as previously defined but is not a proposition, because one cannot assign a truth-value to it without knowing what x is. If x is replaced by 6, the resulting statement is a proposition with a truth-value of 1. If x is replaced by 3, the result is a proposition with value 0. Such a statement is an example of a propositional function. Another propositional function, this one involving two "unknowns" or indeterminates, is

$$x \text{ is a } y ,$$

not an uncommon phrase at all. When a particular meaning is assigned to x and to y, one has a statement which in some context may then be regarded as a proposition. Another such form is

$$(x \text{ is an even integer}) \wedge (x \text{ is greater than } 2) .$$

All of these statements are examples of propositional functions. The following general definition will serve the purposes of this text.

DEFINITION 1.13.1. *A propositional function* is a statement about one or more symbols $x, y, z, \ldots$ which becomes a proposition when a particular meaning is assigned to these symbols. The symbols $x, y, z \ldots$ are called the *indeterminates* of the propositional function.

As always, it is handy to have some notation. We will commonly denote a propositional function having a single indeterminate by $p(x)$ when the indeterminate is denoted by x, (or $p(y)$, or $p(\theta)$ if the indeterminate is denoted instead by y or by θ). We will agree that the particular symbol for the indeterminate is not important. That is,

$$p(x): \; x \text{ is an even integer}$$
$$p(y): \; y \text{ is an even integer}$$

are thought of as the same propositional function because replacing the indeterminate in one of these by the symbol for the indeterminate in the other gives the other.

If a propositional function involves two indeterminates, we will use the notation $p(x, y)$. Thus,

$$p(x, y): \; x \text{ is a } y$$

is a propositional function which is considered to be the same as

$$p(\theta, \gamma): \; \theta \text{ is a } \gamma\,.$$

With the concept of a propositional function we can begin dealing in a logical fashion with a whole new class of propositions, including all those considered up to now in this section, and also including all propositional forms one is likely to encounter in mainstream mathematics.

A propositional function is not a proposition, it has no truth-value. For a given meaning assigned to the indeterminate or indeterminates, the propositional function becomes, by definition, a proposition having a truth-value. But this is *not* the way by which propositional functions justify their existence and utility.

Again let $p(x)$ be the statement

$$x \text{ is an even integer}\,.$$

There are two special *propositions* which can be formed with the use of $p(x)$. These are

$$P_a: \text{ For all } x, p(x)$$
$$P_e: \text{ There exists an } x \text{ such that } p(x)\,.$$

The proposition P_a has the truth-value of 0, the proposition P_e has the value 1, in the sense that while some meanings for x make

$$x \text{ is an even integer}$$

a false statement, there is at least one meaning for x, namely x being the number 2, such that

$$x \text{ is an even integer}$$

is a true statement. To begin formalizing these ideas we make the

DEFINITION 1.13.2. If $p(x)$ is a propositional function, then an *interpretation* of x is a meaning assigned to x for which $p(x)$ is a proposition.

For example, if

$$p(x)\text{: } x \text{ is an even integer}$$

then $x = 4$, $x =$ your professor, $x = \int_0^1 y\,dy$, and $x = \int_0^2 y\,dy$ are all interpretations of x which make $p(x)$ a proposition with truth-values of 1, 0, 0 and 1, respectively.

It would seem reasonable in mathematics to rule out your professor and other irrelevant things as interpretations to be considered in statements like

$$x \text{ is an even integer}$$

without having to say so each time such a propositional function occurs. We do the reasonable thing; we restrict the possible interpretations of the indeterminate to some natural class. For the propositional function above we will, without saying so, understand that the only interpretations of x we care to consider are numbers. If

$$q(x)\text{: } x \text{ is a continuous function}$$

then we will agree only to consider functions as interpretations for x. In general then, along with a given propositional function, there will always be some fixed class of interpretations for its indeterminates.

The following definition is the important one for this section. It concerns propositional functions with only one indeterminate and it is left to you to make a similar definition for situations where there is more than one, if the need arises.

DEFINITION 1.13.3. Let $p(x)$ be a propositional function and let $\mathscr{C}$ be a class of interpretations for x. The statement

$$P_a\text{: For all } x, p(x)$$

is given a truth-value of 1 if $p(x)$ has a truth-value of 1 for each interpretation of x in $\mathscr{C}$; if $p(x)$ has a truth-value of 0 for at least one interpretation of x in $\mathscr{C}$ then P_a is given a truth-value of 0. The statement

$$P_e\text{: There exists an } x \text{ such that } p(x)$$

has a truth-value of 1 if for at least one interpretation of x taken from $\mathscr{C}$, $p(x)$ has value 1; if $p(x)$ has a truth-value of 0 for each interpretation of x in $\mathscr{C}$, then P_e is given a value of 0.

Propositions of the type defined in Definition 1.13.3 occur with such frequency in mathematical discussions that some nearly universal notation is used to express them. Ordinarily, the statement P_a is written

$$\forall x, p(x)$$

and is read "For every x, $p(x)$ is a true statement." The statement P_e is written

$$\exists\, x \ni p(x)$$

and is read "There exists an x such that $p(x)$ is a true statement." Statements in either of these two forms are referred to as *quantified* statements and the symbols "$\forall$" and "$\exists$" are called *quantifiers*. The symbol "$\ni$" is simply a replacement for the words "such that."

Here are some examples. The statement

There is a number greater than 2

might be written

$$\exists\, x \ni x > 2$$

or simply

$$\exists\, x > 2$$

where the indeterminate x is understood to represent a number. And notice that this last abstract, abbreviated statement still has the meaning of the original statement. That's intended of course, but the point is that if you want quantified statements to be of use to you, you must keep their meaning in mind. More importantly you should not let the notation lose the meaning of that which it represents and become only a sequence of symbols.

The statement

All positive numbers less than one have a square root less than one

might be written

$$\forall x\,,\ \text{if}\ 0 < x < 1\,,\ \text{then}\ \sqrt{x} < 1$$

which is a much better way to say this. It is better because it is more precise, and especially because it is in a ***standard logical form.*** The problem of reinterpreting statements into standard logical forms is most important in this section and will soon be considered in more detail.

Finally, the statement

For every positive number, a, there is an integer, N, such that $|1/n - 0| < a$ for all $n \geq N$

could be written

$$\forall a > 0,\ \exists\ \text{an integer}\ N \ni \left|\frac{1}{n} - 1\right| < a,\ \forall n \geq N\,.$$

Notice how the quantifiers are used to place restrictions, or assert the existence of certain qualifications, on the subject of this statement, the propositional function $|1/n - 0| < a$ with two indeterminates n and a. These quantifiers provide a concise way of stating precise *generalities* and *restrictions—for all n* but only greater than some N that *does exist*—on this inequality. That is a way to *think* about the role of quantifiers, to manipulate them requires more precision.

The negations of each of the above statements, in the order given, could be written

$$\forall x, x \leq 2$$

$$\exists x \ni 0 < x < 1 \text{ and } \sqrt{x} \geq 1$$

$$\exists a > 0 \ni \forall N \ni n \geq N \ni \left| \frac{1}{n} - 0 \right| \geq a$$

and you should be wondering where all this comes from, particularly the last negation. The following rules enable one to arrive at these negations in an almost mindless way. The remainder of this section will be devoted to the use of

RULE 1.13.4. Let $p(x)$ be a propositional function concerning some fixed class of interpretations for the indeterminate x

(a) The proposition $-[\forall x, p(x)]$ is logically equivalent to the proposition $\exists x \ni (-p(x))$.

(b) The proposition $-[\exists x \ni p(x)]$ is logically equivalent to the proposition $\forall x, (-p(x))$.

Here is the justification for (a). Let $\mathscr{C}$ denote the relevant class of indeterminates x. Let P denote the proposition $-[\forall x, p(x)]$ and Q the proposition $\exists x \ni (-p(x))$. We are to show that P and Q have the same truth-value. Suppose the truth-value of P is 1. Then, $\forall x, p(x)$ has truth-value 0. Hence by Definition 1.13.3, $p(x)$ must have a truth-value of 0 for at least one interpretion of x in $\mathscr{C}$. For this interpretation, $-p(x)$ then has a truth-value of 1 and hence by Definition 1.13.3 $\exists x \ni -p(x)$ has value 1. Hence, the truth-value of Q is 1. Now suppose the truth-value of P is 0. Then $\forall x, p(x)$ has a truth-value of 1. Thus $p(x)$ has a truth-value of 1 for each interpretation of x and thus $-p(x)$ has at truth-value of 0 for each interpretation of x. Hence, by Definition 1.13.3, $\exists x \ni (-p(x))$ has a truth-value of 0. Thus, Q has value 0 when P has value 0. Hence, P and Q are logically equivalent.

The remainder of this section is devoted to the use of Rule 1.13.4. Since it is rather a simple rule, you might find it surprising that much study should be devoted to it. The reason is simple enough: if there were strict rules for the formation of sentences in mathematics, we could stop this study of logic and grammar right here, Rule 1.13.4 and a little practice would get you through any difficulties that might arise. But this is not the case, and sometimes formidable linguistic difficulties arise in trying to apply these rules. The problem that invariably occurs is that of recognizing how a given statement fits the forms $\forall x, p(x)$ or $\exists x \ni p(x)$. Again one cannot overemphasize how useful it can be to be able to recognize quantified statements and view them in a form which is both clear to you and amenable to logical manipulation according to the rules developed in this and previous sections. With this capability, you will be several times more able to understand and to express precise mathematical statements. And a final suggestion before considering some examples: your goal in this section should be to learn how to use these rules so that you can forget them, and thus perform what logical manipulations you require naturally and correctly with no more effort than in ordinary speaking or writing.

EXAMPLE 1.13.5. Consider the statement

m: For every x in A, x is in B

or equivalently,

m: $\forall x$, if x is in A then x is in B.

Here the natural class of interpretation for x is those things "in A." With $p(x)$ denoting the statement

If x is in A, then x is in B.

Rule 1.13.4 gives

$$-m\colon\ \exists x \ni -p(x)$$

and, by Theorem 1.6.2,

$-m$: $\exists x \ni x$ is in A and x is not in B.

Perhaps you think this is making too much of too little. Consider now

EXAMPLE 1.13.6. Let y denote a fixed number.

m: For all numbers $\varepsilon > 0$, there is a number a such that $y - \varepsilon < a < y$.

which would ordinarily be written

$$m\colon\ \forall \varepsilon > 0 \quad \exists a \ni y - \varepsilon < a < y$$

with the understanding that ε and a are to be thought of as numbers. You will notice that the proposition m involves two quantifiers. You will also notice a symbol y which has no quantifier. It is a good idea when needing to think about or negate such statements to be certain which entities are quantified and which are not.

An initial negation of m, by Rule 1.13.4 would be

$-m$: $\exists \varepsilon > 0 \ni$ not: $\exists a \ni y - \varepsilon < a < y$.

or

$-m$: $\exists \varepsilon > 0 \ni$ it is false that $\exists a \ni y - \varepsilon < a < y$.

There is nothing wrong with this negation, it is only not right enough! It lacks a certain quality which might be called "substance and utility." If, for one reason or another, one wanted to use $-m$ to reach some further conclusion, then as the negation now stands one doesn't have much to reason with. The statement $-m$ does say there is a number $\varepsilon > 0$ for which a certain statement involving ε is false, hardly the kind of hypothesis to inspire confidence in some eventual conclusion! This is a case for carrying the negation of m further—to obtain something with more substance and meaning.

According to Rule 1.13.4, the statement

$$\text{It is false that } \exists a \ni y - \varepsilon < a < y$$

is logically equivalent to

$$\forall a,\ y - \varepsilon < a < y \text{ does not hold}$$

which can be written

$$\forall a,\ y - \varepsilon \geq a \text{ or } a \geq y.$$

Consequently, if

$$m\text{: } \forall \varepsilon > 0\ \exists a \ni y - \varepsilon < a < y$$

then

$$-m\text{: } \exists \varepsilon > 0 \ni \forall a,\ y - \varepsilon \geq a \text{ or } a \geq y.$$

which is a more "thinkable" statement than the inititial negation of m. This statement as a hypothesis for further argument gives one the following two pieces of hard information.[7]

(1) A number $\varepsilon > 0$.

(2) The fact that for *any* number a either $y - \varepsilon \geq a$ or $a \geq y$.

In a subsequent problem it will be seen how the mathematician uses such information to assert all kinds of things.

EXAMPLE 1.13.7. Consider again Braun's statement

No man can be a patriot on an empty stomach.

As soon as one tries to rewrite this statement in a quantified form, he sees how imprecise and logically unclear a form of expression ordinary grammar is. A more logically precise restatement is

Any man with an empty stomach is no patriot

or better yet,

[7] About as hard a piece of information that one usually gets in such a setting.

For every man, if a man has an empty stomach, then he is no patriot.

Certainly a more precise logical form; certainly less ringing to the ear!

Here is how to negate this statement. By Theorem 1.6.2, the negation of an implication $p \rightarrow q$ is $p \wedge -q$. With this in mind, and using Rule 1.13.4(a), the negation is

There is a man such that this man has an empty stomach and is a patriot,

the same negation arrived at intuitively at the start of this section.

Perhaps you're wondering: where did all the formal notions of indeterminates and propositional functions go in arriving at this negation? These were purposefully left out in line with the earlier suggestion that one should try to learn to apply the *sense* rather than the *letter* of the rules in Rule 1.13.4. But if that's getting ahead of you at this time, let the class of indeterminates for this proposition be the class of all men in a given country and let

$$q(x)\text{: } x \text{ has an empty stomach}$$

$$r(x)\text{: } x \text{ is no patriot}$$

$$p(x)\text{: } q(x) \rightarrow r(x)\,.$$

With this notation, Braun's original statement can be written

$$\forall x,\, p(x)$$

with negation

$$\exists x \ni -p(x)$$

according to Rule 1.13.4. By Theorem 1.6.2 $-p(x)$ is logically equivalent to $q(x) \wedge -r(x)$, whence the negation is

$$\exists x \ni q(x) \wedge -r(x)$$

or, in ordinary language,

There is a man such that this man has an empty stomach and is a patriot.

EXAMPLE 1.13.8. The examples to this point have always involved some knowledge of the content and meaning of the given statement. This was necessary because the original statements were not always given in a form amenable to mindless logical manipulation according to the rules. To make the point that you can understand the logical structure of a quantified sentence and obtain its negation even when you don't know what the sentence is about, provided it is given in logical form, consider the statement *m*: For every x in A there exists a neighborhood U of x such that U is contained in A.

Again, this involves two quantifiers and it is good to first decide on those entities they qualify. Evidently "for every" refers to "x's in A" and "there exists" refers to something called a "neighborhood." Applying Rule 1.13.4(a), an initial negation of n is

$-n$: $\exists$ an x in A such that it is false that $\exists$ a neighborhood U of x such that U is contained in A.

Applying Rule 1.13.4(b), this can be negated to a more positive assertion:

$-n$: $\exists$ an x in A such that $\forall$ neighborhood U of x, U is not contained in A.

If you care to be more precise about the method for doing this, let

$$p(x)\text{: } \exists \text{ a neighborhood } U \text{ of } x \ni U \text{ is contained in } A$$

$$q(U)\text{: } U \text{ is contained in } A$$

and rewrite

$$n\text{: } \forall x \quad (\text{in } A), p(x)$$

$$p(x)\text{: } \exists U \quad (\text{a neighborhood of } x) \ni q(U)$$

or, with the interpretations of the indeterminates unmentioned

$$n\text{: } \forall x, p(x)$$

$$p(x)\text{: } \exists U \ni q(U)\,.$$

Then, by Rule 1.13.4(a)

$$-n\text{: } \exists x \ni -p(x)$$

and by Rule 1.13.4(b)

$$-p(x)\text{: } \forall U, -q(U)$$

whence

$$n\text{: } \exists x \text{ in } A \ni \forall u, -q(u)$$

or

$$-n\text{: } \exists x \text{ in } A \ni \forall \text{ neighborhood } U \text{ of } x, U \text{ is not contained in } A$$

where the interpretations of the indeterminates are respecified.

The procedure of the past two examples was to apply the sense of the rules in Rule 1.13.4 and then see that the results and procedures fit the letter of the rules in the form in which they are stated. Such a loose manner of logically operating should not be tolerated in a discussion about logic and its uses *per se*. That is not the aim of this section. Here you should be interested in results, obtained as quickly and as easily as possible, with the only requirement that they be correct. It is left to you to decide just how detailed a look *you* must take at each sentence in order to obtain its correct negation.

Problems.

I. Justify (b) of Rule 1.13.4.

II. Define propositional functions and a relevant class of interpretations such that each of the following statements can be written in the form $\forall x, p(x)$ or $\exists x \ni p(x)$. Then negate each statement according to Rule 1.13.4.

(1) All of us are asleep.
(2) There is an x which is not a y.
(3) There is an x^2 such that $x = -1$.
(4) For any x, $x^2 \geq 0$.

III. State the negation of each of the following statements. Be sure your answer does indeed have an opposite truth-value.

(1) Every x has property P.
(2) $\exists A \ni A \cap B$ is not empty.
(3) $\forall x$ and y in H, xy^{-1} is in H.
(4) If every continuous function on S is bounded, then S is compact.
(5) If every x has property P, then z holds and $\exists x$ with property Q.
(6) $\exists x \ni 0 < x^2 < 1$.
(7) If a and b are numbers and $a > b$, then $a + \varepsilon > b \; \forall \varepsilon > 0$.
(8) If a cause is just, then it will succeed.
(9) If $x > 1$, then $\log x > 0$.
(10) If a man has no rights left, then he is free again.
(11) Freedom is just another word for nothing left to lose.

IV. Consider the following pairs of statements. Is the second the negation of the first? If not, what then is the correct negation?

(1) $\exists$ an open set V containing A such that every open set U which hits V also hits A.
(1′) $\forall$ open set V containing A $\exists$ an open set U such that U hits V and U does not hit A.
(2) $\forall$ open set U containing A $\exists$ a closed set C which hits U and misses A.
(2′) $\exists$ an open set U containing A such that $\forall$ closed set C, C does not hit U and C does not miss A.
(3) If a is a limit point of S, then there exists an infinite sequence $\{x_n\}$ of points in S such that $\lim_{n\to\infty} x_n = a$.
(3′) a is a limit point of A and for every infinite sequence $\{x_n\}$ of points in A, $\lim_{n\to\infty} x_n \neq a$.

V. Is each of the second statements in the following pairs a consequence of the negation of the first?

(1) $\forall \varepsilon > 0 \; \exists \delta > 0 \ni p(\varepsilon)$
(1′) $\exists a > 0 \ni \forall b > 0, -p(a)$

(2) $\forall \varepsilon > 0 \; \exists$ a number a between 0 and 1 such that $y - \varepsilon < a \leq y$
(2′) $\exists \varepsilon > 0 \ni 1/2 \leq y - \varepsilon$ or $1/2 > y$.
(3) $\exists \varepsilon > 0 \ni a - b \geq \varepsilon$
(3′) $a - b < 10^{-10}$.

14. QUANTIFIERS THAT AREN'T THERE AND OTHER SLEIGHTS-OF-MIND

In this section we take up an aspect of mathematical thought and expression which seems to cause far more than its share of difficulties. This is the use of quantified statements without explicit mention of a quantifier. You may recall the occurrence of such statements in Problem III, Section 13, e.g., (8), (9) and (10). It's not that mathematicians wish to be underhanded about things, it's that mathematical expression has evolved in this way and there is not much else to do but to deal with it as it is. On the other hand, it's likely that, when one does well understand what a particular mathematical statement is about, he has no difficulty realizing how the statement is quantified.

Consider the statement

$$\text{If } n \geq N, \text{ then } \left|\frac{1}{n} - 0\right| < a.$$

Somehow one is to come to understand that such a statement means, more precisely,

$$\forall n, \text{ if } n \geq N, \text{ then} \left|\frac{1}{n} - 0\right| < a$$

or

$$\forall n \geq N, \left|\frac{1}{n} - 0\right| < a$$

because in the initial statement there is no other requirement on n except that $n \geq N$. In the initial form one may have been tempted to negate this as

(a) $$n \geq N \text{ and } \left|\frac{1}{n} - 0\right| \geq a$$

but in the latter form one would instead write

(b) $$\exists n \geq N \text{ such that } \left|\frac{1}{n} - 0\right| \geq a$$

and there is a considerable difference between these two. In (a), one could not be sure whether this refers to one, some or all $n \geq N$. In (b) there is no doubt.

The key to avoiding this kind of difficulty is to realize that in statements such as

$$\text{If } n \geq N, \text{ then } \left|\frac{1}{n} - 0\right| < a$$

or

$$\text{If } x \text{ is in } A, \text{ then } x \text{ is in } B$$

where no qualifications, no bounds are put on the indeterminate other than the hypothesis itself ($n \geq N$, x is in A), then what is precisely meant is

$$\forall n, \text{ if } n \geq N, \text{ then} \left|\frac{1}{n} - 0\right| < a$$

or

$$\forall x, \text{ if } x \text{ is in } A, \text{ then } x \text{ is in } B.$$

To confuse things further, these latter statements would usually be written in the less logically precise form

$$\forall n \geq N, \left|\frac{1}{n} - 0\right| < a$$

$$\forall x \text{ in } A, x \text{ is in } B.$$

These are less logically precise because the statements are implications but they do not have that appearance.

You can see from these examples that the point of difficulty is no more and no less than one of "conventions of mathematical expression." Once trained in mathmatics, one is to understand what is precisely meant by any of the above statements. That not being your lot at this time, you must have a few rules to carry you along for a while. Keep the sense of these rules in mind and you will avoid a lot of misunderstanding. Concerning statements of generality, we have the following examples of conventional mathematical expression.

EXAMPLE 1.14.1. Let $p(\varepsilon)$ be a propositional function whose indeterminate ε is to be interpreted as a number. The proposition: *for every number* $\varepsilon > 0$, $p(\varepsilon)$ may appear in any one of the following forms.

(a) $\varepsilon > 0$ implies $p(\varepsilon)$.
(b) If ε is any positive number, then $p(\varepsilon)$.
(c) If ε is a positive number, then $p(\varepsilon)$.
(d) $\forall \varepsilon > 0, p(\varepsilon)$.
(e) Let ε be a positive number. Then $p(\varepsilon)$.
(f) Given $\varepsilon > 0$, then $p(\varepsilon)$.
(g) If ε is a fixed but arbitrary number greater than 0, then $p(\varepsilon)$.

(h) Let $\varepsilon > 0$ be fixed. Then $p(\varepsilon)$.
(i) $p(\varepsilon)\ \forall\ \varepsilon > 0$.

Of these statements, (g) is the most specific and most clearly stated, but also involves the most writing. Consequently it is almost never used. Notice (b) and (c)—the word *a* means *any*. In (a) or (f), since no restrictions are put on ε other than $\varepsilon > 0$ (one is assumed to understand that ε is a number), one takes this to mean the same as (g). Perhaps (h) is the most peculiar, but it places no restriction on ε other than $\varepsilon > 0$ and means that *for all* $\varepsilon > 0$, $p(\varepsilon)$ is true.

As a particular example the statement

> For every positive number, a, there is a positive integer, N, such that $|1/n - L| < a$ for all $n \geq N$

might be written in any one of the following ways
(a) $\forall a > 0\ \exists N \ni \forall n \geq N, |1/n - L| < a$.
(b) $\forall a > 0\ \exists N \ni$ if $n \geq N$, then $|1/n - L| < a$.
(c) $a > 0$ implies $\exists N \ni$ if $n \geq N$, then $|1/n - L| < a$.
(d) If $a > 0$ then $\exists N \ni |1/n - L| < a$ if $n \geq N$,
and so on. The quantifier for the indeterminate n which does not appear in (b), (c) and (d) is what can be called *implicit* in that, since no restriction is put on n, other than $n \geq N$, one is to understand

$$\text{if } n \geq N$$

to mean

$$\text{for all } n \geq N.$$

If the propositional function refers to something other than a number, for example,

$$\forall x, \text{ if } x \text{ is in } A, \text{ then } x \text{ is in } B$$

the proposition might be written
(a) $\forall x$ in A, x is in B.
(b) x in $\digamma$ implies x is in B.
(c) x is in B if x is in A.
(d) x is in $B\ \forall\ x$ in A.
(e) Any x in A is in B.
Again, in each of the statements (b), (c) and (d) there is an implicit quantifier, not an explicit one as in (a) and (c). The general idea is that if no specific restrictions are placed on the indeterminate, one is to understand that the statement refers to *all* intended interpretations of the indeterminate.

EXAMPLE 1.14.2. This example concerns conventions of mathematical expression in existence statements. The situation here is not so varied as that for statements of generality. The proposition

There is a positive number x such that $p(x)$

where $p(x)$ is a propositional function whose indeterminate is interpreted as a number, might appear in any one of the following forms.

(a) $\exists x > 0 \ni p(x)$.
(b) For some $x > 0, p(x)$.
(c) For at least one $x > 0, p(x)$.
(d) There is some $x > 0$ such that $p(x)$.
(e) There is some number x such that $x > 0$ and $p(x)$.

Of these statements, (c) is perhaps most explicit but (a) and (b) are the most common. In general the word *some* means "at least one."

For a particular example, the proposition

$$\forall a > 0 \ \exists \ a \text{ positive integer } N \ni \forall n \geq N, \left|\frac{1}{n} - L\right| < a$$

might be written

(a) $\forall a > 0$ and for some positive integer N, $|1/n - L| < a$ for all $n \geq N$.
(b) If $a > 0$, then there is some positive integer N such that if $n \geq N$ then $|1/n - L| < a$,

and so on. The idea is that one only wishes to assert the existence of at least one positive integer N with a certain property.

Here is a final example, a negation as complicated as one is likely to encounter for some time.

EXAMPLE 1.14.3. One of the most important concepts in mathematics and its applications is that of *uniform convergence* and concerns a property of a sequence of functions $f_1, f_2, f_3 \ldots$, one for each positive integer, n, and all defined on the same set, S. You do not need to know well what this means in order to understand this example.

A sequence of functions $f_1, f_2, \ldots$ defined on a set S is said to *converge uniformly* to a function f on S iff $\forall \varepsilon > 0 \ \exists n \ni$ if $n \geq N$, then $|f_n(x) - f(x)| < \varepsilon$ for any x in S.

We are to decide when such a sequence does *not* converge unifomly. The criterion for uniform convergence bears down on the expression $|f_n(x) - f(x)| < \varepsilon$ with various qualifications on ε, n and x. Notice that the qualification on n is implicit—if $n \geq N$, then $|f_n(x) - f(x)| < \varepsilon$—meaning

$$\forall n, \text{ if } n \geq N \text{ then } |f_n(x) - f(x)| < \varepsilon \text{ for any } x \text{ in } S.$$

To settle the question of non-uniform convergence, we need to be certain of the role of the four quantifiers involved. Working backwards, the last restriction on $|f_n(x) - f(x)| < \varepsilon$ is

$$|f_n(x) - f_n(x)| < \varepsilon \text{ for all } x \text{ in } S$$

or

$$\forall x \text{ (in } S), |f_n(x) - f(x)| < \varepsilon .$$

This in turn is restricted regarding the symbol n—namely

$$\forall n, [\text{if } n \geq N, \text{ then } \forall x \text{ (in } S) \ |f_n(x) - f(x)| < \varepsilon]$$

which is yet qualified by an assertion about N; viz.,

$$\exists N \ni \{\forall n \ [\text{if } n \geq N, \text{ then } \forall x \text{ (in } S) \ |f_n(x) - f(x)| < \varepsilon]\} .$$

Yet the last qualification,

$$\forall \varepsilon > 0[\exists N \ni \{\forall n \ [\text{if } n \geq N \text{ then } \forall x \text{ (in } S) \ |f_n(x) - f(x)| < \varepsilon]\}].$$

Applying Rule 1.13.4(a), (b), (a) and (a) we obtain the negation as

$$\exists \varepsilon > 0 \ni -[\exists N \ni \{\forall n \ [\text{if } n \geq N \text{ then } \forall x \text{ (in } S) \ |f_n(x) - f(x)| < \varepsilon]\}]$$

or

$$\exists \varepsilon > 0 \ni \forall N - \{\forall n \ [\text{if } n \geq N \text{ then } \forall x \text{ (in } S) \ |f_n(x) - f(x)| < \varepsilon]\}$$

or

$$\exists \varepsilon > 0 \ni \forall N \ \exists n \ni n \geq N \ \textit{and} \ -[\forall x \text{ (in } S) \ |f_n(x) - f(x)| < \varepsilon]$$

and finally

$$\exists \varepsilon > 0 \ni \forall N \ \exists n \ni n \geq N \text{ and } \exists x \text{ (in } S) \ni |f_n(x) - f(x)| \geq \varepsilon.$$

This just about does it, but for one final consideration. Given that statements of generality and existence occur over and over again, how does one go about arguing from, or concluding to, such statements? No really general rules now, this being a part of learning mathematics, but a couple of suggestive examples. Suppose that one wishes the truth of the statement

$$\exists a > 0 \ni \forall \text{ (positive integer) } N \exists n \geq N \ni |1/n - 1/2| \geq a.$$

That is, we wish to *conclude to* this statement.

First, comes existence of an a for the inequality $|1/n - 1/2| \geq a$. Since the numbers $1/n$ are further and further from $1/2$ as n increases, $a = 1/4$ is a reasonable guess. Notice that since the statement calls for the *existence* of an $a > 0$, *any one* $a > 0$ will do. The claim now is that

$$\forall N \exists n \geq N \ni \left| \frac{1}{n} - \frac{1}{2} \right| \geq \frac{1}{4} .$$

This a statement of generality, referring to all N. To verify it, according to Definition 1.13.3, one must verify it for each (positive integer) N. The accepted procedure is to argue the truth of this statement for a *fixed* but otherwise *completely arbitrary* positive integer N. Hence, let N denote an arbitrary, but fixed positive, integer. We want to show

$$\exists n \geq N \ni \left|\frac{1}{n} - \frac{1}{2}\right| \geq \frac{1}{4}.$$

Again, the call is for existence, so any *one* $n \geq N$ will do. From the arithmetic of the matter, $|1/n - 1/2| \geq 1/4$ will hold if $n \geq 4$. But we want this to hold for n greater than or equal to the arbitrary, but fixed, N this argument began with. One choice is $n = N + 4$. Then $n \geq N$ and $n \geq 4$. Hence

$$\left|\frac{1}{n} - \frac{1}{2}\right| = \frac{1}{2} - \frac{1}{n} \geq \frac{1}{2} - \frac{1}{4} \geq \frac{1}{4}.$$

We have shown,

$$\forall N \exists n \geq N, \text{ namely } n = N + 4, \text{ such that } \left|\frac{1}{n} - \frac{1}{2}\right| \geq \frac{1}{4}.$$

Hence the statement,

$$\exists a > 0 \ni \forall N \exists n \geq N \ni \left|\frac{1}{n} - \frac{1}{2}\right| \geq a$$

is a true statement because it is true for $a = 1/2$.

Finally, let us consider an example in which we conclude *from* a statement of generality. We will prove a result that lies at the fundation of the calculus and, without which, it would not exist in its present form. Consider proving the following:

If a and b are real numbers and $|a - b| < \varepsilon \; \forall \varepsilon > 0$, then $a = b$.

The roof is by a modified method of *reductio ad absurdum.* Suppose $|a - b| < \varepsilon \; \forall \; \varepsilon > 0$ and $a \neq b$. Since $a \neq b$ then $|a - b| > 0$. By the hypothesis, since $\varepsilon = |a - b|/153 > 0$, then $|a - b| < |a - b|/153$ or $153 < 1$. Since $1 < 153$, this is a contradiction.

The trick then, of using a *for every* hypothesis, is to make a judicious choice of (in this case) a number to which to apply it.[8] In this argument, one could have used $\varepsilon = |a - b|/2$ or $\varepsilon = |a - b|/10^2$ or similar things once one hit on the right idea.

Problems.

1. Restate the following statement in eight equivalent ways. Let m: If x is a fixed but arbitrary element in S, then $p(x)$.

[8] And if you're one of many who keeps thinking that if the hypothesis asserts *for all*, then you must show something *for all*, recall that one does not *show hypotheses* in an implication, only conclusions! One *uses* hypotheses.

II. Restate the following in four different ways. *n*: There is a function f such that $f(x) = f(0) + \int_0^x g(t)f(t)dt$.

III. One most encounters the problem of negations when trying to prove a theorem of the type: if x is a B, then x is a C. For example: if f is a differentiable function then f is a continuous function. If one wishes to prove such a statement by the contrapositive or the method of *reductio ad absurdum* one must formulate a meaning to: x is not a C or x is not a B. In each of the following, a definition is given. For each of these, write down a "useful" statement which serves as a criterion for knowing when the definition is *not* satisfied. For example, in (a): you are write down the meaning of "b is not an upper bound for S."

(1) Let S be a set of real numbers. The number b is called an *upper bound* for S iff $x \leq b \; \forall \, x$ in S.

(2) The set S is *bounded* above iff there is a number b such that b is an upper bound for S.

(3) Let S be a bounded set of real numbers. A number a is called the *least upper bound* of S if:

(i) a is an upper bound for S, and (ii) if b is an upper bound for S, then $a \leq b$.

(Recall that (ii) could be written as: $\forall$ upper bound b of S, $a \leq b$.

(4) Let f be a function from the topological space X to the topological space Y. We say that f is continuous at x if $\forall$ neighborhood V of $f(x)$ $\exists$ a neighborhood U of x such that $f(U) \subset V$.

(5) Let f be a real-valued function defined on the interval (a, b) and let c in (a, b). If L is a real number, then we say that $\lim_{x \to c} f(x) = L$ iff $\forall \varepsilon > 0 \; \exists a \; \delta > 0$ such that $0 < |x - c| < \delta$ implies $|f(x)| - L| < \varepsilon$. What does it mean to say that $\lim_{x \to c} f(x) \neq L$?

(6) A real-valued function f defined on a set S is *bounded* on S iff $\exists$ a number M such that $|f(x)| \leq M \; \forall x$ in S.

(7) The function, f, defined on the set, S, is said to be *uniformly continuous* on $[a, b]$ iff given $\varepsilon > 0 \; \exists \delta > 0 \ni$ for any x, y in S, if $|x - y| < \delta$, the $|f(x) - f(y)| < \varepsilon$. (The problem contains two implicit qualifiers.)

(8) Let S be a set of real numbers. A number x is called a *limit point* of S iff every open interval containing x also contains a number in S different from x.

IV. This problem requires that you find *meaning* in quantified statements and their negations. For each of (1), (2), (3), (6) and (8) in Problem III above, give a specific set S of the open interval (0, 1) and a specific interpretation of the indeterminate which satisfied the given statement. Then give a set S and an interpretation of the indeterminate which satisfies the negation of each of these statements. You choice of sets and interpretations may vary from statement to statement.

V. Show that each of the following interpretations satisfies the negation of the statement referred to.

(1) For statement (5) let $f(x) = x + 1$, $(a, b) = (0, 2)$, $c = 1$ and $L = 3$.

(2) For statement (7), let $S = (0, 1)$, $f(x) = 1/x$.

(3) For statement (8), let $S = (0, 1)$, $x = 3/2$. Show that x is not a limit point of S. Does $x = 1$ satisfy the given statement?

VI. For each positive integer n, let $f_n(x) = x^n$. Let $f(x) = 0$ and let $S = [0, 1]$. Show that these functions satisfy the negation of uniform convergence obtained in Example 1.14.3. For a start, try $\varepsilon = 1/2$.

VII. Mathematics is a highly efficient form of thought. It is so because it unrelentingly avoids the redundant and the extraneous. It wants to take note only of the essentials and in a very careful way. For these reasons the use of convenient notation which conveys the essentials of the matter is extremely important. The lawyer in his legal document may refer to the "party of the first part," "the party of the second part," etc., to keep track of which persons are being referred to in a legal document. The mathematician will be more likely to refer to the first entity of importance in a proof by x_1, or a_1, or f_1, or S_1 (if (say) this entity is an element, number, function or set, respectively), the second by x_2, a_2, f_2, or S_2 and so on, and throughout the remainder of the proof x_1 will stand for the first thing defined (just as "the party of the first part" always mean the same person throughout the legal document). Contrary to what may be popular belief then, notation is only introduced to clarify, never to confuse. The beginning mathematics student generally has great difficulty in knowing when and how to introduce useful notation. He can learn to do this by imitating the work of others and by simply trying to be clear and precise. This will come naturally in time. The following are exercises in the ***introduction*** of ***useful*** notation.

(1) Referring to Problem III(2) above, if b is *not* an upper bound for S, can one conclude the following: there is a number in S greater than b. Let us denote it by x.

(2) Referring to Problem III(3) above, suppose that S is a set which is not bounded above. Can we conclude that: for every positive integer n there is a number in S, which we (might as well) call a_n such that $a_n > n$. That is, there is a number a_1 in $S \ni a_1 > 1$ and a number a_2 in S such that $a_2 > 2$ and a number $a_{10^{21}}$ such that $a_{10^{21}} > 10^{21}$ and so on.

(3) Refer to Problem III(3). Suppose a is an upper bound for S, but a is *not* the least upper bound for S. Can we conclude: there is a number c which is an upper bound for S such that $c < a$. (Again this is a useful negation, if correct, for it gives one something to work with.)

(4) Refer to Problem III(6). If f is not bounded on S, can we conclude that $\forall$ positive integer n $\exists$ an element x_n in $S \ni |f(x_n)| > n$.

(5) Refer now to Problem III(5). Is the following true: if $\lim_{x \to c} f(x) \neq L$, then $\exists \varepsilon > 0$ such that for each number $n > 0$ there is a number x_n such that $|x_n - c| < 1/n$ and $|f(x_n) - L| \geq \varepsilon$.

(6) Let m: $\exists$ a neighborhood U of x such that U contains no elements of A.

Suppose $-m$ is true. Does it follow that for each neighborhood U of x $\exists$ a corresponding element (which we might well call) x_u of A contained in U?

VIII. In this problem we want to put together the above ideas and complete a proof. We will attempt a proof of a theorem about least upper bounds, using the definitions given in Problem III(3). To prove anything about "least upper bounds" one must use some definition of "least upper bound" (Why?) so we will use this one despite some knowledge you might have of alternate definitions. The theorem to be proven is known as an existence theorem. It asserts that something with a certain property in fact does exist (as a consequence of some hypothesis).

Theorem: If a set S has a least upper bound a and if b is a number less than a, then $\exists$ a number in S which is greater than b.

Proof: Suppose $b < a$ and that the conclusion is false. Then, every number in S is less than or equal to b. By Problem III(3), b is an upper bound for S. By Problem III(3), $a \leq b$ since a is the least upper bound for S. This is a contradiction to the assumption that $b < a$. Hence the proof.

Questions:

1. What form of proof is this?
2. Is this a valid proof?
3. Do you believe the theorem?
4. Does the proof convince you of the validity of the theorem?
5. Can you center the proof around a single idea? What idea?

IX. Here we consider a proof which is of the *for every* type. The theorem asserts that something is true for (infinitely) many objects of a type described in the hypothesis.

Theorem: For every pair of numbers a and b, with $a \neq 0$, the equation $ax = b$ has a solution.

Proof: Suppose not. Then there is a pair of numbers a, b with $a \neq 0$ such that $ax \neq b$ no matter what number x is. Since $a \neq 0$, b/a is a number. But then $a(b/a) = b$, a contradiction. Hence the theorem.

(1) Answer questions (1) through (5) in Problem VIII. Consider the following proof of the theorem.

Proof: Let a and b be two arbitrary but fixed numbers, with $a \neq 0$.

The number $x = b/a$ exists since $a \neq 0$. Since $ax = a(b/a) = b$ the equation has a solution and the proof is complete.

(2) Answer questions (1) through (5) in Problem VIII again.

READING ASSIGNMENTS

1. Write a discussion of Chapters I, II and III in R. B. Kershner and L. R. Wilcox: *The Anatomy of Mathematics* (New York: The Ronald Press Co., 1950). This material is related to but unlike that covered in this chapter.

2. Write a discussion of Chapter I in J. B. Rosser: *Logic for Mathematicians* (New York: McGraw Hill, Co., 1953). This is a discussion of logic, mathematics, and the place of symbolic logic in mathematics.

3. Write a discussion of Chapter XIV in G. T. Kneebone: *Mathematical Logic and the Foundations of Mathematics* (New York: D. Van Nostrand Co., 1963).

4. Write a discussion of Chapter XVIII in B. Russell: *Introduction to Mathematical Philosophy* (New York: Humanities Press Inc. 1924).

Both 3 and 4 are discussions of the place of logic in mathmatics and thought with greater depth than herein. Russell is a clever writer with an original approach.

RELATED READINGS

1. Bittinger, M. L.: *Logic and Proof*, Reading, Mass.: Addison-Wesley Publishing Co., 1970. This is a detailed discussion of logic and proof, with many examples and problems.
2. Denbow, C. H., and Goedicke, V.: *Foundations of Mathematics*, New York: Harper and Row, 1959. Chapter VIII consists of a brief discussion of logic and symbolic logic.
3. Dinkines, F., *Introduction to Mathematical Logic*, New York: Appleton-Century-Crofts, 1964. This is a much more detailed, yet still practical, discussion and development of the material herein. If you need more explanation, this would be a good source.
4. Eves, H., and Newsom, C. V.: *An Introduction to the Foundation and Fundamental Concepts of Mathematics*, New York: Holt, Rinehart and Winston Inc., 1969. Chapter IX introduces symbolic logic intuitively but then proceeds to rigorously develop the laws of logic from a list of primitive (or assumed) propositions. Chapter III discusses the method of *reductio ad absurdum* and its place in the development of non-Euclidian geometry.
5. *Insights into Modern Mathematics*, 23rd Yearbook, Washington, D. C.: National Council of Teachers of Mathematics, 1957.
6. Larson, M. D.: *Fundamental Concepts of Mathematics*, Reading, Mass.: Addison-Wesley Publishing Co., 1970. Chapter I is another author's version of the material herein, concentrating on the same material but different in detail.
7. Pfeiffer, J. E.: Symbolic logic, *Scientific American*, December 1950. This is a non-technical discussion of the place and uses of symbolic logic and its relation to other areas.
8. Stoll, R. R.: *Set Theory and Logic*, San Francisco: W. H. Freeman and Co., 1963. More sophistication and detail here.

9. Tarski, A.: *Introduction to Logic and to the Methodology of the Deductive Sciences*, New York: Oxford University Press, 1949. Tarski is a logician, and the entire book is devoted to logic. Chapters I and II are a good supplement to the material herein.
10. Wilder, R. L.: *Introduction to the Foundations of Mathematics*, Chapter IX, Ed. 2, New York: John Wiley and Sons, Inc., 1965. This is a more sophisticated and precise discussion of the material of this chapter, including a partial but rigorous development of the laws of logic from a list of primitive propositions.

chapter 2

> "Some men see things that are and ask—why? I see things that never were and ask—why not?"
> —G. B. Shaw

THE FOUNDATIONS OF MATHEMATICS

1. INTRODUCTION

Mathematics is one of the oldest creations of the human spirit. In the 4,000 years of its recorded history, it has moved from a relatively simple form of thought of immediate use to most men each day, to a structure of thought whose contemporary advances are likely to find application only in the minds of specialists some years after their discovery. At the same time many of its earliest concepts, counting and elementary geometry, are still used by almost everyone each day in small, unnoticed ways. A fundamental mathematical discovery rarely becomes obsolete, and, in quite advanced mathematics, 2,000-year-old ideas such as the Pythagorean theorem are of crucial importance.

An acquaintance with the history and development mathematics is not necessary in order to do mathematics. Yet such an acquaintance can hardly fail to increase one's appreciation of contemporary mathematics, for a sense of its past adds to its present a meaning it could not possibly acquire by mere description and study. In the initial sections of this chapter we will put away the task of building mathematical structures and wander off into the forest of mathematical history and thought. In subsequent sections we will return again to the detailed examination of a few of its more towering trees.

To extend that analogy, there are two general kinds of trees in this allegorical forest of mathematical thought, pure trees and applied trees. It is, of course, the applied, "practical" mathematics, in the sometimes narrow technological sense of that word, that many associate with mathematics and thereby indicate that they have perhaps not been very far into the forest, for a large majority of the trees are pure. Long ago the tree of number was discovered on the edge of this forest, its fruit put to much use. Upon that discovery others entered this forest to find and cultivate more such trees and did so for centuries. But after a while, beginning most notably with the Greek period, mathematical woodsmen became quite captivated by the subtle beauty and pure aesthetic appeal of quite different trees, trees which at the particular time may have had no known use, and began to cultivate these as well. This is what pure mathematics is, the cultivation of mathematical thought for its own sake. In this past century, the aesthetic appeal of the pure trees in the forest of mathematics has been so great and their cultivation has led to so many profound insights, that today they are the most constantly cultivated.[1] Ever so often someone attends to a seedling of one of these that has some "practical" application. Indeed, this happens sufficiently often in such unexpected ways that mathematics has been said to enjoy an almost "unreasonable effectiveness."[2] But the pure trees receive the most constant attention, and in the depths of the forest, at those several points of furthest penetration, these are nearly the only trees that one will find.

Now this is an apt analogy as to the facts of the development of mathematics, but surely it does raise some questions. How did things come to be this way? Is there meaning and relevance to such activity? Do the very real uses of mathematics, when they come, justify the study it receives? Is the course of development of mathematics inseparable from the nature of mathematics, or can and should mathematics be developed with only use in mind?

As might be expected, the answers to all of these questions fall a little on both sides of the coin. As the coming sections develop, the reader will hopefully begin to fully appreciate the questions as well as begin to find satisfactory answers. To make a start, and to indicate by specific example some upcoming assertions, consider the following question: is the mathematics used to design your stereo receiver any more important than the music it produces? One's initial answer might be "no," for after all the music is the thing. But, then, perhaps "yes," for let us go back to the forest. There one will find some very well cultivated trees, pure ones, the familiar e, π, sine and cosine. Long ago someone learned to use these latter three to compute areas and lengths, but the extensive cultivation of their progeny has long since outstripped these initial uses. Then came electronics and the recognition that these "pure trees" could be used as key elements in its mathematical formulization. Biologists, too, happend along and found that e has a neat role in describing population growth. So, in a different sense of course, the mathematics of the stereo receiver, constructed upon the pure study of e, π, since and cosine, having so many uses outside of the design of a receiver, is as important as the music it produces. The point is that mathematics, by nature developed independently of particular use, is thereby useful, more useful than any one of its

[1] If you wish to obtain the view of one of this century's top mathematicians on the meaning of pure mathematics, the little book, *A Mathematicians' Apology* by G. H. Hardy (New York: Cambridge University Press, 1941) is required reading.

[2] E. Wigher: The Unreasonable Effectiveness of Mathematics in the Natural Sciences, p. 123, in *The Spirit and Uses of the Mathematical Sciences*, T. L. Saaty and F. J. Wey (eds.), New York: McGraw Hill Co., 1969.

particular exploitations. This fits well with the proclivity of many a mathematician, who is often quite content to cultivate a pure tree to its ultimate refinement with no special care for what use it might then or someday have.

Others besides engineers and scientists have ventured into the forest of mathematics to observe the activity there. Most philosophers and thinkers of any sort have found at least a brief journey unavoidable. Their journey has not been one in search of technical proficiency. They have recognized that mathematics is pure thought, uncluttered by all redundancy, irrelevance or personal opinion, incarnate. And pure and direct, and consequently simple in structure as it is, it has at times completely gone astray. Some trees have fallen, the supporting roots of others have been completely exposed, and the exposure of one of these, leading to the discovery of non-Euclidean geometry, was noted in quite a few surrounding areas. That event is a key part in the development of mathematics, for it implied a multiplicity of new trees and methods of cultivation that no one had seen or suspected before.

This chapter is an attempt to make clear the nature and development of mathematics and mathematical thought, and to embellish the allusions and metaphors of the above paragraphs with the history and critical events of its long story.

2. MATHEMATICS AS A SYSTEM OF THOUGHT

In a recently popular little novel written years ago, the wandering Siddhartha spends his lifetime searching for a truth that will satisfy him.[3] He comes to a belief not unlike that of the Australian aborigine: a belief in an existing unity among all things. Be this fact or fiction, the manner of mathematical thought is most surely not unrelated to other matters.

It would not be stretching the point at all to say that the growth and course of development of mathematics parallels that of the person whose mind continues to grow thoughout his life, embracing an ever wider variety of thought and experience. Indeed, so closely related to the course and essentials of thought is mathematics that the 20th century philosopher Alfred North Whitehead was led to observe that, "While it would not be correct to say that omitting mathematics from the history of thought would be like omitting Hamlet from the play which is named after him... it would be analogous to cutting out the part of Ophelia... for Ophelia is quite essential to the play, she is very charming—and a little mad."[4] Mathematics, as an organized system of thought, is a model for any deductive system of thought stripped to its bare essentials. It is man's passion

[3] H. Hesse: *Siddhartha*, New York: New Directions Publishing Corporation.

[4] A. N. Whitehead: Mathematics as an Element in the History of Thought, Chapter 2, *Science and the Modern World*, New York: The Macmillan Co., 1925. Interestng reading, if you have the time.

for clear and certain understanding carried to an extreme that could only be called madness in a less idealized world, yet it may have something to say in this unidealized world, as well.

Mathematics in its earliest years took what was around it as given and tried to make sense of it, as most children do. Methods of counting and combining, thoughts of circles, lines and perhaps parallel lines were its earliest interests. Like most children it sought facts and certainty, slogans, formulas and clear truths it could depend on. For 40 centuries mathematics continued on its way, building on a foundation set gradually and imperceptably in antiquity. Its youth was arithmetic, algebra, trigonometry and plane geometry. Its adolescent, so to speak, was calculus, an attempt to invent and understand really new concepts but in an ancient framework, and often accompanied by contradiction and uncertainty. Its manhood began in 1829, when it was brought to look back upon itself and recognize its meaning and existence for what it really was.

The cataclysmic event was the discovery of non-Euclidean geometry, later hailed as "primate among the emancipators of the human intellect," not only for what it was, but for what it implied about man and his attempts to understand his world. The details will follow, but, in essence, men of mathematics found that the certainties of mathematics in its youth were not certainties after all, but mere assumptions about the way things seemed to be, having no *absolute* validity. Like an individual who at some life's moment realizes that the certainties of youth are no more, that there is more to it, mathematics confronted itself and lost its certainty. This indeed was much to lose, but in its stead came a freedom that has been worth the price. If its most straightforward, naive certainties were mere assumptions and not absolutes, then there was no need to cling to them merely because of their past, for the future was yet to come. In short, mathematics became free to discard, modify or reassert its certainties and with a better awareness of their proper role. This realization gave mathematics a nearly unique freedom for development, and, as the course of mathematics since 1829 has shown, had mathematics continued to cling to the conventional wisdom of the then established mathematical order with its reassuring familiarity and apparent certainty, it would have suffered a tragic death of spirit at far too early an age. Men of mathematics came to realize that as one cannot define all words, one cannot prove all things, that one must begin somewhere and there lies a mere assumption, not an absolute, and a way to see things anew.

The general mathematical method is the search for, and identification of, the most basic and admitted assumptions which explain a given situation, followed by a precise, abstracted and detailed discourse of how matters may be explained by these assumptions. This attitude of precisely detailed explanations is largely restricted to mathematics, but the attitude of working from assumptions is not. The study of history is in great part the search for assumptions that would explain past happenings and perhaps indicate future events. Various schools of economics are predicated on various assumptions about economic forces. Atmospheric science today is largely concerned with finding assumptions that would better explain the workings of the atmosphere. The Bill of Rights is a list of assumptions that in theory govern the relationship between man and the state in the United States. The list goes on and on: philosophy and physics, religion and psychology, foreign policy and internal affairs. But in none of these is it admitted so clearly as in mathematics that one is working with pure assumptions. In none of these does the role of assumptions and the consequent structure of thought and development from them stare one so full in the face. And when things go awry in a system, or study or policy, and matters have proceeded in a more-or-less understood fashion all along, then a prime place to look for weakness is in those assumptions which are the foundation of the entire edifice. And, as the discovery of non-Euclidean geometry indicates, it is essential to recognize all the assumptions, particularly those unconscious preconceptions which can color the entire structure and are thus hardest to see.

All these things aside, there were other consequences of the discoveries of 1829 especially germane to mathematics. Mathematicians today find themselves as free to create in mathematics as the poet, artist or musician is in his area of interest. Because of this, mathematics for many mathematicians is as much an art as a science. But mathematics is scientific, so some clarification is needed. Science is constrained by the demand that it study what matters in the universe. Mathematics on the other hand is constrained only by the spirit of its practioners, as are music and painting. But mathematics differs from these (and it is this which makes it scientific) in that its technique and form is extremely rigid, more so than the scientific method, whereas art forms are thankfully free. The remaining pages of this text are intended to help the reader understand and operate within its rigid techniques and forms, and hopefully to generate an appreciation of them.

3. A VERY BRIEF HISTORY OF MATHEMATICS

Quite often, general and individual advances in knowledge occur in a manner not unlike that of the famous laboratory mouse working his way through some rather intricate maze. Drawn by some chance, for he could conceivably have been a field mouse, he is encouraged by events to attempt the maze. His journey proceeds smoothly at first but as he is drawn deeper into the maze he becomes a bit confused, becomes involved in detours and side journeys. And it is often that only an accumulation of dead-ends eventually leads him to a promising route and smooth passage; only after being lost and confused does he find the right path. The acquisition of understanding often proceeds in much the same way: smoothly at first and throughout much of its development. But often enough one's incomplete knowledge leads him into a quagmire of doubt, confusion and, in retrospect, plain error. Amazingly enough, it is often that from such a quagmire the truest knowledge, the most sparkling advance, follows. And he who follows the pathfinder, he who comes afterward in the quest for understanding, should at least view the quagmire if he is to appreciate those paths that have been found. One is reminded of Santayana's comment—"Those who cannot remember the past are condemned to relive it." This and subsequent sections are a brief chronicle of mathematics' successes, but much emphasis will also be laid on the errors, ignorance and contradictions that were manifest at various key points in its development, both to lead into the manner of mathematical thought that has thereby resulted and to point out the fallibility of even an idealized thought world. The discovery of irrational magnitudes, the realization that Euclidian geometry was not the absolute truth of the physical world, the contradictions of the calculus, the Russell paradox, were all disasters in their time, but todays understanding would not be ours had they not occurred.

Ancient Greece was the setting for advances in understanding and knowledge that has had few rivals before or since. In that time of relative peace and plenty, borne on the idea that understanding and the cultivation of knowledge for its own sake was quite the thing, Grecian learning in mathematics and other matters flourished and its influence extends even into this century. Though mathematics is one of the oldest creations of the human spirit, until the Greek period it consisted of only computation and measurement. Parallel lines were thought of as wagon tracks, abstraction had not yet occurred. There are no recorded attempts in ancient mathematics to demonstrate (prove) that formulas or procedures were in truth independent of their uses. Mathematical texts were not unlike instructions yet used in elementary mathematics courses today: instead of stating and verifying a general procedure to solve a whole class of problems, the text simply gave examples of the solutions of each of several particular problems of a single type.[5]

It is in the work of the Greek Thales that one first finds any attempt to demonstrate the general validity of a procedure or method, but not until the later work of Euclid is a consistent attempt made to prove the validity of what was known or believed. For the first time someone tried to show why, for example, any diameter of any circle bisects the circle, and with this the mathematical method was born. Two other outstanding names of the period were Pythagoras and Archimedes, the latter one of the greatest mathematicians and physical scientists of all time. The Pythogorean doctrine, disproved in its own time, was that whole numbers were at the root of many things, mystical as well as mathematical, and that every length must be a whole number multiple of a fraction of some unit length.

[5] H. Eves: *An Introduction to the History of Mathematics*, p. 61. New York: Holt, Rinehart & Winston, 1969.

Greece in time was absorbed by the Roman empire and Greek mathematics, with its concern with "why," did not meet the Roman interest in "how," and ceased to develop. The fall of Rome saw the advent of the Dark Ages and for over 600 years a period of intellectual stagnation, constant violence and the rule of superstition put a stop to nearly all advances in knowledge. Parts of Greek mathematics managed to be preserved only in the monasteries of Europe and on the southern side of the Mediterranean, where the Moslem empire arose and provided a sufficiently stable climate for learning to continue if not to flourish.

The European Renaissance began in the 15th and 16th centuries. In the 3 centuries previous to this time things had already begun to stir a bit, however, influenced in part by the Moslem presence in Spain. In this period Fibonacci's work is found, the sole significant contribution to number theory between that of the Greek Diophantus and the later work of Fermat. Here, too, one finds for the first time the concept of fractional exponents, and, in the early 16th century, the use of $+$ and $-$ signs, the symbol for equality, and the first use of letters to represent unknowns. Copernicus suggested his revolutionary thesis that the earth, despite our presence, was not the center of the universe and needed a better trigonometry to do his work. One should also note that before and during this period mathematics was also done in India, the Orient, and, of course, the Moslem world. The word *algebra* is an Arabic word.

Modern mathematics begins in the 17th century. In this period Napier introduced logarithms, Descartes invented analytical geometry, and Fermat began modern number theory. Galileo and Kepler contributed to mathematics by way of their investigations in astronomy. A new geometry, *projective geometry* was created by Desargues, and Pascal and Fermat began a theory of probability. At the close of the century Isaac Newton and Gottfried Leibnitz independently invented the calculus. It must be noted that all of these inventions were presaged by other works going back, in some cases, for centuries. Integral calculus was nearly discovered by the Greeks, notably Archimedes with his applications of the "method of exhaustion" due to Eudoxus. But Zeno's paradoxes, concerning the divisibility of a given magnitude into a sufficiently large number of sufficiently small parts, formed a river across the road to calculus that would not be successfully crossed for nearly 2,000 years.[6]

Mathematics of the 18th and subsequent centuries is not elementary. The great names of the 18th century were Euler (who established the notations: $f(x)$, $\sum$, $i = \sqrt{-1}$, among many other things) and Lagrange, who was the first real mathematician to take serious note of the quicksand-like foundations of calculus. The Bernoulli family made numerous contributions. Dirichlet introduced the function concept taught in calculus today. Taylor published his *Taylor's Series* while paying no heed to the possible

[6] See Section 2.8.

divergence of these infinite series.

Coming to the 19th century one can make brief mention of the names, if not the actual discoveries, of other mathematicians who played a dominant part in 19th century developments. Laplace published fundamental works in the fields of celestial mechanics, probability theory and differential equations. Gauss, ranked with Newton and Archimedes as one of the three greatest mathematicians of all time, anticipated the discovery of non-Euclidean geometry and gave the first good proof of the fundamental theorem of algebra: that any polynomial of degree n with complex coefficients has at least one complex root. Fourier studied heat conduction and forced upon us a need for yet better concepts of *function* and *convergence*. Riemann, near the close of the century, accelerated the tendency toward abstract generalization in analysis, presaged topological considerations in that field, and did much to improve the concept of integrability, leading to the 20th century development of the Lebesque integration; his ideas have not yet all been completely developed. In 1831 Galois cast the spotlight of importance on the specially beautiful theory of groups and essentially settled age-old problems concerning the algebraic solution of polynomial equations. Cayley and Hamilton originated the revolutionary, but now highly useful, algebras of matrices and quarternions in which the commutative law of multiplication $(a \cdot b = b \cdot a)$ did not hold. And, it was in this century, too, that Boole developed the symbolic treatment of logic that the earlier Leibnitz had once attempted.

The 20th century has seen the invention of topology which, apart from its own importance, has transformed analysis and clearly resolved the difficult limit concept. The topic of functional analysis, growing out of the notion of an abstract vector space, now unifies many of the concerns of classical analysis mentioned above. Work in algebraic topology is at this time very important. A relatively new theory, category theory, has arisen within the past 20 years in an attempt to unify many of these newer areas. And of course there has been the rise of computer science, though it has had, as yet, little effect on pure mathematics. And the continued and continuing study of the old reliables, geometry, differential equations, integration theory, applied mathematics, has gone on, influenced by and influencing in all of the newer areas just mentioned.

In summary, the work of the 19th and 20th centuries involves yet more and more notable mathematics and mathematicians, a few of whom we will later encounter. The work of these centuries can be said to have two essential parts. One was the ever more rapid continuation and development of general mathematical knowledge on what is now a truly vast scale. The other was the resolution of the paradoxes, discrepancies, errors and things left undone amid the work of the 20 or so preceding centuries. Without question, this latter activity had as much subsequent influence as any of the successful discoveries preceding and concomitant with it, and it is to these crises that we now turn.

There have been three great crises in the evolution of mathematics. It is notable that the first and second of these were not resolved until the 19th century, though the first occurred among the Greeks nearly 2,400 years ago. Beginning with a single fixed unit of length the Greek scholars of the Pythagorean school were able to make sense of multiples of this length as well as fractional parts of these multiples. It was believed that given two lengths, one must be a multiple of some fraction of the other. But then came the Pythogarean theorem which implied that in a right triangle two sides of which were of the unit length, the third must be a length whose square was two unit lengths. The proof that this length *could not be* a multiple of a fraction of the unit length (see Section 1.10, Problem (V)) demolished certain parts of the Pythogorean theory of mathematics. Efforts even were made to keep the matter a secret. Men of mathematics were, in this instance, as protective of their disasters as politicians! The difficulty of the existence of irrational numbers was not decisively settled until the work of Richard Dedekind in 1872, although his work is similar to that of Eudoxus in the fifth book of Euclid's *Elements*.

The second great crises occurred in the 17th, 18th and early 19th centuries and was concerned with calculus. The essential problem was with its logical and philosophical basis, but the upshot concerned difficulties of a more practical nature—the methods of calculus were shot through with errors and its conclusions could be unreliable. The very good mathematicians who developed and used calculus did not employ the rigorous methods of definitions and proofs used today. The difficulties and contradictions which resulted took some time in becoming apparent because these men were capable mathematicians who had a good sense of how things should be, but calculus could become useless in the hands of the less gifted. Two underlying reasons for the difficulties that did occur were the lack of a general function concept as well as a too great a reliance on geometrical considerations and visual proofs. Early 19th century mathematicians even used physical phenomena to justify new mathematics, an approach unheard of today.

The history of calculus is in part the reason for the insistence on the rigorous proof of assertions in mathematics today. If one takes the time to read a small part of the history of mathematics during this period, he will find quotations from the leading mathematicians of the day which sound remarkably like the kind of answer he might have been inclined to give in reply to a hard question posed by his calculus instructor. He might then perhaps better understand the admonition that his reply was not quite a good one. Here is a quote from Johann Bernoulli's introduction to differential calculus in 1691: "A quantity which is diminished or increased by an infinitely small quantity is neither diminished nor increased."

These difficulties resided in the need for a good concept of number and limit. Witness L'Huilier: "If a variable quantity at all stages enjoys a certain property, its limit will enjoy this same property" despite the presently known fact that an irrational number is the limit of a sequence of rational numbers, and the then apparent belief that a circle (a non-polygon) is the limit of inscribed polygons. This mistaken hope about limiting cases persisted even into the 19th century.[7] But one should not forget one thing: the calculus was an astoundingly successful and powerful tool of mathematical analysis.

The French mathematician Augustin Cauchy introduced hard headed thinking in calculus and analysis with the publication of his lectures in analysis in 1821.[8] Cauchy's method of good complete proofs following from careful definitions took some time to gain acceptance. Even Gauss, perhaps the greatest mathematician of all time, is said to have referred to analysis as the "science of the eye." The school of mathematicians which relied on visual perception and other intuitive methods in the study of continuity, differentiability and limits was destroyed by the work of Karl Weierstrass in 1854. By indisputable methods, Weierstrass produced a continuous curve having a tangent line at *none* of its points, a decidedly difficult thing to visualize and quite a shock to the intuitively founded conventional mathematical wisdom.

Weierstrass, according to Bell, deserves the title of the "father of modern analysis."[9] More technically, Weierstrass replaced the geometrical foundation of calculus and analysis with an arithmetical one, founded on a detailed knowledge of the real number system. Weierstrass, following Cauchy, also firmly established the need for viewing the idea of the limiting processes of analysis not as a *dynamic* one—"*$f(x)$ approaches L as x approaches a*"—but as a *static* one—given arbitrarily small but *fixed* interval around L there is an interval around a such that for x in this interval around a, $f(x)$ is in the given interval around L—this is the $\varepsilon - \delta$ notion used today. The first is great for intuition and for getting ideas, the second is the best yet found to certify them.

The third great crisis in mathematics occurred in the late part of the 19th century with the advent of Cantor's theory of sets and of the actual, as opposed to the potential, infinite. Some years after Cantor's definition of a set as "any collection into a whole of definite and separate objects of our intuition or thought" came Russell's paradox and a shock to mathematical thinking so severe that its ripples are still being felt. The alarming thing about this discovery is that, as will be seen, (1) the notion of a set is, on its surface, about as simple a thing as one can come by, (2) much of mathematics can be made to rest on set theory and (3) operations with sets are essentially equivalent to our most fundamental reasoning processes. It must be noted, however, that difficulties in set theory occur either as a result of careless or circular definition or in the context of infinite sets or

[7] C. B. Boyer: *The History of the Calculus and its Conceptual Development*, Chapter VI, New York: Dover Publications, 1959.

[8] E. T. Bell, *Men of Mathematics*, p. 286, New York: Simon & Schuster, 1937.

[9] *Ibid.*, p. 406.

infinite collections of sets—a context which is nevertheless unavoidable! Although the difficulties arising from the theory of sets have not been decisively settled to this day, mathematics has expanded in all directions using the theory of sets, and, despite the many different and independent ways in which set theory is used, no insoluable contradictions have resulted—yet.

A large part of the remainder of this text is devoted to a study of the mathematics that has resulted from the difficulties that gave rise to these crises. In this chapter we will soon take up the real number system, and touch lightly ancient difficulties involving the limit concept. In later chapters we will consider a theory of sets and functions and infinite sets and close with a completely general study of modern concepts of limit and continuity. But at this point it is proper to turn to that discovery which led to the methods and outlook capable of meeting all these crises and successes.

4. THE DEVELOPMENT OF NON-EUCLIDEAN GEOMETRY

In the three periods of mathematical crisis just considered it appeared that the whole of certain parts of the subject might collapse into a useless collection of supposed theorems, void of meaning, signifying nothing. Indeed this did happen, but not in any essential sense. In each case the subject was spared this embarrassment by the brilliance and inventiveness of its adherents. It is now time to look at the underlying mathematical method which enabled the recovery—the foundation of the house of mathematics. This is the axiomatic method lying quietly and almost unnoticed at the base of ancient geometry for 2,300 years, only to be thoroughly shaken by the discoveries of Gauss, Bolyai, Lobachevsky and Riemann in the early 19th century, and made to rise up and out of their work to become the most acceptable foundation yet for all of mathematics.

In the year 1800 mathematics consisted of Euclidean geometry, arithmetic, algebra, and calculus. In the year 1970 the subject classification index of the American Mathematical Society listed 58 primary areas of mathematics, each containing an average of at least 15 subareas! The average mathematician may be highly proficient in only a few of these subareas and without some effort two average mathematicians in different primary areas are unlikely to be able to tell each other what they are doing.[10] Whereas in the year 1800 every prominent mathematician might legitimately describe his particular interest as mathematics *per se*, only a very, very few of the mathematicians who consistently produce original mathe-

[10] You may find it interesting to spend a few minutes looking over a recent copy of the *Mathematical Reviews*, which is a collection of mere abstracts of published mathematical papers isssued bimonthly. This will give you a good idea of the size of today's mathematical activity. The Reviews are published by the American Mathematical Society, Providence, R. I.

matics today would dare claim so much. Evidently, something happened in between. This happening was the release of mathematics from any dependence on direct observation of the physical world, beginning with the discovery of non-Euclidean geometry in 1829. Before 1829 mathematicians depended on the logic of their senses, after 1829 only on the logic within themselves, the logic of their minds.

It is impossible to study mathematics and not be influenced by the Greek mathematics done in the years 600 to 300 B.C. During this time the *Book of Elements* was compiled by Euclid and, as noted, the first consistent attempts at demonstration—proof—of mathematical statements were made. The *Elements* was concerned mainly with geometry and geometry was thought to be the absolute truth on which man's understanding of the physical world must be based. This view was wrong in principle and in fact, the principle being that absolute truth does not apparently exist and the fact that even an idealized universe is not a concrete realization of Euclidean geometry. Nevertheless, this view was held for over 2,000 years.

The celebrated role of Euclid and the *Elements* in mathematics is due to two things. The first is that these works contain the theorems of (Euclidean) geometry and their proofs, these proofs being abstract and independent of any direct observation of the physical world to which they might be applied. The second is that Euclid recognized that these theorems and their proofs ultimately depended on unproven assumptions which he called axioms and postulates. Some examples were (1) the whole is the sum of its parts, (2) through every two points there is a unique straight line (a straight line being the locus of points transcribing the shortest distance between these two points) and (3) through any point p not on a straight line L there is a unique straight line through p which does not meet L—the parallel postulate.[11]

As Euclid pointed out, the whole of geometry rested on these and his other postulates and all were unproven. But they seemed to be obviously valid assumptions, and up to the year 1829 the word axiom was synonymous for a "self-evident truth, beyond all logical opposition."[12] There was, however, the recurring feeling that one of these, the parallel postulate, ought not to be an axiom, that it ought really to be provable from the other axioms; this feeling rested on the belief that as long as the interpretation of a straight line was that of the shortest distance between the points on it, then the parallel postulate should be a consequence of further deduction from the remaining axioms. It was so intuitively obvious!

There were three people who began to suspect that there might be more to it than that. These were Carl Friedrich Gauss (1777–1855), Janos Bolyai (1802–1860) and Nicolai Lobachevsky (1793–1856). Their suspicions were founded in their own work as well as the earlier work of others,

[11] This is not precisely Euclid's statement, but it is easier to discuss.
[12] Bell, *Men of Mathematics*, p. 249–306.

particularly Saccheri (1733), who had attempted to prove the parallel postulate by the method of ***reductio ad absurdum*** but had never been able to reach a contradiction! Because Lobachevsky was the first to publish his work, we will concentrate on what he did.

Einstein said, "Lobachevsky challenged an axiom," and, Bell added, "An axiom is a prejudice sanctified by thousands of years."[13] At the turn of the 16th century Francis Bacon had set forth a then unheard way of thinking. "As the foundation (in studying nature)," Bacon wrote, "we are not to imagine or suppose but to *discover* what nature does or may be made to do" and the scientific method was born. Almost concurrently, Galileo was amassing the data to show that the earth did indeed revolve around the sun as Copernicus had suggested. In the centuries that followed many another ancient article of faith was annulled, culminating with Darwin's discovery on a 13,000-foot high Andes mountain the same fossilized sea shells he had earlier found only a few feet above the seashore. Einstein in the next century was to follow Bacon's advice. Unable to satisfy his sense of how things should be without questioning the idea that two events could happen in different places *at the same time*, Einstein came to the discovery of relativity and a better understanding of the universe, the atom, time and space. Yet despite all these discoveries, despite all the doubters who have followed Bacon, among all those who have challenged the conventional wisdom of their time, it is Lobachevsky that Bell calls "the Copernicus of all thought" because he was able to "knock the eternal truth out of geometry" when after over 2,000 years the postulates of geometry had "taken on the aspect of hoary and immutable necessary truths, revealed to mankind by a higher intelligence as the veritable essence of all material things."[14] Perhaps it was because, whereas all the other discoveries following the Dark Ages were concerned with quite complex phenomena, ordinary plane geometry was the essence of simplicity, its assumptions so plainly true!

Just what did Lobachevsky do? In brief, he gave an example of a consistent geometry in which the interpretation of point and straight line were the "same" as in Euclidean geometry and in which all of the postulates of Euclid were valid except for the parallel postulate. This postulate was replaced by the postulate that through a point not on a straight line, L, there is more than one straight line not meeting L. As a corollary he forced on us the unavoidable necessity of proving the obvious before we know it is true, for Euclid's parallel postulate is very "obvious" but by virtue of Lobachevsky's example it is not even true (as a consequence of the remaining axioms of geometry)!

But more had better be said about the ensuing example lest the reader think this whole discussion is merely specious. In the broadest sense the

[13] *Ibid.*, p. 305.
[14] *Ibid.*, p. 305.

matter resides within that enduring subject of poets, philosophers, and other writers, more so than scientists: this is the matter of sense awareness, as it is sometimes called. In simplistic terms it has two parts, one inner directed and one outer directed. For the inner, there is what your senses can perceive of the world outside you. If your eyes had evolved a receptiveness to infrared radiation rather than ordinary light, the world would be a very different place in appearance to you; it would nevertheless be the same world.[15] For the outer, there is the appearance one imposes on what his senses receive. Often this amounts to placing an order or structure on what he senses. Thus the physicist supposes an order and structure for the atom, the biologist finds DNA, the repositor of the genetic code, and says it looks like a spiral staircase. He does this to better understand and explain it. But once this is done one has placed a structure, or an appearance upon a reality and one's perception of reality itself is altered. And here is the point that concerns us: one then perceives through the appearance imposed and perhaps no longer senses the whole, possibly less ordered, or differently structured reality. What then happens (and this is the "proof" of the last assertion) is that in further study, through the structure assumed, one begins to notice discrepancies between what the supposed structure implies and the reality emits; it is then time to change the model, change the supposed structure. This is how the present day model of the atom has come about, this is in part what led Einstein to relativity. This is why our decendents will not see things as we do. It has, for example, been an assumption for centuries of physics that one can analyze the whole by analyzing its separate, simpler parts. This appears to be an untenable approach in ecological studies and seems inadequate for the life sciences in general.[16]

Now suppose yourself to be a mildly knowledgeable student of Euclidian geometry in the year 1829. You have come to think of geometry as a collection of truths about points, lines, planes, angles, congruence, etc., which are but logical consequences of certain initial statements about points, lines, etc. For example, "Through two points there is one and only one straight line." You recognize these initial statements as being unproven, but thoroughly believable on the basis of intuition, except possibly for the

[15] An interesting point here. We are so complex, perceive so much, it is hard to see that we may not perceive all there is. To extend an example of the naturalist Loran Eiseley in *The Immense Journey* (New York: Vintage Books of Random House, Inc., 1957) consider the common, ordinary frog. It has a very primitive nervous system and because of this has managed to survive relatively unchanged for eons of time. Much of the world, the wind, trees, flowers, does not exist for a frog. It is attuned almost solely to small, rapidly moving, flying invertebrae—in short, thoroughly alive insects. Put a captured frog in a cage with a feast of deceased flies and it will starve to death. Put a hundred thousand frogs in a hundred thousand separate cages with a feast for each—maybe one would eat. Call him Lobachevsky, or Newton, or Einstein!

[16] T. L. Saaty and F. J. Weyl (eds.): *The Uses and Spirit of the Mathematical Sciences*, p. 203, New York: McGraw-Hill, 1969.

parallel postulate which several well known geometers believe ought to be a consequence of the remaining axioms anyway, but have been unable to prove. Nevertheless, this postulate seems true as you stand on a flat plane and imagine the parallel to L through p extending out into the mist at infinity. Here is where the point of sense awareness comes in: you have come to see the geometry of the world through its axioms, through an order imposed some 2,000 years earlier. That is, reality has been replaced by only a description of reality, but so thoroughly imbibed with this view as you are, you might consider anyone who would suggest it to be only a supposition to be only a disruptive outsider out to infect your thoroughly healthy beliefs.

Unless you were a Gauss, Bolyai, or Lobachevsky! These mathematicians showed that if a slightly different interpretation or meaning of "straight line" is taken, one that is coincidently more in accord with physical reality on the surface of the earth, then one finds that all the axioms of Euclidian

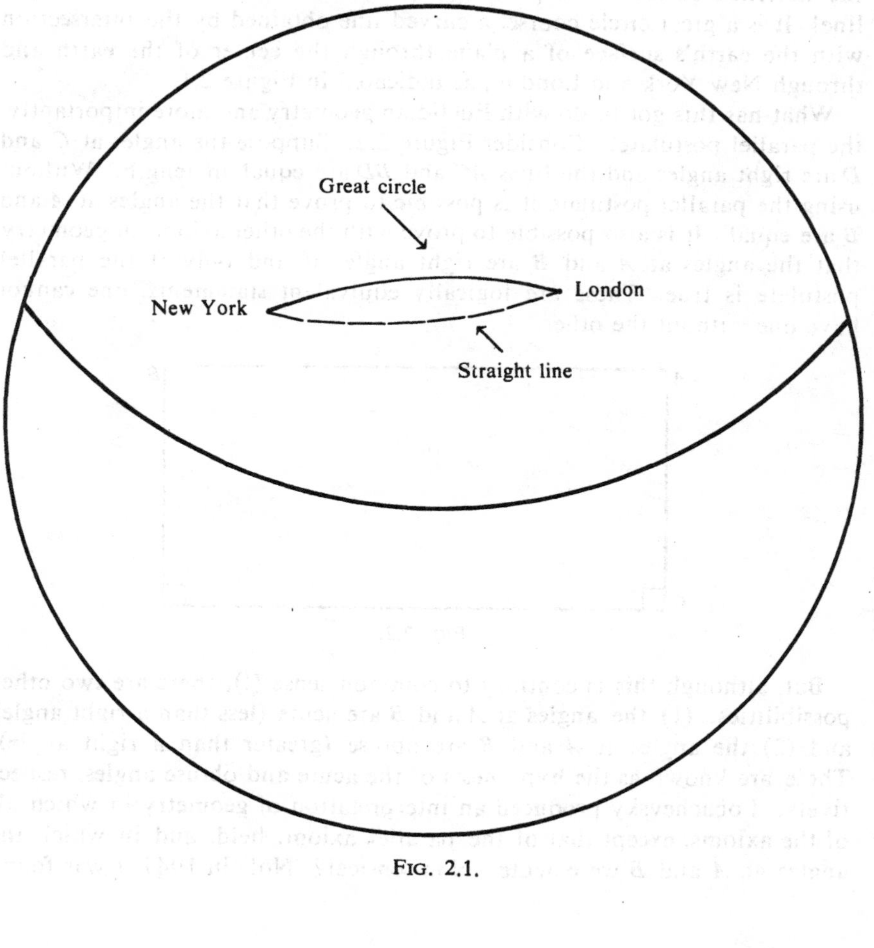

FIG. 2.1.

geometry remain true, save for one: the parallel postulate. Implication one: the parallel postulate can never be proven from the remaining axioms. Implication two: even in the context of all the remaining axioms about points, lines, etc. and despite the "obvious" character of the parallel postulate in this context, it is not only not obvious, it is not even true. It only seems obvious because one has come to think in terms of what one considers a straight forward common sense interpretation of straight line.

How did this example escape the notice of scholars for 2,000 years? One cannot know with certainty, but perhaps it was because of a view of the world around us so naturally and gradually accepted that it had never been recognized as crucial to the whole interpretation of geometry. This unconscious viewpoint probably had its roots in the uses of geometry for surveying, construction and for sailing about the Mediterranean, and reinforces again the point of sense-awareness. For a moment think of that line which represents the shortest distance between New York and London on the flattened surface of a perfectly spherical earth—it is not a straight line! It is a great circle course, a curved line obtained by the intersection with the earth's surface of a plane through the center of the earth and through New York and London, as indicated in Figure 2.1.

What has this got to do with Euclidian geometry and more importantly, the parallel postulate? Consider Figure 2.2. Suppose the angles at C and D are right angles and the lines AC and BD are equal in length. Without using the parallel postulate it is possible to prove that the angles at A and B are equal. It is also possible to prove with the other axioms of geometry that the angles at A and B are right angles if and only if the parallel postulate is true. These are logically equivalent statements, one cannot have one without the other.

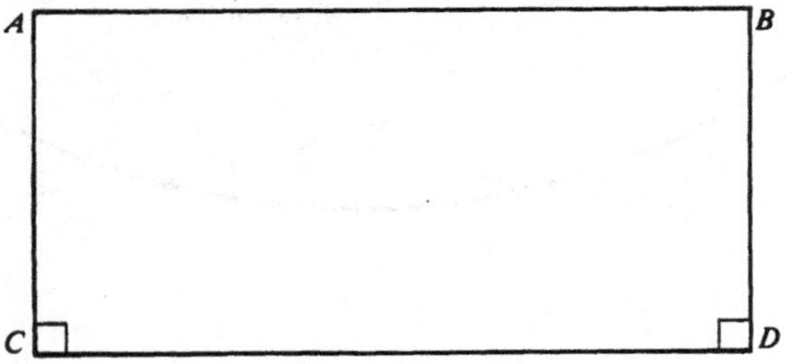

FIG. 2.2.

But, although this is contrary to common sense (!), there are two other possibilities: (1) the angles at A and B are acute (less than a right angle) and (2) the angles at A and B are obtuse (greater than a right angle). These are known as the hypothesis of the acute and obtuse angles, respectively. Lobachevsky produced an interpretation of geometry in which all of the axioms, except that of the parallel axiom, held, and in which the angles at A and B were acute. Non-sensical? No! In 1947 it was found that Lobachevskian non-Euclidian geometry provides a better description of visual space than Euclidian geometry.[17] It is easier to believe in Lobachevsky's example if we first go back to the surface of a sphere and consider a simpler example not due to Lobachevsky.

Keep in mind that we want all of the usual interpretations of line and point and the other axioms of geometry to hold, except for the parallel postulate. In particular we want a straight line to be the shortest distance between two points. Thus let us think of geometry on the surface of a half-sphere without edge on one side, a straight line between two points being a *geodisic*, a great circle course as defined above. Now let us draw the "rectangle" in Figure 2.3, on a sphere.

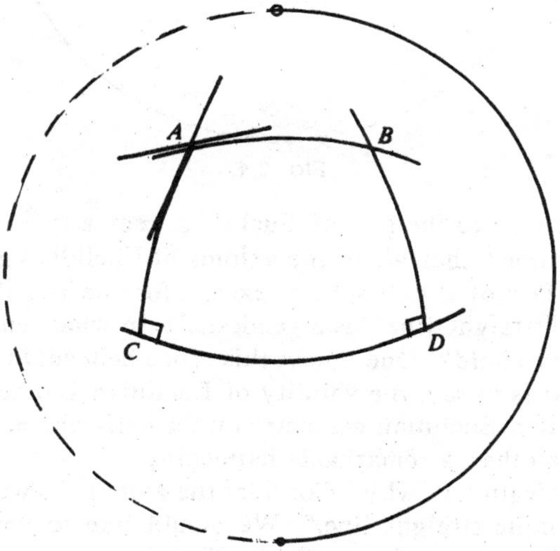

FIG. 2.3.

It is not too hard to believe that the angles at A and B (defined to be the angles between the tangents T and T') are obtuse. Notice also Figure 2.4, where it is clear that there is no parallel to the equator which passes through the point p, recalling that a straight line must be a great circle if all the other axioms of geometry are to hold. (More about the other axioms shortly.) In fact, one can argue without resort to pictures that the hypothesis of the obtuse angle is equivalent to the existence of no parallels through a point not on a given line. However, this conclusion would seem less than satisfactory were there no examples for which the hypothesis of the obtuse angle did hold.

[17] R. K. Luneburg: *Mathematical Analysis of Binoucular Vision*, Princeton: Princeton University Press, 1947. This reference is taken from H. Eves and C. V. Newsom: *Introduction to Fundations and Fundamental Concepts of Mathematics*, p. 78, New York: Holt, Rinehart and Winston, Inc., 1965.

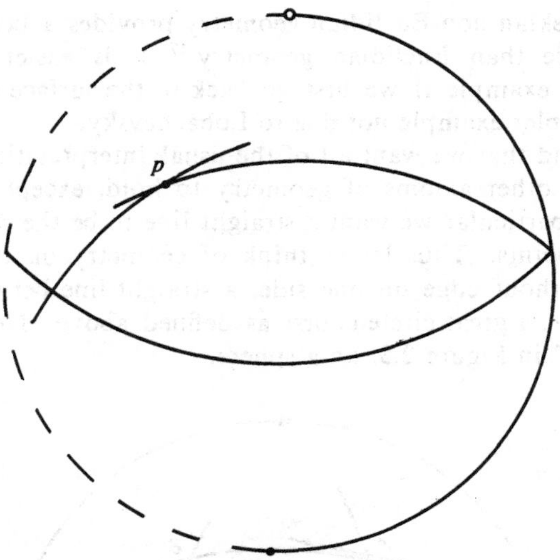

FIG. 2.4.

Let us return now to the role of Euclidian geometry in the face of all this. It was claimed that all of the axioms of Euclidian geometry would hold on the surface of a half-sphere, except for the parallel postulate, if one interprets "straight line" as a geodesic.[18] In what sense, or how does one know, these to hold? One knows this if one believes in solid Euclidian geometry! That is to say, the validity of Euclidian geometry implies the validity of the non-Euclidian geometry on the half-sphere. One can't have one without the other, a remarkable happening.

Here is an indication of why. Consider the axiom: "Two distinct points determine a unique straight line." We would like to verify this or the surface of a half-sphere without edge.[19] That is, we would like to know that any two distinct points on the half-sphere lie on one and only one geodesic. Consider the center of the completed sphere. It and the two given points are non-colinear. From solid Euclidian geometry, three non-colinear points determine exactly one plane. The intersection of this plane and the half-sphere is the unique straight line (geodesic) through the given points. This is an indication of how Euclidian geometry validates non-Euclidian geometry. In the same way Euclidian geometry can be used to show the existence of no parallels to a given point not on a given line. A line L is the intersection of a plane and the half-sphere through its

[18] Actually another one of Euclid's assumptions, this one unrecognized by him and not stated formally, fails to hold in the model for non-Euclidean geometry just given. Euclid assumed that any straight line could be extended indefinitely and, of course, this does not hold on the half-sphere. But the lack of this property only would seem to make the existence of a parallel on the sphere even easier to obtain, yet it is still impossible.

[19] Notice that this is not true on a whole sphere where any two antipodel points are on infinitely many geodesics.

center. Any line through a given point *p* determines another such plane. These are not parallel planes since they both contain the center, nor are they coincident since *p* is not on the line *L*. These planes intersect in a straight line lying in both planes. This straight line intersects both geodesics in a common point. Hence, there is no parallel to *L* through *p*.

It is equally possible to prove that the hypothesis of the acute angle is equivalent to the existence of more than one parallel through a point not on a given line. Lobachevsky published a geometry in which this occurred —and later Beltrami (1868) published a model for Lobachevskian geometry, the pseudosphere of Figure 2.5, having constant negative curvature (a sphere has constant positive curvature), obtained by rotating the curve *T* (called a tractrix) about the horizontal line *H*.

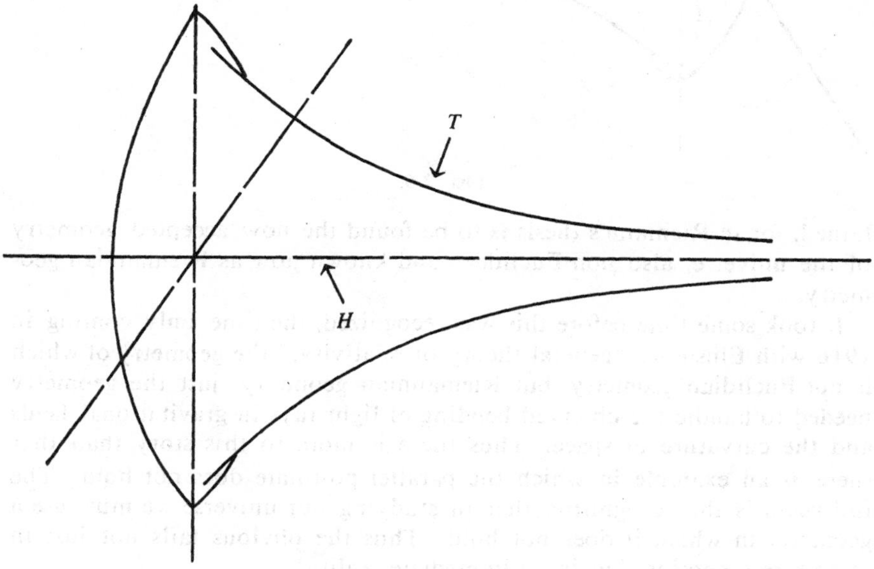

FIG. 2.5.

The geodesics of the pseudosphere which are to be used as straight lines in the geometry on it are not easy to describe. In Figure 2.6 some examples of the geodesics are given.

This then is what Lobachevsky did. He provided a logically consistant geometry in which all of the postulates of Euclid's geometry held, with the same interpretation of point and straight line, with the exception of the parallel postulate. This meant that the parallel postulate could never be proven from the remaining postulates and that its obvious truth was not only not obvious but in fact not even there. Not long afterward (1854) Riemann published his doctoral thesis laying to rest forever the necessary truth of Euclidian geometry and by implication the kind of thinking that led to the belief in such a kind of truth. But even more than this was ob-

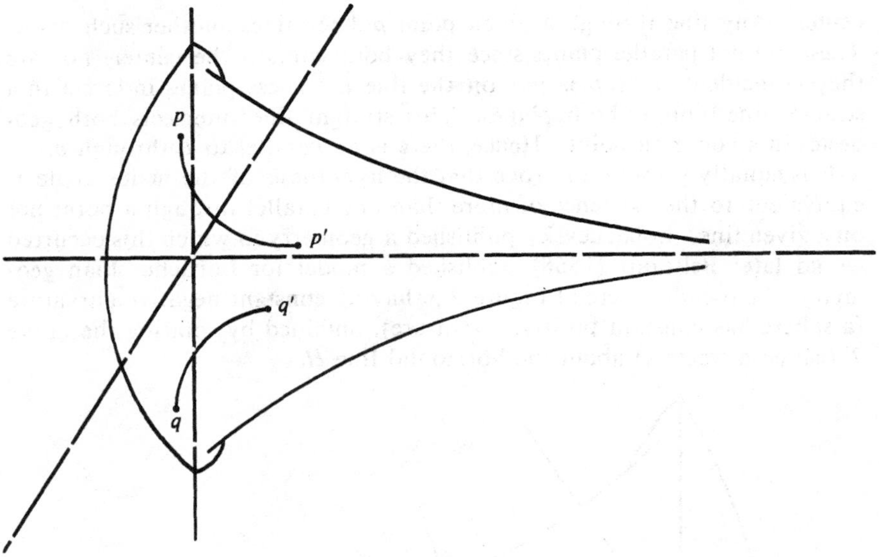

FIG. 2.6.

tained, for in Riemann's thesis is to be found the now accepted geometry of the universe, also non-Euclidean and known now as Riemannian geometry.

It took some time before this was recognized, the time only coming in 1916 with Einstein's "general theory of relativity," the geometry of which is not Euclidian geometry, but Riemannian geometry, just the geometry needed to handle the observed bending of light rays in gravitational fields and the curvature of space. Thus there is more to this story than that there is an example in which the parallel postulate does not hold. The full point is the recognition that in studying our universe we must use a geometry in which it does not hold. Thus the obvious fails not just in some rare exception, but in an immediate reality!

There are several things to be drawn from all this. We have already noted some of these in a general sense. Because of the work of Riemann and Einstein one can add another: the obvious may be far from a better truth. We want to examine a third implication, the most important one if one is to understand what mathematics is. The implication is 2-fold: (1) valid, self-consistent (even useful) theories of mathematics can exist without meeting with "common sense," which is only environmentally induced knowledge anyway, and (2) mathematics has need of no interpretation in the physical world, for the uses of non-Euclidean geometry to describe reality in the universe did not really occur until over 50 years after its discovery, and Euclidean geometry, studied and developed for thousands of years did not, in the end, refer to anything real at all, but only to an ideal plane surface. These observatons set mathematics free from direct observation of any reality in the world about us and are what have led to its present diversity. But this is not to say that mathematics should not have an interpretation in reality, only, to repeat, that it needs none. Indeed, the most prolific areas of mathematics are those that were at least initially motivated by the need to solve real world problems.

With Lobachevsky's work and the ancient example of Euclidean geometry at hand, the idea soon got around that a mathematical theory could be constructed on a set of given conditions (axioms) which were taken to be true without regard to any reality. There is one clear requirement for such a system of assumed conditions—that they be consistent in themselves and not eventually lead to contradiction. We will discuss this in more detail later on. The constructions on this set of axioms are called theorems and are obtained by deduction from the axioms using a fixed system of logic which itself is constructed from a set of axioms. There is, of course, clear beauty and great simplicity in such an approach, but, even more, there is an economy of effort—for whenever the axioms are true about a given instance of reality, then all the theorems are true about that reality when the same logic used in the proofs is used in discussing the reality. Thus, the Pythagorean theorem gives correct information about *any* particular right triangle confronted in a setting in which the axioms of Euclidean geometry correctly describe reality (such as a plane region), whereas a theorem from Riemannian geometry gives correct information in *any* particular setting where its axioms hold in reality. It is not that either is a true or correct geometry, it is that each is true or more useful in a particular setting. There can be tremendous problems involved in knowing when the axioms for one or the other are true in a particular instance but this is legitimately the physicist's problem, not the mathematician's. His problem is: given the axioms, what must follow from them?

Now there is also a converse to this and it is a most important one. Let us suppose one is given a particular question to decide, whether it be in physics, or biology or sociology or whatever. Suppose that one can set down the basic facts about this problem which take into account all the things one deems essential to the study, and in such a fashion that they have meaning without reference to the original problem, so that it now only supplies an instance of when these statements are actually true. These may take the form of equations (e.g., Maxwell's equations describing electro-magnetic fields) or statements (e.g., the second axiom for special relativity: the speed of light is independent of the speed of its source) or both (the laws of thermodynamics). One may then attempt to draw further conclusions using *only* these *assumptions* by abstract reasoning with no further regard to the physical entities on which they are based. Having these new conclusions one can interpret their meaning back in the original setting. This is no mean feat, being perhaps more difficult than obtaining the conclusions themselves. If these interpretations give consistently correct results, then the axioms were well chosen; if not, perhaps they can be modified or changed completely.

This last is what happened with the axioms of mechanics. These were first given by Isaac Newton; the familiar laws governing force, gravity, and action and reaction. These gave good results until they were applied to the study of masses, forces inside the atom or to velocities near the speed of light. They then failed to give good answers and were replaced by the assumptions which form the basis for relativity in conjunction with the quantum mechanics of atomic particles. (Incidentally, quantum field theory is almost wholly extremely abstract mathematics—the study of linear transformations of infinite dimensional spaces of vectors.) This points up again the place of laws of axioms. It is not that they are true or false, right or wrong. It is that they can be right in one place and wrong in another—of use here, but not perhaps there.

The search for assumptions, stated mathematically, which might explain a given interest, is known as the setting up of a mathematical model of what one wishes to study. One tries to obtain a faithful representation of what one is studying in abstract mathematical terms which have meaning independent of the particular thing they are derived from. The task of reaching new conclusions and obtaining new insight and new information then becomes to a great extent one of pure mathematics, with the exception of the problem of interpretation mentioned above. The search for a good mathematical model is of constant interest in such diverse fields as physics, biology, economics, sociology, computer science, meteorology and in pure mathematics itself. The most remarkable thing about this is that a mathematical model for a problem in biology may be exactly like one in sociology or in economics or in physics. The abstract structure of both problems may turn out to be the same, and a conclusion abstractly arrived at beginning with the same assumptions but with no regard to either of the intended applications can then be applied to both!

A rather recent example is that of group theory. A group is an abstraction of the set of integers 0, ± 1, ± 2, $\cdots$ with the operation of addition. This abstraction has many other realizations, totally different in their individual appearance. Although groups were first studied by mathematicians before 1900, the theory has recently been used by physicists to predict the existence of subatomic particles.[20]

Having seen briefy then the relationship of the axiomatic method to the uses of mathematics, we have not yet looked at its use for mathematics itself. This is the real concern here. Following Lobachevsky's work, mathematicians in other areas of mathematics began to look for and to set out the assumptions underlying their own areas of interest. These had been

[20] *Mathematics in the Modern World*, p. 249–262, San Francisco: W. H. Freeman & Co. 1968. This book contains a wealth of interesting material about mathematics and its uses. The details of even a non-technical explanation of the uses of group theory as applied to physics are beyond the intent of this book.

there all along but perhaps had never been explicitly delineated. An axiom system was given by Peano for the natural number system, the properties of which had been in development since man began counting but, not surprisingly, had never been given an explicit foundation. Following this, axiom systems were given for the real number system. It was also found that the complex numbers could be defined, using the idea of ordered pairs, in terms of the real numbers, and that in fact the whole number system could be constructed upon only the natural number system. Whitehead and Russell in the *Principia Mathematica* attempted to show that all of mathematics could be founded on an axiom system for logic. The axioms were to be a collection of "primitive" tautologies such as: $p \vee -p$, $p \wedge q \rightarrow p$, and so on. The question of an acceptable axiomatic structure for all of the set theory one might wish to use is still not settled.

Algebraic properties in use for centuries were recognized as axioms and with this came an extension of Lobachevsky's discovery. If Lobachevsky could use another version of the parallel postulate and get a consistent and interesting geometry, why not try an algebra in which (say) the commutativity properties of addition and multiplication $(a + b = b + a)$ and $(a \cdot b = b \cdot a)$ were not assumed? This is what Cayley and Hamilton did and the algebraic structures so obtained have been of essential use in both pure and applied mathematics. With that began a stampede off in all directions leading to a plurality of mathematical structures all undreamed of even a century ago. And in our time there is increasing discussion as to whether this proliferation can continue to be worthwhile along with much hard work going on to collect and unify many of the resulting areas.

The conclusions then are these. Mathematics requires no regard for, and no justification in, physical reality. Indeed, within certain limits, the less regard it pays to reality the better are its chances for wide application. In this view pure mathematics is the most useful mathematics. In many a view it is the most soul satisfying.

Problems. Each of the following problems calls for a written report on some related reading matter. Try *not* to make your report a mere list of topics covered. Rather pick out those parts that interest you and write up your own thoughts about these.

(1) Write a three-page report on Chapters I, II and III in E. T. Bell's, *Mathematics: Queen and Servant of Science.* (New York: McGraw Hill, 1951.)

(2) Write a two-page report on E. R. Stabler's *Introduction to Mathematical Thought,* p. 36–42, 115–119, and 170 173. (Reading, Mass.: Addison-Wesley, 1953.)

(3) Pick out a newspaper column or a short magazine article which sets out conclusions and examine it for its "mathematical" structure. That is, identify its axioms (assumptions stated as facts without justification) its proofs (reasonings from its assumptions) and theorems (the conclusions it reaches). Take careful note, too, of those

undefined words whose meaning may be crucial to the article. If its conclusions do not agree with what you think is a fact, what must you believe about its assumptions? If its assumptions seem invalid to you, could its conclusions still be correct?

(4) Write an essay on the nature of mathematics, truth and mathematical truth. You may find one or more of the following comments (taken from E. T. Bell's *Men of Mathematics*) helpful in this regard:

"I regret that it has been necessary for me in this lecture to administer such a large dose of four-dimensional geometry. I do not apologize, because I am really not responsible for the fact that nature in its most fundamental aspect is four-dimensional. Things are what they are····"—A. N. Whitehead (*The Concept of Nature*, 1920).

"Strange as it may sound, the power of mathematics rests on its evasion of all necessary thought and on its wonderful saving of mental operations."—Ernst Mach.

"Mathematics is the tool specially suited for dealing with abstract concepts of any kind and there is no limit to its power in this field. For this reason a book on the new physics, if not purely descriptive of experimental work, must be essentially mathematical."—P. A. M. Dirac (*Quantum Mechanics*, 1930).

"As I proceeded with the study of Faraday, I perceived that his method of conceiving the phenomena [of electromagnetism] was also a mathematical one, though not exhibited in the conventional form of mathematical symbols. I also found that these methods were capable of being expressed in the ordinary mathematical forms, and thus compared with those of the professed mathematicians."—James Clerk Maxwell (*A Treatise on Electricity and Magnetism*, 1873).

"How can it be that mathematics, being after all a product of human thought independent of experience, is so admirably adapted to the objects of reality?"—Albert Einstein (1920).

"Every *new* body of discovery is mathematical in form, because there is no other guidance we can have."—C. G. Darwin (1931).

"In my opinion a mathematician, in so far as he is a mathematician, need not preoccupy himself with philosophy—an opinion, moreover, which has been expressed by many philosophers."—Henri Lebesgue (1936).

"Mathematics is the most exact science, and its conclusions are capable of absolute proof. But this is only so because mathematics does not *attempt* to draw absolute conclusions. All mathematical truths are relative, conditional."—Charles P. Steinmetz (1928).

"It is a safe rule to apply that, when a mathematical or philosophical author writes with a misty profundity, he is talking nonsense."—A. N. Whitehead.

"As far as the laws of mathematics refer to reality, they are not certain; and as far as they are certain, they do not refer to reality."—Albert Einstein.

"Thus mathematics may be defined as the subject in which we never know what we are talking about, nor whether what we are saying is true."—Bertrand Russell.

Finally, you may also wish to look over the comments of others on this subject as found in H. Eves and C. V. Newsom: *Introduction to the Foundations and Fundamental Concepts of Mathematics*, Ed. 2, p. 188, New York: Holt, Rinehart and Winston Inc., 1965.

5. THE AXIOMATIC METHOD IN MATHEMATICS

The great proponent of formalism in mathematics was David Hilbert, an early 20th century mathematician whose insight and influence extends throughout the mathematics of this century. Hilbert once described mathematics as "a game played with meaningless marks on paper." The intent of this section is to lend meaning and substance to that remark. Following this section we will consider an axiomatic foundation for the natural number system and for the real number system. Here we will first be content with a thoroughly useless and then a thoroughly useful axiom system which will get the key points across.

Any discussion, any system of thought must begin somewhere, and this extends even down to the level of the individual words used in the discussion. It is impossible to define all words in a given language system. For example, imagine a language consisting of (say) three words and try to define all these words *within* that language. In a very few areas of mathematics the use of any words at all might introduce unwanted shades of meaning and all technical discussions in such systems are carried on only in symbols. In most systems of mathematics two kinds of words are used. One kind consists of the words of ordinary language used in ordinary ways, most often to state the important sentences of the system in a clear fashion subject to logical meaning and manipulation. Such words as *a, the, if* ··· *then, therefore, hence* and so on are examples of these. The remaining words are the technical words of the discussion and are, in the end, the *only* subject of the discussion. Not all of these can be defined, because one must begin somewhere. Thus these fall into two types: (1) undefined technical words which are admitted initially as simply undefined and (2) technical words defined in terms of the underfined words. For example if *point, locus, between, transcribing* and *shortest distance* are taken as undefined technical words in geometry, then the technical word *straight line* is defined as the *locus of points transcribing the shortest distance between any of its points*. The words *of, the, its* are undefined, non-technical words whose importance is not critical to the definition.

A body of mathematical thought is a systematic study of the undefined technical terms with which the study begins. This study must also begin with some facts about these terms. These are the *axioms* of the study: the statements which are given a truth-value of *true*. The axioms are statements about (that is, relating) the undefined terms, no more and no less. They need not be true about any particular thing that exists anywhere in this universe, they are only taken as true. If they happen to have an interpretation one regards as real, then, theoretically, that is just a happy coincidence. The idea is that although these are statements about underfined meaningless words, they provide a sequence of given logical connections between these words that enable one to obtain further and hopefully deeper logical connections between these same words.

Along with the axioms, undefined technical terms and a non-technical language basis, an overiding system of logic is also imposed on the structure. This system of logic provides a set of stated rules, the rules of the game, for manipulating the given statements of the theory. Ordinarily this logic is that studied in Chapter 1. Although we did not establish this logic axiomatically it is possible to do so. It is known for example that beginning with the undefined symbols $\vee$ and $-$ and the undefined concept *proposition*, one can derive all of the laws of logic found in Chapter I from the following axioms, which, being taken as true, may be thought of as primitive or initial tautologies. These are

I. For p and q propositions, $p \vee q$ and $-p$ are propositions.
II. $-(p \vee p) \vee p$
III. $-q \vee (p \vee q)$
IV. $-(p \vee q) \vee (q \vee p)$
V. $-(-q \vee r) \vee [-(p \vee q) \vee (p \vee r)]$

These are taken from Russell and Whitehead's *Principia Mathematica* and, despite their appearance, really do lead to the laws of logic. Logic, too, then has an axiomatic basis though this is not an important point in understanding the axiomatic method.

At this point it is not unlikely for the reader to have thought of the following questions. How does one know how to choose the undefined words and axioms? Where do these come from? They do not in practice come first. Roughly what happens is this. A rather systematic but hazily founded study or theory is developed concerning some matter of interest. If the study proves useful and provides insight, then one feels that it is worth the effort to discover and delineate a sure foundation on which it can be made to rest. In other words, knowing what one wants, one goes back and searches for reasons and underlying justifications. This is a technique probably not unfamiliar but care should always be taken to admit that the ultimate foundation is but a collection of assumptions. The mathematician is not guided by a search for truth in the statement of his axioms, and admits they are assumptions but which, *if accepted, must* lead to this or that stated conclusion. He is vitally interested in the wherewithal of the word *must*! His only claim is that if his axioms and logic are agreed on as fact in a given concrete setting, then his conclusions must be accepted as fact, even though, and because, he paid no attention to that particular setting in reaching them.

Suppose that a collection of undefined technical terms in given along with some statements (axioms) about them along with a system of logic. The *formal* development of the body of mathematics (further statements about these undefined terms) is then purely a matter of combining the statements of the axioms according to the rules of composition given by the logic one is using. When one has obtained a new statement about the undefined terms which he deems sufficiently important to make a permanent note of, he commonly calls this statement a ***theorem***. If it is of lesser importance it may be called a *lemma*. The combinations of the statements in the axioms, done according to the rules of the logic, which led up to this statement now called a theorem, is know as a *proof* of the theorem. Since the axioms were given a truth-value of *true* and the theorem is obtained using the assumed logic by inference from the axioms, the theorem has a truth-value of *true*. Then one has the axioms and the theorem to use as true statements to derive further true statements. And the game goes on!

Perhaps it would now be a proper time to discuss again the role of *definitions* in such a structure. A definition is a statement neither true nor false; the only requirement for a definition is that it be consistent within itself. The need for a definition might occur in the following way. In developing the theory based on the axioms, the mathematician gets the idea that it would be nice to have a single concept which seems to gather together several recurring techniques or devices he has used to think about and prove theorems to that point. He then tries to define this concept using what is at hand—the undefined terms, the axioms, the logic, and the theorems already proven. This done, he can make use of the new concept in a precise and convenient way, for more than likely it wraps together several ideas into a single one. Furthermore, he may wish to prove new statements about the thing just defined, which, because the definition goes back to the undefined terms, is still only a new "truth" about the undefined terms. Here is an important point: a definition (say of the limit concept) may be motivated by an intuitive idea, it may give one an idea or a way of thinking about the thing defined, but that is *not* its role! The role of a *definition is* to give one *a rule* or criteria *for identifying* the thing defined. Until new facts are proven about the thing defined it is the *only* way to tell when something is or is not the thing defined. If any other way is used to determine this, then in effect the definition has been changed to a different one in which this "other way" is now the rule for identifying.

We have been describing the skeleton of a deductive system of thought. To put some flesh on its bones, let us take up an example and review what has been said. Given below is a set of axioms about four undefined terms. More than likely they will seem virtually meaningless—they are! Nevertheless, we will see that it is possible to play the mathematical game and obtain conclusions from these ultimately meaningless statements. We will then see how the statements can be given meaning and how the theorems give new information about the meaning assigned to the undefined terms and axioms.

We suppose a *collection* of *spirits* and *substances* and a notion of *possesses* satisfying the following axioms (the undefined words are those italicized).

AXIOM 1. There is at least one *spirit*.

AXIOM 2. Each *spirit possesses* a *substance*.

AXIOM 3. Given a *substance* there is a *spirit* not *possessing* that *substance*.

In this example one almost immediately needs a definition before anything can be proven.

DEFINITION 1. Two spirits are *distinct* if there is a substance such that one spirit possesses this substance and the other does not.

One can now prove the theorem.

THEOREM 1. There are at least two distinct spirits.

Proof. By Axiom 1 there is a spirit, call it s. By Axiom 2, s possesses a substance, S. By Axiom 3, there is a spirit, t, not possessing the substance S. By Definition 1, s and t are distinct spirits.

What has been done? Not much, but at least let us analyze that. It has been shown that the three axioms guarantee the existence of at least two spirits in the sense of Definition 1 only, whereas the axioms themselves do not directly state this.

Let us make a further difinition.

DEFINITION 2. Two substances are *distinct* if there is a spirit possessing one of the substances but not the other.

With this, another theorem can be proven.

THEOREM 2. There are at least two distinct substances and there are also distinct spirits which possess each of these substances.

Proof. By Axiom 1, there is a spirit, s, which by Axiom 2 possesses a substance, S. By Axiom 3 there is a spirit, t, not possessing S. By Axiom 2 there is a substance, T, such that t possesses T. Since t possesses T and t does not possess S, then S and T are distinct by Definition 2. Since s possesses S and t does not possess S, then s and t are distinct by Definition 1. Since s possesses S and t possesses T, we are through.

If one were absolutely stuck for anything to do with his time he could carry this further. One could, for example, ask the question: Are there three distinct spirits? No matter how long one tried to prove that the answer is "yes" he would never succeed. How does the author know this? Mostly because he planned it that way, but also because he has in mind a specific realization (interpretation) of the axioms in which, for the meaning assigned to spirit, there are only two spirits.

To see what is meant by this we must take up the idea of an interpretation or "real-world" realization of an axiom system. The essence of this topic is again well indicated by Hilbert. One of Hilbert's more memorable statements on the role of undefined terms, axioms and interpretations was to the effect that: "Whatever is true about points on lines, must be true about beer mugs on tables." Hilbert here was referring to the idea that because the undefined terms are given no meaning in proving the theorems, then when a meaning is assigned to these terms so that the axioms are satisfied, then the theorems will be true about the meaning assigned to the terms.

Let us give a meaning to, or interpretation of, the axiom system above. Let spirit mean "beer mug" possesses mean "is on" and substance mean "table." In order to have an interpretation of the axiom system, we suppose ourselves to now be in a beer hall. To satisfy Axiom 1, there must be at least one beer mug. To satisfy Axiom 2, every beer mug must be on a table with none left behind the bar. To satisfy Axiom 3, it is necessary that, given a table, there is a beer mug which is not on that table. The picture of the beer hall appears as shown in Figure 2.7.

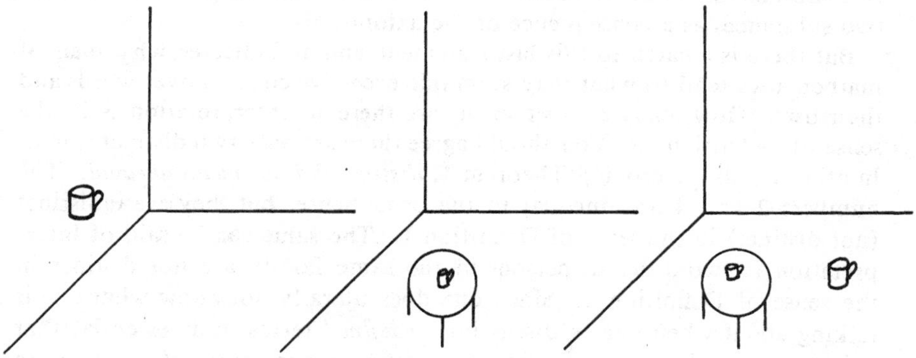

FIG. 2.7.

Theorem 1 says there must be at least two distinct beer mugs as is already apparent from the figure. Theorem 2 says there are at least two distinct tables; this is also apparent from Axiom 2 applied to our last drawing. Thus our latest interpretation is shown in Figure 2.8, and this beer hall satisfies all the axioms. Do you now see why one can never prove there are three distinct spirits from the axioms?

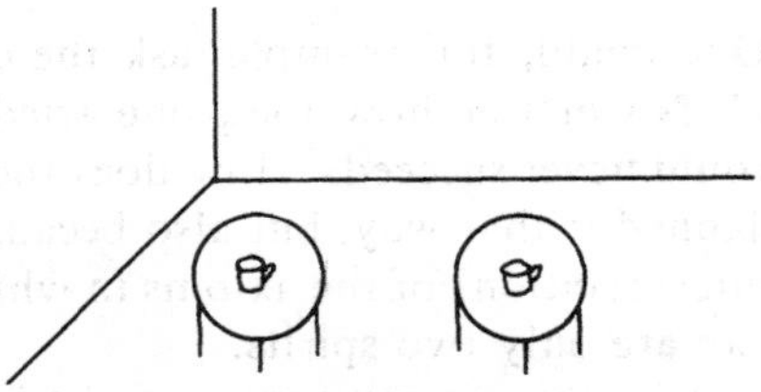

FIG. 2.8.

With that somewhat lightheaded realization of this axiom system, let us return to more sober matters and illustrate the power of the axiomatic method. Here are three more totally different realizations of this axiom system, given in tabular form.

Interpretations	Spirit	Substance	Possesses
1	Persons	Family	Is a member of
2	Point	Line	Is on
3	A positive, whole number	{1, 3, 5, · · ·}[21] or {2, 4, 6, · · ·}	Is in

In each of these instances, the axioms are true statements. Thus the theorems are true statements; not very profound but still true. Notice the third example. Here there are infinitely many spirits (apparently) but only two substances. Thus we could also not prove that there are more than two substances as a consequence of the axioms.

But there is a catch to this last statement and it indicates why men of mathematics tend to what may seem like excessive concern over words and their uses. How many *distinct* spirits are there in Interpretation 3 in the sense of Definition 1? You should agree there are only two distinct spirits. In other words, in proving Theorem 1, *distinct did not mean unequal.* The numbers 2 and 4 are unequal in the usual sense, but they are indistinct (not distinct) in the sense of Definition 1. The same can be said of Interpretation 1; two different persons in the same family are not distinct in the sense of Definition 1. Since one does literally not know what one is talking about when one is discussing *undefined* terms, it is essential that *only meaning taken from the axioms or definitions is acceptable.* These are *all* that one can *allow* himself to know.

The only worth of the axiom system given above that is apparent is its value as an example. Before we go on to some thoroughly useful axiom systems, a summary of what we have seen is in order: It is possible to reason about meaningless entities (spirit and substance) on the basis of a given set of statements about these entities. One can make definitions which

[21] This denotes the collection of numbers 1, 3, 5, etc., thought of as a single entity.

attempt to refine the properties inherent in the statements. One can assert and prove further "facts" in terms of the definitions and meaningless words. When one comes upon an interpretation in the real world of the axioms and undefined terms, the conclusions obtained apply to this interpretation. One could have equally well reasoned out each conclusion in each of the interpretations separately, but in very little time would have seen that the arguments used had the same abstract structure that appears in the proof of Theorems 1 and 2. And there lies the economy and the nature of mathematical thought and the axiomatic method. *There is the essence of mathematical thought.* This example, though it be trivial, is what was meant by the earlier claim that when a problem (say) biology has the same abstract structure as one in (say) economics or in another area of pure mathematics, then any conclusion arrived at by reasoning without regard to any one of these particular interpretations then applies to all of them. This is the worth of mathematical thought. This is why it is useful, and is a great source, no doubt, of its "unreasonable effectiveness."

Let us now take a look at one of the most useful yet initially simple branches of pure mathematics, the theory of groups. A group is the abstraction of a fundamental algebraic structure found first in arithmetic but then in several other areas. Consider the operation of addition on the numbers $0, \pm 1, \pm 2, \cdots$. Among other properties, the following seem related to the essentials at hand:

(1) A sum of two numbers is a single number.
(2) $a + (b + c) = (a + b) + c$
(3) $a + b = b + a$
(4) There is a number, z, such that $a + z = a$ for all numbers a, namely $z = 0$
(5) For any number, a, there is a number, b, such that $a + b = z$, namely $b = -a$.

with the exception of (3), which is a rather unnatural requirement in much of mathematics, these "truths" of addition of numbers, once abstracted, reappear in an almost endless variety of mathematical settings. Here are these properties abstracted, the axioms of group theory.

A group G is a *collection* of *elements* and a symbol, $\circ$, with the following properties

(1) For a, b, c and d elements of G, with $a = b$ and $c = d$, one has that $a \circ b$ and $c \circ d$ represent the same element in G.
(2) For a, b, c in G, $a \circ (b \circ c) = (a \circ b) \circ c$.
(3) There is an element e in G, called an identity, such that $a \circ e = a$ for all a in G.
(4) For each a in G there is an element a' in G, called an inverse for a, such that $a \circ a' = e$.

Here "$=$" merely means the same element of G and is not an undefined term.

There are two models for this axiom system immediately at hand. One is obtained by taking G to be the collection of numbers $0, \pm 1, \pm 2, \cdots, 0$ the operation of addition, e the number 0 and a' the number $-a$. Another is obtained by taking G to be the collection of all fractions p/q where p and q are whole numbers with $p \neq 0$, and $q \neq 0$, 0 the operation of multiplication, e the number 1 and a' the number q/p when a is the number p/q. Before trying to prove some general results of abstract group theory let us consider a point that may not yet be clear.

The allusion has been made to mathematics as a game played with meaningless words or symbols beginning with assumed truths about these. Given that analogy, the logic assumed and the axioms given are the rules of the game. To use anything else to play the game is cheating. To introduce a new rule or method of playing the game is to either add an axiom to the system or to the logic. Thus to play the game of abstract group theory is to deduce new facts about a collection G with an operation $\circ$ satisfying (1), (2), (3) and (4) using the usual logic, using only these and *nothing more*. Reasoning on the basis of an axiom system then answers one valid and common question: "What do I use to prove theorems?" The axioms, the previous theorems deduced from these, and *only* these. The mystery is removed; the difficulty of proving the theorem remains!

In the case of group theory we have a very good, concrete example in mind to give us ideas about what may be true, namely addition on the set of integers. One thing which is true in this model is that $a + c = b + c$ implies $a = b$. Here is a proof that this property holds in any abstract group.

THEOREM 1. If $a \circ c = b \circ c$, then $a = b$.

Proof. Since $a \circ c = b \circ c$, then by Axiom 1, $(a \circ c) \circ c' = (b \circ c) \circ c'$ where c' exists and has the property guaranteed by Axiom 4. By Axiom 2, $a \circ (c \circ c') = b \circ (c \circ c')$. By Axiom 3, $c \circ c' = e$ and, hence, by Axiom 1 applied twice,

$$a \circ e = a \circ (c \circ c') = b \circ (c \circ c') = b \circ e .$$

Since $a \circ e = a$ and $b \circ e = b$, this means $a = b$ and the proof is complete.

At this point it should be evident that the constant repetition of the symbol $\circ$ in this discussion can be avoided if we simply define the symbol ab, for any $a, b \in G$, to mean $a \circ b$. Thus, Theorem 1 says that $ac = bc$ implies $a = b$. Henceforth, we will use this new notation, but its properties are only those it has by virtue of its definition: xy means $x \circ y$.

A thoughtful one, reviewing the axioms and the principal model might be led to ask: does $ca = cb$ imply $a = b$? Since we do not assume $ca = ac$, the matter is neither clear nor trivial. The crux of the proof of Theorem 1 is that $cc' = e$ and, if we attempted to use this on the equation $ca = cb$, we would obtain $(c'c)a = (c'c)b$. But we have no information about $c'c$. As it turns out $c'c$ is e, but that requires proof.

THEOREM 2. For any c in G, $c'c = e$.

Proof. Reasoning as in Theorem 1, we have $c'(cc') = c'(e) = c'$. Let b denote c' and consider b', which exists by Axiom 4. From the equation $c'(cc') = c'$ and Axiom 1, we have $[c'(cc')]b' = c'b' = bb' = e$. Hence, using Axiom 2, $e = [c'(cc')]b' = c'[(cc')b'] = c'[c(c'b')] = c'[c(bb')] = c'[ce] = c'c$. That is, $e = c'c$ and the proof is complete.

We have here begun the theory of what is a very important and extensive area of modern mathematics and some exercises shortly will follow. To some, group theory is the epitomy of pure mathematics. Yet group theory has been of use in settling age old problems concerned with the solution of polynomial equations, the problem of squaring the circle, the more contemporary problems of analysis and topology and, as was mentioned earlier, to predict the existence of fundamental particles in nuclear physics. This is truly a generous span of application covering problems thousands of years old up to problems conceived only within the last decade. This is remarkable! This also suggests something else, that perhaps goes like this. There is nothing in the axioms of group theory that would lead one to suspect that the results of group theory could have such a variety of uses and interpretations. Not only that, with some of these applications in mind, one well suspects that he might never be able to think of what theorems one could actually prove in the setting of a group. He would be confused by *irrelevant details in the model.* That is the power of mathematics—it draws "necessary conclusions which are exceedingly unobvious" to combine the words of two different authors, and these conclusions can have meaning in such disparate situations that one might not think them to be true at all if the applications rather than theory were first in his mind. The case just made then is that it is sometimes easier and more straightforward to think pure mathematics than to think applications. A further possibility is that it may be more profitable with regard to the amount of time spent to achieve a broad range of application.

There is yet another point which touches on this viewpoint. The applied technical student, in need of a kind of manipulative proficiency in mathematics, is known to protest that the mathematics he is being taught, being so abstract, cannot be of much use. "This is too theoretical. Why don't we do something practical rather than all this theory?" This viewpoint, though it has firm basis in his own experience, does ignore the nature of mathematics, as well perhaps as the meaning of being learned as well as trained. Learning is for designing the machine, refining its workings, or perhaps questioning the need for even more machines. It is the nature of what mathematics attempts to do that the theoretical is at the same time the more practical. Within certain limits of course! One does have the eminent G. H. Hardy's comment that "the most utilitarian mathematics is perhaps the most uninteresting." The reader will find in taking more advanced courses in mathematics that what is studied is usually as abstract as the group theory we have been looking at. Nonetheless, most of the advanced mathematics studied today had its origins in "practical" problems, the traces of these often long lost once the problem was put in a suitable mathematical setting. Perhaps it will help a little to recall this when he has cause to wonder "What possible use could this esoteric exercise in mental gymnastics have?"

Problems. In the setting of an abstract group G,

(1) Prove that $ea = a$ for any a in G beginning with the equation $ea = (aa')a$.

(2) Prove that c' is the only element a in G such that $cc' = e$. To do this, suppose $b \in G$ and $cb = e$, and show that $b = c'$.

(3) Prove that the equation $ax = b$ has a solution in G for any given (fixed) choice of a and b. That is, show that there does exist in G an element c such that $ac = b$. Think a minute. All you're given is a and b. If you tried to solve $2 + x = 3$ your answer would be $x = 3 - 2$.

(4) Prove that e is unique. That is, if $ab = a$ for all $a \in G$, then $b = e$.

(5) Prove that $(a')' = a$ for any a in G. To do this, let $b = a'$. Show $b' = a$. Recall from (2) that b' is the only element of G such that $b'b = e$.

(6) Suppose you are given the following conjecture to settle: there are two trees in the state of Missouri having the same number of leaves. Set up an axiom system, made up of at most two or three axioms and two undefined terms from which an answer of "yes" would follow from your axioms. Try to choose these axioms so that they are believable in a real world sense. If you approach this correctly, various interpretations of this system would yield the following conclusions: (a) there are two books in the library with the same number of pages, (b) there are two men in New York with the same number of hairs on their heads and (c) there are two brooms with the same number of straws. For a further hint, if you believe these statements, try to *abstract* the reasoning behind your belief to apply to all these examples, To do such is to do mathematics, in its fullest and most meaningful sense.

6. THE NATURAL NUMBER SYSTEM

The natural numbers are what we commonly denote by 1, 2, 3, · · ·. In this section we inquire as to the basis for these symbols and the most common operations with them. And despite the position that mathematics is independent of the physical world, we begin by asking, if not definitively answering: do these have a basis in something more than the imagination? While these number concepts lie at the basis of almost all mathematics, they are thought to be so universally apparent that it has been suggested that the first message one could hope to communicate to intelligent nonterrestial beings would be the idea: $1 + 1 = 2$. This would seem to indicate a suspicion that these number concepts are more than a product of man's imagination, that they must have some realization in a common universe of experience. No doubt without especially meaning to, the Noble prize winning biologist, Professor George Wald, in an address entitled "The Human Experience," obliquely suggested that $1 + 1$ was just about the first thing there ever was.[22]

In brief there was once, and is now in other parts of the universe, a dense cloud of hydrogen atoms, each with a single nucleus, which by gravitational and other forces were brought closer and closer together increasing the temperature within the cloud to some five million degrees at which point four hydrogen nuclei combine to form a single helium nucleus. At even higher temperatures, in what are called Red Giants, these helium nuclei, known also as α-particles, began to combine. One plus one helium combine to form an atom of beryllium, then one more nucleus combines to form carbon. Then oxygen, when one carbon captures one helium. So went and so still goes the birth of a star, and after some happening on our own star, the sun, came this earth. Whereupon, with the accidental collision of molecules in some warm primordial sea, here we are, mostly carbon, nitrogen, oxygen and hydrogen, all *integral* multiples of hydrogen, and the blood coursing through our veins is still most chemically like warm sea water. So goes a small part of the view of the scientist on our origins.

Kronecker, in an earlier century, said it another way: "God made the integers, all the rest is the work of man." He was speaking of mathematics, not the origins of the universe, but the two points are well taken: the essence of the positive whole numbers has been around a longer time than we and all mathematics follows after them. It is not surprising that no one thought to examine the simplest notions on which our ideas of whole numbers are based until the mid-19th century when all of mathematics was being questioned as to its ultimate foundation. The credit for providing an axiomatic foundation for the natural number system is given to G. Peano, who provided a set of axioms in 1889. Peano decided that the most basic thing about the natural numbers is not addition, or the notion of counting, but something even simpler: the idea that there is a first number and that each number is followed by exactly one more number. That is, each number has a "successor." The work of Peano and other mathematicians has established that all the facts we use about the natural numbers are logical consequences of the following four axioms.

2.6.1. We suppose a collection $\mathscr{N}$ whose elements are called *num-*

[22] A repetition of Professor Wald's main points cannot be repeated here. An article of his, "The Origin of Life," appeared in the August, 1954 issue of *Scientific American.* Along these lines, the very literary yet scientifically founded writings of Loren Eiseley in ***The Firmament of Time***, ***The Immense Journey*** and ***The Unexpected Universe*** (New York: Harcourt Brace & World, Inc., 1969) are vivid and challenging to the scientifically oriented reader.

bers and a relation of *successor* with the following properties.[23]

(1) There is a number, denoted by 1, which is not the successor of any other number.

(2) Every number has a unique successor.

(3) Every number except 1 must be the successor of exactly one number.

(4) If M is a collection of natural numbers and

  (a) 1 is in M

  (b) the successor of any number assumed to be in M is also in M,

  then M is the collection $\mathcal{N}$.

To better relate this to what you already know, think of Axiom 1 as saying that there is a first number, Axiom 2 as saying that there always is exactly one more number, Axiom 3 that each number except (possibly) the first number follows one and exactly one number. And now for Axiom 4. One might wish to view it as a special principle of logic reserved for discussions of the natural numbers. It is most often referred to as the principle of mathematical induction. It can be visualized, somewhat depressingly, by thinking of one of the current facets of the human experience: the stand-still bumper-to-bumper traffic of the afternoon rush toward home. If the last car in the line is hit from behind with sufficient force, collision of each car with its successor in the traffic will continue one after another.[24] There being no loss of energy due to friction between successive integers, the induction principle says that whatever is true about the first number 1, *and* is true about the successor of each number for which it is *supposed* true, must be true of all the numbers. Take particular note of the latter part of this statement; if one did not know that for every car bumped from behind it does indeed bump its successor, then one could not conclude that all will be bumped. That is, to argue by induction one must argue that whatever is true for an arbitrary number must be true of its successor.

All of what you know about the natural numbers and considerably more than that can be derived from these axioms by ordinary logic. It is not the purpose of this section to derive even the properties of addition from these. We will, however, see how addition arises in this context and study Axiom 4 in more detail, in particular proving the equivalence of it with a more intuitively apparent statement about the natural numbers. In subsequent sections we will assume a familiarity with those basic properties of the natural numbers that the reader has long made use of since learning arithmetic. Subsequent developments will, however, require special familiarity with the induction principle.

[23] Strictly speaking, both the terms *collection* and *element of* are undefined technical words. We will use them intuitively in this and the next section, and clear up the matter in the sequel.

[24] As recalled from newspaper reports, the record for such a sequential collision is 265.

First a reformulation of the induction principle that might better make clear its relation to logic. Most reasoning is deductive, p implies q, *from a stronger* or general condition p *to a weaker* or more specific conclusion q. Inductive reasoning is referred to as reasoning from the specific to the general; reasoning from a large number of special cases to the general case. The principle of induction is a formal statement to this effect about the natural numbers. More specifically, suppose $P(n)$ is a propositional function whose indeterminate n is to be interpreted as a natural number. We have seen that to assert the truth of the proposition $[\forall n, P(n)]$ one must argue the truth of $P(n)$ for an arbitrary natural number. The induction principle furnishes a specific procedure for doing this and permits the introduction of an additional hypothesis in such an argument. Stated formally,

DEFINITION 2.6.2. *The principle of inductive reasoning.* Let $P(n)$ be a propositional function concerning an indeterminate n and let the class of interpretations of n be natural numbers. Then, if

(a) $P(1)$ is true

(b) If $P(n)$ is true, then $P(n')$ is true, where n' is the unique successor of n,

then the proposition $[\forall n, P(n)]$ is true. That is, $P(n)$ is true for all natural numbers n.

Proof. By the definition of a propositional function, for each n, $P(n)$ is either true or false. Let M be the collection of all those numbers n for which $P(n)$ is true. By Definition 2.6.2(a), 1 is in M. By Definition 2.6.2 (b), if n is in M, then so is its successor. By the principle of induction, M is the collection $\mathscr{N}$—reinterpreted: $P(n)$ is true for all numbers n.

The strength of inductive reasoning resides in Definition 2.6.2(b). It allows one to suppose the truth of $P(n)$, *giving one something to work with* to conclude the truth of $P(n')$ where n' is the successor of n. That is, to know all cars will be bumped from behind, it suffices to know that if a car is assumed bumped, then it bumps its successor.

Now for addition. The definition of the addition of the number 1 to a given number is quite straightforward and has already been indicated.

DEFINITION 2.6.3. For a given natural number n, define $n + 1$ to be the unique successor of n (which exists by Axiom 2).

The problem that remains is that of defining $n + m$ for any number m. The key idea is that, by Definition 2.6.3, $n + 1$ is defined, and $(n + 1) + 1$ is defined, and $[(n + 1) + 1] + 1$ is defined, and so on. But this can go on forever and not be finished! Clearly, induction is the technical device needed to make the definition in a finite number of steps.

DEFINITION 2.6.4. If m is a natural number and m is the successor of k and $n + k$ is defined, then define $n + m$ to be $(n + k) + 1$.

THEOREM 2.6.5. For a given number n and for all numbers m, $n + m$ is defined.

Proof. If $m = 1$, then $n + 1$ is defined by Definition 2.6.3. Suppose $n + k$ is defined and k' is the successor of k. By Definition 2.6.4, $n + k'$ is defined (to be $(n + k) + 1$). These two sentences, reinterpreted, say that if M is the collection of all numbers m such that $n + m$ is defined, one has (1) 1 in M (2) if k is in M then the successor k' of k is in M. By Axiom 4, M is the collection $\mathcal{N}$. That is, $n + m$ is defined for all m.

With addition defined we can get to the most important result of this section. This is the thoroughly intuitive belief that any collection of natural numbers has a first or smallest number.

DEFINITION 2.6.6. If k and n are numbers, define $k < n$ to mean $n = m + k$ for some number m. We will say that "k is less than n" if $k < n$. If $k < n$ or $k = n$, we will write $k \leq n$.

Before establishing the main result we need some preliminary facts to aid in its proof. These are

LEMMA 2.6.7. For natural numbers n, m, and k

(a) $n + 1 = 1 + n$

(b) $n + m = m + n$

(c) $(n + m) + k = n + (m + k)$

which can be proven by induction on n, m, and k, respectively, using the definition of addition. Additionally,

LEMMA 2.6.8. If $m < n$, then $m + 1 = n$, or $m + 1 < n$.

which can be proven using Definition 2.6.6 and Lemma 2.6.7.

Here is the result we wish.

THEOREM 2.6.9. If K is a non-empty collection of natural numbers, then there exists a number k in K such that $k \leq n$ for all n in K.

This is an existence theorem. It asserts the existence of some thing with some property. It is often the case that it is easier to prove existence by showing that the denial of existence leads to a contradiction. This is the plan of the proof. Note the introduction of symbols in the proof to aid its exposition.

Proof of Theorem 2.6.9. Suppose K is a non-empty collection of natural numbers and there *does not exist* a k in K such that $k \leq n$ for all n in K. Let M be the collection of all those n in $\mathcal{N}$ less than every number in K. We claim M is the collection $\mathcal{N}$, a contradiction to the assumption that K is non-empty.

According to Axiom 4 we must show (a) 1 is in M and (b) if n is in M, then so is the successor of n. Suppose that 1 is not in M. Then, by definition of M, this must mean that 1 is not less than every number in K. Hence there must be a number k in K such that 1 is not less than k. If k is not 1 then k is the unique successor of exactly one number k' by Axiom 3. But by Definition 2.6.3, this means $k = k' + 1$. By Definition 2.6.6, this means $1 < k$, a contradiction. Hence, $k = 1$. But then 1 is in K and, by repeating the argument just made for any k in K, we get $1 \leq k$ for all k in K. However, $1 \leq k$ for all k in K is contrary to the supposition of our first paragraph. Therefore, we must conclude that 1 is in M.

Suppose now that n is in M and let n' be the successor of n. If j is in K, then $n < j$ by definition of M. Hence, $n + 1 = j$ or $n + 1 < j$ by Lemma 2.6.8. That is, $n' = j$ or $n' < j$. Hence, $n' \leq j$ for any j in K. Thus, n' cannot be in K by the assumption on K. Hence, $n' < j$ for all j in K, since $n' \leq j$ and $n' \neq j$. By definition, n' is in M. By Axiom 4, M is the collection $\mathscr{N}$, a contradiction to the fact that K is non-empty.

The reader may be wondering: "Is the matter any clearer or more certain than before? I've always believed, more-or-less, that any collection of natural numbers has a first or smallest number." A good question. We now know that this belief can be founded on the assumption of the axiom of induction, a statement a good deal less clear or intuitive! And that is the key to a decent reply to the question. It turns out that Theorem 2.6.9 and the concept of addition implies in turn the induction axiom. In other words, the less intuitive induction axiom is logically equivalent to the more intuitively reasonable Theorem 2.6.9; this result is left for the reader to manage a proof of.

We will not carry this study any further in this text. Subsequent sections and chapters will assume the reader's familiarity with the simplest properties of addition and multiplication of the natural numbers.

Problems.

(1) Restate Axiom 4 and Definition 2.6.2 in terms of addition.

(2) Let s be a real number. Prove that for any natural number, n, $1 + s + s^2 + \cdots + s^n = (1 - s^{n+1})/(1 - s)$.

(3) Prove or disprove: $n^2 - n + 41$ is evenly divisible only by itself and 1 for any n in N.

(4) Prove Lemma 2.6.7.

(5) Prove Lemma 2.6.8.

(6) Assuming the properties of addition in Lemmas 2.6.7 and 2.6.8, prove Axiom 4, restated as in (1), under the assumption that any non-empty collection of natural numbers has a first element. That is, suppose M is a collection of numbers satisfying (a) and (b) of Axiom 4 and show that M is the collection $\mathscr{N}$, assuming Theorem 2.6.9. For a hint, suppose M is not $\mathscr{N}$ and let K be the collection of all natural numbers not in M.

(7) Prove that $1 + na \leq (1 + a)^n$ for all n, where a is a fixed real

number greater than or equal to 1.

(8) What is wrong with the following proof?

Claim: Let $P(n)$: Any collection of n buttons consists of buttons of the same color. Then $P(n)$ is true for every n.

Proof: Clearly $P(1)$ is a true statement. Suppose $P(n)$ is a true statement and consider $P(n+1)$. Let S_{n+1} be a collection of $n+1$ buttons and let S_n be this collection with one button removed. Because $P(n)$ is true, S_n must consist of buttons of the same color, call it c. Now let S_n^* consist of S_n with one button removed and replaced by that button in S_{n+1} first removed. Again S_n^* has n buttons and because $P(n)$ is true, S_n^* must consists of buttons of the same color, c. Putting all buttons together again they all have color c. Hence $P(n+1)$ is true.

(9) A decreasing sequence of n natural numbers is a collection $a_1, a_2, \cdots, a_n$ of natural numbers such that $a_1 > a_2 > a_3 > \cdots > a_n$.

(a) Show that for every natural number, n, there is a decreasing sequence of n natural numbers.

(b) Show that there cannot be any infinite decreasing sequence of natural numbers.

(10) State, and prove by induction, a formula for the sum of the first n natural numbers for any natural number, n.

(11) Make up an interesting problem or formula which you then settle by inductive proof. Which do you think is the more significant and/or difficult process: (a) deciding upon what might be true for all n or (b) proving it by induction for all n?

7. THE REAL NUMBER SYSTEM

Real number properties are the end point of discussion in much of mathematics and and its applications. Unlike the natural numbers, the existence and meaning of what are called real numbers is not obvious, intuitive or otherwise there to behold almost without trying. Furthermore, there are different kinds of real numbers each of which presents different problems of existence and meaning.

The number zero was a long time in coming. It was quite a point of difficulty to conceive and discuss well a number that measured of nothing in a time when the very concept of number was closely tied to physical, either geometric or quantitative, existence. One can imagine a similar difficulty with the concept of a negative whole number if the idea of number is intertwined with physical existence. The existence of positive rational numbers, fractions, was not really a problem to early Greek mathematicians, who hardly separated the concept of number from that of geometric configuration. Given a line L it was easy to imagine it to be made up of parts each of length $1/q$, as we now call it, where q is a whole number. From that it was easy to believe in p of these parts of length $1/q$

laid end-to-end and hence a line of what we now call length p/q. Moreover, all lines were thought to be of this kind. But then, with the Pythagorean theorem, it was shown that the hypotenuse of a right triangle with two sides of unit length could not be a whole multiple of a fractional part of this unit length. Here was geometric existence in contradiction to a whole theory of number existence. As noted earlier, Eudoxus tried to make sense of the matter and Dedekind definitely did over 2,000 years later.

The historical developent of the real number system is quite in contrast with today's method of development. Historically, things began with only the natural numbers and Euclidian geometry. It is now known that one can, with a certain minimum of set theory, begin with the natural numbers and define the negative integers, rational numbers and then the real numbers in a very orderly and rigorous way.[25] While this ties the existence and meaning of the real numbers to that of the natural numbers in a direct manner, this development does not parallel the historical one. Moreover, while all this is possible, in situations where one is interested primarily in the development of some body of mathematical knowledge that requires the use of real numbers, one commonly assumes the existence of real numbers as undefined objects whose properties are given axiomatically. It is the purpose of this section to set out such an axiomatic foundation for the real number system and to derive from these axioms those properties most in need of understanding. Before giving these axioms it might be well to better mention what we are after. We seek an axiomatic foundation for all of the usual arithmetical, algebraic and order properties of the natural numbers, integers, rational numbers, and irrational numbers including those numbers thought of as infinite decimal expansions. We are not interested in complex or "imaginary" numbers. In subsequent sections a working knowledge of the real numbers will then be assumed.

We suppose a collection $\mathscr{R}$ of *numbers* along with three undefined symbols $+$, $\cdot$, and $=$, having the following properties.

A.1. For any numbers a and b, $a + b$ is a number and, if $a = c$ and $b = d$, then $a + b = c + d$.

A.2. For any a and b, $a + b = b + a$.

A.3. For any a, b and c, $a + (b + c) = (a + b) + c$.

A.4. There is a number, to be denoted by 0, such that $a + 0 = a$ for any number a.

A.5. For each number a there is a number, denoted by $-a$, such that $a + (-a) = 0$. Moreover $-a$ is unique: if b is a number such that $a + b = 0$ then $b = -a$.

M.1. For any numbers a and b, $a \cdot b$ is a number and if $a = c$ and $b = d$ then $a \cdot b = c \cdot d$.

[25] For a brief development see H. Eves and C. V. Newsom: *Introduction to the Foundations and Fundamental Concepts of Mathematics*. For a thorough, and thoroughly good, development, see E. G. H. Landau: *Foundations of Analysis*. (New York: Chelsea Publishing Co, 1951).

M.2. For any a and b, $a \cdot b = b \cdot a$.
M.3. For any a, b and c, $a \cdot (b \cdot c) = (a \cdot b) \cdot c$.
M.4. There is a number, to be denoted by 1, such that $a \cdot 1 = a$ for any number a, and $1 \neq 0$.
M.5. For any number a except 0, there is a number, denoted by $1/a$ such that $a \cdot 1/a = 1$. Moreover, $1/a$ is unique: if $ab = 1$ then $b = 1/a$.
D.1. For numbers a, b and c, $a \cdot (b + c) = a \cdot b + a \cdot c$.

These eleven axioms are enough to derive all of the usual finite arithmetical operations on real numbers that one desires, excluding root extraction, using only the usual logic of mathematics. But these assumptions do not characterize the real numbers, nor do they provide sufficient information for the development of those parts of mathematics that deal with real number concepts. More axioms are needed but the reasons are not obvious. We will develop these reasons through a series of problems. The next problem shows that something much less than the real numbers meets the assumptions above.

Problems. In parts (1) and (2) of this problem, two models or interpretations of the above axiom system are suggested. Show that these satisfy all the axioms.

(1) Let $\mathscr{R}$ consist of two symbols, any α and β, and define $+$ and $\cdot$ by the tables

$+$	α	β
α	α	β
β	β	α

$\cdot$	α	β
α	α	α
β	α	β

You might first decide which of these "numbers," α or β, plays the role of 0 in A.4 and 1 in M.4.

(2) Assume the existence and the usual arithmetic of the rational numbers. Let $\mathscr{R}$ consist of all rational numbers and $+$ and $\cdot$ be ordinary addition and multiplication, respectively.

(3) Show that one cannot assign a meaning to division by 0 which is consistent with the above axioms. Begin with the definition that the quotient of a number, b, by a number, a, is that number c such that $b = ac$. Similarly show that division of 0 by 0 cannot be consistent.

This problem indicates that more assumptions are needed to characterize the real number system. No doubt you have noticed that there is no mention in the axioms of that familiar and useful number concept: comparative magnitude, size or order. Here then are additional axioms.

AXIOM O.1. There is a fixed non-empty collection P of numbers, each of which is called *positive*, such that if a and b are positive, then so are $a + b$ and $a \cdot b$.

AXIOM O.2. If $a \neq 0$, then either a or $-a$ is positive, but not both. The number 0 is not positive. Any number $a \neq 0$ which is not positive will be called *negative*.

These axioms of course introduce a new undefined term, *positive*, and in seeking models for this enlarged axiom system one is free to interpret *positive* as he wishes, but must obtain properties of Axioms O.1 and O.2 for this interpretation. One model for this extended system is that of Problem 2 above, where one interprets positive in the usual sense. The following problem shows that these two axioms bring one closer to an axiomatic basis for the real numbers by precluding at least one interpretation.

Problems.

(1) Show that there is *no* interpretation of *positive* in Problem 1 above for which the assumptions of Axioms O.1 and O.2 are valid. Hint: begin by showing, say, that the collection P consisting of only β cannot satisfy both Axioms O.1 and O.2. Then try the collection consisting of α only. There remain two other possibilities for P.

(2) Show that the assumption $1 \neq 0$ in M.4 is necessary in order to have more than one real number. That is, show that a collection $\mathscr{R}$ with a single element 0 and $+$ and $\cdot$ defined by $0 + 0 = 0$, $0 \cdot 0 = 0$ satisfies Axioms A.1–A.5, M.1–M.5, D.1, O.1 and O.2 with the exception of M.4.

It turns out that one more axiom is needed to obtain a number system with all the properties one ordinarily desires for the real number system. It is more difficult to use and more sophisticated in nature, for it introduces into the system $\mathscr{R}$ properties of a special infinite character. It is a most important axiom, for it makes the calculus possible in its present form. It is the axiom that gives meaning to the notion of an unending decimal representation of number. It makes root extraction meaningful and possible. And, going back to the problem of Greek mathematics, it fills out the gaps in the number line, asserting, by implication, the existence of a unique number corresponding to the line segment whose length is that of the hypotenuse of a given right triangle. For example, in Figure 2.9, if one begins with a given line segment of length called 1 (*a*), lays it end-to-end upon itself (*b*), constructs a right triangle of sides 1 (*c*), rotates its hypotenuse back onto the extended line (*d*), then this final axiom asserts the existence of a number corresponding to the point p. In the axiom system thus far given, since the rational numbers are an interpretation and this hypotenus cannot be of rational length, there is no guarantee of such a number without an additional assumption. This assumption is appropriately called the completeness axiom. We will eventually show that $\mathscr{R}$

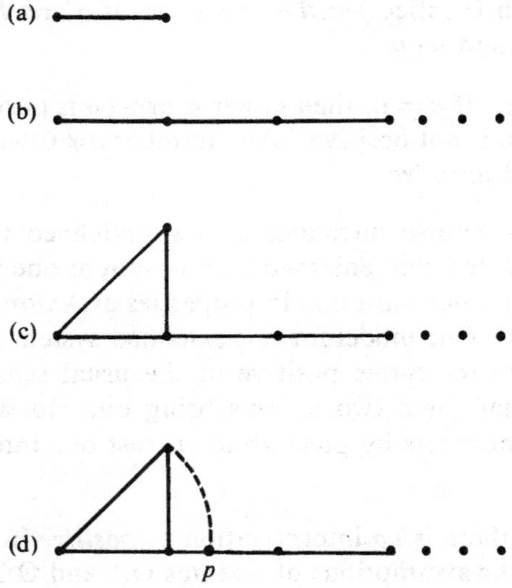

FIG. 2.9.

with this axiom is a "connected" set, having no "gaps."

Two final points. With the axioms given previously one is confined to the study of the algebra and arithmetic of numbers. The completeness axiom transports one into the domain of mathematical analysis, whose origins are the calculus and the concept of a limit. Finally, and most importantly, this axiom asserts the mathematical existence of an entirely new kind and order of infinity, distinct from that of the natural numbers, but the proof of this assertion must await the substantial theoretical development to be undertaken in subsequent chapters.

The completeness axiom cannot even be easily stated without the introduction of additional terminology in the context of the axioms thus far admitted.

DEFINITION 0.1. If a and b are numbers we will say that a is *greater than* b, written $a > b$, if $a + (-b)$ is a positive number. If $a > b$ or $a = b$, we will write $a \geq b$. We will say that b is *less than* a, written $b < a$, if $a > b$. Similarly, $b \leq a$, if $b = a$ or $b < a$.

DEFINITION 0.2. A collection M of numbers is said to be *bounded above* if there is a number b such that $x \leq b$ for *all* numbers x in M. The number b is called an *upper bound* for M.

DEFINITION 0.3. The number b is called a supremum of a collection M of natural numbers if

(1) b is an upper bound for M.

(2) If a is any upper bound for M, then $b \leq a$.

If both conditions (1) and (2) hold for a collection M and a number b, we will write $b = \sup M$.

This last definition is the familiar definition of the least upper bound of a set M of real numbers. The terminology *supremum* of M is more common today and that is the reason for its use. It will be subsequently shown that the number $\sup M$, if it exists, is unique.

Of course, one can define something without knowing that it exists. For example, all the definitions above can be made in any model of our present axiom system for $\mathscr{R}$, notably the rational numbers, but there is no guarantee that one is defining something that actually exists in the model. In the rational number interpretation for $\mathscr{R}$, the collection M of all rational numbers p/q such that $(p/q)^2 < 2$ would not have a supremum, because if it did have a supremum, call it b, one would (eventually!) have $b^2 = 2$, which cannot hold for the *rational* number b.[26] Thus the final axiom

C.1. *The Completeness Axiom.* Any non-empty collection of numbers in $\mathscr{R}$, which is first of all bounded above, has a supremum in $\mathscr{R}$.

The brunt of this supposition is that a supremum must *exist in* $\mathscr{R}$.

Before considering some preliminary problems within this now complete axiom system for $\mathscr{R}$, we will prove two theorems in order to indicate methods of proof in this axiom system and to contrast arithmetic with analysis. First an arithmetical result.

THEOREM 2.7.1.

(a) For any real number a

(1) $-(-a) = a$.

(2) $a \cdot 0 = 0 \cdot a = 0$.

(3) $(-1) \cdot a = -a$.

(b) 1 is a positive number.

Proof. (a).

(1) According to Axiom A.5, $-(-a)$ is the *only* number c with the property that $-a + c = 0$. By Axioms A.2 and A.5, $(-a) + a = a + (-a) = 0$. Hence a must equal $-(-a)$.

(2) From Axioms A.4 and D.1, $a \cdot 0 = a \cdot (0 + 0) = a \cdot 0 + a \cdot 0$. From this and Axioms A.5, A.3, A.1 and A.4, one then has $0 = a \cdot 0 + -(a \cdot 0) = (a \cdot 0 + a \cdot 0) + -(a \cdot 0) = a \cdot 0 + [(a \cdot 0) + -(a \cdot 0)] = a \cdot 0 + 0 = a \cdot 0$. Hence, $0 = a \cdot 0 = 0 \cdot a$ by Axiom M.2.

(3) From Axioms M.4, M.2, D.1 and A.5 and from (2) above, $a + (-1) \cdot a = a \cdot 1 + (-1) \cdot a = a \cdot 1 + a \cdot (-1) = a \cdot [1 + (-1)] = a \cdot 0 = 0$. By Axiom A.5, as in (1) above, one must have $(-1) \cdot a = -a$.

[26] See Problem 4 at the close of this section.

(b) Either 1 or -1 is positive by Axioms O.2 and M.4. If -1 were positive then, by Axiom O.1, $(-1)\cdot(-1)$ would be positive. But by (1) and (3), $(-1)\cdot(-1) = -(-1) = 1$, a contradiction to O.2.

The next theorem is non-arithmetical and is concerned with the supremum. It says that sup M can only represent one real number and hence that one can speak of *the* supremum of a collection M.

THEOREM 2.7.2. If a non-empty set M has a supremum, then sup M is unique.

Proof. Suppose that two numbers, a and b, are both supremums of a single set M. By 0.3, part (1), both a and b are upper bounds for M. By part (2), one then has $a \leq b$ and, on the other hand $b \leq a$. Hence, according to Problem 8(b) below, $a = b$,

Problems.

(1) For numbers a and b, prove
 (a) $1/(1/a) = a$ for $a \neq 0$.
 (b) $-(ab) = (-a)b = a(-b)$.
 (c) $(-a)(-b) = ab$.
 (d) $-(a+b) = (-a)+(-b)$.
 (e) $-(1/a) = 1/(-a)$ for $a \neq 0$.

(2) Define $a - b$ to mean $a + (-b)$. Why can one not prove the theorem $a - b = b - a$?

(3) For $b \neq 0$ define a/b to mean $a \cdot (1/b)$. Prove that

$$\frac{a}{b} + \frac{c}{d} = \frac{ab + bc}{bd}.$$

(4) Show that $-0 = 0$ and $1/1 = 1$.

(5) Show that if a and b are positive, then $-a \cdot b$ is not.

(6) Use Theorem 2.7.1(b) to give a short proof of Problem 1 following the statement of Axioms O.1 and O.2.

(7) Show that if a is in P, then so is $1/a$.

(8) For numbers a, b, and c, show, possibly using the results in Theorem 2.7.1, that
 (a) Either $a \leq b$ or $b \leq a$ for any choice of a and b.
 (b) If $a \leq b$ and $b \leq a$, then $a = b$.
 (c) If $a \leq b$ and $c \geq 0$, then $ac \leq bc$.
 (d) If $a \leq b$ and $c \leq 0$, then $ac \geq bc$.
 (e) If $a \leq b$, then $a + c \leq b + c$.
 (f) If $a \leq 0$ and $b \leq 0$ then $a \cdot b \geq 0$.

(9) For a real number a define $|a|$ to be a if a is positive and to be $-a$ if a is not positive. Prove that
 (a) $|a| + a$ is positive for $a \neq 0$.
 (b) $a - |a|$ is not positive.

(c) $|ab| = |a|\,|b|$
(d) $|-a| = |a|$
(e) $|a - b| = |b - a|$

(10) Show that for numbers a and b,
(a) $|a| < b$ iff $-b < a < b$.
(b) $-|a| \leq a \leq |a|$
(c) $|a + b| \leq |a| + |b|$
(d) $|a| - |b| \leq \big||a| - |b|\big| \leq |a - b|$

(11) Thinking of the number system as a line with each point representing a number—viz.

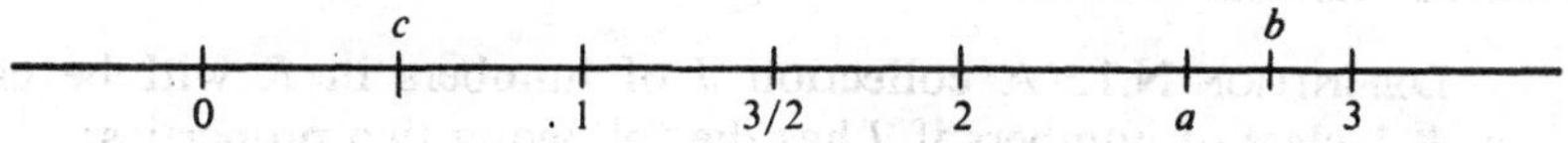

express the following statement in notation alone: The number a is within a distance c of the number b.

(12) Show that if $b = \sup M$ and $c < b$, then there is a number x *in* M such that $c < x \leq b$. An indirect proof may be the best method here. This is perhaps the most useful property of the number $\sup M$.

(13) Let M be a collection of numbers which is bounded above. Let a be a positive real number and let aM denote the collection of all products $a \cdot b$ where b is in M. Let $c = \sup M$ and show that $ac = \sup aM$. What happens if a is not positive?

8. THE NATURAL NUMBERS AS A PART OF $\mathscr{R}$

How do the natural numbers fit into this scheme of things? Do the axioms for $\mathscr{R}$ imply the existence of a subclass of $\mathscr{R}$ which satisfies Peano's axioms for the natural numbers? It is clear that most likely such a class would contain the numbers 1, $1 + 1$, $1 + 1 + 1$, and so on. So on, indeed! How does one give meaning to "so on" in the context of the axiom system for $\mathscr{R}$? One could say that a natural number in $\mathscr{R}$ is a number of the form $1 + 1 + \cdots + 1$, but then one must explain in terms of what is at hand, the axioms for $\mathscr{R}$, what this expression means without resort to whole number concepts less he precede in logical circles.

A way out of this dilemma is provided by a class, rather than an individual, approach to the problem. We will shortly consider a theory of classes or sets, and this exercise in the definition of those numbers in $\mathscr{R}$ to be thought of as natural numbers indicates the utility as well as the spontaneity of set or class concepts in mathematics. In this section we will continue to use the class or collection notion intuitively, which is not really being mathematically honest, for we are then not stating *all* the assumptions needed for the development of the real numbers. We are stating all the non-set-theoretical assumptions. At any rate, the approach is to define a certain class of numbers in $\mathscr{R}$, the individual numbers in this class to be called natural numbers, rather than attempt a definition of a single number as a natural number. But the reader no doubt well believes in the natural numbers and their properties as a subclass of the real numbers $\mathscr{R}$, so what is the point of this exercise? Simply stated, to indicate the tricky and sometimes impressive reasoning that must be employed in the initial development of an axiom system, but in a relatively familiar setting. Think of these axioms as defining the set of real numbers. The goal is to isolate, using only these assumptions, those numbers in $\mathscr{R}$ that should properly be called "natural numbers." We begin with

Definition N.1. A collection J of numbers in R will be called a $+$ 1-class of numbers if J has the following two properties:

(a) 1 is in J.

(b) If x is in J, then $x + 1$ is in J.

The name $+$ 1-class is merely a suggestive label attached to a collection of numbers with properties (a) and (b), if any such collections exist! But notice how these properties are most certainly desired for the natural numbers. That is, if we hope to define a class of numbers in $\mathscr{R}$ to be thought of as the natural numbers, then this class had better have properties (a) and (b) or else we're on the wrong track.

We would also be engaging in useless mental exercises if there were no $+$ 1-classes to begin with. Hence

There is a $+$ 1-class .

Proof. Consider the collection P of positive numbers which exists by Definition 0.1. By Theorem 2.7.1 (b), 1 is in P and (a) holds for P. For property (b), if x is in P then since 1 is in P, Definition 0.1 asserts that $x + 1$ is in P. Hence P is a $+$ 1-class.

But it is intuitively clear that P is ***not*** what we would want to call the natural numbers. For example 1 is in P and letting $2 = 1 + 1$, 2 is in P. Hence, by Problem 5 above, $1 \cdot (1/2)$ which we ordinarily call 1/2, is also in P. Thus P is too *large* for our purposes. And there is the key! All $+$ 1-classes are apparently large enough to contain what we should like to call the natural numbers. By definition, 1 is in *any* $+$ 1-class. Also by definition, part (b), $1 + 1$ is then in *any* $+$ 1-class, and for the same reason $1 + 1 + 1$ is *any* $+$ 1-class. Hence the definition

Definition N.2. A number n in R will be called a ***natural number*** if n is in *any* and *every* $+$ 1-class.

As an indication that we are on the right track with this definition, we have

THEOREM 2.8.1. If a is in R and $a < 1$, then a is *not* a natural number.

Proof. Let J be the class of numbers in R greater than or equal to 1. Since $a < 1$, then a is not in J. Moreover, J is a $+$ 1-class, for 1 is in J and, if x is in J, then $x \geq 1$ and, hence, $x + 1 \geq 1 + 1 \geq 1$ by Definition 0.1 and Theorem 2.7.1 (b). Since J is a $+$ 1-class and a is not in J, then, by definition, a is not a natural number.

Before going further, stand back a little and try to see what's going on. Our definition of natural number in $\mathscr{R}$ does at least yield 1 as a natural number and no number less than 1 as a natural number. But more is needed for conviction that Definition N.2 does isolate only those numbers such as $1 + 1$, $1 + 1 + 1$, etc., as natural numbers. Two additional theorems further the cause.

THEOREM 2.8.2. If n is a natural number, then so is $n + 1$.

Proof. Suppose n is a natural number. Then n is in every $+$ 1-class by Definition N.2. By definition N.1, $n + 1$ must then also be in every $+$ 1-class. By Definition N.2 again, this says $n + 1$ is a natural number.

THEOREM 2.8.3. If n is a natural number and $n \neq 1$, then $n - 1$ is a natural number, where $n - 1$ is defined to be the number $n + (-1)$.

Here is a good example of the use of class or set concepts in proof.

Proof. Let J be the collection of all numbers $m + 1$, where m is a natural number, or $m = 0$. Then J is a $+$ 1-class. For, 1 is in J where m is taken to be 0 and if $k = m + 1$ is in J, then, by Theorem 2.6.5, k is a natural number and so by definition of J, $k + 1$ is in J. That is, $(m + 1) + 1$ is in J and so by Definition N.1, J is a $+$ 1-class.

Now let n be any natural number with $n \neq 1$. Since by Definition N.2, n is in any $+$ 1-class, then n is in the class J. Since $n \neq 1$ and J consists only of the number 1 or numbers of the form $m + 1$, then n must equal $m + 1$ for some natural number m. But then $m = n - 1$ and hence $n - 1$ is a natural number.

The following problem completes the job of defining the proper natural number class in R.

Problems.

(1) The point of this problem is to show that Peano's axioms for the natural numbers hold in R for the class of numbers defined in Definition N.2. Let $\mathscr{N}$ be the collection of all natural numbers in $\mathscr{R}$ as defined by Definition N.2 and if n is in $\mathscr{N}$ define the successor of n to be the number $n + 1$. Show that Peano's axioms are satisfied in this contest. You may wish to use the theorems above. The

meaning of this is that the assumption of the existence of the class $\mathscr{R}$ with the properties (axioms) assumed about $\mathscr{R}$, the existence of mathematical logic, and the use of the concept of classes of numbers implies in turn the existence of the natural numbers. Notice, however, that the completeness axiom plays no role in the existence of the natural numbers.

(2) Show that if n is a natural number and a is a real number such that $n < a < n + 1$, then a is not a natural number in the sense of Definition N.2.

From here on out in this section we assume the natural numbers as thoroughly familiar members of $\mathscr{R}$, with all the usual properties assumed without further proof. We will also assume total familiarity with the usual finite arithmetical and order properties of the elements of $\mathscr{R}$ to be thought of as real numbers. That is, we assume a familiarity with addition, subtraction, multiplication and division and their properties relative to the order properties as long as only finitely many numbers are involved. Under such assumptions we wish to investigate some implications of the completeness axiom and the real number properties it yields. In particular we will consider the existence and meaning of infinite decimal expansions as well as that of root extraction.

These matters deserve discussion in a larger context. Up to this point we have almost exclusively operated in the domain of arithmetic and algebra, which involves none of those processes usually termed *limit processes*. These latter processes bring one into the domain of analysis and, to a certain extent, topology. These processes are what separate calculus and more broadly, analysis, from arithmetic and algebra, and limit operations can be handily discussed only in the context of the completeness axiom or an equivalent assumption—that is, in the entire real number system. The rational numbers just will not do. The whole matter is highly complicated and it will hardly be fully elucidated here, but one cannot pursue a full understanding without a good grasp of those topics we do consider here. In subsequent chapters we will often find ourselves involved with these matters in a natural way and will come to appreciate more and more of their substance. But, in short, analysis by and of limiting processes is intimately involved with the completeness axiom and, as noted earlier, this inherent division between algebra and analysis was not fully recognized for many centuries, finally being squarely faced only by Cauchy, Dedekind and Weierstrass in the 19th century. Weierstrass did it most definitively, relating limiting processes to already well understood arithmetical-algebraic operations through devices (concepts) such as the completeness axiom. The broad purpose of this and the latter part of the next section is to indicate the nature of this relationship and we begin with a theorem, which by its name, indicates its ancient origins.

THEOREM 2.8.4. *The Archimedian Property.* If a is a real number greater than 0 and b is a real number, then there exists a natural number n such that $na > b$.

Geometrically this theorem asserts that if a line of length b is given, along with a line of length a, then enough (n) copies of the line of length a, laid end-to-end will give a line of length greater than b. In Figure 2.10, $n = 5$ would appear to suffice. Here now is a proof of this theorem.

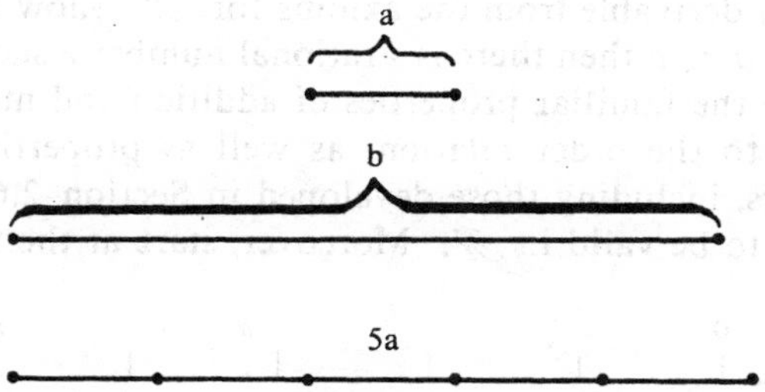

FIG. 2.10

Suppose that the conclusion is false. Then for all natural numbers, n, $na \leq b$. If M is the collection of all numbers of the *form* na, where n is a natural number, this says that b is an upper bound for M. Hence M has a supremum, call it c. Thus, for any natural number n, $na \leq c$. But for any natural number n, $n + 1$ is a natural number and $(n + 1)a \leq c$ and thus $na \leq c - a$. But then $c - a$ is an upper bound for M and moreover $c - a < c$ since $a > 0$. This is a contradiction to the property of c that it be less than or equal to *any* upper bound for M. Hence the proof by *reductio ad absurdum.*

With this single additional property, certain fundamental results about the real number system can be verified. These are detailed in a subsequent problem but as an indication of the fundamental nature of the Archimedian property and its manner of use we include a proof of the fundamental

THEOREM 2.8.5. As n gets larger, $1/n$ gets smaller and approaches 0. Precisely: if ε is any positive number, then there is a natural number N such that if $n \geq N$, then $1/n < \varepsilon$.

Proof. Since $\varepsilon > 0$, then by the Archimedian property there is a natural number N such that $N\varepsilon > 1$. If $n \geq N$ then $n\varepsilon \geq N\varepsilon > 1$ and hence $1/n < \varepsilon$.

So the obvious has been proven again! But is it really obvious? To what extent does the reader believe "$1/n$ approaches 0" only because this was once required to get along in the calculus and has since become familiar? Such might suffice for every day purposes, but it is no road to really profound ideas, no road in this case to the mathematical infinite. That is one reason for working through the following problems, to obtain a grasp of the essential notions which underly one's real number concepts. Better understanding them, one may be able to do more with them.

Problems.

(1) A number r in $\mathscr{R}$ is called *rational* if $r = p \cdot (1/q)$ where p, q are natural numbers and $q \neq 0$. You may wish to use the more familiar notation $p/q = p \cdot (1/q)$, but the properties of this notation are only derivable from its definition as $p \cdot (1/q)$ whose properties in turn are only derivable from the axioms for $\mathscr{R}$. Show that if a, b are in $\mathscr{R}$ and $a < b$ then there is a rational number r such that $a < r < b$. Assume the familiar properties of addition and multiplication with respect to the order relation, as well as properties of the natural numbers, including those developed in Section 2.6 which are now known to be valid in $\mathscr{R}$. Moreover, stare at the figure below!

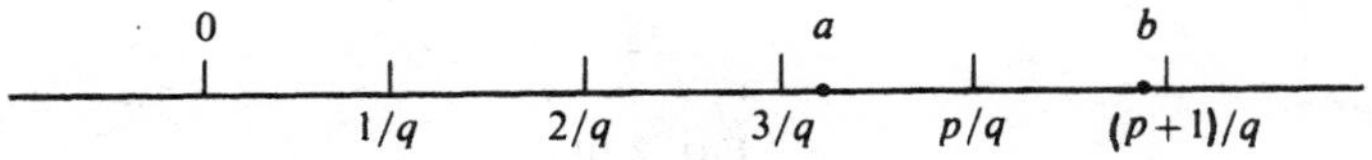

(2) Using Theorem 2.8.4, show that if a is a real number and $a \geq 0$ and $a < \varepsilon \; \forall \varepsilon > 0$, then $a = 0$.

(3) Using Theorem 2.8.4, show that if x is a real number, $x \geq 0$ and $x^2 < 2$, then $\exists$ a natural number n such that $(x + x/n)^2 < 2$. Similarly, show that if $x^2 > 2$ then $\exists n$ such that $(x - x/n)^2 > 2$.

(4) Show that $\exists$ a real number c such that $c^2 = 2$. For a hint, show that the collection M of real numbers x such that $x \geq 1$ and $x^2 < 2$ is non-empty and bounded above, and try $c = \sup M$. Once done, one usually denotes $\sup M$ by $\sqrt{2}$. The existence of square roots of other positive numbers b can be obtained by abstracting your proof for $b = 2$.

(5) Show that there does not exist a real number c such that $c^2 = -1$.

(6) Prove that for any real number x, $(1 - x)(x + x^2 + x^3 + \cdots + x^n) = x^n - x^{n+1}$ for any natural number n. Hence for $x \neq 1$,

$$x + x^2 + \cdots + x^n = \frac{x^n - x^{n+1}}{1 - x} \quad \text{and}$$

$$x + x^2 + \cdots + x^n \leq \frac{1}{1 - x} \quad \text{for } 0 \leq x < 1 .$$

(7) Once one has assumed the existence of the real numbers, again using a bit of intuitive set theory, one can construct the complex numbers. These are sometimes referred to as "imaginary" numbers but they are no more imaginary than the real numbers. Without the use of set theory one can define the complex numbers as follows. Suppose the real numbers $\mathscr{R}$ are given and i is a symbol. Let C be the collection of all forms $a + bi$ where a and b are real numbers.

Define

(a) $a + bi = c + di$ iff $a = c$ and $b = d$

(b) $(a + bi) + (c + di) = (a + c) + (b + d)i$

(c) $(a + bi) \cdot (c + di) = (ac - bd) + (bc + ad)i$.

Show that the axioms A.1–A.5, M.1–M.5 and D.1 are satisfied. Denote $1i$ by i and show that $i^2 = -1$. Finally show that Axioms O.1 and O.2 cannot be satisfied (i.e., that no sense of positivity can be placed on the complex numbers). Hint: suppose there is a collection P satisfying Axioms O.1 and O.2 By Theorem 2.7.1(b), 1 is in P. Then either i or $(-1)i$ in P and from this obtain a contradiction.

9. INFINITY

"But if we are to have censorship, then who will censor the censor, and who will censor the censor's censor and . . .?" Almost everything one ordinarily encounters in this world is, for most purposes, of finite extent, including any attempt at censorship, and that may be why it is a dangerous idea. But mathematics is not the real world (!) and even with its "unreasonable effectiveness" in solving certain real world problems, mathematics has forever been confronted by problems with infinity. It has come away from each of these confrontations, refreshed, with new and better understanding and a greater insight that has always increased its worth. And mathematicians themselves have been ever amused by the irresistible mental charm that infinity has, always trying to put a finger on it and nowadays mostly succeeding. Many a new student of mathematics has no doubt wished that infinity would just go away while surely sensing it would take much of interest with it.

Why is it that infinity does indeed occur so often in mathematical thought? Is it simply an irresistibly interesting concept that one could dismiss were it not so inviting to muse over, or is it essential to mathematical thought? The answer must come down hard on the side of the latter. Few ideas have so challenged the mind of man. And few ideas have been so necessary to mathematical man, for the concept of infinity imparts a certain freedom to mathematical thought and speculation that is only slowly appreciated. But that is mere rhetoric. A key reason for the unremitting appearance of infinity is the search for simplicity: one needs to understand the complex and does so through the medium of approximation by the simple. Long before the formal establishment of the calculus, mathematical man tried to understand the complex as the limiting case of an *infinite* number of ever slightly better approximations, each a matter of greater simplicity than the whole. As it turns out, that is also the key to understanding and using the notion of infinity itself in a consistent way. Always, the crux of infinite approximation has been: precisely how does the complex end relate to the infinite sequence of approximations to it?

A particular example of infinite approximation is that of finding the area of a circle. One could approach this problem by approximating the circle by polygonal paths the area of each easily computed from the formula for the area of a triangle. Such approximations gain more interest of course when the complex end is not so well known as a circle and it is there that a good understanding of the *methods* of approximation become crucial.

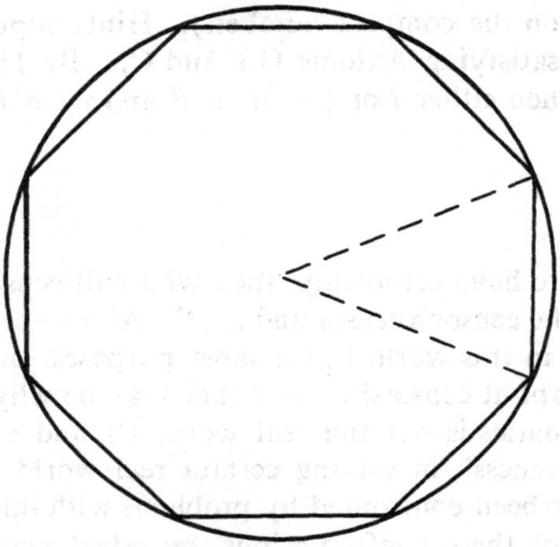

The great difficulty the student of mathematics encounters with infinity is that he thinks he knows something about it. He does intuitively. But there are few areas of thought where intuition can so lead one astray, and as this text unfolds the reader will evermore so appreciate that remark. The thing about infinity is that one begins with a straightforward common sense notion, and rapidly finds himself in a mire of contradiction unless he is extremely careful of his path. It is not a matter of thought to be handled lightly. This section will present no axioms for dealing with infinity. But this discussion is appropriate to this chapter. The section contains only one idea. See if you can make it a part of your thinking about infinity. It can carry you far and without it you will always be a little lost.

Properly speaking there are at least two intuitive ways whereby one encounters the notion of infinity. One is that of an infinite set, the other is the idea of "approaching infinity," the infinitely large or infinitely small. We will be concerned only with the latter in this section but will, in time, confront the former. Someone once wrote that "you can't sing the blues if you've never suffered" and that is the plan of this section: first to try to dispel any sure thoughts the reader may have about "approaching infinity," then try to put these back together again in what will hopefully be a more certain form.

First of all, infinity is not a number or anything so concrete as that. It is best thought of as a concept—roughly the concept that there is always one more, that one is never quite there. This is the concept of the *potential* infinite. This is what one intuitively means by infinity. Cantor introduced an *actual* infinite to mathematics, but it is a separate and essentially more profound kind of thing. We are here concerned only with the potential infinite and begin by examining the phrase "as x approaches infinity." The difficulty that one must face when using this notion to do anything in a reliable way is that ordinary logical thought processes do not always clearly lead to the conclusion one expects, and so may be useless when one does not know what to expect! Once this is realized one is left with the choice of either creating a special logic to handle infinity or to make precise definitions of the relevant ideas in terms of finite, hence previously understood, concepts. This last is the approach usually taken in calculus courses in the discussions of limits and an appreciation of this approach should be the aim in this section.

Consider the difficulty in walking across a room 10 feet long at the speed of 5 feet per second. Before one can reach the other side he must reach the middle; this takes 1 second. Assuming one accomplishes this, one still cannot reach the other side until he covers one-half the remaining distance; this takes an additional 1/2 second. If he has managed to get this far, he still has one-half the remaining distance to cover and this requires an additional 1/4 second. Continuing in this way we find that it has taken $1 + 1/2 + 1/4 + \cdots + 1/2^n$ seconds to cover all but $10/2^{n+1}$ of the length of the room. Now no matter how we continue to reason this way we will not be able to conclude that one can indeed make it all the way across the room. We could begin to prattle phrases like, "as n approaches infinity the remaining distance becomes zero," but this is too much like the politician's argument that "I am confident of your support because all reasonable men will support the right policy" in that both avoid the substance of the issue. The substance of the issue here is that if there is a remaining distance, then it is plainly *not* zero and the room has not been crossed, and there will be a remaining distance no matter how large n is. Furthermore, no matter how large a fixed value of n we chose to use in deciding how much of the room is left to be walked we have the sorry conclusion that n is much closer to the number *one* than to what we think of as infinity! Thus the notion of "approaching infinity" is meaningless in this intuitive sense, and if the reader's thought processes on this have been thoroughly shaken then there is some hope of clearing things up definitively. The solution is to get the problem into the context of previously understood finite concepts.

Thus, on the basis of an ability to add finitely many numbers and ordinary logic let us try to resolve the above problem in a manner consistent with reality—reality being that within our concept of time one can walk across the room in 2 seconds, because it is 10 feet long and he is traveling 5 feet per second. We want to take our reasoning above, which after all is correct and commonplace as it stands, and resolve it with this reality.

As noted above, after n such bits of reasoning and walking, we find that in an amount of time $t_n = 1 + 1/2 + \cdots + 1/2^n$ one has covered all but $10/2^{n+1}$ of the remaining distance. We then have

$$(1) \quad (1 - 1/2)t_n = (1 - 1/2)(1 + 1/2 + \cdots + 1/2^n)$$
$$= 1 + 1/2 + \cdots + 1/2^n - 1/2 - 1/4 - \cdots - 1/2^{n+1}$$
$$= 1 - 1/2^{n+1}.$$

so that

$$(2) \quad t_n = (1 - 1/2^{n+1})/(1 - 1/2) = 2(1 - 1/2^{n+1}) = 2 - 1/2^n.$$

We now define the entire phrase "t_n approaches L as n approaches ∞" to mean the following: no matter how small a number $\alpha > 0$ we are given, there is always an integer N_α (this is to indicate that N_α might depend on α) such that if $n \geq N_\alpha$ then $|L - t_n| < \alpha$. First of all the reader should note that this definition involves only finite concepts—inequalities, finite algebraic expressions, and the existence of certain finite numbers (N_α) for each fixed number $\alpha > 0$. Secondly, and this is the key to the matter, the definition replaces the notion of "t_n approaching L" by the requirement that the distance, $|t_n - L|$, between t_n and L satisfies *infinitely many* fixed conditions—namely, be less than *any* $\alpha > 0$ (if n is large enough)—there being infinitely many choices of α for which this must be satisfied. This is the idea of always better approximation. Finally, if this definition is to square with reality, then $L = 2$ had better satisfy it and only $L = 2$ should satisfy it. Moreover, we must be able to obtain this conclusion without resort to "infinite" concepts. In fact, the formula $t_n = 2 - 1/2^n$ is enough to do this. Let us see how.

Suppose we are given $\alpha > 0$. We claim there is a number N_α such that if $n \geq N_\alpha$ then $|2 - t_n| < \alpha$. Let us choose N_α to be a fixed integer greater than $\log(1/\alpha)/\log 2$. If $n \geq N_\alpha$ then $n > \log(1/\alpha)/\log 2$ and hence $n \log 2 > \log 1/\alpha$ so that $\log 2^n > \log 1/\alpha$. From familiar properties of the log and exponential functions we obtain $2^n > 1/\alpha$ or $1/2^n < \alpha$ for any such value of n. Thus, we have

$$(3) \quad 0 \leq 2 - t_n = 2 - (2 - 1/2^n) = 1/2^n < \alpha$$

which is what we wanted to prove, because this means $|2 - t_n| < \alpha$ for $n \geq N_\alpha$.

Now let us see that $L = 2$ is the *only* number that will work in this definition. Suppose we try a number $L \neq 2$; suppose further that $L > 2$. Let $\alpha = (L - 2) > 0$. Then

$$L - t_n = L - (2 - 1/2^n) = (L - 2) + 1/2^n = \alpha + 1/2^n.$$

Thus $L - t_n$ cannot be less than α no matter what n is since

$$L - t_n = \alpha + 1/2^n > \alpha$$

for all n.

Now suppose that $L < 2$. Let $\beta = 2 - (L + 2)/2 > 0$. From what we have just shown (3) there is an N_β such that $n \geq N_\beta$ implies $0 \leq 2 - t_n < \beta$ and hence

(4) $$2 - t_n < 2 - (L + 2)/2$$

or

$$t_n > (L + 2)/2$$

which means

(5) $$t_n - L > (L + 2)/2 - L = (2 - L)/2 > 0$$

for $n \geq N_\beta$.

If we then chose $\alpha = (2 - L)/2$, there can be no N_α such that $n \geq N_\alpha$ implies $|L - t_n| < \alpha$. For suppose N_α could be found. If we take $n = N_\alpha + N_\beta$, then $n \geq N_\beta$ and hence $t_n - L > (2 - L)/2 = \alpha$ from (5).

Of course, the reader who has been through a study of infinite series in calculus will find something at least vaguely familiar about the above. The point here is not to study infinite series. The point is that ordinary logic can be used to obtain correct results in a reliable way if one makes careful definitions of "t_n approaches L as n approaches ∞" or "as x approaches infinity," or the like, *in terms of finite verifiable expressions*. Once such definitions are made, one can reach such conclusions, as we did above, in finitely many steps involving ordinary logic and finite algebraic expressions. There is no need to even mention infinite sums like $1 + 1/2 + 1/2^2 + \cdots$ in order to obtain the correct answer of 2, even though our initial reasoning at first seems to leave us with the dilemma that only by "letting n approach infinity" could one get across the room.

The difficulty encountered above is a takeoff on a paradox that goes all the way back to the Greek philosopher Zeno (who wondered how an arrow could even reach its target if the usual conception of time was assumed). This difficulty remained to plague mathematics, particularly calculus and more generally the mathematical analysis of infinite processes, until the 19th century. The work of the French mathematician, Cauchy, and of the German, Weierstrass, resolved the difficulty essentially as above. Both of these men can justifiably be called the founders of modern analysis because of their contribution to this matter of infinity.

Problems.

(1) Here is another example of how knowledge of an event of "finite type" can yield knowledge about an event of infinite type. Suppose you are given a road without end. Suppose that by walking down the road you can reach *any* given point on the road; to reach any one point requires only a finite amount of time. Finally suppose that no matter how far you do go down the road you observe that a yellow stripe is painted down the middle. Can you conclude that

the yellow stripe must be painted down the entire length of the road? Why or why not?

(2) Prove that if the difference between two given numbers a and b can be made smaller than any given positive number, then $a = b$. Is your reasoning analogous to that in (1)?

(3) In our discussion above we decided to pick on the phrase "as x approaches infinity." We could have equally well dwelt on "as x approaches zero." Let us take a look at how the humble genius and founder of the calculus, Isaac Newton, handled this problem. It should be encouraging but for the fact that Newton lived in the 17th century and possessed a brilliant mind whose discoveries even today seem remarkable. (Yet Newton was to say of himself: "I do not know what I may appear to the world; but to myself I seem to have been only like a boy playing on the seashore, and diverting myself in now and then finding a smoother pebble or a prettier shell than ordinary, whilst the great ocean of truth lay all undiscovered before me." Giving credit to those who had gone before him, Newton was also to say that, if he had seen further than other men, it was because he had stood on the shoulders of giants.)

Newton wanted to compute the derivative of $y = x^2$ by algebraic processes and did not have at hand the good definition of a limit later given by Cauchy. Newton of course knew what he wanted and set out to get it as best he could.

Let Δx be an increment in x; i.e., Δx is a number different from zero. Then $((x + \Delta x)^2 - x^2)/\Delta x$ is the average rate of change in y over the interval from x to $x + \Delta x$. We should think then that the rate of change of y *at* x would be obtained by letting "Δx approach zero." Since $(x + \Delta x)^2 - x^2 = 2x(\Delta x) + (\Delta x)^2$ then

$$\frac{(x + \Delta x)^2 - x^2}{\Delta x} = 2x + \Delta x\,.$$

Hence letting "Δx approach 0" we see that the rate of change of y at x is $2x$.

But obviously Newton had let $\Delta x = 0$ in the last expression and the philosopher (George) Bishop Berkeley would not let him forget it. Berkeley pointed out that in the step just previous to this Newton had divided by Δx. That's okay as long as $\Delta x \neq 0$. But in the following step Newton essentially set $\Delta x = 0$. "Make up your mind Isaac," Berkeley may have said. At any rate the logic of Newton was certainly unsatisfactory. But the answer was correct; the question that was to plague lesser mathematicians (in particular, the methodical mathematical plodder who wished to use the calculus) for 2 centuries was "Will such reasoning with these infinitesimals like Δx always be correct?" (Infinitesimals was just a fancy word—Berkeley called them "ghosts of departed quantities" and calculus was described as the study of quantities which are "barely something and almost nothing"). Can you make sense of Newton's idea? That is, can you use ordinary logic with finite quantities either zero or not zero and obtain $2x$ as the only answer?

(4) Mathematicians up through the 18th century were known to perform formal algebraic operations on expressions of "infinite extent" as though they were ordinary finite algebraic expressions. Resolve the paradox posed by the following sequence of formal algebraic operations:

$$\begin{aligned} 0 &= 0 + 0 + \cdots + 0 + \cdots \\ &= (1-1) + (1-1) + \cdots + (1-1) + \cdots \\ &= 1 - (1-1) - (1-1) - \cdots - (1-1) - (1-1) - \cdots \\ &= 1 - 0 - 0 - \cdots - 0 - 0 - \cdots = 1\,. \end{aligned}$$

Using the central idea of this section, can you assign a single numerical value to the formal sum

$$1 - 1 + 1 - 1 + 1 - 1 + \cdots$$

which is consistent with ordinary arithmetic on finite sums?

(5) What is wrong with the following argument that $2 = 1$?

$$\begin{aligned} 1 &= 1 - 2/3 + 2/3 - 3/5 + 3/5 - 4/7 + 4/7 - 5/9 + \cdots \\ &= 1/3 + 1/3\cdot 5 + 1/5\cdot 7 + 1/7\cdot 9 + \cdots \\ &= \frac{1 - 1/3}{2} + \frac{1/3 - 1/5}{2} + \frac{1/5 - 1/7}{2} + \cdots \\ &= (1/2 - 1/6) + (1/6 - 1/10) + (1/10 - 1/14) + \cdots \\ &= 1/2 - 1/6 + 1/6 - 1/10 + 1/10 - 1/14 + \cdots \\ &= 1/2\,. \end{aligned}$$

Hence, $2 = 1$.

(6) If you begin with the algebraic expression $1/(1-x)$ and formally divide $1-x$ into 1 you will find that the division processes does not end and that the quotient, as a matter of purely formal manipulative thought, is $1 + x + x^2 + \cdots + x^n + \cdots$. If you stop dividing after n steps you obtain a quotient of $1 + x + x^2 + \cdots + x^n$ and remainder of x^{n+1}. That is,

$$\frac{1}{1-x} = 1 + x + \cdots + x^n + \frac{x^{n+1}}{1-x}\,.$$

Now here is the question. For, say, $x = 2$, $1/(1-x)$ is defined and equals -1. What about $1 + x + x^2 + \cdots$? What is the remainder after n steps? What is

$$1 + 2 + 2^2 + \cdots + 2^n + \frac{2^{n+1}}{1-2}$$

(the sum of quotient and remainder) equal to? Can you, from some viewpoint, make sense of what's going on here? Examine this process of "infinite" division for the numbers $x = 0$, $x = -1$, and $x = -1/2$ as well.

(7) Show that $1 + 1/2 + 1/3 + 1/4 + \cdots$ does not add up.

The contradictions and absurdities that could arise in "infinite" computations were not settled until the creation of a good definition for the convergence of an infinite series along with good convergence tests. Much of the early work in this area is due to Weierstrass. We will not take these particular matters up, but will instead consider one part of the "arithmetization of analysis," this part having to do with infinite decimal expansions. Our purpose is to confront the mix of arithmetic, analysis and infinity in a somewhat familiar setting.

We wish to consider the meaning, an exact meaning, amenable to computation, that should be placed on expressions such as

$$.500 \cdots$$

or

$$.4999 \cdots$$

or

$$.1212212221 \cdots + .0808808880 \cdots.$$

At this point we might agree that .5000 means 1/2. We should also agree that the latter two expressions are rather unclear as to exact numerical value. What could $.4999 \cdots$ mean? There is no way to *actually compute* by finite arithmetical means a specific numerical value for $.4999 \cdots$. There is no way to actually compute the product

$$(1/4)(.4999 \cdots)$$

in a finite number of steps. Not even the largest computer could yield an exact value for this expression, It is true that one could compute

$$1/4(.49) = .122500 \cdots$$

$$1/4(.499) = .1247500 \cdots$$

$$1/4(.4999) = .12497500 \cdots$$

and so on. In fact, one could compute

$$1/4(\underbrace{.499 \cdots 900}_{n} \cdots)$$

given any specific $n - 1$ digits of 9's. It is, peculiarly like walking across that room! We can compute $1/4(.499 \cdots)$ to any desired degree of approximation, as close as we wish, but we cannot, in our short lives, compute by arithmetic an exact value for

$$1/4(.499 \cdots) .$$

This discussion sets the problem up. We wish to assign a meaning to an expression such as

$$.4999 \cdots$$

or, more generally, an expression

$$.a_1a_2a_3a_4 \cdots$$

where a_i is called the i-th *digit* of this expression (a_i can be any one of the numbers $0, 1, 2, 3, \cdots, 9$). This meaning we assign must be amenable to additive and multiplicative operations on such expressions by means of *finite* arithmetical operations—for such operations are all that one can actually perform. We should also like to know, with mathematical certainty, whether or not

$$.4999 \cdots = .5000 \cdots = 1/2 .$$

We remind the reader that all we have at hand are the axioms of Sections 7 and 8 and facts drawn from them to work with in this problem. Note that all of those axioms expressly deal with only finite sums and products, except for one, the completeness axiom, which has no clear relationship to arithmetic.

Consider the expression

$$.4999 \cdots$$

and act as though it really has the usual meaning. This decimal expansion represents a number which is greater than all the numbers

$$.4, .49, .499, .4999, \cdots, .\underbrace{4999 \cdots 9}_{n}$$

for any $n - 1$ number of 9's. In fact, we generally think of $.4999 \cdots$ as *exactly* the *smallest* number *greater than all* of these approximations. That is the key to the definition of an infinite decimal expansion. We will define it to be the *smallest number greater than every one* of its approximating finite decimal expansions. And we will now see how this can be done in the context of the axioms for $\mathscr{R}$.

Though we have not yet said it, by

$$.49, .499, \cdots, \underbrace{499 \cdots 9}_{n}$$

we of course mean the finite arithmetical sums

$$4/10 + 9/10^2$$
$$4/10 + 9/10^2 + 9/10^3$$
$$4/10 + 9/10^2 + 9/10^3 + \cdots + 9/10^n$$

which are defined and exist by the axioms for $\mathscr{R}$. In general we define the expression

$$.a_1a_2a_3 \cdots a_n$$

to be the rational number

$$a_1/10 + a_2/10^2 + a_3/10^3 + \cdots + a_n/10^n$$

for any n, where $a_1, a_2, a_3, \cdots, a_n$ are digits chosen from the numbers 0, 1, 2, 3, 4, 5, 6, 7, 8, 9 and only these.

With these numbers a_i it is then apparent that

$$a_1/10 \leq 9/10,\ a_2/10^2 \leq 9/10^2,\ \cdots,\ a_n/10^n \leq 9/10^n.$$

Hence

$$.a_1a_2a_3 \cdots a_n = a_1/10 + a_2/10^2 + \cdots + a_n/10^n \leq 9/10 + 9/10^2 + \cdots + 9/10^n$$

or

$$.a_1a_2a_3 \cdots a_n \leq 9/10(1 + 1/10 + \cdots + 1/10^{n-1}).$$

According to Problem 6 at the close of Section 8,

$$1 + 1/10 + \cdots + 1/10^{n-1} \leq 10/9$$

and, hence, for any n number of digits a_i, we have

$$.a_1a_2 \cdots a_n \leq 9/10(10/9) = 1.$$

Hence the collection M of all the numbers

$$.a_1,\ .a_1a_2,\ .a_1a_2a_3,\ \cdots,\ .a_1a_2 \cdots a_n,\ .a_1a_2 \cdots a_na_{n+1},\ \cdots$$

for any *finite* number of digits has an upper bound. By the completeness axiom, M has a supremum—we call that supremum the infinite decimal expansion

$$.a_1a_2a_3 \cdots a_n \cdots.$$

Thus $.a_1a_2a_3 \cdots$ is the smallest number larger than or equal to all the numbers

$$.a_1,\ .a_1a_2,\ .a_1a_2a_3,\ \cdots,\ .a_1a_2a_3 \cdots a_n$$

no matter how large n is.

So we have a definition, a rule for identifying, a meaning for $.4999\cdots$ or more generally $.a_1a_2a_3\cdots$. Let us see how to use it to prove something with this definition. Let us, for example, show that

$$1/2(.499\cdots) = .25 = 1/4\,.$$

For sake of discussion, let $a = .499\cdots$. We must show that $(1/2)a = 1/4$. First of all, $(1/2)a \leq 1/4$ because for any n number of digits,

$$\underbrace{.499\cdots 9}_{n} = 4/10 + 9/10^2 + \cdots + 9/10^n \leq 5/10 = 2(1/4)\,.$$

Hence $2(1/4)$ is an upper bound for all the numbers $\underbrace{.499\cdots 9}_{n}$ and hence $2(1/4) \geq a$ since, *by definition*, a is the supremum of the collection of all these and is smaller than or equal to any upper bound. Thus $1/4 \geq (1/2)a$. On the other hand, because $\underbrace{.499\cdots 9}_{n} \leq a$, no matter how large n is, we have

$$\begin{aligned}0 \leq 1/4 - (1/2)a &\leq 1/4 - 1/2(\underbrace{.499\cdots 9}_{n})\\ &= 1/4 - 1/2(4/10 + 9/10^2 + \cdots + 9/10^n)\\ &= 1/4 - 5/10(4/10 + 9/10^2 + \cdots + 9/10^n)\\ &= 1/4 - (2/10 + 4/10^2 + 9/10^3 + \cdots + 9/10^n + 5/10^{n+1})\\ &= 5/10^{n+1}\end{aligned}$$

by a whole lot of arithmetic. Finally

$$0 \leq 1/4 - (1/2)a \leq 5/10^{n+1} \leq 5/n$$

or

$$n(1/4 - (1/2)a) \leq 5 \text{ for } any \text{ number } n\,.$$

By the Archimedean property $1/4 - 1/2a$ must be 0. Hence, $1/4 = (1/2)a$. Incidentally, it follows from this that $.4999\cdots = a = 1/2 = .5000\cdots$.

That is a lot of bother! That is also a way to validate computation with infinite decimal expansions. As a theoretical tool it is what enabled late 19th century mathematicians to prove with certainty (given the axioms for $\mathscr{R}$) the rules of addition, multiplication, division and approximation of such decimal expansions that you were encouraged to accept and learn some years ago.

Our interest in this matter is one of having a theoretical tool, a logically acceptable means, without babbling about "infinity" and "using enough decimal places," of handling infinite decimal expansions when they prove to be useful. To summarize: (1) one can give a real number meaning to the expression $.a_1a_2a_3\cdots$ using the concepts of arithmetic (the sums $a/10 + \cdots + a_n/10^n$) and the completeness axiom and (2) it is possible to ascertain facts about the real number $.a_1a_2a_3\cdots$ so defined by means of finite arithmetical operations. This is quite in analogy to proving that one can get across the room. We are forced to define $.a_1a_2a_3\cdots$ in terms of approximate stages, $.a_1a_2\cdots a_n$, as we could, for the sake of argument, consider crossing a room by approximate stages. But as we noted, to get there involves something extra—here it is the completeness axiom.

Having defined the expression $.a_1a_2a_3\cdots$ we still have not defined numbers like $26.7521\cdots$. This is simple enough: if N is a natural number, define $N.a_1a_2a_3\cdots$ to be the number

$$N + .a_1v_2a_3\cdots .$$

Let us now prove a general result to indicate a general method of proof.

THEOREM 2.9.1. $1/10(.a_1a_2a_3\cdots) = .0a_1a_2a_3\cdots$.

Proof. Let $a = 1/10(.a_1a_2a_3\cdots)$. Then by definition of $.a_1a_2a_3\cdots$,

$$a \geq 1/10(.a_1a_2\cdots a_n) = 1/10(a_1/10 + a_2/10^2 + \cdots + a_n/10^n) .$$
$$= a_1/10^2 + a_2/10^3 + \cdots + a_n/10^{n+1} = .0a_1a_2\cdots a_{n+1} \text{ for all } n .$$

Hence, by definition of $.0a_1a_2a_3\cdots$, we have

$$a \geq .0a_1a_2a_3\cdots .$$

Suppose it were true that $a > .0a_1a_2a_3\cdots$. Then,

$$.a_1a_2a_3\cdots > 10(.0a_1a_2a_3\cdots)$$

and hence, because $.a_1a_2a_3\cdots$ is the smallest upper bound for all the numbers $.a_1a_2\cdots a_n$ for any n, there must be an n for which

$$.a_1a_2a_3\cdots a_n > 10(.0a_1a_2a_3\cdots) .$$

Since $.a_1a_2a_3\cdots a_n = a_1/10 + a_2/10^2 + \cdots + a_n/10^n$, this yields, upon division by 10,

$$a_1/10^2 + a_2/10^3 + \cdots + a_n/10^{n+1} = .0a_1a_2\cdots a_n$$
$$> .0a_1a_2a_3\cdots ,$$

a contradiction. Hence the proof.

The reader should note that all arithmetical operations are performed only on finite sums. Theorem 2.9.1, despite its rather specific nature, does indicate the general approach to proving those arithmetical operations with decimal expansions that the reader learned to perform long ago. It is not our purpose to develop the proofs of any more of these, but to prove that any real number has an infinite decimal expansion, and in some sense only one decimal expansion. These results will add more meaning to these unending series of digits now defined by the above to represent numbers.

First an example to motivate a method of proof. Suppose $c = .487 \cdots$. Geometrically, this means c lies to the right of four-tenths of the interval from 0 to 1, plus eight-one hundredth's to the right of that four-tenths, plus seven-one thousandth's to the right of this point, and so on. The idea of the proof is to narrow down on the exact position of an *arbitrary* but fixed number c by breaking up the interval into smaller and smaller parts.

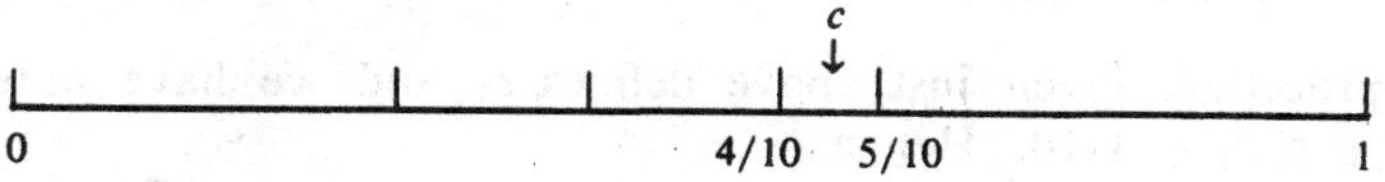

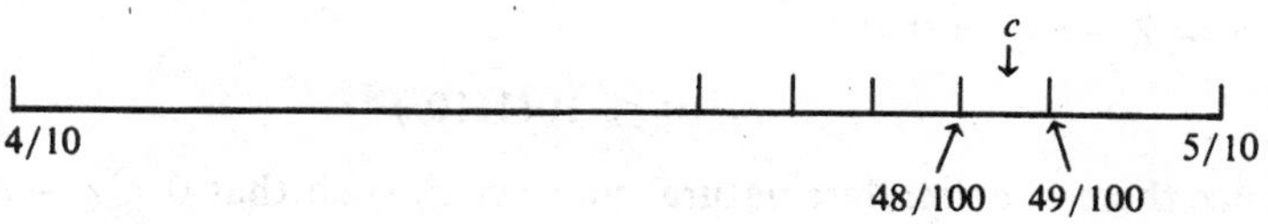

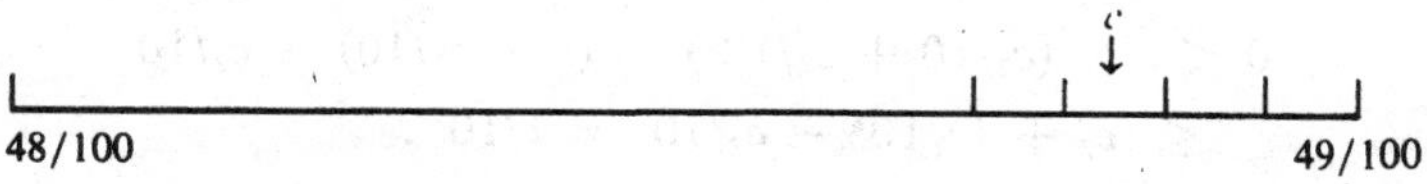

Suppose now that c is an arbitrary number with $0 < c \leq 1$. We will give a procedure for locating the first digit in a decimal expansion for c.

Since $c \leq 1$, then there is a natural number n such that $c \leq n(1/10)$—namely $n = 10$. According to Theorem 2.6.9, there is then a *smallest* natural number, let us call it d_1, such that $c \leq d_1(1/10)$. From this and the previous sentence, and the fact that $0 < c$, we know three things: (1) $d_1 \neq 0$, (2) $d_1 - 1 \leq 9$ and (3) $(d_1 - 1)(1/10) < c \leq d_1(1/10)$. Let $c_1 = d_1 - {}_1$. Then, we know from (3) that c lies in the interval of numbers from $c_1/10$ to $(c_1 + 1)/10$ and hence that c_1 must be the first digit in the decimal expansion for c. If, for example, $c = .487$, we would find by the above procedure that $d_1 = 5$, $c_1 = 4$ and $4/10 < .487 < 5/10$. A geometric view of this argument appears below.

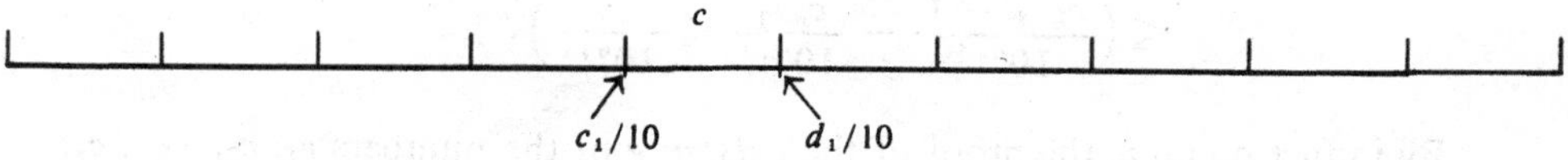

Here is the theorem we wish to prove, and its proof.

THEOREM 2.9.2. If $0 < c \leq 1$ is a real number, then for each natural number n there is a number c_n equal to one of the numbers 0, 1, 2, $\cdots$, or 9, such that

$$c > .c_1c_2c_3 \cdots c_k \qquad \text{for all k}$$

and

$$c = .c_1c_2c_3 \cdots.$$

Proof. We claim that for every natural number n there is a number c_n, either 0, 1, 2, $\cdots$, or 9, such that

$$(6) \qquad 0 < c - (c_1/10 + c_2/10^2 + \cdots + c_n/10^n) \leq 1/10^n.$$

The procedure given just above defines c_1 and we have noted that $c_1/10 < c \leq c_1 + 1/10$. Hence,

$$0 < c - c_1/10 \leq 1/10.$$

The above claim is then true for $n = 1$.

For $n = 2$ we note that

$$c - c_1/10 \leq 10(1/10^2)$$

and hence there is a smallest natural number d_2 such that $0 < c - c_1/10 \leq d_2(1/10^2)$. Let $c_2 = d_2 - 1$. Note that $c_2 \leq 9$ and that $c_2/10^2 < c - c_1/10$. Moreover,

$$\begin{aligned} 0 < c - (c_1/10 + c_2/10^2) &= (c - c_1/10) - c_2/10^2 \\ &\leq (c_2 + 1)/10^2 - c_2/10^2 = 1/10^2. \end{aligned}$$

Hence (6) holds for $n = 2$.

We complete the proof of existence of the numbers c_n by induction on n. Suppose numbers $c_1, c_2 \cdots, c_n$ have been found satisfying (6) above. Let $b = c - (c_1/10 + c_2/10^2 + \cdots + c_n/10^n)$. Then $0 < b \leq 1/10^n$ and hence $0 < b \leq 10(1/10^{n+1})$ so that, as before, there is a smallest number d_{n+1} such that $0 < b \leq d_{n+1}(1/10^{n+1})$. Let $c_{n+1} = d_{n+1} - 1$. Then

$$c_{n+1}/10^{n+1} < b \leq (c_{n+1} + 1)/10^{n+1}$$

and hence

$$0 < c - \left(\frac{c_1}{10} + \frac{c_2}{10^2} + \cdots + \frac{c_n}{10^n} + \frac{c_{n+1}}{10^{n+1}}\right)$$

$$\leq \left(\frac{c_{n+1} + 1}{10^{n+1}} - \frac{c_{n+1}}{10^{n+1}} \leq \frac{1}{10^{n+1}}\right).$$

By induction on n, the proof of the *existence* of the numbers $c_1, c_2, \cdots, c_n$, one for each natural number n and satisfying (6), is complete.

We will now show that $c = .c_1c_2c_3 \cdots$. From (6) we first note that, for all n, $c > .c_1c_2 \cdots c_n$ and hence that $c \geq .c_1c_2c_3 \cdots$. If $c > .c_1c_2c_3 \cdots$, then by the Archimedian property (2.8.4) there is an n such that $n(c - .c_1c_2c_3 \cdots) > 1$ or, using (6) again

$$1/10^n > c - .c_1c_2 \cdots c_n \geq c - .c_1c_2c_3 \cdots \geq 1/n .$$

This is a contradiction. Hence, $c = .c_1c_2c_3 \cdots$ and the proof is complete.

That *any* positive real number has at least one decimal expansion is left for the reader to prove as a corollary to Theorem 2.9.2 and the Archimedian property.

Let us turn now to the matter of uniqueness of infinite decimal expansions. We have already noted that 1/2 has the two decimal representations $.5000 \cdots$ and $.4999 \cdots$. But these are distinguishable in an abstract sense: While, for any n number of digits, $.499 \cdots 9 < 1/2$ one has $.500 \cdots 0 = 1/2$. In Theorem 2.9.2 we saw that any number c, with $0 < c \leq 1$, has at least one expansion $.c_1c_2c_3 \cdots$ with the property: $.c_1c_2 \cdots c_n < c$ for any n. It is to our purpose to prove that such a number c has *only one* expansion with this latter property. This result, as well as being informative, will be of some use in the study of infinite sets.

THEOREM 2.9.3. The number c with $0 < c \leq 1$ has *only one* decimal expansion $.c_1c_2c_3 \cdots$ such that

$$.c_1c_2 \cdots c_n < c \qquad \text{for all } n .$$

Proof. Suppose that c has two decimal expansions, say $.c_1c_2 \cdots c_n \cdots$ and $.c_1'c_2' \cdots c_n' \cdots$ such that

$$.c_1c_2 \cdots c_n < c \quad \text{and} \quad .c_1'c_2' \cdots c_n' < c \qquad \text{for all } n .$$

We claim that $c_n = c_n'$ for all n.

If not, then, there being some n for which $c_n \neq c_n'$, there must then be a smallest such n, call it n_0. Hence, $c_1 = c_1'$, $c_2 = c_2'$, $\cdots c_{n_0-1} = c'_{n_0-1}$ but $c_{n_0} \neq c'_{n_0}$. Thus either $c_{n_0} < c'_{n_0}$ or $c_{n_0} > c'_{n_0}$. Let us suppose $c_{n_0} < c'_{n_0}$ and that $d = c'_{n_0} - c_{n_0}$ is their difference.

We claim that $c > .c_1'c_2'c_3' \cdots$ which will be a contradiction. Noting that $c > .c_1c_2 \cdots c_{n_0}$ by hypothesis, let $\delta = c - .c_1c_2 \cdots c_{n_0} > 0$. Since $c_i = c_i'$ for $i \leq n_0 - 1$ and $d = c'_{n_0} - c_{n_0}$ we have,

$$\begin{aligned} c - .c_1'c_2' \cdots c'_{n_0} &= c - .c_1c_2 \cdots c_{n_0-1}c_{n_0} + d/10^{n_0} \\ &= \delta + d/10^{n_0} . \end{aligned}$$

Hence, $c - \delta - d/10^{n_0} = .c_1'c_2' \cdots c'_{n_0}$ and for all $n > n_0$

$$(7) \qquad \begin{aligned} c - \delta - d/10^{n_0} + c'_{n_0+1}/10^{n_0+1} + \cdots + c'_n/10^n \\ = .c_1'c_2' \cdots c'_{n_0+1} \cdots c_n' . \end{aligned}$$

Now,

$$\begin{aligned} c'_{n_0+1}/10^{n_0+1} + \cdots + c_n'/10^n &\leq 9/10^{n_0+1}(1 + 1/10 + \cdots + 1/10^{n-(n_0+1)}) \\ &\leq 9/10^{n_0+1} \frac{1}{1 - 1/10} = \frac{1}{10^{n_0}} \leq \frac{d}{10^{n_0}} \end{aligned}$$

by Theorem 2.9.3. Combining this inequality with (7) we obtain

$$c - \delta - .c_1'c_2' \cdots c_{n_0}'c_{n_0+1}' \cdots c_n' \geq 0.$$

Hence,

$$c - \delta \geq .c_1'c_2' \cdots c_n' \qquad \text{for all } n,$$

and, by definition of $.c_1'c_2'c_3' \cdots$,

$$c - \delta \geq .c_1'c_2'c_3' \cdots$$

or

$$c - .c_1'c_2'c_3' \cdots \geq \delta > 0, \quad \text{a contradiction.}$$

This completes the proof.

To this point we have dealt only with decimal expansions in *base* 10. The above theorem is proven by successively splitting the interval in first 10 parts of length 1/10, then 100 parts of length 1/1000, and so on. It is equally possible to operate in base 2 or 3 or 5 or whatever positive number one choses. Base 2 does, however, have certain advantages both theoretically and in applications.

Decimal representations in base 2 for the numbers 1/2, 1/4 and 3/4, respectively, would be .1, .01 and .11. For,

$$1/2 = 1/2^1$$
$$1/4 = 1/2^2 = 0/2^1 + 1/2^2$$
$$3/4 = 1/2^1 + 1/^2.$$

In base 2 one would in general define $.a_1a_2 \cdots a_n$ to mean

$$a_1/2 + a_2/2^2 + a_3/2^3 + \cdots + a_n/2^n$$

where, *in this base*, each number a_k is either 0 or 1. One would then define $.a_1a_2 \cdots a_n \cdots$ to be the supremum of all the numbers $a_1/2+a_2/2^2+\cdots+a_n/2^n$, for all n. This kind of representation is particularly adapted to computer operations because a switch is either open (0) or closed (1) and a number can be stored in the computer's memory simply by directing that a series of switches be open or closed in a definite order.

One can similarly operate in base 3, where the possible digits are 0, 1 and 2. In base 3, one would find

$$1/3 = .1000 \cdots$$
$$7/9 = .2100 \cdots = 2/3 + 1/3^2$$
$$1/4 = .020202 \cdots.$$

The key to the computations that verify the last equality is given by the next theorem. The computations throughout this section have hinted at this result and its purpose is to wrap up into one package the reoccurring techniques of these computations. It is useful in any base one chooses to operate within and is one of the most useful approximation facts of mathematics.

THEOREM 2.9.4. Let x be a real number with $0 \leq x < 1$. The supremum of the collection M of numbers $1, 1 + x, 1 + x + x^2, \cdots, 1 + x + x^2 + \cdots + x^n, \cdots$ is $1/(1 - x)$.

Proof. According to Problem 6 at the close of Section 2.8

$$1 + x + x^2 + \cdots + x^n \leq 1/(1 - x) \qquad \text{for all } n.$$

Hence $1/(1 - x)$ is an upper bound for M. Thus, M has a supremum, let us call it y. We claim that $y = 1/(1 - x)$ and already know that $y \leq 1/(1 - x)$.

Suppose that $y < 1/(1 - x)$ and let $d = 1/(1 - x) - y$ be their difference. Since $x < 1$, then $1/x = 1 + a$ with $a > 0$. By the Archimedian property (Theorem 2.8.4) there is a natural number n such that

$$[1/(1 - x)d - 1] < na.$$

Hence,

$$1/(1 - x) < d(1 + na)$$

and by Problem 7, Section 2.6, $1 + na \leq (1 + a)^n = (1/x)^n$. Hence, $x^n/(1 - x) < d$ for this choice of n.

By Problem 6, Section 2.8.

$$1 + x + x^2 + \cdots + x^{n-1} = \frac{1 - x^n}{1 - x} = \frac{1}{1 - x} - \frac{x^n}{1 - x}$$
$$> \frac{1}{1 - x} - d = y.$$

But this is a contradiction. Hence $y = 1/(1 - x)$. This completes the proof.

The reader may recognize the expression $1 + x + \cdots + x^n + \cdots$ as a *geometric series*; Theorem 2.9.4 is the proof that its "sum" is $1/(1 - x)$ when $0 \leq x < 1$. There is no need to deal in these matters with infinite sums. Theorem 2.9.4, the completeness axiom, and arithmetic is all that one needs.

Problems.

(1) Write down base 2 representations for the numbers

$$13/16, 23/64 \text{ and } 3/8.$$

Interpret these representations geometrically on the number line.

(2) Write down decimal representations for 1/9 in bases 2, 3 and 10, respectively.

(3) In base 10, prove that
(a) $.023 + .091 = .114$
(b) $.023 + .09111 \cdots = .11411 \cdots$

(4) Suppose the real number a has decimal representation $.aaa \cdots$ in base b, b a natural number greater than 1. (Think of b as 2, 3 or 10.) Show that $0 \leq a - .a_1a_2a_3 \cdots a_n \leq 1/b^{n-1}$. This means that $1/b^{n-1}$ is an upper estimate on the degree of approximation of $.a_1a_2 \cdots a_n$ to $.a_1a_2a_3 \cdots$.

(5) Compute exact fractional valus for $.1111 \cdots$ in bases 2, 3 and 10, respectively.

(6) Without using Theorem 2.9.4 prove
(a) In base 10, $1 = .999 \cdots$
(b) In base 2, $1/2 = .0111 \cdots$

(7) Using Theorem 2.9.4 prove in base 10 that
(a) $.aaaa \cdots = a/9$
(b) $.aa'aa'aa' \cdots = 10a/99 + a'/99$.
More generally, in base b, prove that
(c) $.aaa \cdots = a/(b-1)$
(d) $.aa'aa'aa' \cdots = ab/(b^2-1) + a'/(b^2-1)$

(8) Suppose that $c = .c_1c_2c_3 \cdots$ and that after the kth digit the digits for c repeat in the same order. That is, suppose

$$c_1 = c_{k+1} = c_{2k+1} = c_{3k+1} = \cdots,$$

$$c_2 = c_{k+2} = c_{2k+2} = c_{3k+2} = \cdots c_k = c_{2k} = c_{3k} = \cdots.$$

Hence

$$c = .c_1c_2 \cdots c_kc_1c_2 \cdots c_kc_1c_2 \cdots c_kc_1 \cdots.$$

Show that c is the rational number

$$\left(\frac{c_1}{10} + \frac{c_2}{10^2} + \cdots + \frac{c_k}{10^k}\right)\left(\frac{1}{1-(1/10^k)}\right)$$

Note that this is but a generalization of Problem 7(c), (d).

(9) Suppose that $0 < c < 1$ and that c is a rational number. By Theorem 2.9.2, $c = .c_1c_2c_3 \cdots$. Show that $\exists$ a k such that

$$c_1 = c_{k+1} = c_{2k+1} = \cdots, \ c_2 = c_{k+2} = c_{2k+2} = \cdots,$$

$$\cdots c_k = c_{2k} = c_{3k} = \cdots.$$

The import of (8) and (9) is that the rational numbers are only those real numbers whose decimal expansions repeat after a certain k-number of digits.

(10) Classify the following as rational or irrational (not rational)
(a) $.232323 \cdots$
(b) $.1010001000000010 \cdots$ where $a_n = 1$ if $n = 2^k - 1$ for some

k and $a_n = 0$ if $n \neq 2^k - 1$ for all k.
(c) $.10100100010 \cdots + .01011011101 \cdots$
(d) $.232323 \cdots + .474747 \cdots$
(e) $.1212212221 \cdots + .0808808880 \cdots$

(11) Discuss the rational or irrational status of $a + b$ where a and b are real numbers. Prove your assertions.

(12) Given $a < b$, is there an irrational number x such that $a < x < b$?

(13) Prove that any positive real number has a base 10 decimal representation of the form $N.a_1a_2a_3 \cdots$ where N is a natural number. How would you handle negative numbers?

(14) Prove that 0 has only one decimal representation.

(15) Let $b = .b_1b_2b_3 \cdots$. Suppose that a is a number such that
(a) $a \geq .b_1b_2b_3 \cdots b_n$ for all n
and
(b) $a - .b_1b_2b_3 \cdots b_n \leq c/n$ for all n where c is some constant.
Prove that $a = b$.

(16) Prove the analogue of Theorem 2.9.2 for base 2.

(17) Prove the analogue of Theorem 2.9.3 for base 2.

10 SOME PROBLEMS AND PROPERTIES OF AXIOM SYSTEMS

One just does not sit down and wildly write out a list of assumptions from which he hopes conclusions will follow! All other previous impressions aside, the process is just the reverse. Knowing what one wants, what conclusions one wishes, one sets out to find specific assumptions from which they will follow. Rationalizing, it is sometimes called; or, looking for facts to support one's conclusions. That kind of reasoning is often, and justly, criticized. But the tenor of the criticism misses a key point. There is nothing really wrong in looking for believable "truths" that will support the conclusions one already has. The hang-up with such a process is that the same assumptions that lead to what one wants, can at times equally well lead to what one does not want, does not suspect, and perhaps wishes not to admit. In mathematics such unexpected results especially occur when one begins with assumptions that extend to the infinite the same ideas which work extremely well for the finite. We will encounter such unsettling results when we consider the topic of infinite sets. And mathematicians are still wrestling with a perfectly reasonable axiom of set theory, the "axiom of choice," which quite frankly leads to almost unbelievable conclusions and yet, as an assumption, seems perfectly acceptable. We will consider this axiom when we have considered enough structure to make its special form meaningful. People who do mathematics know no reputations will be lost, in fact, some new ones may be made when unsuspected conclusions result from a given set of assumptions. They also know that much mathematical progress has occurred in this way.

This section is concerned with certain desirable properties and undesirable problems of any axiom system. These are: *equivalence, consistency, completeness, categoricalness* and *independence.* Of these, consistency presents the greatest problem, while completeness turns out to be a will-o'-the-wisp. The study of axiomatic systems as entities in themselves is known as *meta-mathematics.*

The property of *equivalence* is not a point of interest in the presence of a single axiom system, two at least are needed. Suppose that S_1 is a collection of axioms about a collection of undefined terms T_1, and S_2 is a collection of axioms about a collection of undefind terms T_2. The axiom systems (S_1, T_1) and (S_2, T_2) (we write it this way to remind you that T_1 relates to S_1, T_2 relates to S_2) are said to be equivalent iff the terms in T_2 are definable in terms of those in T_1 and the axioms in S_2 can be proven from the axioms in S_1 and, conversely, the axioms of S_1 can be proven from those in S_2 when the undefined terms in T_1 are defined in terms of those in T_2.

The concept of equivalent axiom systems goes back to earlier attempts to prove the parallel postulate. "Proofs" were given, but on closer examination these proofs were seen to make use of statements which were but unadmitted assumptions. Attempts were then made to show that such an assumption could be used in place of the parallel postulate as an axiom of Euclidian geometry. But then it was invariably found that geometry with the parallel axiom implied geometry with this new assumption in place of it. In other worlds, unsuspected assumptions used to prove the parallel postulate were in fact assumptions equivalent to it, thus giving geometries equivalent to ordinary geometry.

Quite conceivably two different axiom systems can arise in two different contexts and not be recognized as equivalent for many years. If they are both sufficiently obscure, perhaps never! If they are both important, then at the slightest hint of equivalence, someone would set out to settle the question. But there is another aspect to the concept of equivalence. One could for example take an axiom system with five axioms and give an equivalent axiom system with one axiom wherein the five were combined into one huge assumption. Or, conversely, one could expand the five into twenty-five simpler assumptions. After all, the number eight can be defined as three more than the first prime which exceeds by one the smallest number greater than two having an integral square root; eight is also seven plus one. In other words, equivalence is a property of axiom systems which is somewhat a matter of taste.

Problems. Here is a problem on equivalence which has some meaning and importance. Suppose the existence of a collection H of *elements* called a *herd* and an undefined symbol $*$ with the following properties

(a) For a and b in H, $a*b$ is an element of H and if $a=c$ and $b=d$ then $a*b = c*d$.

(b) For a, b, c in the herd, $a*(b*c) = (a*b)*c$.

(c) Given a and b in the herd, there exists an element x in the herd such that $a*x = b$ and an element y in the herd such that $y*a = b$.

(1) Show that the following is a model for a herd: H is the collection consisting of 0, ± 1, ± 2, $\cdots$ and $*$ is the operation $+$.

(2) Show that the axiom system for a herd H and its undefined operation $*$ is equivalent to the axiom system for a group G and its undefined operation $\circ$ when H and G are identical collections of elements. For a hint, consider the solution to the equation $a*x = a$. Call it e. Show that $c*e = c$ for *all* c in the herd. This gives the identity for H. There remains the problem of inverses.

The property of *independence* of an axiom system is not a mandatory one. Simply stated, an axiom system is independent if it is not redundant, if it is the bare minimum of assumptions needed to found the whole theory, no assumptions being repeated in different guises. As you might suspect, it may not be at all obvious whether a given axiom system is independent. Yet independence is a desirable property of an axiom system, desirable for several reasons. The axioms are shorter and often simpler. In a particular instance where one wishes to apply the conclusions of the theory, one has less assumptions to verify as valid before doing so. An independent set of axioms is more esthetically satisfying; one knows that these are the minimum assumptions on which the entire theory can be built, no fewer will do and no more are needed.

A formal definition of independence is a rather impractical one to apply. An axiom system S is said to be independent if for each axiom A of S, one *cannot* prove A from the remaining axioms of S. How does one decide that one *cannot* prove something—not suspect it, but know it? For the method, in this setting, we first consult history.

The most famous example of independence is again the parallel postulate. Thought for 2,000 years to be surely dependent on the remaining axioms of geometry, a logical consequence of them, it was finally shown by the work of Lobachevsky, Bolyai, and Gauss to be independent of the remainder when these men gave models of the system in which all the axioms held but for the parallel axiom. This is the method yet used today for determining the independence of an axiom system.

Suppose an axiom system S has five postulates. One can show these are independent by finding five interpretations of the undefined terms such that in each interpretation, one of the axioms does not hold and the remainder do hold; of course, one interpretation is needed for each axiom one is to show is independent. The idea is that, if four axioms hold in a model and the other does not, then it cannot be proven from the four because that would contradict the fact of the model.

We have already considered the problem of independence without saying so. The first eleven axioms for the real numbers, given in Section 7, were shown in the problems that followed to be independent of the next two,

0.1 and 0.2. There a model was given in which the first eleven axioms held but the next two did not. Hence these two are logically independent of the first eleven. Moreover, Axiom 0.2 is independent of 0.1. For if we take the rational numbers as an interpretation of $\mathscr{R}$ and take P to be *all* the rational numbers then Axiom 0.1 holds but 0.2 does not. More importantly, the irrationality of $\sqrt{2}$ shows that the completeness axiom is independent of the remaining axioms for the real numbers.

As it turns out, the first five axioms for the real numbers lack independence. Notice that the Axioms A.1, A.3, A.4 and A.5 are the axioms of Section 5 for a group with $\circ$ the operation $+$. Notice that the uniqueness of the identity in a group is a consequence of the four axioms for a group by Problem 4 in Section 2.5. Yet, in $\mathscr{R}$ with the operation $+$, which is a group, we stated uniqueness of the identity as part of the Axiom A.4. This part of Axiom A.4 is not then independent. This brings up a point where lack of independence is desirable. In beginning mathematics courses, in order not to bore or lose the student with nit-picking details he is not likely to appreciate, it is standard practice to assert much more in the way of assumptions than are strictly needed, in order to get on with the main body of the theory.

Problems. Show that each of the following models verifies the independence of the indicated axiom.

(1) For a group G with the operation $\circ$, let G consist of the number $0, 1, 2, 3, \cdots$ and $\circ$ be addition. Show that the first three axioms hold in this model but the fourth does not.

(2) Consider Peano's axioms for the natural numbers. Find a model in which all axioms, except for the induction axiom, hold.

(3) Consider the axioms for a herd given above and let H be the set of real numbers greater than 0 and less than 1. Show that the first two axioms for a herd are satisfied when $*$ is interpreted as ordinary multiplication, but the third axiom does not hold.

An axiom system is said to be *categorical* if there is essentially only one model of it, and that is no precise definition. To remove a little of the imprecision, if a system is categorical, then between any two models of the system one can define a one-to-one correspondence between the elements of the models and between the interpretations of the undefined terms. It is in general easier to decide that a system is not categorical than to prove that a system is categorical even when it is. The axiom systems for the real numbers and for Euclidian geometry are both categorical. But, for example, the first eleven axioms for the real numbers do not constitute a categorical system. The two models introduced in the problem set at the beginning of Section 2.7, one having finitely many numbers, and the other having infinitely many, show that these eleven axioms are not categorical: one cannot obtain a one-to-one correspondence between the constituent parts of these models. The axiom system for group theory is not categorical, nor will the axiom system for a topological space be categorical. As far as it concerns us herein, if a syatem is not categorical, one should be cautious in guessing at generalizations of a truth in a model, extended to the abstract system itself, for in yet another model this may not be a truth and hence could not be provable from the axioms.

Now for the difficult, most interesting and most demanding properties of axiom systems, *completeness* and *consistency*. We will consider consistency first. It is an essential; in the sense remarked on in Chapter 1, consistency is the whole ball game. An axiom system S is said to be consistent if it does not imply contradictory conclusions. That is a really useless definition! It gives no feasible rule for determining consistency. The theorems deducible from an axiom system for Euclidian geometry seem endless. How is one to know that someday someone will not prove a statement contradictory to one already proven? In truth, one does not know.

Mathematicians have attempted to handle problems of consistency through, again, the use of models. In many cases, particularly where one can find a model with only finitely many parts, this is a feasible and accepted approach. One can simply verify that all the assumptions of the system are true for each of the finite number of cases possible in the model. Since one believes in the consistency of the model, if it is simple and finite, one then feels assured of the consistency of the abstract system.

Problem. Show that the collection G consisting of two marks 0 and 1 and the symbol $\circ$ defined by the table is a model for the axiom system for a group. Do this by

$\circ$	0	1
0	1	0
1	0	1

checking out all the possible cases of $a \circ b$, $a \circ (b \circ c)$, $(a \circ b) \circ c$, $a \circ c = a$, $a \circ a' = e$ for a, b, c being either 0 or 1. This finite model constitutes an accepted method for verifying the consistency of the group axioms.

When it comes to axiom systems which demand a model with infinitely many parts, consistency is a real problem. Here mathematicians have been content with the notion of *relative consistency*. For example, it can be shown that Euclidian geometry is consistent *if* the real number system is consistent. Unfortunately no one has ever shown the latter! So where does that leave one? The real numbers, in one form or another, have been a constant object of study for over 2,000 years. No one has yet proven one statement in contradiction to another. That is all that can be said with certainty. With less certainty, one can take the natural numbers as given and properly axiomatized by Peano, and from these construct, using some set theory and the purest logic, a model satisfying all the axioms given in

Section 7. But then this model relies on set theory, and nothing so naively there, yet so profoundly disturbing to mathematicians, has yet been found. Even the simplest of set theories is very difficult to completely and rigorously axiomatize.

This is what the problem of consistency of the basic structures of mathematics, Euclidian geometry and the real number system, come down to: the natural numbers and a basic theory of sets. We will take up a theory of sets in the following chapter.

Now for completeness, and a real surprise of sorts. In the latter part of the 19th and early part of the 20th centuries, the axiomatic method was looked upon as a really certain way of founding all of mathematics. But for all its virtues it does have one outstanding defect, discovered in 1931 by Kurt Gödel. Roughly speaking, Gödel proved that in an axiomatic theory with "enough" assumptions, there must exist a statement which can be stated within the system, but which cannot be decided upon within the system. That is, one cannot prove using the axioms and the theory that the statement is either true or false. Or, in other words, there will always be an unanswerable question in any extensive theory. This discovery doomed the hopes of some that all mathematical and scientific knowledge could be placed on a universal axiomatic foundation in which all questions could be answered. Gödel's discovery is one of the most, perhaps the most, outstanding discovery of this century.[27] Moreover, Gödel also discovered that proofs of consistency must fall back of logical principles which are themselves subject to speculative doubt as to consistency. We have already mentioned a case somewhat it point, the "law of the excluded middle," applied to a statement about more than a finite number of things.

What then is the ultimate validity of mathematics? This is quite a deep matter of course and very much a philosophical one. Let us ignore the philosophical aspect and instead consider only the real situation in mathematics as it exists today. It appears that mathematics cannot hope for a proof of absolute consistency, or absolute validity, in large part because of the necessity of dealing with concepts of infinite extent and the accompanying lack of models in the real world. While one can interpret certain real world possibilities in the context of infinite concepts, and this is extremely fruitful in applying mathematics to this world, one is then but assuming the consistency of infinite reasoning and using it in applications. In such use one does arrive at consistent and useful "practical" results. But since the implication of a true conclusion by a false hypothesis is an integral part of mathematical reasoning, this observation offers little succulence.

[27] A non-technical article which will give you the flavor of Gödel's work can be found in the article by E. Nagel and J. R. Newman: Gödel's proof, *Scientific American*, June 1956. For a brief discussion of undecidable propositions in the familiar setting of $\mathscr{N}$, see M. Kac and S. Ulam: *Mathematics and Logic*, pp. 124–128, New York: F. A. Praeger, Publishers, Inb., 1968.

An answer to the ultimate validity of mathematics is perhaps best offered by the plain *existence* of mathematics and nothing more. Moreover, mathematics exists not as merely a stale rehashing of past truths discovered, but as a constantly growing, innovative monument to man's aesthetic sense. This is apparently the view of that eminent corporate mathematician of recent (1949) vintage, Nicholas Bourbaki.[28] Bourbaki has undertaken the immense task of compiling and verifying all of the mathematics that man has found continually useful, beginning with its most basic principles. For Bourbaki, mathematics is what is produced by mathematicians working in a tradition critically evolved through 25 centuries. As such, Bourbaki is not concerned with the *eternal* consistency of mathematics, admitting that contradiction may yet arise; Bourbaki does believe that the existence of interesting mathematics, of wide ranging use within and without mathematics, is justification enough for its validity. For Bourbaki, mathematics is the study of structures, patterns of relationships, each structure being founded on a collection of axioms. The chief existent structures are algebraic, topological and ordinal, although he admits more may yet arise. The mathematical researcher either recognizes in the essential relations of a problem he faces one of these established structures and hence studies it in the context of established mathematics, or abstracts from his problem a new structure which incorporates the essentials of his problem. Subject to the accepted procedures, or *procedures he can gain acceptance of*, his results then become mathematics. There are few areas of endeavor where one is so free to create, so free to strike new directions unbounded by any pseudo-norms that have lost their usefulness. All one need do to accomplish something new in mathematics is to do it definitively.

Time to close this chapter. The reader has been presented with but a glimpse of the foundations and fundamental questions of mathematics as though through a knot hole in the ball park wall. Only a brief glimpse, granted as much to provoke a little interest and present a few perhaps unsuspected "truths" of mathematics, as to present any hard and certain knowledge. Though the author habitually scorns such naive metaphors, it is now time to get into the ball park and play the game. That is, to do some particular mathematics. The mathematics that follows has as its only goal an introduction to the fundamental concepts and manner of thought relevant to all mathematics above the elementary level.

Related Readings

1. Baker, A.: *Modern Physics and Anti-Physics.* Reading, Mass.: Addison-Wesley Pub. Co., 1970. An account of physics and scientific thinking for the non-physicists

[28] N. Bourbaki is not a person but has some 20 volumes of mathematics to his name. For a brief period, Bourbaki was something of an inside joke and trade secret among mathematicians. See "Nicholas Bourbaki" by P. R. Halmos in *Mathematics in the Modern World* (San Francisco: W. H. Freeman & Co., 1968) or in the May 1957 issue of *Scientific American* for an insider's biography.

with dialogues between humanist and physicist on the meaning and worth of science, and some good explanations of some key problems and viewpoints of physics.
2. Bell, E. T.: *Mathematics: Queen and Servant of Science*, New York: McGraw-Hill, 1951. Bell has a talent for making the reading of mathematics really interesting.
3. Bell, E. T.: *Men of Mathematics*, New York: Simon & Schuster, 1937. One book on the human side of the world's great mathematicians.
4. Bochner, S.: *The Role of Mathematics in the Rise of Science*, Princeton: Princeton University Press, 1966. This is a book not "in mathematics but about mathematics," to quote the author, and provides some depth and detail related to the ideas presented in this Chapter and others besides. If you find interest or worth in an overall view of science and mathematics, this is a book worth looking over, for it attempts to "explicate how deep is the mystery ··· of the entry of mathematics into science."
5. Boyer, C. B.: *History of Calculus and Its Conceptual Development*, New York: Dover Publications, 1959.
6. Boyer, C. B.: *A History of Mathematics*, New York: John Wiley & Sons, Inc., 1968. A good history of mathematics that almost reaches the present.
7. Dinkines, F.: *Abstract Mathematical Systems*, New York: Appleton-Century-Crofts, 1964. This book presents an axiomatically founded study of the fundamental algebraic structures, groups, rings and fields, with the intent of introducing the user to modern mathematics.
8. Eves, H.: *An Introduction to the History of Mathematics*, Ed. 3, New York: Holt, Rinehart & Winston, 1969. Another good history.
9. Eves, H., and Newsom, C. V.: *An Introduction to the Foundations and Fundamental Concepts of Mathematics*, New York: Holt, Rinehart & Winston, 1965. Chapter VI describes the axiomatic method and related aspects, including some detailed examples.
10. Grattan-Guinness, I.: *The Development of the Foundations of Mathematical Analysis from Euler to Riemann*, Cambridge: MIT Press, 1970. This is a deeper and technically more demanding account of the ideas we have but mentioned, and the problems we have encountered in Section 2.8.
11. Kerschner, R. B., and Wilcox, L. R.: *The Anatomy of Mathematics*, Ed. 2, New York: The Ronald Press Co., 1973.
12. Kneebone, G. T.: *Mathematical Logic and the Foundation of Mathematics*, New York: D. Van Nostrand Co., 1963. A more sophisticated discussion of mathematics and its applications.
13. Kramer, E. E.: *The Nature and Growth of Modern Mathematics*, New York: Hawthorn Books, Inc. 1970. An awesome encyclopedic account of mathematics and its development.
14. Lieber, L. R.: *Infinity*, New York: Holt, Rinehart & Winston Inc., 1953. Mathematics need not always be technical and slow in order to be interesting mathematics! This is a beautiful math book, and a pleasure to read, but sometimes just a bit corny—1953 you know!
15. Manheim, J. H.: *The Genesis of Point-Set Topology*, New York: The MacMillan Co., 1964. A tour of the mathematics and physics of the previous centuries which led to the invention of the 20th century field of point-set topology. Again, the reader will obtain a broad view and some details of how mathematics develops, and also learn a good bit about the concepts of calculus, no doubt coming to a better understanding of them. This book and topology itself is a study of limit processes.
16. *The Mathematical Sciences: A Collection of Essays*, Cambridge: The MIT Press, 1969. Published for the National Research Council of the National Science Foundation, this is a collection of articles by working mathematicians attempting to explain the context and import of present day mathematics, with no assumption of familiarity or technical ability on the reader's part—but such ability helps!
17. *The Mathematical Sciences: A Report*, Washington, D.C.: National Academy of Sciences publication 1681, 1968. The socio-economic, academic-political and technological aspects of mathematics in the United States—near the date it was published and already under dispute.

18. *Mathematics and the modern World*, San Francisco: W. H. Freeman & Co., 1968. A really good book about mathematics and its uses, providing an answer to the complaint: "What's this good for?"
19. Meschowski, H.: *Evolution of Mathematical Thought*, San Francisco: Holden-Day Inc., 1965. This book will give one a good idea as to what mathematics is about and what its philosophical problems are.
20. Newman, J. R. (Ed.): *The World of Mathematics*, New York: Simon & Schuster, 1956. Four volumes on mathematics in almost all its aspects—mathematics in music, art, literature, games, warfare, logic, infinity, probability, the physical world, etc.
21. Russell, B.: *Introduction to Mathematical Philosophy*, New York: Simon & Schuster, 1971. Any literate person should read something Russell has written, sometime. It need not be on mathematics!
22. Saaty, T. L. (Ed.): *Lectures on Modern Mathematics*, 3 Vols, New York: John Wiley & Sons, Inc., 1967. A series of introductions to the mainstream areas of modern mathematics.
23. Saaty, T. L., and Weyl, F. J. (Eds.): *The Uses and Spirit of the Mathematical Sciences*, New York: McGraw-Hill, 1969. This is a collection of articles, all nontechnical, which attempt to provide a feeling for the uses and purposes of mathematics. These articles should give one an idea of the behind-the-scenes mathematical activity in our culture and provide some ideas as to what training in mathematics is good for.
24. Stabler, E. R.: *An Introduction to Mathematical Thought*, Reading, Mass.: Addison-Wesley, 1953. A nice little book.
25. Stein, S. K.: *Mathematics: The Man Made Universe*, San Francisco: W. H. Freeman & Co., 1963. A freshman level introduction to the spirit of mathematics.
26. Stoll, R. R.: *Set Theory and Logic*, San Francisco: W. H. Freeman & Co., 1963. This text as a whole is a more mathematically sophisticated and formal discussion of logic, axiom systems and the theory sets than presented herein.
27. Tietze, H.: *Famous Problems of Mathematics*, Baltimore: The Graylock Press, 1969.
28. Ulam, S.: *Problem in Modern Mathematics*, New York: Interscience, 1960. This one goes deep, but you will get an idea of how much mathematics there simply is, by scanning its pages and reading here and there. It's a shame to say it this way, but this book will cause you to realize the little bit of the subject you are actually aware of.
29. Kac, M., and Ulam, S.: *Mathematics and Logic: Retrospect and Prospects*, New York: Frederick A. Praeger Publishers, Inc., 1968. The title is misleading in the sense that this book is a tour of many aspects and problems of mathematics. Worth looking over is the chapter entitled "Relations to Other Disciplines."
30. Whitehead A. N.: *Science and the Modern World*, New York: The Macmillan Co., 1925. Old, but still good, Chapter 2 discusses mathematics in the history of thought.
31. Wilder R. L.: *Introduction to the Foundations of Mathematics*, New York: John Wiley & Sons, 1952. Chapters I and II present the axiomatic method and examine it in more sophisticated detail.
32. Witter, G. E.: *Mathematics: The Study of Axiom Systems*, New York: Blaisdell Publishing Co., 1964.

PART 2

"The only doubt is whether I shan't someday suddenly be overwhelmed by the passion for things that are eternal and perfect, like mathematics. Even the most abstract political theory is terribly mundane and temporary." Lord Bertrand Russell, shortly before his dismissal from Trinity College and imprisonment for opposition to British war policies, 1916

THE STRATEGIC ATTACK IN MATHEMATICS

At the International Congress of Mathematicians held in Paris in 1900, the eminent German mathematician David Hilbert presented the now famous list of 23 questions he believed to be fundamental to the development of mathematics in this century. Confronting these questions and armed with Hilbert's view that mathematics should ever be forward looking and completely autonomous of all disciplines, absolutely free to choose new concepts and procedures with only the requirement of consistency, mathematicians changed the face, emphasis, and approach of mathematics from that of all previous centuries. Mathematics in this century has undergone such fundamental changes in interest and method as to easily rival that of any still viable institution. As a result there has been a veritable explosion of mathematical knowledge, and despite the discriptive language of conflict, once the momentary trauma induced by Cantor's theory of sets had subsided, these changes took place quietly and with decorum, yet completely altering the form and fabric of mathematical exposition. In a word, mathematics became a study of abstract *structure*, an art of structuring reason, the fundamental tool being Cantor's theory of sets. Instead of studying a particular kind of number, equation or function, mathematicians now habitually undertake such studies in the context of sets of numbers, sets of possible solutions to an equation, sets of functions. In a single phrase, one *structures* his reason about the particular in the *context* of the general.[1]

That in itself is not new! In ancient geometry a particular question about (say) a right triangle was considered in a *context* larger than the points and lines of the triangle itself, involving ideas of lines, circles, etc., and, in such a context, answered. What is new to the 20th century is that this approach became common in almost all areas of mathematics.

[1] This approach is, according to P. R. Halmos, like judging an individual by the company he keeps.

Wishing to know something about a particular kind of continuous function, today's mathematician may choose to view this function as but a single element in the collection of all functions which are continuous. This viewpoint has its origins in the dissertation of Maurice Fréchet in 1906. Fréchet introduced the idea of an *abstract space* whose points could be numbers, points in ordinary Euclidian space, functions, lines or—almost—whatever. Up to that time, the only useful notions of space had been those closely aligned to that of the ordinary three-dimensional space we claimed for so long to reside in. Fréchet's theory was a study of those properties of a completely arbitrary collection of points in a "space" which hold *independently of their particular nature*, whether they be number, points, functions, etc. From this resulted the concept of a topological space that we will study. As will be seen, topology is a rather neat abstract theory of space which provides a precise context for the study of ideas that could not fit into a concept of space in the rigid Euclidian mold.

In part 2 of this text we are going to parallel, in microcosm, the development of mathematics in this century as best we can. We will begin with a single main question, pose another closely related to it, and try to answer these in the best possible way. In seeking the answers we will develop several fundamental mathematical structures which underlie much contemporary mathematics. And that is the real goal, to generate an introduction to the fundamental concepts and methods of mathematics.

Here is the first of these questions. Consider the unit interval $S = [0, 1]$ of all real numbers greater than or equal to 0 and less than or equal 1. You were encouraged in ealier mathematics courses to think of S as a "connected" line of length 1 as indicated in the Diagram.

Remove the point at 1/2 and let $S_1 = [0, 1] - \{1/2\}$. Surely, the length, $\ell(S_1)$, of S_1 is still 1 for it could not be less than 1. Now remove the point at 1/4 and let $S_2 = [0, 1] - \{1/2, 1/4\} = S_1 - \{1/4\}$; surely, $\ell(S_2) = 1$. Now remove the point at 3/8, letting $S_3 = S_2 - \{3/8\}$; surely $\ell(S_3) = 1$. Remove now the points at 5/16 and 7/16 from S_3 and let $S_4 = S_3 - \{5/16, 7/16\}$. Surely $\ell(S_4) = 1$. Let $S_5 = S_4 - \{9/32, 11/32, 13/32, 15/32\}$; surely $\ell(S_5) = 1$. We can *conceive* of continuing this process, for $n = 6, 7, 8, \ldots,$ times, each time removing the midpoint of each segment of S_{n-1}. Let S_∞ denote what is left when all the points removed in the indicated manner are gone. Here is a version of the question we intend to ask: should S_∞ appear as indicated in the diagram? If so, we find $\ell(S_\infty) = 1 - (1/2 - 1/4) = 3/4$, whereas $\ell(S_1) = \ell(S_2) = \ldots = \ell(S_n) = \ldots = 1$!

The reader who is familiar with the material in the latter part of Section 2.8 knows that the answer to this particular question is "no" because all the numbers removed are rational and hence there must be some, the irrational ones, remaining between 1/4 and 1/2. The question we wish to raise is more general than the above and considers the *process* by which the set S results.

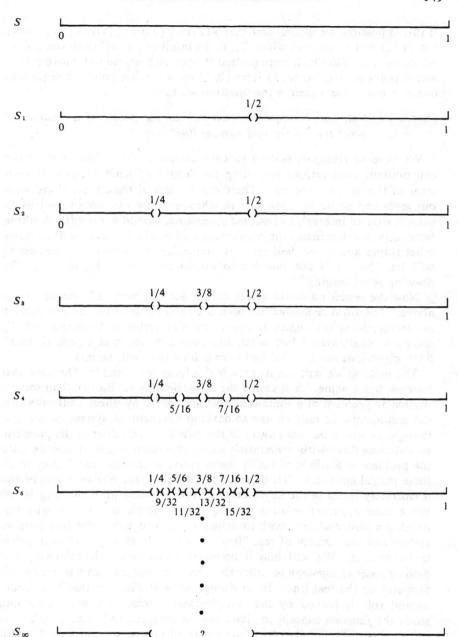

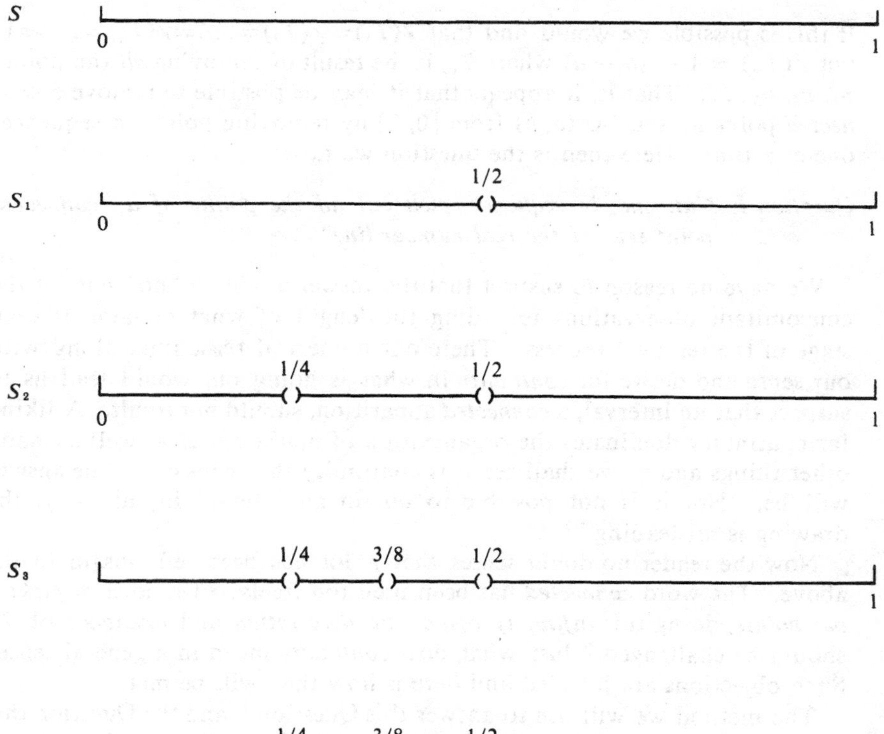

The pictures above indicate that it might be possible to begin removing points $p_1, p_2, p_3 \ldots$ from the set S obtaining sets $T_1 = S - \{p_1\}$, $T_2 = S - \{p_1, p_2\}$, $T_3 = S - \{p_1, p_2, p_3\} \ldots$, in such a way that, when we are through, we will have removed an entire interval (a, b) from $S = [0, 1]$.

If this is possible we would find that $\ell(T_1)=\ell(T_2)=\ldots=\ell(T_n)=\ldots=1$, but $\ell(T_\infty)=1-(b-a)$ where T_∞ is the result of removing *all* the points $p_1, p_2, p_3 \ldots$. That is, it appears that it may be possible to remove a *connected* point set such as (a, b) from $[0, 1]$ by removing points in sequence, one at a time. Here then is the question we raise:

Question I: Can one, in sequence, pick out all the points of a "connected point set" of the real number line?

We have no reason to suspect that the answer might be "no" but for the concomitant observations regarding the length of what remains at each stage of the removal process. There our numerical reasoning, along with our sense and desire for *continuity* in what is going on, would lead us to suspect that an interval, a *connected* apparition, should not result. A liking for continuity dominates the organization of mathematics as well as many other things and as we shall see it is continuity that wins out. The answer will be, "No, it is not possible to obtain an interval in this way, the drawing is misleading."[2]

Now the reader no doubt senses that a lot has been left unsaid in the above. The word *connected* has been used too freely. The idea of *picking out points*, doing this *infinitely often*, the *description* and *existence* of T_∞ should be challenged. Just what does *continuity* mean in a general sense? Such objections are justified and here is how they will be met.

The method we will use to answer this Question I, and the Question that follows, has a name. It is called the *strategic attack*. Rather than solve a particular problem of a somewhat abstract nature by direct confrontation, the mathematician may choose to develop a structured system of abstract thought in which he picks away at the essential difficulties of the problem, surrounding those with penetrating ideas, characterizing their essence until the problem as a whole virtually melts away, becoming "easy" prey to all these partial advances. To describe it another way, one very slowly and deliberately boxes in the key aspects of the problem until nothing is left but a most *apparent solution*. Using the "strategic attack" we will first develop a *theory of sets*, both to define T_∞ and to study the real number system and the concept of real "line" that we use as a geometrical representative of it. We will find it necessary to introduce the relatively new field of *pointset topology* in order to define "connected" and to study this property on the real line. In so doing, we will discover that the fundamental role is played by the completeness axiom. We will define and study the *function* concept to give a precise meaning to "picking out." As to the idea of "infinitely often," encountered before, we will find it useful to study it from a new perspective, Cantor's beautiful theory of *infinite*

[2] This conclusion is one of the most interesting of Cantor's original theory of sets. Although it can be settled (quickly in fact) by arguments involving infinite decimal expansions, we will intentionally take a longer and hopefully more illuminating approach.

sets and *cardinal numbers*. In short, just as mathematics in this century witnessed the rise of extensive systems of thought in the hope of answering Hilbert's brief list of questions, we will follow the same road and hope to give the reader a fair idea of the manner and method of contemporary mathematical thought.

There is another noteworthy aspect of contemporary mathematical thought which we will illustrate by asking a second question. When an extensive abstract system of mathematical thought is developed it is often possible to raise and answer, or simply to answer, questions for which the system was not even specifically designed. Such is the case here. Question I raises the question of "constructing" a connected set by a picking out process. Since we will have a concept of connectedness at hand in dealing with this question, we will ask another question, related to it, but in an opposite direction.

Question II: *Given a connected object, to what extent can it be changed, distorted or transformed in such a way that the resulting object will still be connected*?

An answer to Question II will not require much more mathematics than Question I, and its solution will fall out in an even easire manner though its discussion will be carried on a much more abstract level. An answer to Question II will entail a discussion and broad conceptualization of the one word used in discussing Question I that we have not yet mentioned again: *continuity*. Continuity, although used in generating Question I, is not material to its solution. But continuity is one of the most studied concepts in mathematics, an understanding of it is quite worthwhile, and in answering Question II we will come to a completely general description of this fundamental concept simply by putting together a good deal of the material needed for Question I.

There is a pedagogical point to all of this, too. The reader will witness the natural emergence of quite an abstract body of mathematics from the need to answer some particular questions that can be viewed as rather concrete. It is often not apparent in many a mathematics course that the theory originated from an intended application, though it almost surely did. The limits of time and the sheer volume of what one needs to know often preclude a clear exposition of how the polished theory arose from its origins. Without faith, one could regard the theory as but more abstraction of no particular use, and a decisive aspect of mathematics has been lost. (A good example is the subject of linear algebra, a subject of extremely wide practical application, but exceedingly abstract in appearance.) As noted in Chapter 2, mathematics is abstract because that is a very good way to do useful mathematics; hopefully, the problems we consider will provide a good example of the natural course to abstraction that mathematics does habitually follow.

The intent of subsequent chapters is to introduce the reader to the fundamental concepts and methods of mathematics. Question I, and subsequently Question II, only provide motivation and a link between the various ideas. Regarding the methods of modern mathematics, one aim is to induce the reader to prove some of the results needed to answer these questions. The proofs of all the essential results which answer Question II are left to the reader as well as many of those for Question I. The path to finding these proofs is 2-fold: push and develop an intuitive feeling for the essential concepts, and verify your intuition by mathematical logic. Impose the logic of your mind on the logic of your sences to obtain a certain conclusion. You should find that neither aspect can be slighted at the expense of the other. Regarding the concepts to be discussed, the universal importance of set theory and the abstract notion of a function between sets will cause us to explore these topics largely without mention of the original questions, and similarly, for a large part of the study of counting and infinite sets. Important, too, will be a host of smaller investigations

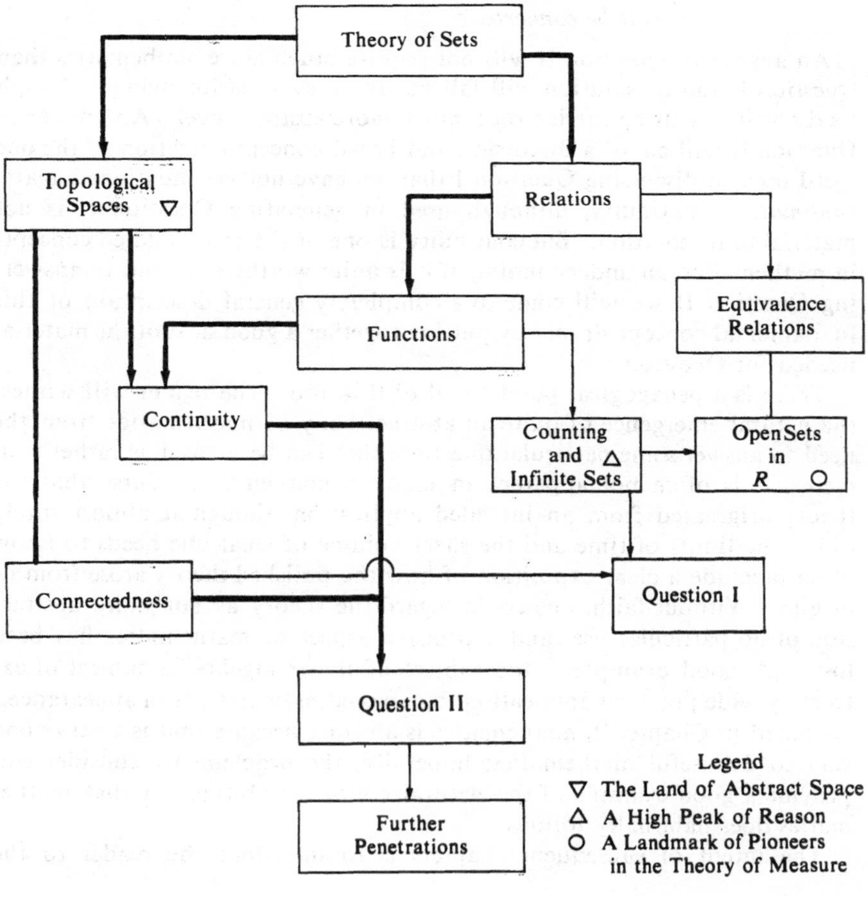

undertaken in structuring this body of knowledge. In consequence, in order to understand the whole story, it will be necessary to stand off a bit from time-to-time and look at the whole of what is going on—to get out from under the trees of sets, functions, infinity, connectedness, separation, relations, closures and the other topics we will consider and look at the whole forest of mathematical thought that will be growing around you. If not, it will form truly an impenetrable forest in your mind. A map for the wayfaring stranger, with the main roads more heavily marked, is provided in the accompanying diagram.

chapter 3

"We need elucidation of the obvious more than investigation of the obscure."—Judge Oliver Wendell Holmes

A FORMALLY INFORMAL THEORY OF SETS

1. INTRODUCTION

The impact of the theory of sets on all of mathematics bids fair support for Holmes' observation. Now considered to be perhaps the most fundamental concept of mathematics, the concept of a set originated only in the 1890s with the work of the German mathematician Georg Cantor with his definition of a set as "a collection of definite distinguishable objects of one's intellect, thought of as a whole." Such an obvious thing: a *collection* of separate things *to be thought of as an entirety.* The investigation of this obvious idea, spurred on by Russell's paradox, has done more to advance mathematics at a single blow than perhaps any other concept yet discovered. The theory of sets is still unsettled to this day and, as Russell showed, this "obvious" concept can lead to quite a contradiction.

There is another interesting historical point about set theory. As abstract as it can be at times, and must be to avoid contradiction, it arose from Cantor's study of Fourier series, a topic rich in the applications of mathematics. The French mathematician Jacques Fourier invented the series that bear his name in order to study heat conduction in solids. Today they are an indispensable tool in electrical engineering. To understand the essential nature of Fourier series approximation (Question I indeed is related to certain problems of Fourier approximation), one must make use of a quite extensive theory of sets, as well as several other theories which would not exist without the theory of sets.

The use of set theoretic concepts pervades all of mathematics and in a most natural way. In Sections 2.6, 2.7 and 2.8 the concept of a collection of numbers was nearly indispensable to an organization and establishment of much of the theory of the real and natural numbers. Moreover, the use of collections of numbers occurs somewhat naturally once one has the idea of using collections at all. Recall especially the process for obtaining the natural numbers as a subclass of the axiomatized real numbers using the concept of a $+1$-class. In that instance, the notion of classes or sets is more than an organizational-technical tool, becoming an integral part of the whole idea of the natural numbers.

But the profound influence of set theory on present day mathematics does not lie solely in its all pervasive use. Perhaps its greatest influence is the result of its initial simplicity in conflict with the almost bottomless pits of nonsense it can lead to if one is not careful. Mathematicians, in trying to avoid a useless theory, have been driven to standards of rigor and precision of expression undreamed of less than a century ago. This in turn has led to deeper understanding and, more importantly, insight into the essential problems of mathematics. Since understanding mathematics is all that mathematicians do, this has been all to the point.

It is not the purpose of this text to evolve an absolutely uncontradictory theory of sets upon an axiomatic basis.[1] Its purpose is to evolve a theory of sets sufficient unto mainstream mathematics, in particular the subsequent subject matter of this text, upon a mildly intuitive axiomatic basis. There will be a hint of the rigor of modern set theory, but that is all. If you desire a thorough approach, or would simply like to know more about set theory, see the lists of textbooks at the close of this chapter. A rigorous discussion and development of modern set theory would entail a degree of abstraction that is simply too much for the novitiate in mathematical thought. One great difficulty in a rigorous development of set theory is that its basic concepts so emulate the ordinary use of language and logic, that the use of any words at all in the theory can throw it off. For example, the mindless trivia

All dogs are animals

can be exalted to an incalculably more impressive, if no more profound place in the sun, by restatement in the form

$$x \text{ in } D \to \text{ in } A$$

where D is the set of dogs and A the set of animals! In a thoroughly rigorous theory one cannot be too careful in using sets, lest one lose track of where grammar ends and set theory begins and vice versa.

We will not get involved in such difficulties because we will not insist upon a completely rigorous theory of sets. For our purposes, the word *set* will be an undefined term which we will think of as a collection of things—points, numbers, lines, triangles, etc.—drawn from some fixed class of "distinguishable concepts of one's intellect"—a ***universal set*** for the particular discussion. For example, if we wish to consider sets of real numbers, then a universal set for such a discussion will simply be the set consisting of all the real numbers. This requirement in itself avoids some of the simpler paradoxes, as long as one does not allow the universe to be the set of all

[1] A most interesting, informative, and brief, discussion of axiomatic set theory, with a complete list of axioms can be found in the article "Non-Cantorian Set Theory" by P. Cohen and R. Hersch in the December 1967 issue of *Scientific American*.

sets, whatever that is![2] We are more concerned with using set theory than in developing it, and when we will have need of it we will already have at hand a fixed collection (set) that concerns us at the moment. This collection will be a universal set within which the set theory we do develop can be applied. We will shortly make this convention more precise.

What good does it do to have a collection unless one can discuss the things in it? None, lest these collections be simply left to vegetate. Thus we suppose a second undefined notion of "belonging to," "is in" or "being in" a collection. More precisely, we suppose an undefined term *element* and an undefined symbel ε, to denote the concept of "is in," and we speak of an "*element* which *is in a set.*" More precisely still, if we use a letter S to denote a set, and a letter x to denote a element, we will write

$$x \in S$$

read, "x is in S," and thought of as

"*x is* an *element in* the *set S.*"

You can see from this discussion how simply and naively a set theory can and did arise. If one wishes to think of a collection S consisting of the numbers 1, 2 and 3 one is likely to wish to think of 1, 2 or 3 as being *in* the collection.

Finally we impose on all of this the logic discussed in Chapter 1, and are ready to state the first two axioms of our informally formal set theory.

AXIOM 1. There exists a set U, called the universal set, such that $\forall$ set S, if $x \in S$ than $x \in U$.

In plain English, the elements of all sets to be discussed are to be elements of a supposed "universal" set.

AXIOM 2. For $x \in U$ and S a set, the statement $x \in S$ is a propositional function—i.e., it has a truth-value of 0 or 1 and not both, for any element x.

Again, in plain English, if S is a set and x is an element, then either x is in S or x is not in S and not both; a small matter, but surely fundamental! If the statement $x \in S$ is false, we will use the the notation $x \notin S$, and this statement is read "x is not in S" which we will regard as expressing the true statement: not $x \in S$.

The second part of Cantor's notion of set, and the whole point of set theory, is that the collection of definite distinguishable concepts is to be "thought of as a whole." In other words, a set is determined solely by

[2] See H. Eves and C. V. Newsom: *Introduction to the Foundation and Fundamental Concepts of Mathematics*, p. 296 (New York: Holt, Rinehart and Winston Inc., 1969) or R. L. Wilder: *Introduction to the Foundations of Mathematics*, p. 104 (New York: John Wiley and Sons, Inc., 1952).

what's in it—solely by its elements. If two sets, S and T, have the same elements, then they are to be considered *equal*—a new undefined term to be denoted by $=$, read as usual. In precise, if not plain, English and logic

AXIOM 3. For sets S and T,

$$S = T \text{ iff } (\forall x)[(x \in S \rightarrow x \in T) \wedge (x \in T \rightarrow x \in S)$$

or equivalently

$$S = T \longleftrightarrow \forall x, (x \in S \longleftrightarrow x \in T)\,.$$

This does say things precisely, and, moreover, here is a key point: this axiom gives a rule for deciding when two sets S and T are equal. One must be able to show that the two implications $x \in S \rightarrow x \in T$ and $x \in T \rightarrow x \in S$ are true implications, and makes the matter of set equality one of logic about the relation of "belonging to."

One immediate consequence of this axiom is that in listing the elements of a set one does not care either to repeat any element in the list or to list the elements in any particular order. Thus the sets $S = \{1, 3, 7, 5\}$ and $T = \{3, 1, 5, 7\}$ are equal; one also regards the set $V = \{1, 3, 1, 5, 3, 7\}$ as being equal to both S and T. All are equal because all have the same elements, or, in accord with Axiom 3, if $x \in S$ then $x \in T$ and $x \in V$. And if $x \in V$ then $x \in T$ and $x \in S$. Finally, if $x \in T$ then $x \in S$ and $x \in V$, and hence it is consistent with Axiom 3 to write: $S = T = V$.

Let us take another example. Suppose U is the collection $\mathscr{N}$ of all natural numbers, S is the collection of all odd numbers and T is the collection of all numbers of the *form* $2n + 1$ where $n \in \mathscr{N}$. Then if $x \in S$, then x is odd. If one divides x by 2, one must have a remainder and this remainder must be 1. If n is the quotient, then $x = 2n + 1$ by definition of quotient and remainder. Hence $x \in T$. Conversely, if $x \in T$, then $x = 2n + 1$ for some choice of n and hence 2 cannot divide x evenly. Hence x is odd, or $x \in S$. Thus both implications $x \in S \rightarrow x \in T$ and $x \in T \rightarrow x \in S$ are true and one would be playing by the rule of Axiom 3 to write: $S = T$.

You may have noticed certain sleights of mind in that example. Specifically: S is the collection of all odd numbers, T is the collection of all numbers of the form $2n + 1$. These statements assert the *existence* of sets, with no stated justification. This will not do! Why? The Russell paradox is a good example of what casual existence can lead to. If one does not admit the *existence* of the set of all ordinary sets, one has no paradox to speak of!

As a matter practical to this text, existence of sets comes down to this. Given a universe of discussion, some statements may perhaps be true of elements in that universe. We need only assume that, for each statement a set consisting of all those elements for which the statement is true does exist. For example, if $\mathscr{N}$ is the universe and $p(n)$ the statement

$$p(n)\text{: } n \text{ is an even number}$$

we wish to assume that a set E consisting of all n such that $p(n)$ is true does exist—in short, a set E consisting of all even numbers. This is the nature of the final axiom we need for a simple theory of sets.

Before stating it, here is some conventional notation. The set E above would be described as

$$E = \{n\text{: } n \text{ is an even number}\}$$

or in terms of $p(n)$,

$$E = \{n\text{: } p(n)\}$$

read "E is the set of *all* $n \in \mathcal{N}$ such that n is an even number." The "{ }" are read "set of all," the ":" is read "such that." The sets S and T defined above would be written

$$S = \{n\text{: } n \text{ is an odd number}\}$$

$$T = \{k\text{: } k = 2n + 1 \text{ for some } n\}$$

or

$$T = \{k\text{: } k \text{ is of the form } 2n + 1 \text{ for some } n \in \mathcal{N}\},$$

this last more clear, but also more lengthy. Here now is

> AXIOM 4. If $p(x)$ is a propositional function whose indeterminate is to be interpreted as an element of U, then $\exists$ a set M consisting of those, and only those elements of U for which $p(x)$ is true. More precisely, $\exists$ a set M such that $x \in M \leftrightarrow p(x)$ has a truth-value of 1.

Along with this axiom we adopt the conventional notation $M = \{x\text{: } p(x)\}$, read "M is the set of all x (in U) such that $p(x)$ is true." If it is necessary to specify the universal set, this will be written $M = \{x \in U\text{: } p(x)\}$: "the set of all $x \in U$ such that $p(x)$ is true."

There are several clinkers in this axiom, some subtle and some not so subtle. The subtle ones are concerned with the kinds of propositional functions to be admitted for use in forming sets—again the Russell paradox is a liability to keep in mind. As we will use set theory, such subtleties will not be a problem, but do not be misled. Our definition of a propositional function in Chapter 1 was not a model of rigor, and despite the appearance of rigor in Axiom 4, it is no more rigorous than the concept of a propositional function.[3] For a less subtle difficulty, suppose $U = \mathcal{N}$ and

$$p(n)\text{: } n \text{ is an even number which is the sum of two primes.}$$

Deciding what is in and what is not in $M = \{n\text{: } p(n)\}$ would be equivalent

[3] For a discussion of this difficulty, which plagued those who first sought a firm foundation for set theory, see the nice little text by P. R. Halmos: *Naive set Theory* (New York: Van Nostrand, Reinhold Co., 1960).

to solving the 2-century-old Goldbach conjecture, that every even number is the sum of two primes. You can verify that 2, 4, 6, 8 and more even numbers yet are in M, since $2 = 1 + 1$, $4 = 2 + 2$, $6 = 3 + 3$, $8 = 5 + 3$, and so on, but no one has yet determined whether or not M contains all the even numbers. In other words, Axiom 4 assumes the existence of sets whose elements may not be determinable, or have not yet been determined.

It also asserts the existence of a bit of nothingness, the empty set. Like zero, this set is there and counts, but not really. It is more a handy concept, convenient of use, simplifying communication. The idea is that of a set containing no elements at all, a seeming denial of what one would think a collection should be. But such are the consequences of abstraction, so here it is.

THEOREM 3.1.1. *If there exists a set, then there exists a set $\square$ having no elements at all. In other words, $x \in \square$ is a false statement for any $x \in U$.*

Proof. Suppose a set exists and call it M. Let $p(x)$ be the statement $x \in M \wedge -(x \in M)$. By Axiom 2, $p(x)$ is false for each $x \in U$. Let $\square = \{x\colon p(x)\}$. This set exists by Axiom 4, and also by Axiom 4, $x \in \square$ iff $p(x)$. That is, $x \in \square$ is false for any x.

Now, as we have been pained to point out, and you have perhaps been pained to witness, mathematics is abstract. Suppose $q(x)$ is a proposition about the elements of U which, like the statement $p(x)$ used to define the empty set $\square$ in Theorem 3.1.1, is always false. Do the rules of the game say that $\square = \{x\colon q(x)\}$? This latter set exists by Axiom 4, so the question is legitimate. Moreover it is important. If for example $U = \mathscr{N}$ and $q(n)$: $n + 2 = 1$, and $p(n)$: $n + 3 = 1$, one would certainly like it that $\{n\colon n + 2 = 1\}$ and $\{n\colon n + 3 = 1\}$ be the same empty set, if the concept is to be of any use. This is the point of the next result, which is left for you to prove, with the reminder that you must use Axiom 3 to determine set equality; and perhaps a convention of formal logic.

THEOREM 3.1.2. *If $q(x)$ as well as $p(x)$ are propositional functions which are both false for all interpretations of x in U, then $\{x\colon q(x)\} = \{x\colon p(x)\}$.*

In plain English, there is only one empty set, which we henceforth denote by $\square$.

This section will close with a listing of a few other conventions of set notation. If a set A has (say) exactly 10-elements it is most natural to denote these elements by $a_1, a_2, \ldots, a_{10}$ and write $A = \{a_1, a_2, \ldots, a_{10}\}$. We do not mean here to imply that any order is given to the elements of A. We could equally well write $A = \{a_7, a_6, a_{10}, a_9, a_8, a_1, a_2, a_3, a_4, a_5\}$. If A has only one element, call it a, then $A = \{a\}$ and A is called a singleton set. Notice that a and $\{a\}$ are conceptually different things; the first is an element, the second is a set and trouble will come if they are treated as the same—again the Russell paradox bears witness to the importance of such conceptual distinctions. Thus writing $a \in \{a\}$ is correct, but writing $a = \{a\}$ will bring on the wrath of the gods of mathematics, for this last is mixing apples and pears, oil and water, and similar things. Finally, along these lines, if A is a set and $x_1 \in A, x_2 \in A, \ldots$, and $x_n \in A$, we will write $x_1 x_2, \ldots, x_n \in A$. If A consists only of the elements $x_1, x_2, \ldots, x_n$ (here n is a natural number), we will as before write $A = \{x_1, x_2, \ldots, x_n\}$. These subscripts are no more than a convenient way of keeping track of things.

Problems.

(1) Let U be the collection $\mathscr{R}$ of real numbers. For each statement $p(x)$ below use the notation adopted above to describe the set $A = \{x: p(x)\}$ in at least two ways.
 (a) $p(x): x^2 = x$
 (b) $p(x): (x - 1)^2 > 1$
 (c) $p(x): x^2 < 0$

(2) Let the universe of discussion be the set U of real numbers between 0 and 1 used in the statement of Question I. Describe the sets $T_1, T_2, T_3, \ldots, T_n$ therein using Axiom 4. Can you describe the set T_∞ using Axiom 4?

2. FUNDAMENTAL SET OPERATIONS

Ninenty-nine percent of the manipulative uses of set theory involve the notation and concepts of this section. In other words, time spent on this section will pay off well. The main ideas are concerned with obtaining new sets from old. All of the axioms play a fundamental role in defining these new sets and relating them to one another. These may be formed quite differently, yet in the end be the same, and such is the way the mathematician avoids unnecessary thought, by establishing the important relations, once and for all.

The three principle operations for forming new sets may be described as putting together, cutting down and throwing out. On the plane of accepted appearance and exposition, which every institution must put up with, these are known as union, intersection and complimentation. These words then replace the idea, it all sounds more imposing, etc., but the idea is what counts. The union of sets, the putting together, is just what the latter words imply. The elements of two or more sets are put together to make a new, perhaps larger set. The intersection, the cutting down of two sets, is that set which results when one considers as a set only the elements in both sets simultaneously. The compliment of the set B in the set A is that set which results in throwing out of A those of its elements which are also in B. If none of the elements of A are in B, nothing is thrown out. More precisely,

DEFINITION 3.2.1. Let A and B be sets. The sets $A \cup B$, $A \cap B$ and $A \backslash B$, read "A union B," "A intersect B" and "A slash B" or "the complement of B in A" are defined

(a) $A \cup B = \{x: x \in A \vee x \in B\}$
(b) $A \cap B = \{x: x \in A \wedge x \in B\}$
(c) $A \backslash B = \{x: x \in A \wedge x \notin B\}$.

The validity of these definitions and the existence of the sets defined resides in Axioms 2, 3, and 4. Axiom 2 justifies the notion that $x \in A \vee x \in B$, $x \in A \wedge x \in B$, $x \in A \wedge x \notin B$ are propositional functions. Axiom 4 says that the sets so defined exist. Axiom 3 makes the respective equalities meaningful. For example, $x \in A \cup B$ iff $x \in A$ or $x \in B$. Thus $A \cup B$ is the putting together of the elements of A with those of B into a collection $A \cup B$ to be thought of as a whole. Here now is a special definition for a special case of complimentation.

DEFINITION 3.2.2. If A is a set, the *complement* of A is the set $U \backslash A$, where U is the universal set.

Thus $U \backslash A$ is the set of all those elements in U which are *not* in A; or, $U \backslash A$ is U with A thrown out. Here now is a theorem and its proof.

THEOREM 3.2.3. For any two sets A and B,

$$A \cap (U \backslash B) = A \backslash B .$$

Proof. According to Axiom 3 one must show $x \in A \cap (U \backslash B) \leftrightarrow x \in A \backslash B$. Let us first show that $x \in A \cap (U \backslash B) \rightarrow x \in A \backslash B$. If $x \in A \cup (U \backslash B)$ then by Definition 3.2.1(b), the definition of intersection, we have $x \in A$ and $x \in U \backslash B$. By Definition 3.2.1(c), we have $x \in U$ and $x \notin B$. Since $x \in A$ and $x \notin B$, then by Definition 3.2.1(c), $x \in A \backslash B$. This completes the proof of the desired implication.

Now let us show that $x \in A \backslash B \rightarrow x \in A \cap (U \backslash B)$. If $x \in A \backslash B$, then $x \in A$ and $x \notin B$. Since $x \in U$ then $x \in U \backslash B$ by Definition 3.2.1(c). Hence $x \in A$ and $x \in U \backslash B$ and therefore $x \in A \cap (U \backslash B)$ by Definition 3.2.1(b). Hence the proof.

There is a visual way to think about such results of set theory, known as a Venn diagram. If we picture U as all there is, A and B as portions of U, we have a picture as in Figure 3.1. Notice that $A \backslash B$, the doubly crossed portion is exactly those things in $U \backslash B$ (shaded upward) and in A (shaded downward) as expressed in Theorem 3.2.3. Returning to Definition 3.2.1(a), the union of two sets A and B is the entire shaded portions of Figure 3.2, and $A \cap B$ the doubly shaded portion. But such pictures do not *prove* anything; such devices were cast out in the early 19th century, and no one has seen any need to bring them back as *methods of proof*. As explanation and motivation fine, but enough contradictions have resulted so as to demand that one must use the logic of the mind rather than the logic of the senses

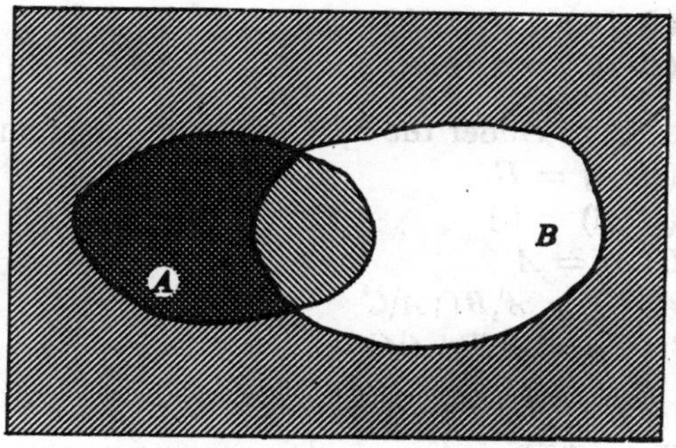

FIG. 3.1.

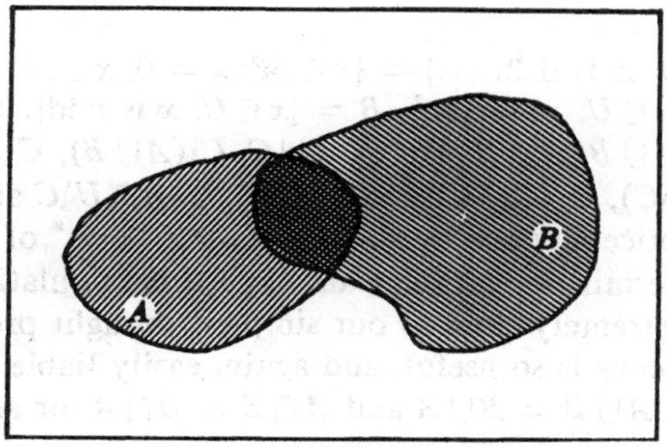

FIG. 3.2.

in proof. On this matter of proof, study carefully the *form* of the proof of Theorem 3.2.3. Note not the logical form, but the organizational form. Notice how everything goes back to some definite else, an axiom or a definition of the terms involved is used. Inscribe in stone the thought: the definition of a term must be used to learn something new about it. There is nothing else to appeal to, for the definition provides the only meaning assigned to the term. If you ignore this, you will not know where to start in proving a theorem.

With the axioms and Definition 3.2.1 and the proof of Theorem 3.2.3 as a model, the following theorems are left to the reader to prove. It is suggested that you draw Venn diagrams to get things straightened out in your mind and then make particular use of Axiom 3 to determine these set equalities.

THEOREM 3.2.4. Let U be the universal set, A, B and C be sets. Then

(a) $A \cap (B \cap C) = (A \cap B) \cap (A \cap C)$
(b) $A \cap (B \cup C) = (A \cap B) \cup (A \cap C)$
(c) $A \cup (B \cap C) = (A \cup B) \cap (A \cup C)$
(d) $A \cup (B \cup C) = (A \cup B) \cup (A \cup C)$

(e) $A \cup (B \cup C) = (A \cup B) \cup C$
(f) $A \cap (B \cap C) = (A \cap B) \cap C$

THEOREM 3.2.5. Under the hypotheses of Theorem 3.2.4,
(a) $A \cup (U \backslash A) = U$
(b) $A \cap (U \backslash A) = \square$
(c) $U \backslash (U \backslash A) = A$
(d) $A \backslash (B \cup C) = A \backslash B \cap A \backslash C$
(e) $A \backslash (B \cap C) = A \backslash B \cup A \backslash C$

THEOREM 3.2.6. For any set A
(a) $A \cup \square = A \backslash \square = A$
(b) $A \cap A = A \cup A = A$

Problems.

(1) Let $U = \{0, \pm 1, \pm 2, \ldots\} = \{x \in \mathscr{R}: x = 0, x \in \mathscr{N} \text{ or } (-x) \in \mathscr{N}\}$. Let $A = \{x \in U: x \text{ is even}\}$, $B = \{x \in U; x \text{ is odd}\}$, $C = \{x: x < 5\}$. Describe $A \cup B$, $A \cap B$, $A \cap C$, $A \cup C$, $C \backslash (A \cup B)$, $C \backslash A \cup C \backslash B$, $C \backslash A \cap C \backslash B$, $A \backslash (A \backslash C)$, $U \backslash (A \cap C)$, $U \backslash A \cup U \backslash C$, $U \backslash A \cap U \backslash C$ and $A \cap B \cap C$.

(2) Do you notice any analogies between the "laws" of sets in Theorem 3.2.6 and certain laws of logic and logical manipulations? Set operations are extremely close to our simplest thought processes. That is why set theory is so useful, and again, easily liable to error.

(3) Prove that $A \cup B = B \cup A$ and $A \cap B = B \cap A$ for any sets A and B.

3. SUBSETS

What with Axiom 4 and Definition 3.2.1, we see that in a given universe there are likely to be lots of sets lying around waiting to be discovered and related to other sets. Sometimes there is no apparent relation, but sometimes an extremely deep and non-intuitive one may be present. For the time being, we will be concerned only with sets that are either equal or not equal and among the latter, try to refine their relationship a bit. For this we need the

DEFINITION 3.3.1. Let A and B be sets. We will say that A is a ***subset*** of B, and denote this by $A \subset B$, if the following implication holds: $x \in A \rightarrow x \in B$. If, in addition, there is an $x \in B$ such that $x \notin A$, we will say that A is a ***proper subset*** of B and denote this by $A \subsetneq B$. If $A \subset B$, we might also write this as $B \supset A$ as well as $B \supsetneq A$ if $A \subsetneq B$.

The idea of subset is *sub* or part of: $A \subset B$ means the elements in A are also in B. A as a whole is a sub-part of B. With that, a pause for reflection. Notice that although the key idea of set theory is to think of a collection of things as a whole, the definitions always come down to element level propositions. A set A is a subset of B if each element of A is an element of B. This is in the nature of the axioms, of course, but more yet to

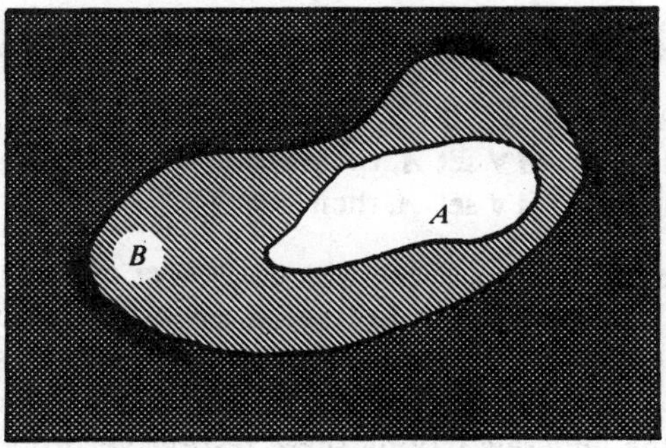

FIG. 3.3.

the point. One may wish to think of A as a part of B, where both are thought of as a whole, but the practical, the straightforward way to initially determine the validity of this thought is on the element level—verify that each element of A is an element of B. Once done, one then can think of A as a part of B with justification. But more yet: as the proofs of the above theorems indicate, *the way* to determine the relationship between sets at this elementary level is on the element level. A good point to keep in mind, it will save you much time and materially further your progress in the sequel.

Again some Venn diagrams. Figure 3.3, where $A \subset B$, suggests a couple of relationships involving the subset concept. These are

THEOREM 3.3.2. If A and B are sets, then
(a) $A \subset B$ iff $U \backslash B \subset U \backslash A$
(b) $A \subset B$ iff $A \cap U \backslash B = \square$

Here is a proof of a part of (a). Suppose $A \subset B$. We must show $U \backslash B \subset U \backslash A$. According to Definition 3.3.1, it must be argued that $x \in U \backslash B \rightarrow x \in U \backslash A$. If $x \in U \backslash B$ then $x \notin B$. Since $A \subset B$ then by 3.3.1, $y \in A \rightarrow y \in B$ and by the contropositive, $y \notin B \rightarrow y \notin A$. Since $x \notin B$ then we know $x \notin A$ and hence $x \in U \backslash A$. Thus $U \backslash B \subset U \backslash A$.

Here is part of the proof of (b), and an indication of how a mathematical theory proceeds and used previous results. We will show that $A \subset B$ implies $A \cap U \backslash B = \square$. Since $A \subset B$ then $x \in A \rightarrow x \in B$. Hence $x \in A$ and $x \notin B$ is false. By Theorem 3.1.2, $\square = \{x\colon x \in A \wedge x \notin B\}$ and by Definition 3.2.1(c), this last set is $A \backslash B$. Hence $A \backslash B = \square$. Substituting now in Theorem 3.2.3, we obtain $A \cap U \backslash B = \square$ which is the desired end.

Here is a basic "truth" of set theory, left for you to prove.

THEOREM 3.3.3. If B is a set, then $\square \subset B$.

This study of basic set theory concludes with the following theorems, the first "obvious" from a Venn diagram, the second likely to cause difficulty.

THEOREM 3.3.4 If $U = A \cup B$ and $A \cap B = \square$, then $A = U \backslash B$.

THEOREM 3.3.5. Let B be a fixed set in the universal set U.
(a) If $A \cap B = A \;\forall$ set A, then $B = U$.
(b) If $A \cup B = A \;\forall$ set A, then $B = \square$.

This latter theorem characterizes the empty set and the universe among all possible sets. The *converses* of these statements are trivial. For the first, if $B = U$ then $A \cap B = A \cap U = A$ for any set A. For the second, if $B = \square$ then $A \cup B = A \cup \square = A$ for any set A. The theorem claims that $B = U$ and $B = \square$ are the *only* sets having these respective properties. Here is a proof of (a), you are left to prove (b).[4]

To prove (a) one is given that $A \cap B = A$ and this is supposed to be true no matter what set A is. One is to conclude that B must be the set U. Suppose $B \neq U$. Then $U \backslash B \neq \square$. Apply the hypothesis that $A \cap B = A$ for *any* set A, to the particular set $A = U \backslash B$. One obtains $U \backslash B \cap B = U \backslash B$. But, according to Theorem 3.2.5(b), $U \backslash B \cap B = \square$. Hence $\square = U \backslash B$. Hence the proof by *reduction ad absurdum*, for we have $U \backslash B \neq \square$ and $U \backslash B = \square$.

Problems.
(1) Prove the remaining parts of Theorem 3.3.2.
(2) Prove Theorem 3.3.3.
(3) Prove that the following are logically equivalent:
(a) $A \subset B$.
(b) $A \cap B = A$.
(c) $A \cup B = B$.
Does it suffice to prove (a) $\rightarrow$ (b) $\wedge$ (b) $\rightarrow$ (c) $\wedge$ (c) $\rightarrow$ (a)? Reminder: Definition 3.3.1 must be used.
(4) Prove that for two sets, A and B, $A = B$ iff $A \subset B$ and $B \subset A$.
(5) Prove that $A \cap B \subset A \cup B$ for any sets A and B.
(9) Prove Theorem 3.3.4
(7) Prove Theorem 3.3.5(b).

4. SET THEORY: SECOND FLOOR

The advantages and uses of the concept of a collection of objects thought of as a whole need not be, and are not, limited to a fixed universal set and subsets of it. It is both natural and especially useful to make use of the idea of a set of sets or, to help make things clear, a collection of subsets of a given universal set. But such things are not subsets of U; their elements, being sets, are not elements of U. These collections of sets occur at what might be called another conceptual level. Consider an example.

Let U be the set $\mathscr{N}$ of natural numbers. Then $\mathscr{N} = A_1 \cup A_2 \cup A_3$ where

[4] Proving a theorem of this type is discussed in more detail following Example 1.14.3.

$A_i = \{n \in \mathscr{N}: n = 3k + i \text{ for some } k \in \mathscr{N}\}$. That is, $A_1 = \{1, 4, 7, \ldots\}$, $A_2 = \{2, 5, 8, \ldots\}$, and $A_3 = \{3, 6, 9, \ldots\}$. For one reason or another one may wish to consider as a set the collection $\mathscr{A}$ of sets A_1, A_2, A_3. That is, $\mathscr{A} = \{A_1, A_2, A_3\}$. Despite the appearance of rigor here, there is really nothing in the axioms to justify calling such a grouping a set or to use set-theoretic notation; $\mathscr{A}$ is of another species. Properly speaking, $A_1 \in \mathscr{A}$ and $A_1 \subset \mathscr{N}$. Moreover, for example, the number $4 \in A_1$ and it is clearly not consistent with the meaning (the concept) of $\mathscr{A}$ to write $4 \in \mathscr{A}$. What is consistent is: $4 \in A_1 \in \mathscr{A}$, $A_1 \subset \mathscr{N}$ and $4 \in \mathscr{N}$. Nor would it be conceptually accurate to write $A_1 \subset \mathscr{A}$.

The key point here is that $\mathscr{A}$, properly speaking, occurs at a third level of thought. There is level one: the elements of a universal set. Then there is level two: sets consisting of elements. Now there is level three: collections consisting of sets. But all this properly requires an axiom. (Incidentally, we are not idly beating the bushes looking for a generalization of things: the concept of a collection of sets is fundamental to the definition of a topological space.) Thus, this next axiom forces our set theoretic thoughts to a new level, above that of the basic set theory of the previous sections.

AXIOM 5. For a given universal set U, there exists a *collection* denoted by 2^U which, thought of as a second universal set, satisfies the previous four axioms of set theory under the conditions:

(a) The underfined word *set* is replaced by the undefined word *collection*.

(b) The notion of belonging to is defined as:

$$A \in 2^U \quad \text{iff} \quad A \subset U.$$

The main point of this axiom is that it asserts the existence of the set or better, the collection, of all subsets of U, this being the collection 2^U. We regard the word *collection* as an undefined term here, and we will always use the wording "collection of sets" rather than "set of sets" to help keep matters clear. Thus there are now three levels: ***elements, sets, collections***, each having the property of "belonging to" in the sense of the first four axioms, in that: elements belong to sets and sets belong to collections.

Some further remarks are in order. The first concerns the idea of regarding 2^U as a second universal set to which the previous axioms apply under the definition of "ε" stated in Axiom 5: $A \in 2^U$ iff $A \subset M$. This is just a neat way of stating the assumption that set theory is to apply to collections of sets as well, that all previous definitions and theorems will carry over to a setting in which "element" is replaced by "set" and "set" replaced by "collection." For example, if $U = \mathscr{N}$, and $A = \{1, 2\}$, $B = \{2, 4, 6\}$, $C = \{1, 7, 5\}$, $D = \{n: n \text{ is prime}\}$, and $E = \{n: n \text{ is even}\}$, then one has, for example, these collections of sets: $\mathscr{A} = \{A, B, E\}$ and $\mathscr{B} = \{C, D, E\}$. One can speak of $\mathscr{A} \cup \mathscr{B} = \{A, B, C, D, E\}$ and $\mathscr{A} \cap \mathscr{B} = \{E\}$ and

$\mathscr{A}\backslash\mathscr{B} = \{A, B\}$. Here $1 \in A \in \mathscr{A}$, $7 \in C \in \mathscr{B}$ and while, for example, $\{7\} \subset C$ one does *not* have $\{7\} \in \mathscr{B}$.

A second point involves, again, the Russell paradox and other paradoxes of set theory and reinforces the above discussion of notation and "belonging to" concepts. In the Russell paradox, one begins with elements, defines ordinary sets, and then considers the collection of all ordinary sets. But to define an ordinary set, one must speak in terms of propositional functions of the form $M \in M$ where M is a set. Our axioms do not admit such as propositional functions. We have only elements "ε" sets, sets "ε" collections—provided one does not regard set and collection as synonymous. And that is the point of saying that 2^U is a second universal set whose elements are to be the subsets of U, not to be confused with the elements of U itself. In other words, the relation ε is defined only between elements and sets and not between sets and sets. This paragraph is an extremely native and blurred introduction of Russell's *theory of types*, which Russell created to avoid his famous paradox. Hopefully it suffices for the reader. It will suffice for the remainder of this text. One should consult other works for a more rigorous development.

The next definition introduces some special set notation, used only for these particular sets in a particular setting. This notation is standard and will be used throughout this text.

DEFINITION 3.4.1. In the set $\mathscr{R}$ of real numbers, with the order relation $<$ and $\leq$ (2.7, Definition 0.1) define the special subsets of $\mathscr{R}$ known as open, closed and half-open, half-closed intervals of $\mathscr{R}$ as follows: For $a, b \in \mathscr{R}$ and $a \leq b$, let

$$(a, b) = \{x \in \mathscr{R}: a < x \wedge x < b\} \quad [a, b] = \{x \in \mathscr{R}: a \leq x \wedge x \leq b\}$$

$$[a, b) = \{x \in \mathscr{R}: a \leq x \wedge x < b\} \quad (a, b] = \{x \in \mathscr{R}: a < x \wedge x \leq b\}$$

Finally the indefinite intervals of $\mathscr{R}$ are defined as follows

$$[a, \infty) = \{x \in \mathscr{R}: a \leq x\} \qquad (-\infty, a] = \{x \in \mathscr{R}: x \leq a\}$$

$$(a, \infty) = \{x \in \mathscr{R}: a < x\} \qquad (-\infty, a) = \{x \in \mathscr{R}: x < a\}$$

Problem. Let $U = \mathscr{R}$. Let $\mathscr{A} = \{[0, x)$: the third digit in the nonterminating decimal expansion for x is $7\}$. Thus $[0, 1.25760\ldots) \in \mathscr{A}$. Let $\mathscr{B} = \{[x, y): y \in \mathscr{R}$ and the first digit for x is $0\}$. Thus $[1, 1.065\ldots\} \in \mathscr{B}$. Which of the following statements are conceptually correct? In addition, which are valid set theoretic statements? For example, $[2, 2.0280\ldots) \subset [0, 1.02570\ldots)$ is conceptually correct but set theoretically invalid.

(a) $\mathscr{B} \subset \mathscr{A}$
(b) $\mathscr{A} \subset \mathscr{B}$
(c) $[1, 1.687200\ldots\} \in \mathscr{A}$
(d) $[1, 1.687200\ldots\} \in \mathscr{B}$
(e) $[0, 1.00700\ldots) \subset \mathscr{A}$

(f) $\mathscr{A} \cap \mathscr{B} = \mathscr{A}$
(g) $[0, 0) \in \mathscr{A}$
(h) $1/2 \in [0, 1.5070) \subset \mathscr{A}$
(i) $1/2 \in \mathscr{A}$
(j) $1/4 \in [0.0120\ldots, 2) \in \mathscr{B}$
(k) $[0, 1.02700\ldots] \in \mathscr{A}$
(l) $[0, 1.02700\ldots] \subset [0, 1.0370\ldots) \in \mathscr{A}$.

Some fundamental results in the theory of measure and probability are the result of arguments about collections of sets. The following is not one of them, but it does indicate how conditions placed on a class of sets implies further conditions on the class.

THEOREM 3.4.2. Let $\mathscr{A}$ be a collection of subsets of a universal set U, for which $\mathscr{A}$ has the three properties
(1) $\mathscr{A}$ is not empty; i.e., $\mathscr{A}$ contains at least one set.
(2) If $A \in \mathscr{A}$ then $U \backslash A \in \mathscr{A}$ also.
(3) If $A \in \mathscr{A}$ and $B \in \mathscr{A}$, then $A \cup B \in \mathscr{A}$.
Then $\mathscr{A}$ has the three additional properties.
(1′) $U \in \mathscr{A}$
(2′) $\square \in \mathscr{A}$
(3′) If $A \in \mathscr{A}$ and $B \in \mathscr{A}$ then $A \cap B \in \mathscr{A}$.

You may have difficulty in knowing where to start a proof. First of all you are to use any one of (1), (2) and (3) or a combination of these to prove each of (1′), (2′) and (3′). To get an idea of what's going on, you should make up a particular example. For example, let $U = \{1, 2, 3\}$. Try to make up a set $\mathscr{A}$ having properties (1), (2) and (3). You might begin with (1). Suppose for example that the set $\{1, 2\} \in \mathscr{A}$. What other sets must be in $\mathscr{A}$ in order for (2) and (3) to be satisfied? You now have an example of an $\mathscr{A}$. Does it in fact have properties (1′), (2′) and (3′)? Try to see why. Having done this you then have an example of the theorem. You should be ready now to try a proof.

Here is an example of how one might state a proof of (1′).

Proof of (1′). By (1), $\mathscr{A}$ is non-empty. Thus suppose a set, let us call it B, belongs to $\mathscr{A}$. By (2) the set $U \backslash B \in \mathscr{A}$. Since $B \in \mathscr{A}$ and $U \backslash B \in \mathscr{A}$, then by (3) $B \cup (U \backslash B) \in \mathscr{A}$. By Theorem 3.2.5(a), $U = B \cup (U \backslash B)$ and thus $U \in \mathscr{A}$ and (1′) is proven.

You are left to prove (2′) and (3′).

There is one final relation between two sets that enables set theory to serve its purposes. If two sets, A and B, are given, one has the possibilities: $A = B$, $A \subsetneq B$, $B \subsetneq A$, $A \cap B = \square$ and $A \cap B \neq \square$. This last is simply there and doesn't say anything particular about an arbitrary element of A relative to B; it may be in B or may not be. The second to last does, however, say at least a little: that no element of A is in B and no element of B

is in A. Such sets are called disjoint and the formal definition is as follows.

DEFINITION 3.4.3. If A and B are sets, then A and B are said to be *disjoint* if $A \cap B = \square$. If $\mathscr{A}$ is a collection of sets, then the sets in $\mathscr{A}$ are said to be ***mutually disjoint*** if every pair of sets $A, B \in \mathscr{A}$ are disjoint, and $\mathscr{A}$ is called a ***mutually disjoint collection.***

For example, if $\mathscr{A} = \{[0, 1], (1, 2], (2, 3]\}$ then $\mathscr{A}$ is a mutually disjoint collection because $[0, 1] \cap (1, 2] = [0, 1] \cap (2, 3] = (1, 2] \cap (2, 3] = \square$.

Problems.

(1) Show that the following are logically equivalent: A and B are disjoint, $A \subset U \backslash B$, $B \subset U \backslash A$.

(2) Show that for any two sets, A and B, the sets A and $B \backslash A$ are disjoint and $A \cup B = A \cup B \backslash A$. Any union of two sets is a union of two disjoint sets!

(3) Prove the remainder of Theorem 3.4.2.

(4) Give an example of a mutually disjoint collection having more than a finite number of sets in it.

(5) Give an example of a collection $\mathscr{A}$ such that it is false that $B \in \mathscr{A}$ and $A \subset B \rightarrow A \in \mathscr{A}$.

5. CROSS PRODUCTS OF SETS

We will shortly take up set theoretic operations on arbitrarily large collections of sets. In this sections we confine ourselves to a final method of forming new sets from old involving only two sets. This is the notion of a cross product of sets, and ultimately relies on Axiom 5 for existence.

You are familiar with a special cross product of sets, the cartesian coordinate plane wherein analytic geometry is carried on. You were encouraged to think of the cartesian plane as a collection of points, each specified uniquely by a pair (a, b) of real numbers, a being the distance of the point from the "y" or vertical axis, b the distance from the "x" or horizontal axis of the plane. Really, some extraneous notions were used to define such ordered pairs of numbers, in particular the distance notion. Given two sets A and B, consisting of numbers or *whatever,* one is not lost to speak of a pair (a, b), where $a \in A$ and $b \in B$. That is, one can think of pairs of elements (a, b) ordered by the requirement that $a \in A$ and $b \in B$. One might then wish to think of a set of ordered pairs. Simple enough concepts in themselves, but purely for the sake of an example of the lengths mathematical man has gone to in the past half-century to establish all of set theory on a firm foundation, we will take an abstract and certainly non-intuitive approach to the definition of an ordered pair of elements. It will not add one whit of meaning to your concept of ordered pair. This approach will only show how the existence and meaning of ordered pairs can be made to rest on the set theory of the previous sections.

But first a word on the essential difficulty. Suppose one begins with a universal set U and two subsets A and B. Now, one can write down a symbol (a, b) where $a \in A$ and $b \in B$, and some sort of meaning is attached. Enough meaning is present so that one does realize that this symbol (a, b) and what it represents is a new kind of animal, not an element of A or of B or *even the universal* set U! For example, if $U = \mathscr{R}$ then $(1, 2)$ in no reasonable or natural sense represents a real number. It is an element of a new and totally different kind of set, different from $\mathscr{R}$.

The questions that this matter presents to the mathematician who wishes to have a fixed and certain foundation for set theory are these: where do such pairs come from? What justifies their theoretical existence and in what context do they lie? Do they, for example, belong to any set already present? Such questions are essential in theory if not in particular practice. The following definition and subsequent theorem provide all the answers. First of all, for the sets A and B and the elements $a \in A$ and $b \in B$, the sets $\{a\}$ and $\{a, b\}$ exist, this by Axiom 4; e.g., $\{a, b\} = \{x\colon x = a \vee x = b\}$. Then, by Axiom 5, the collection $\{\{a\}, \{a, b\}\}$ exists. It is not an element of U, not a subset *of* U, but rather a subset of 2^U. Here now are answers to the above questions.

DEFINITION 3.5.1. Let A and B be subsets of U. The *ordered pair* where $a \in A$ and $b \in B$ is defined to be the collection $\{\{a\}, \{a, b\}\}$. The collection of all such ordered pairs (a, b) is denoted by $A \times B$ and called the *cross product* of A by B.

As noted earlier, this would not add one whit of meaning to the symbol (a, b) that was not intuitively already there. Conceivably it may remove some! But it does give a precise meaning to the symbol (a, b) in terms of things *already present*: $\{\{a\}, \{a, b\}\}$. Here is the theorem which justifies the term *ordered* in the terminology *ordered pair*.

THEOREM 3.5.2.
(a) $(a, b) = (c, d)$ iff $a = c$ and $b = d$.
(b) $(a, b) = (b, a)$ iff $a = b$.

This theorem asserts, simply, that the definition of (a, b) in Definition 3.5.1 implies an order to the pair. Two such pairs are equal only if their "first" elements are equal and their "second" elements are equal. Here now is a partial proof.

Proof of (*a*). First the sufficiency. Suppose $a = c$ and $b = d$. We claim $(a, b) = (c, d)$. We must show, according to Definition 3.5.1, that $\{\{a\}, \{a, b\}\} = \{\{c\}, \{c, d\}\}$. Since $a = c$, then $\{a\} = \{c\}$. Since $b = d$ also, then $\{a, b\} = \{c, d\}$ and hence $(a, b) = \{\{a\}, \{a, b\}\} = \{\{c\}, \{c, d\}\} = (c, d)$.

Now suppose $(a, b) = (c, d)$ We will show $a = c$. Since $\{a\} \in \{\{a\}, \{a, b\}\}$ and we are supposing $\{\{a\}, \{a, b\} = \{\{c\}, \{c, d\}\}$, then $\{a\} \in \{\{c\}, \{c, d\}\}$ and hence either $\{a\} = \{c\}$ or $\{c, d\}$. If $\{a\} = \{c\}$ then $u \in \{c\}$ and hence $a = c$

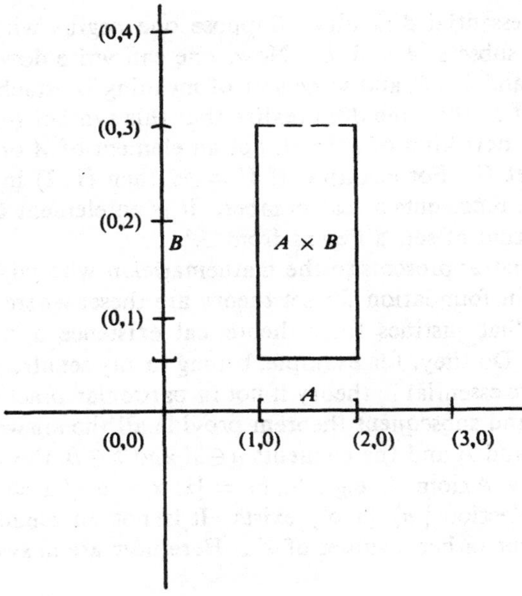

FIG. 3.4.

and we are through. If $\{a\} = \{c, d\}$, then $c \in \{c, d\} = \{a\}$ and again $c = a$. Hence $a = c$ in either case. Similarly, $b = d$.

Now for some examples of how one thinks about ordered pairs and cross products. Suppose $U = \mathscr{R}$, $A = [1, 2]$, $B = [1/2, 3)$. Then one thinks of $\mathscr{R} \times \mathscr{R}$ as the coordinate plane, and in turn thinks of $A \times B$ as that subset of the coordinate plane indicated in Figure 3.4. Suppose now that $U = \mathscr{N}$, $A = \{1, 3, 5, \ldots\}$ and $B = \{1, 2, 3\}$. Here one thinks first of $\mathscr{N} \times \mathscr{N}$ as a subset of the cartesian plane, which, as already indicated, is $\mathscr{R} \times \mathscr{R}$. The points of $\mathscr{N} \times \mathscr{N}$ are sometimes called the *lattice points* of the plane. Here $A \times B$ is thought of as the x'd points indicated in Figure 3.5.

One can easily conceive of cross products for which there are no pictures to go along. For example, if U is a set and we regard 2^U as the universe for this discussion and $\mathscr{A} \subset 2^U$, $\mathscr{B} \subset 2^U$ then $\mathscr{A} \times \mathscr{B}$ is defined to be all pairs of sets (A, B) where $A \in \mathscr{A}$ and $B \in \mathscr{B}$. No pictures for this one. Purely mental conception grounded in notation.

Problems.

(1) Give an example of sets A and B such that $A \times B \neq B \times A$.

(2) How would you define $A \times B \times C$ for three sets A, B and C? Or, how would you define an ordered triple? How about an ordered quadruple?.

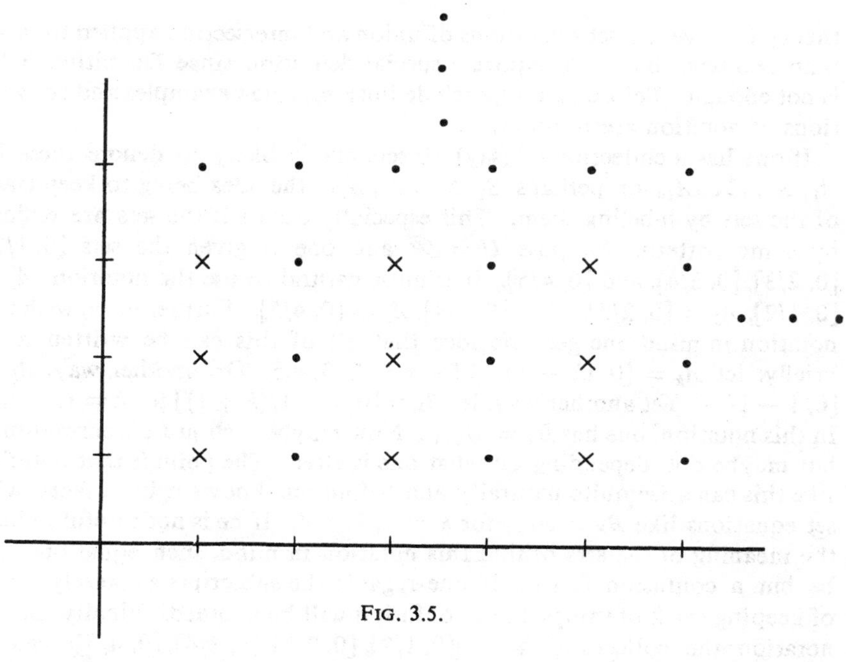

FIG. 3.5.

(3) Let $U = \mathscr{R}$, $A = [1, 3/2)$, $B = \{x\colon x = \sin y \text{ for some } y \in \mathscr{R}\}$. How do you interpret $A \times B$ as a subset of the cartesian plane $\mathscr{R} \times \mathscr{R}$.

(4) Let $U = \mathscr{R} \times \mathscr{R}$, $C = \{(x, y)\colon x^2 + y^2 = 1\}$, $D = \{(x, y)\colon x^2 + y^2 = 1/4\}$. Think of C and D as circles of radius 1 and 1/2 centered at the origin(0, 0). Can you picture $C \times D$ as a subset of

$$(\mathscr{R} \times \mathscr{R}) \times (\mathscr{R} \times \mathscr{R})\,?$$

(5) Let A, B, C, D be sets. Prove or disprove

$$(A \times B) \times (C \times D) = [(A \times B) \times C] \times D$$

without getting too involved with the proper universal set for this problem!

(6) Prove the remainder of Theorem 3.5.2(a); that is, that $b = d$. Then use (a) to conclude (b).

6. OPERATIONS WITH ARBITRARILY LARGE COLLECTIONS OF SETS

For most purposes, collections of sets can be separated into two kinds: those that are finite, having only some n-number of sets in the collection where $n \in \mathscr{N}$ and those that are not finite. We will later on define the terms finite and infinite exactly. For the time being, finite means only so many, 2, 10 or 20, or n for some $n \in \mathscr{N}$. Many important uses of set theory involve the set operations of union and intersection applied to more than two sets. But such require a special definition since Definition 3.2.1 is not enough. Before making such definitions, some examples and conventions of notation are in order.

If one has a collection of (say) 10 sets one is likely to denote these by $A_1, A_2, \ldots, A_{10}$ or perhaps $S_1, S_2, \ldots, S_{10}$, the idea being to keep track of the sets by labeling them. This especially occurs if the sets are defined by some pattern. Suppose $U = \mathscr{R}$ and one is given the sets [0, 1/2], [0, 2/3], [0, 3/4], and [0, 4/5]. It is most natural to use the notation: $A_2 = [0, 1/2]$, $A_3 = [0, 2/3]$, $A_4 = [0, 3/4]$, $A_5 = [0, 4/5]$. Furthermore, with this notation in mind one gets the idea that all of this can be written more briefly: let $A_k = [0, (k-1)/k]$ for $k = 2, 3, 4, 5$. Or, another way, $A_k = [0, 1 - 1/k]$. Yet another way: let $B_k = [0, 1 - 1/(k+1)]$ for $k = 1, 2, 3, 4$. In this notation, one has $B_k = A_{k+1}$. Now maybe such just obscures things, but maybe not, depending on what one is after. The point is that notation like this can arise quite naturally and before one knows it he is faced with set equations like $B_k = A_{k+1}$ for $k = 1, 2, 3, 4$. If he is not careful to keep the meaning of the sets behind this notation in mind, such equations will be but a confusion factor. If one regards the subscripts as merely a way of keeping track of things, most confusion will be avoided. Finally, in this notation the collection $\mathscr{A} = \{[0, 1/2], [0, 2/3], [0, 3/4], [0, 4/5]\}$ can be described as

$$\mathscr{A} = \{A_k : k = 2, 3, 4, 5\}$$

or

$$\mathscr{A} = \{B_k : k = 1, 2, 3, 4\}\,.$$

And a word about this last expression too. Here the propositional function defining the collection $\mathscr{A}$ is not really used properly. A more rigorous method for describing $\mathscr{A}$ would be as follows. In $2^{\mathscr{R}}$, let $p(A)$ be the propositional function $p(A)$: $A = A_2 \vee A = A_3 \vee A = A_4 \vee A = A_5$. Then $\mathscr{A} = \{A : p(A)\} = \{A_2, A_3, A_4, A_5\}$. But as these things are usually done in mathematics, when no misunderstanding will occur, one simply writes $\mathscr{A} = \{A_k : k = 2, 3, 4 \text{ or } 5\}$ because this is more natural and straightforward.

The general story then on a finite collection $\mathscr{A}$ of sets $A_1, A_2, \ldots, A_n$ is this. One usually writes $\mathscr{A} = \{A_1, A_2, \ldots, A_n\}$ or $\mathscr{A} = \{A_k : k = 1, 2, \ldots, n\}$ or $\mathscr{A} = \{A_i : i = 1, 2, \ldots, n\}$. All these notations get the idea across that $\mathscr{A}$ consists of n sets whose labels are $A_1, A_2, \ldots, A_n$, or simply A_k for $k = 1, 2, \ldots, n$. As will soon be seen, such notation allows one to handily use inductive reasoning in arguing about finite collections of sets.

Mathematics becomes most interesting when more than a finite number of things is involved. We can easily think of collections of infinitely many sets. For example, there is the collection $\mathscr{A}$ of all open intervals of $\mathscr{R}$. There is the collection $\mathscr{B}$ of all intervals of the form $[0, x)$ for $x > 0$. There is the collection $\mathscr{C}$ in $\mathscr{R} \times \mathscr{R}$ of all sets of the form $\{(x, y) : y =$

$x + n\}$ where $n \in \mathscr{N}$. This last consists of parallel lines of slope one intersecting the vertical axis at the points 1, 2, 3, ... Then there is the collection of all subsets of $\mathscr{N}$. One can go on and on.

If $\mathscr{A}$ is a collection of sets, finite or not, we will have need to consider two sets naturally related to $\mathscr{A}$. One is the set of all those elements belonging to at least one set in $\mathscr{A}$, the other, the set of all those elements belonging to every set in $\mathscr{A}$. These ideas are but the extension of the ideas of union and intersection to more than two sets. To make these ideas useful and precise we need some notation.

If the first example above of a collection $\mathscr{A}$ of sets there were only four sets and we could write $\mathscr{A} = \{A_2, A_3, A_4, A_5\}$ or $\mathscr{A} = \{A_k: k = 2, 3, 4, 5\}$. In the collection $\mathscr{B}$ described above we apparently cannot get by this way for $\mathscr{B}$ contains all intervals of the form $[0, x)$ for $x > 0$. For example, $[0, 1) \in \mathscr{B}$, $[0, 1/2) \in \mathscr{B}$, $[0, \sqrt{2}) \in \mathscr{B}$, for every $n \in \mathscr{N}$, $[0, n) \in \mathscr{B}$ and so on. In such a case one adopts the notation $\mathscr{B} = \{[0, x): x \in \mathscr{R}$ and $x > 0\}$. Properly speaking one would define $\mathscr{B}$ using propositional functions as follows. Let $p(B)$: $B = [0, x)$ for some $x \in \mathscr{R}$ with $x > 0$. Then $\mathscr{B} = \{B: p(B)\}$. But it is more convenient, and just as precise, to write as before $\mathscr{B} = \{[0, x): x \in \mathscr{R}$ and $x > 0\}$ or at most $\mathscr{B} = \{B$: B is a set of the form $[0, x)$ for $x \in \mathscr{R}$ and $x > 0\}$. Here now are the formal definitions of union and intersection for a collection of sets.

DEFINITION 3.6.1. Let $\mathscr{A}$ be a collection of subsets of a given universal set. The union of the sets in $\mathscr{A}$ is defined to be the set

$$\{x \in U: \exists\, A \in \mathscr{A} \ni x \in A\}$$

and is denoted by $\bigcup_{A \in \mathscr{A}} A$. The intersection of the sets in $\mathscr{A}$ is defined to be the set

$$\{x \in U: x \in A \ \forall \text{ set } A \in \mathscr{A}\}$$

and is denoted by $\bigcap_{A \in \mathscr{A}} A$.

Returning to the set $\mathscr{B} = \{[0, x): x \in \mathscr{R}, x > 0\}$ we see that

$$\bigcup_{B \in \mathscr{B}} B = \{y: \exists\, B \in \mathscr{B} \ni y \in B\} = \{y \in \mathscr{R}: y \geq 0\} = [0, \infty)$$

and

$$\bigcap_{B \in \mathscr{B}} B = \{y: y \in B \ \forall\, B \in \mathscr{B}\} = \{0\}\,.$$

We do see this! Because, if $y \in \mathscr{R}$ and $y \geq 0$ then $y \in B = [0, 2y)$ and $[0, 2y) \in \mathscr{B}$. On the other hand, if $\exists\, B \in \mathscr{B} \ni y \in B$, then since this set is of the form $[0, x)$, then $y \in [0, x)$ and hence $y \geq 0$. Thus $\bigcup_{B \in \mathscr{B}} B = \{y \in \mathscr{R}: y \geq 0\}$. For the matter of intersection it is clear that $0 \in B \ \forall\, B \in \mathscr{B}$. On the other hand if $y \in \bigcap_{B \in \mathscr{B}} B$, then $y \in B \ \forall\, B \in \mathscr{B}$. Hence $y \geq 0$. Suppose $y > 0$. Then we would have $y \in [0, y/2)$ since this last *is* a set in

$\mathscr{B}$. But this is contradictory and hence $y = 0$. Thus $\bigcap_{B \in \mathscr{B}} B = \{0\}$.

The matter really need not get so complicated in many cases. If a collection of sets $\mathscr{A}$ is given, think of $\bigcup_{A \in \mathscr{A}} A$ as the set of all elements that occur in at least one set in $\mathscr{A}$ and think of $\bigcap_{A \in \mathscr{A}} A$ as the set of all elements that occur in *every* set in $\mathscr{A}$.

Now to specialize matters a bit. If a collection $\mathscr{A}$ consists of only finitely many sets $A_1, A_2, \ldots, A_n$ it is customary to use the notations

$$\bigcup_{k=1}^{n} A_k = \bigcup_{A \in \mathscr{A}} A$$

and

$$\bigcap_{k=1}^{n} A_k = \bigcap_{A \in \mathscr{A}} A$$

Going along with notational conventions considered earlier one commonly writes,

$$\bigcup_{k=1}^{n} A_k = \{x \in U\colon \exists\, k = 1, 2, \ldots, n \ni x \in A_k\}$$

$$\bigcap_{k=1}^{n} A_k = \{x \in U\colon \forall\, k = 1, 2, \ldots, n,\ x \in A_k\}.$$

We now prove a general result which makes theorectical discussions of unions and intersections of such collections of sets amenable to inductive reasoning.

THEOREM 3.6.2. If $A_1, A_2, \ldots, A_n, A_{n+1}$ are any $n + 1$ sets, then

(a) $\displaystyle\bigcup_{k=1}^{n+1} A_k = A_{n+1} \cup \Big(\bigcup_{k=1}^{n} A_k\Big)$

(b) $\displaystyle\bigcap_{k=1}^{n+1} A_k = A_{n+1} \cap \Big(\bigcap_{k=1}^{n} A_k\Big)$

Proof. Let $\mathscr{A} = \{A_1, \ldots, A_n\}$ and $\mathscr{A}' = \{A_1, \ldots, A_n, A_{n+1}\}$. We are to show

$$A_{n+1} \cup \Big(\bigcup_{k=1}^{n} A_k\Big) = \bigcup_{k=1}^{n+1} A_k$$

or, by definition of this notation

$$A_{n+1} \cup \Big(\bigcup_{A \in \mathscr{A}} A\Big) = \bigcup_{A \in \mathscr{A}'} A.$$

According to Axiom 3 we must show

$$x \in A_{n+1} \cup \Big(\bigcup_{A \in \mathscr{A}} A\Big) \quad \text{iff} \quad x \in \bigcup_{A \in \mathscr{A}'} A$$

or, equivalently,

$$x \notin \bigcup_{A \in \mathscr{A}'} A \quad \text{iff} \quad x \notin \Big[A_{n+1} \cup \Big(\bigcup_{A \in \mathscr{A}} A\Big)\Big].$$

Now

$$x \notin \bigcup_{A \in \mathscr{A}'} A = \{y \in U: \exists\, A \in \mathscr{A}' \ni y \in A\} \quad \text{iff} \quad \forall\, A \in \mathscr{A}',$$

one has $x \notin A$. This holds iff $x \notin A_{n+1}$ and $x \notin A \;\forall\, A \in \mathscr{A}$ and this, iff, $x \notin A_{n+1}$ and $x \notin \bigcup_{A \in \mathscr{A}} A$, and therefore iff $x \notin A_{n+1} \cup (\bigcup_{A \in \mathscr{A}} A)$. Thus the proof.

One further case. Suppose one is given a collection $\mathscr{A}$ of sets $A_1, A_2, \ldots, A_n, \ldots$, one set for *every* natural number n. Thus

$$\mathscr{A} = \{A_1, A_2, \ldots\} = \{A_k: k = 1, 2, \ldots\} = \{A_k: k \in \mathscr{N}\}$$

in the usual notations for such things. Such a collections is called a *sequence* of sets. The usual notation adopted for the union and intersection of such a collection is $\bigcup_{k=1}^{\infty} A_k$ and $\bigcap_{k=1}^{\infty} A_k$. That is, these symbols are defined as

$$\bigcup_{k=1}^{\infty} A_k = \bigcup_{A \in \mathscr{A}} A = \{x \in U: \exists\, A_k \in \mathscr{A} \ni x \in A_k\}$$

and

$$\bigcap_{k=1}^{\infty} A_k = \bigcap_{A \in \mathscr{A}} A = \{x \in U: \forall\, A_k \in \mathscr{A}, x \in A_k\}\,.$$

More often one writes

$$\bigcup_{k=1}^{\infty} A_k = \{x \in U: x \in A_k \text{ for some } k = 1, 2, \ldots\}$$

and

$$\bigcap_{k=1}^{\infty} A_k = \{x \in U: x \in A_k \;\forall\, k = 1, 2, \ldots\}\,.$$[5]

If the sets A_k are given specifically, such as $A_k = [0, 1/k]$, then for $\bigcup_{k=1}^{\infty} A_k$ and $\bigcap_{k=1}^{\infty} A_k$ one would instead write

$$\bigcup_{k=1}^{\infty} [0, 1/k] \quad \text{and} \quad \bigcap_{k=1}^{\infty} [0, 1/k]\,.$$

Finally, referring back to the definition of these as $\bigcup_{A \in \mathscr{A}} A$ and $\bigcap_{A \in \mathscr{A}} A$ respectively, where $\mathscr{A} = \{A_k: k \in \mathscr{N}\} = \{[0, 1/k]: k \in \mathscr{N}\}$, one has

$$\bigcup_{k=1}^{\infty} [0, 1/k] = [0, 1]$$

and

$$\bigcap_{k=1}^{\infty} [0, 1/k] = \{0\}\,.$$

The final result of this section is a useful set theoretical equation which mirrors perfectly the negation laws for quantified statements. These equations are known as De Morgan's laws and are but extensions of Theorem 3.2.5(d) and (e) to operations on collections of sets.

[5] There is no set A_∞! Only a set A_k for each *number* $k \in \mathscr{N}$.

THEOREM 3.6.3. Let $\mathscr{A}$ be a collection of sets and let B be a set. Then

(a) $B \backslash \bigcup_{A \in \mathscr{A}} A = \bigcap_{A \in \mathscr{A}} (B \backslash A)$

and

(b) $B \backslash \bigcap_{A \in \mathscr{A}} A = \bigcup_{A \in \mathscr{A}} (B \backslash A)$.

Before a partial proof, a word about the notation of the theorem which has not yet been entirely defined. Specifically, $\bigcap_{A \in \mathscr{A}} (B \backslash A)$ means the natural thing: $\{x \in U: x \in B \backslash A \ \forall A \in \mathscr{A}\}$. Similarly, $\bigcup_{A \in \mathscr{A}} B \backslash A$ means $\{x \in U: \exists A \in \mathscr{A} \ni x \in B \backslash A\}$.

Proof of (a). By Axiom 3 we are to show $x \in B \backslash \bigcup_{A \in \mathscr{A}} A$ iff $x \in \bigcap_{A \in \mathscr{A}} B \backslash A$. If $x \in B \backslash \bigcup_{A \in \mathscr{A}} A$, then $x \in B$ and $x \notin \bigcup_{A \in \mathscr{A}} A$. By Definition 3.6.1 this can only mean that $\forall A \in \mathscr{A}, x \notin A$. Hence $\forall A \in \mathscr{A}$ one has $x \in B$ and $x \notin A$, or by Definition 3.2.1 $x \in B \backslash A$. But this is exactly the definition of $\bigcap_{A \in \mathscr{A}} B \backslash A$. That is, $x \in \bigcap_{A \in \mathscr{A}} B \backslash A$.

Conversely, if $x \in \bigcap_{A \in \mathscr{A}} B \backslash A$, then, by definition, $x \in B \backslash A \ \forall A \in \mathscr{A}$. Thus, $x \in B$ and $x \notin A \ \forall A \in \mathscr{A}$. Hence, again by Definition 3.6.1, $x \notin \bigcup_{A \in \mathscr{A}} A$. But this means $x \in B \backslash \bigcup_{A \in \mathscr{A}} A$ by Definition 3.2.1. Hence the proof.

This completes our study of basic set theory and set theoretical operations. The next section is the final one on set theory *per se*, being a primitive introduction to the "axiom of choice."

Problems. Compute $\bigcup_{A \in \mathscr{A}} A$ and $\bigcap_{A \in \mathscr{A}} A$ for each given collection $\mathscr{A}$.

(1) Let $\mathscr{A} = \{[0, x)$: the third digit in the non-terminating decimal expansion of x is $7\}$.

(2) Let

$$\mathscr{A} = \{[0, 1/n]: n \in \mathscr{N}\}$$
$$= \{[0, 1], [0, 1/2], [0, 1/3], \ldots, [0, 1/k], \ldots\}.$$

(3) Let $\mathscr{A} = \{[0, 3], [1/2, 2], [1, 3/2)\}$.

(4) Let

$$\mathscr{A} = \{(1 - 1/n, 1 + 1/n): n \in \mathscr{N}\}$$
$$= \{(0, 2), (1/2, 3/2), (2/3, 4/3), (3/4, 5/4), \ldots\}$$

(5) In $\mathscr{R} \times \mathscr{R}$, let

$$\mathscr{A} = \{[0, x) \times (y, 1]: x, y \in \mathscr{R}, 2 > x > 0 \text{ and } 0 < y \leq 1\}.$$

Hint: Draw a picture of an arbitrary element of $\mathscr{A}$.

(6) In $\mathscr{N}$ let

$$\mathscr{A} = \{\{1\}, \{2, 4\}, \{1, 3, 5\}, \{2, 4, 6, 8\}, \{1, 3, 5, 7\}, \ldots\}$$
$$= \{A: A \text{ consists of either the first } n \text{ even numbers or the first } n \text{ odd numbers}\}$$

$= \{A: A = \{1, 3, 5, \ldots, 2k-1\}$ or $A = \{\{2, 4, 6, \ldots, 2k\}, k \in \mathscr{N}\}\}$.

(7) Prove Theorem 3.6.2(b).

(8) Let $A_k = [0, 1 - 1/k]$ for $k = 1, 2, \ldots, 23$. Describe each of $\bigcap_{k=1}^{23} A_k$ and $\bigcup_{k=1}^{23} A_k$ as an interval of real numbers.

(9) Consider the Venn diagram which suggests the equation $\bigcup_{i=1}^{4} A_i = \bigcup_{i=1}^{4} B_i$ where $B_1 = A_1$, $B_2 = A_2\backslash A_1$, $B_3 = A_3\backslash(A_1 \cup A_2)$ and $B_4 = A_4\backslash(A_1 \cup A_2 \cup A_3)$ and $B_i \cap B_j = \square$ for $i \neq j$. Prove the general result: If $A_1, \ldots, A_n$ are n sets and $B_1 = A_1$, $B_2 = A_1\backslash A_2$ and in general $B_k = A_k\backslash\bigcup_{i=1}^{k-1} A_i$ for $k = 3, \ldots, n$, then

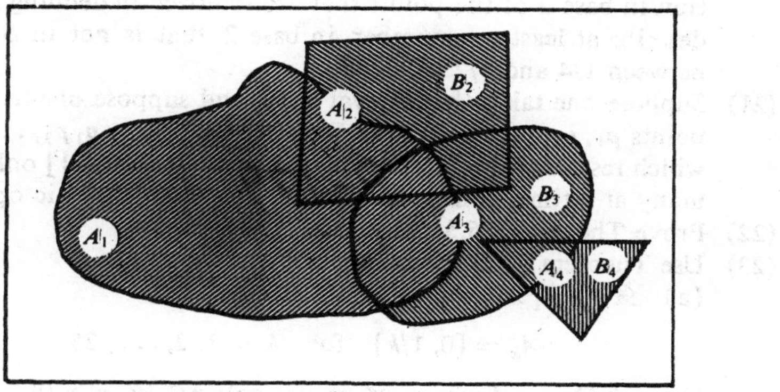

(a) $B_i \cap B_j = \square$ for $i \neq j$

and

(b) $\bigcup_{i=1}^{n} B_i = \bigcup_{i=1}^{n} A_i$.

In the set $\mathscr{R}$ of real numbers

(10) Let $A_n = (1, 1 + 1/n)$ for each whole number n. What is $\bigcap_{k=1}^{\infty} A_k$ and $\bigcup_{k=1}^{\infty} A_k$?

(11) Let $B_n = (1 - 1/n, 1 + 1/n)$. What is $\bigcap_{k=1}^{\infty} B_k$?

(12) Let $C_j = [1, 1 + 1/n]$. What is $\bigcap_{i=1}^{\infty} C_i$?

(13) Let $D_k = [k, k + 32]$. What is $\bigcap_{j=1}^{\infty} D_j$?

(14) Let $E_i = [i, \infty) = \{x: x \geq i\}$. What is $\bigcap_{m=1}^{\infty} E_m$?

(15) Let $F_k = [-1 + (-1)^k/k, 1 + (-1)^k/k]$. What is $\bigcap_{j=1}^{\infty} F_j$ and $\bigcup_{i=1}^{\infty} F_i$?

(16) Give an example of a sequence of non-empty sets A_n, $n = 1, 2, \ldots$, such that $A_{n+1} \subsetneqq A_n$ for all n and $\bigcap_{k=1}^{\infty} A_k = \square$.

(17) Give an example of a sequence of finite sets whose union is not finite. A finite set is roughly one whose elements can be counted such that the tally eventually stops.

(18) Let $A_n = [0, 1 - 1/n]$. Define new sets B_n, $n = 1, 2, \ldots$, such

that $B_i \cap B_j = \square$ for $i \neq j$ and $\bigcup_{k=1}^{\infty} B_k = \bigcup_{k=1}^{\infty} A_k$. Notice that $A_n \subset A_{n+1}$ for all n.

(19) In $\mathscr{R} \times \mathscr{R}$ let

$$A_k = [0, 1 + 1/2 + 1/2^2 + \cdots + 1/2^k] \times (-1, 2^k]$$

and compute $\bigcup_{k=1}^{\infty} A_k$ and $\bigcap_{k=1}^{\infty} A_k$.

(20) Let $S_1, S_2, S_3 \ldots$ be the sets described in the statement of Question I in the introduction to Part 2. Argue that the indicated set S_∞ is in fact $\bigcap_{k=1}^{\infty} S_k$. Writing the numbers 1/4, 1/2, 3/8, 5/16, 7/16, 9/32, etc., in base 2 (see Problem (1) (after Theorem 2.9.4)) we have $1/4 = .01$, $1/2 = .1$, $3/8 = .011$, $5/16 = .0101$, $7/16 = .0111$, $9/32 = .01001$, $11/32 = .01011$ and so on. Give a general description in base 2 of the points that are removed in defining S. Then, describe at least one number, in base 2, that is not in S, but lies between 1/4 and 1/2.

(21) Suppose one takes the interval [0, 1] and suppose one has selected points $p_1, p_2, p_3 \ldots$ from it. Describe $[0, 1] \backslash \{p_1, p_2\, p_3, \ldots\}$ as a set which results from (1) picking the points p_i for [0, 1] only finitely many at a time, combined with (2) some set theoretic operation.

(22) Prove Theorem 3.6.3(b).

(23) Use Theorem 3.6.3 to compute:

(a) $\mathscr{R} \backslash \bigcup_{k=1}^{25} A_k$ where

$$A_k = [0, 1/k] \quad \text{for} \quad k = 1, 2, \ldots, 25\,.$$

(b) $\mathscr{R} \backslash \bigcap_{A \in \mathscr{A}} A$ where $\mathscr{A} = \{[x, \infty): x \in \mathscr{R}, x > 0\}$.

(c) $\mathscr{R} \backslash \bigcup_{k=1}^{\infty} [-k, k]$.

(d) $\bigcap_{n=1}^{\infty} (1 - 1/n, 1 + 1/n) = \bigcap_{n=1}^{\infty} \mathscr{R} \backslash [(-\infty, 1 - 1/n] \cup [1 + 1/n, \infty)]$.

7. THE AXIOM OF CHOICE

Mathematicians are yet skirmishing over this one. Basically it can be viewed as a special principle of logic to be applied to infinite classes. In fact, *on its surface* it appears to be so patently logical, or reasonable, that it was not even recognized as an assumption when first used.

In 1904 the mathematician Zermelo proved what is called the Well-Ordering Theorem. Consider the order relation "$<$" on the set of real numbers $\mathscr{R}$. It has the properties:

(1) If $a \neq b$, then $a < b$ or $b < a$.

(2) If $a < b$, then $a \neq b$.

(3) If $a < b$ and $b < c$, than $a < c$.

Now consider the subset $S = (0, 1)$ of $\mathscr{R}$. This set does *not* have a first or smallest element! For, $0 \notin S$ and if $x \in S$ then $x/2 \in S$ and hence there is no $s \in S$ such that $s \leq x$ for *all* $x \in S$. Recall that in the set $\mathscr{N}$, on the other hand, every non-empty subset does have a smallest element. This indicates that there is something different about ordering the real numbers as opposed to ordering the natural numbers. Zermelo's well ordering theorem says that given *any* set T, not necessarily a set of numbers, there *can* be defined between the elements T an order relation "$<$" having the abstract properties (1), (2) and (3), and the additional property that for any subset S of T, *S does have a smallest element* in the ordering defined. Such a set T is said to be well ordered.

Unfortunately Zermelo's theorem does not say how to define the order relation, but only that one can be defined. No one has yet defined an order relation on $\mathscr{R}$ for which (say) the set $S = (0, 1)$ has a smallest element. From the discussion above, an order relation with this property would necessarily be different from the usual order relation of $a < b$. At any rate, the whole idea was a bit too much for mathematicians to take and they began examining Zermelo's proof of the "well ordering" theorem. It was discovered that the proof depended on the following assumption, which Zermelo did not recognize or admit as an assumption, probably because it seemed so natural.

The Axiom of Choice. Let $\mathscr{A}$ be a collection of mutually disjoint non-empty subsets of a universal set U. Then there exists a set B consisting of exactly one element taken from each set $A \in \mathscr{A}$.

The axiom of choice says, no more and no less, that one can pick an element out of each set $A \in \mathscr{A}$ and collect these elements together into a new set. For example, let $U = \mathscr{N}$, $\mathscr{A} = \{\{1, 3\}, \{5, 7, 9\}, \{2^k: k = 1, 2, \ldots\}\}$. The axiom of choice says a set $B = \{1, 5, 4\}$ exists. No big thing there! This set could also be obtained from Axiom 4. For, let $p(x)$: $x = 1 \vee x = 5 \vee x = 4$. Then $B = \{x: p(x)\}$. The catch comes when the set $\mathscr{A}$ contains infinitely many sets. Then one may not be able to form a propositional function to define the set B for, literally, too much would have to be written to form the propositional function! But there is a better explanation, due, again, to Russell.

Problem. Suppose there are infinitely many pairs of black shoes in the universe and infinitely many pairs of black socks to go with them.

(1) Let U consist of all black shoes, $\mathscr{A}$ consist of each pair taken as a set. Show that the set B of all left shoes exists as a consequence of Axiom 4.

(2) Now let U consist of all black socks, $\mathscr{A}$ consists of all pairs of socks, each taken to be a set in $\mathscr{A}$. Why is Axiom 4 not sufficient to assert the existence of a set B consisting of one sock from each pair? What assumption would suffice to assert the existence of a set consisting of exactly one sock from each pair?

(3) Suppose that in (2) one could take each pair of socks from $\mathscr{A}$ and label one sock from this pair with a white mark. Let $p(x)$: $\exists A \in \mathscr{A}$ such that $x \in A$ and x has a white mark. Does the set $B = \{x$: x is

a sock with a white mark} exist by Axiom 4? Now, what assumption would allow one to make such a labeling to begin with?

The proof of the well ordering theorem from the axiom of choice is a rather esoteric, but important, thing in mathematics. We will not present it, but we can obtain a little insight into the importance of the matter from the following consideration. Consider again the set $\mathcal{N}$. The induction axiom is fundamental to the study and use of the natural numbers, and as has been shown the induction axiom is equivalent to the statement that every non-empty subset of $\mathcal{N}$ has a smallest number. In other words the fact that $\mathcal{N}$ is well ordered is equivalent to inductive reasoning. To carry the analogy a bit further, the axiom of choice implies the ability to well order *any* set. In subsequent mathematics courses where the axiom of choice is well nigh indispensible, one finds that the use of this axiom arises naturally when one must deal with collections of sets that are too large to be subscripted by the numbers in $\mathcal{N}$. Recognizing that statement as very vague, here is another way to say it: when one attempts to generalize certain theorems about a collection $\mathcal{A}$ of the form $\mathcal{A} = \{A_1, A_2, \ldots, A_n\}$ to a collection of the type found in Problem 3.6(1), which cannot be subscripted with the natural numbers as we shall later see, one is likely to make use of the axiom of choice. That is yet a rough approximation to the truth of the matter but it does, hopefully, give a little insight.

After it was pointed out that the "believable" axiom of choice implied the surprising and non-intuitive well ordering theorem, a natural question was: does the converse hold? Here is a sketch of a proof that it does. Assume that any set can be well ordered. Let $\mathcal{A}$ be a mutually disjoint collection of non-empty subsets of a set U. Each set $A \in \mathcal{A}$ can be well ordered and hence, in this ordering, each set A has a smallest element. This amounts to labeling one x out of each set A, namely the smallest x in the well ordering for A. For a given $x \in U$, the statement $q_x(A)$: x is the smallest element in A, is then a true or false statement, and not both, for each $A \in \mathcal{A}$. Hence $\exists\, A \in \mathcal{A} \ni q_x(A)$, is a statement which is either true or false about each $x \in U$. Let $p(x)$: $\exists\, A \in \mathcal{A} \ni q_x(A)$. Then $B = \{x: p(x)\}$ exists by Axiom 4. But B consists of exactly one element taken from each set $A \in \mathcal{A}$. Hence the assumption of well ordering implies the axiom of choice. The rather believable and the rather unbelievable are equivalent!

But more to this tale yet. Large parts of modern mathematics could not exist without the axiom of choice. This is particularly true of topology and the areas of measure theory and functional analysis, though not so true of modern algebra. Moreover, a large part of the mathematical theory of quantum physics depends on the axiom of choice. The results of this theory, known as quantum field theory and mathematically realized in terms of operations on infinite-dimensional spaces of vectors, gives useful results with predictable accuracy to the physicist who cares to use them. One is led to think then that the axiom of choice is all right, since its implications do coincide with the physicist's view of reality.

But the mathematicians won't let it rest there, for there's is not to be concerned with reality. Mathematicians have shown that the axiom of choice, is, by ordinary logic, logically equivalent to several well known and often used statements of set theory besides the well ordering theorem, none of which is very readily believable. These are too technically complicated to elucidate here, but the equivalence of this axiom and a certain result of Euclidian geometry can be given, and it might well shake the readers faith in the whole esoteric but extremely useful business.

The Banach-Tarski Paradox.[6] In Euclidian three-dimensional space, if S_1 and S_2 are any two spheres of possibly different positive radii, then there exists subsets $A_1, \ldots, A_n$ of S_1 and $B_1, \ldots, B_n$ of S_2 such that

(a) $S_1 = \bigcup_{i=1}^{n} A_i$ and $S_2 = \bigcup_{i=1}^{n} B_i$.
(b) $A_i \cap A_j = \square$ and $B_i \cap B_j = \square$ for $i \neq j$.
(c) Each A_i is congruent to B_i(!).

In plain English again, if one has a sphere the size of a pea and another the size of the sun, then one can decompose the pea into *finitely* many parts which by translations and rotations in three-space can be made to fill out the sun! Such is a good example of an unwanted, unsuspected and certainly mind-blowing consequence of an otherwise reasonable assumption.

The Banach-Tarski paradox is so difficult to conceive of you may wish to take the view that any assumption that would lead to it cannot be accepted in a reasonable mathematics. Take such a view if you will, many eminent mathematicians have doubted the axiom of choice and refused to use it, but in discarding it you also discard a great part of the useful and interesting mathematics of the past half-century. So again a bit of reassurance. In 1940 Gödel was able to show that *if* set theory without this axiom is consistent, then set theory with the axiom of choice is also consistent.[7]

Related Readings

1. Bourbaki, N.: *Elements de Mathematique, Livre I: Theorie des Ensembles,* Paris: Hermann, 1954.
2. Dinkines, F.: *Elementary Theory of Sets*, New York: Appleton-Century-Crofts, 1964. This book would serve well for additional explanation and examples of the theory of sets.
3. Eves, H., and Newsom, C. V.: *An Introduction to the Foundations and Fundamental Concepts of Mathematics.*, New York: Holt, Rinehart and Winston Inc., 1965.
4. Halmos, P. R.: *Naive Set Theory*, Princeton: Van Nostrand, 1960. This is a nice little book authored in an interesting way.

[6] This is actually a special form of the paradox. For a fuller discussion of it see L. M. Blumenthal: A paradox, a paradox, a most ingenious paradox, *American Mathematical Monthly*, *47*: 346–353, 1940.

[7] P. Cohen and R. Hersch: "Non-Cantorian Set Theory" (*Scientific American*, December 1967) is again a good further reference on this matter.

5. Hayden, S., and Kenison, J. F.: *Zermelo-Fraenkel Set Theory*, Columbus, O.: Charles E. Merril Publishing Co., 1968. This presents an axiomatic theory of sets that is a bit more rigorous than that presented herein.
6. *Insights into Modern Mathematics*, 23rd Yearbook, Washington, D. C.: The National Council of Teachers of Mathematics, 1957.
7. Stoll, R. R.: *Set Theory and Logic*, San Francisco: W. H. Freeman & Co., 1963.
8. Suppes, P.: *Axiomatic Set Theory*, New York: Dover Publications, Inc., 1972.
9. Vilenkin, N. Y.: *Stories About Sets*, New York: Academic Press Inc. (paperback), 1968.

chapter 4

"Although this may seem a paradox, all exact science is dominated by the idea of approximation."—B. Russell

TOPOLOGY AND CONNECTED SETS

1. INTRODUCTION

This chapter will put some life into the abstract set theory so far developed. That is to say, we will make use of it for something other than its own sake. In this *first* instance, to use it to develop and study the concept of a topological space and relate it to both Question I and Question II. Regarding Question I, we will show that $\mathscr{R}$ is a connected topological space. Concerning Question II, we will consider the general idea of connectedness and attempt to define and characterize its essence as seen by that relatively new specise of mathematicians known as topologists. But, to be candid about it, you should know that the author is an analyst, and this development of topology will no doubt be strongly influenced by the way an analyst looks at topology: an extremely beautiful part of mathematics which is very useful in analysis! Thus the more esoteric, to an analyst, parts of topology will be largely avoided, and most discussions and examples will be carried on in the setting of "nice" subspaces of ordinary Euclidian space. But the flavor of abstract point set topology will be most certainly there.

The concept of a topological space is a rather recent one in mathematics. First known as *analysus situs*, a name given to it by Henri Poincaré in 1895, topology was at first concerned with those properties of a spatial form which are not lost when the form is changed (deformed) in a continuous manner. Question II asks whether "connectedness" is such a property. Because of its subject matter, topology became known as "rubber-sheet" geometry, but its present connection with geometry is now recognized as mostly a historical one. Topology is now an extremely important, interesting and independent area of mathematics in itself. It has two principle forms: point set topology and algebraic topology. The first received its greatest development in the years 1920–1950 and has now been well worked over as a research area, only the hardest problems remaining. The latter is yet an active research area. It is interesting to note that while algebraic topology initially drew on the concepts of modern algebra to do topology, it now stands as a tribute to the oneness of mathematics that things have come full circle and, on the frontiers of algebra, one finds the methods of algebraic topology indicating new directions and possibilities.

We will be concerned solely with point set topology. What then is the concern of point set topology? Think of a set of points in the two-dimensional space $\mathscr{R} \times \mathscr{R}$. It may have length or area, a certain size or shape, such as a triangle, curve or square. Let us call these properties *quantitative.* They can be expressed and determined in terms of numbers, ordered pairs, equations, etc., as in analytical geometry. Such point sets also have other properties. A triangle or square along with its insides are "connected" point sets in the intuitive sense. These also have an inside, an outside, a boundary. These are *qualitative* properties that have little to do with precise shape, size or length. Such properties are the concern of topology. Topology rarely seeks numerical descriptions of point sets, it is more concerned with the nature of the relationship between the various points in a point set which, for example, determine that a square, triangle or circle are connected point sets despite their dissimilar appearance. To a topologist, a square, circle and triangle are indistinguishable. Having been long trained to make strict distinctions between such objects in order to learn much about them, this attitude may seem to be of dubious utility, but consider it from another angle. If all that one is interested in at the moment is "connectedness," then one should *not* wish to concern himself with the fact that a square has corners in very special places or that a circle does not have any at all! These are immaterial to the appearence of "connectedness" in these figures. Conclusion: a useful abstract theory of connectedness would be one that pays no regard to such extraneous details, one that picks on the essence of "connectedness" and nothing more. Thus, an abstract theory of connectedness, and in particular the theory of topological spaces as a whole, generally ignores those properties of a point set subject to algebraic or geometrical description in the usual sense. It concentrates on making just as precise the more nebulous qualitative properties of a point set such as how the points in a point set are attached, connected, to the point set as a whole. Let us begin then with a rough look at just this: the connectedness of a point set, and attempt to find a context in which it can best be discussed.

Consider Figure 4.1 made up of two point sets A and B with $A \cap B = \{x\}$. It satisfies our intuitive sense of a connected point set. Now despite the fact that we do conceive of connectedness as a global property, a property of the whole of a point set, it is apparent that the crux of the connectedness of A to B comes down to the apparent connectedness of A, and of B, and, most importantly, the existence of the point x in $A \cap B$. A topologist attempting to draw some abstract criteria from this last observation about x might begin with the general guideline that connectedness, despite

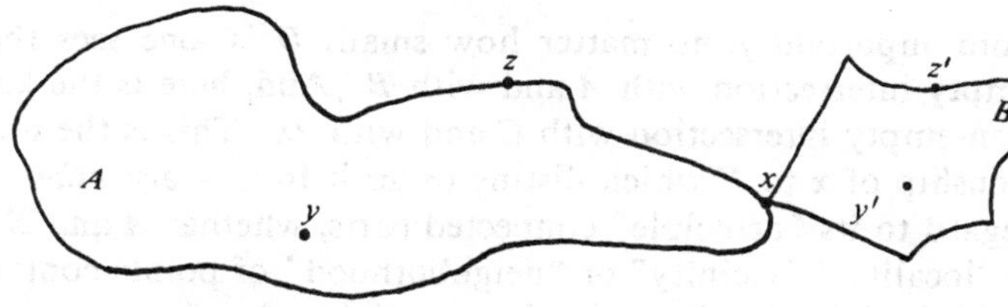

FIG. 4.1

its global character, must turn on details of local character as well. That is to say, connectedness does appear to involve details about each point of a point set.

But details about single points of a point set won't do either! Here is why. Let us try to qualitatively describe the relationship of the crux point x, to the set $X = A \cup B$, as opposed to the relationship of y, z, y' or z' to X. At first this seems easy, x is the only point in both A and B. But this judgment as to the essentials at hand is quickly seen to be illusory. Why? It is too dependent on drawings and pictures to indicate an abstract method of characterizing connectedness. For let $C = A$ and $D = B/\{x\}$. Now $C \cup D$ is the same "connected set" X, and $C \cup D$ is still connected by the crux point x, but our easy criteria that x be in both of the parts C and D making up the connected set $C \cup D$ has been *lost by a trivial change in the designation*, the naming, of the parts which make up the whole of the set $X = A \cup B = C \cup D$. Because mathematics cannot function when such trivial modifications in the description of a problem destroy the essential criteria it would try to use in solving the problem, in this case the criteria that x be in both of the individually connected parts of the set X, something better is needed.

But we seem to be at an impasse. We have more or less decided that the connectedness of X cannot be well described in terms of the universe of discussion, the whole point set X. We have also seen that if one concentrates on what appears to be the crucial point, x, "connecting" A to B, or C to D, and x alone, one obtains nothing definitive. Yet there is something special about the point x relative to A and B, and to C and D. This something special does involve the only set theoretic concept left between element and universe, that of a subset, in particular certain subsets of X which contain the point x.

In Figure 4.2, consider any "neighborhood" U of x. That is, let U be the interior of a non-degenerate circle centered at x. No matter how large,

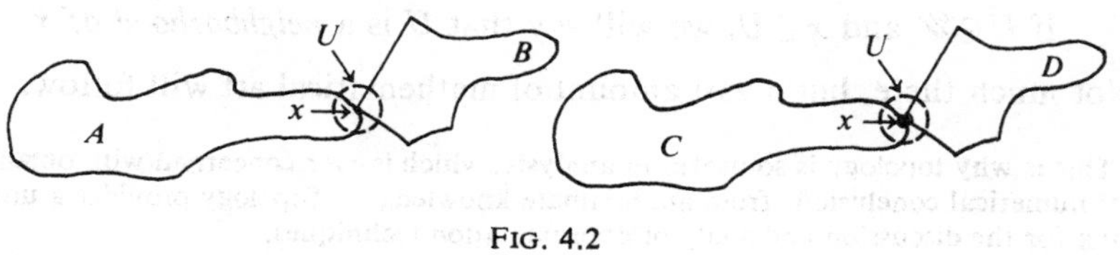

FIG. 4.2

but *much* more importantly, no matter how small, U is, one sees that U has a non-empty intersection with A and with B. And, here is the key, U also has a non-empty intersection with C and with D. This is the quality of the relationship of x to X which distinguishes it from *every other point* in X with regard to its "principle" connected parts, whether A and B or C and D: any "locality," "vicinity" or "neighborhood" of points containing x on the "inside" of this locality, vicinity or neighborhood of x, must hit, intersect, both of the principle parts. Such a view is the essence of the topologists view of space, point sets, connectedness and the like. He is interested in drawing conclusions from hypotheses about what is true in a neighborhood of a point, as opposed to what is true at a point. Implicit in such a view is the use of approximate knowledge.[1] For example, the important point is that any "neighborhood" of x does hit both A and B, both C and D; where it hits—the exact points of the intersection—are not really material to the essentials of the matter. That may strike you as leaving a bit too much to chance, a little too imprecise to reach precise conclusions. But have you noticed the *any* qualification in that criteria: *any* neighborhood of x hits both A and B, both C and D? As in the earlier section on infinity, the quantifier *for all* makes up for the indefiniteness of what is being hypothesized.

At this point the reader has at best a vague idea of what topology is concerned with. It does concern sets, thought of as sets of points, and their relationship to other sets. The relationships have to do with subsets, called "neighborhoods" of the whole which contain a single point, and their relationship to the whole. It took years of research before topologists decided on a proper and minimal axiomatic foundation for their corner, indeed cornerstone, of mathematics. We will not try to see at this time why they chose as they did, but will simply begin with the axioms for a topological space, which amount to set theoretic statements, pure and simple. Here are these axioms.

A topological space is a pair $(X, \mathscr{U})$ where X is a set whose elements are called the *points* of the topological space and $\mathscr{U}$ is a fixed collection of subsets of X called *neighborhoods* with following properties:

AXIOM 1. If $x \in X$, then $\exists U \in \mathscr{U}$ such that $x \in U$.

AXIOM 2. If $U, V \in \mathscr{U}$ and if $x \in U \cap V$, then $\exists W \in \mathscr{U}$ such that $x \in W$ and $W \subset U \cap V$.

If $U \in \mathscr{U}$ and $x \in U$, we will say that U is a *neighborhood of* x.

Not much there, but a vast amount of mathematical art will follow. Let

[1] This is why topology is so useful in analysis, which is ever concerned with obtaining exact numerical conclusions from approximate knowledge. Topology provides a unified setting for the discussion and study of approximation techniques.

us begin with some examples and suggestions as to how to think of a topological space. First of all it does take two things before one has such a creature. One requires a set X, *and* one requires a designated collection $\mathscr{U}$ of subsets of X. The set $\mathscr{U}$ is called a topology for X (meaning: there may be other topologies around). A point x is in X, not in $\mathscr{U}$. A neighborhood U is in $\mathscr{U}$, not in X. A point x is in a neighborhood U which is in turn in $\mathscr{U}$. A neighborhood U is a subset of X. Thus one has: $x \in U \in \mathscr{U}$, $x \in U \subset X$, $\mathscr{U} \subset 2^X$, in the usual notation.

Typically one thinks of X as a point set, or universe of discussion, as in Figure 4.3. About each point x in X one imagines a subset of points $U \in \mathscr{U}$ containing x. Axiom 1 says that every point must have at least one neighborhood. Axiom 2 says that in a drawing such as in Figure 4.4 things must appear roughly as indicated. Or, to say it another way, if two neighborhoods intersect at all, then their intersection must contain an entire neighborhood.

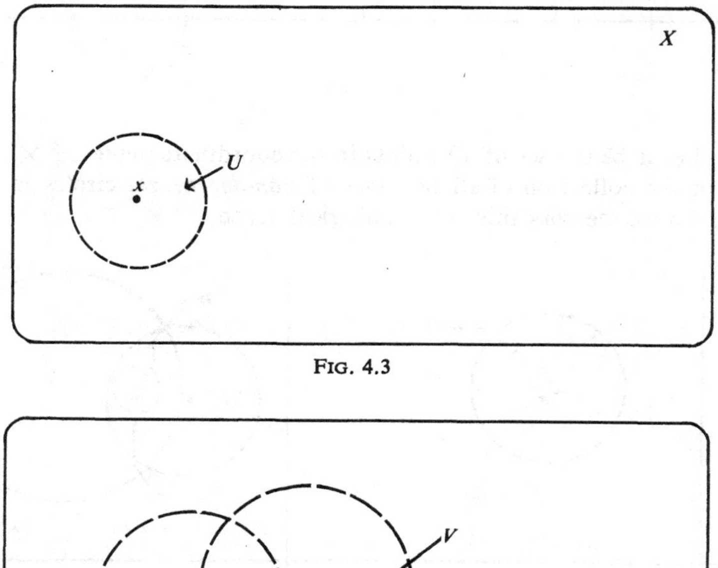

FIG. 4.3

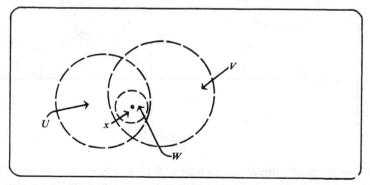

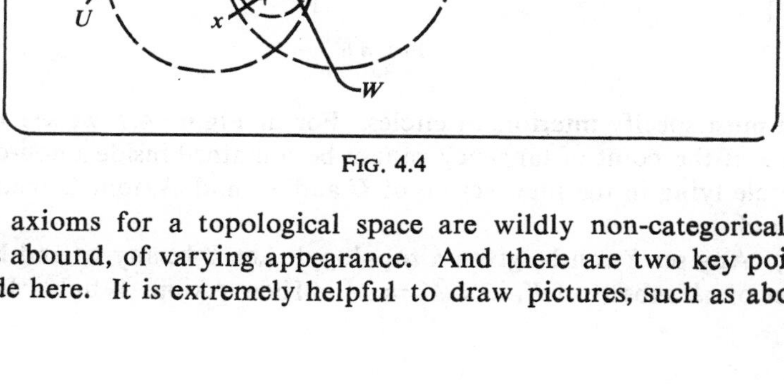

FIG. 4.4

The axioms for a topological space are wildly non-categorical. Examples abound, of varying appearance. And there are two key points to be made here. It is extremely helpful to draw pictures, such as above, in order to motivate one's topological thoughts and keep one's ideas straight. It is extremely useless to *prove* anything in topology by drawing pictures because, at the least, no picture could possibly encompass all the varied interpretations of a topological space. One must use the logic of the mind and not of the senses to *verify* conclusions. One should also use the logic of the senses to *guess* at possible conclusions. Here now are the most prominent examples of topological spaces we will consider in the sequel.

EXAMPLES. (a) Let X be the set $\mathscr{R}$ of real numbers and let $\mathscr{U}$ consist of *all* open intervals (a, b) of real numbers. Axiom 1 is satisfied because if $x \in X$, then $x \in (x - 1, x + 1)$, which is an open interval. Axiom 2 is satisfied because, if $x \in (a, b) \cap (c, d)$, then $x \in (a', b')$ where a' is the larger of a and c and b' is the smaller of b and d. Figure 4.5 is a visualization of what has been said.

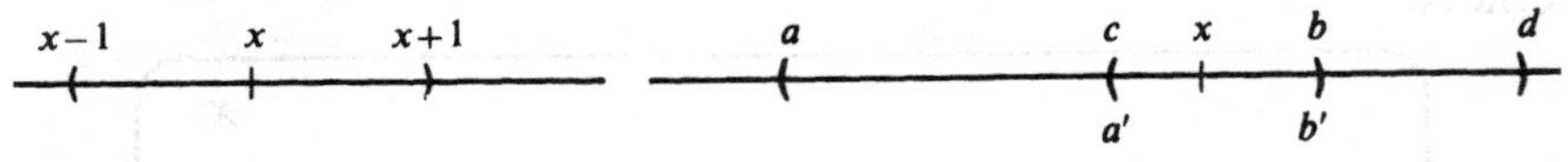

FIG. 4.5

(b) Let X be the set of all points in the coordinate plane $\mathscr{R} \times \mathscr{R}$ and let $\mathscr{U}$ be the collection of all *interiors* of *non-degenerate* circles in X. In Figure 4.6 we see why this is a topological space.

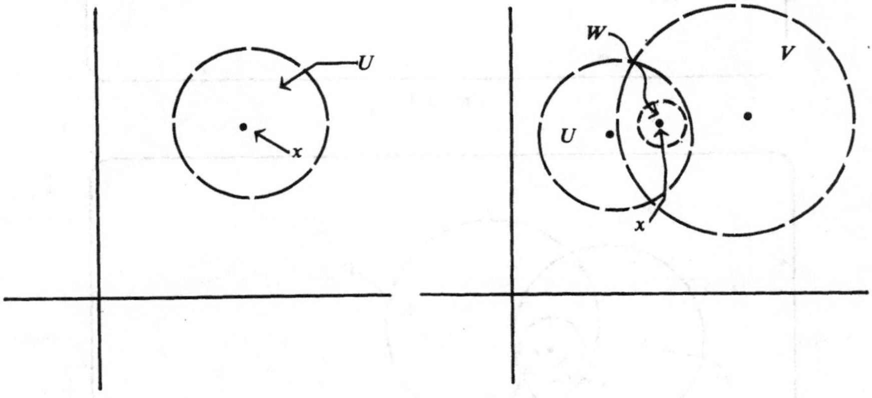

FIG. 4.6

We must specify interiors of circles. For in Figure 4.7 we see that the point x at the point of tangency cannot be contained inside a non-degenerate circle lying in the intersection of U and V, and Axiom 2 would then not hold.

(c) Any set X can be given a topology! Let X be any set and let $\mathscr{U}$ be the set of *all* subsets of X, i.e., $\mathscr{U} = 2^X$. Then Axiom 1 holds for given

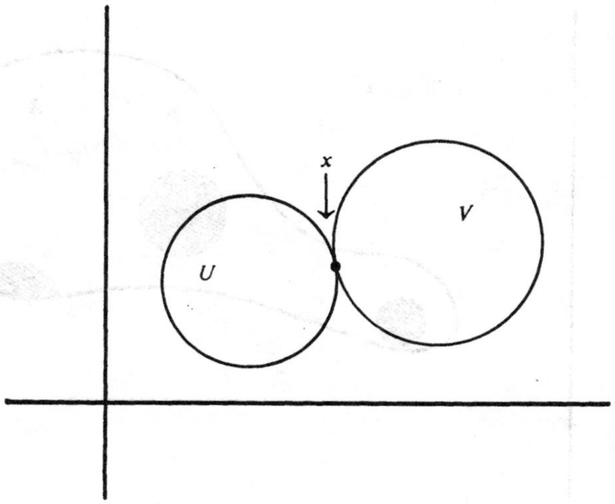

FIG. 4.7

$x \in X$, then $x \in \{x\}$ and $\{x\} \in \mathscr{U}$. Also, if $x \in U \cap V$ then $W = U \cap V$ is a subset of X and hence $W \in \mathscr{U}$. Since $x \in W \subset U \cap V$, (2) holds. The topology $\mathscr{U}$ is called the "discrete" topology on X.

(d) Any set be given two topologies! Let X be a set, $\mathscr{U} = \{\square, X\}$. Then Axiom 1 holds because $x \in X \in \mathscr{U}$ and Axiom 2 holds vacuously. The topology $\mathscr{U}$ is called the "indiscrete" topology on X.

Let us consider Example (a) again, where $X = \mathscr{R}$. The topology given there will be henceforth referred to as the *usual topology* on $\mathscr{R}$. Similarly, the topology in Example (b) will be called the *usual topology* of the plane. Each of these sets has other topologies. For example, with $X = \mathscr{R}$, one has both the discrete topology $\mathscr{U}$ of all subsets of $\mathscr{R}$, the indiscrete topology $\mathscr{U}$ consisting of only $\square$ and $\mathscr{R}$, as well as the usual topology. We will be most interested in the topology $\mathscr{U}$ of all open sub-intervals of $\mathscr{R}$ and and for this reason call it the usual topology of $\mathscr{R}$. Here are further examples.

Problem. Verify that the following pairs $(X, \mathscr{U})$ are topological spaces using either diagrams, such as in Example (b) or analytical arguments, such as in Example (a), when this is not too involved.

(1) Let X be the set of real numbers and let $\mathscr{U} = \{(a, \infty): a \in X\}$.

(2) Let $X = (0, 1)$ and let $\mathscr{U}$ be the set of all open intervals in X.

(3) Let $X = [0, 1]$ and let $\mathscr{U}$ be the set of all subsets of X obtained by intersecting X with all possible open intervals of numbers. For example, $(1/3, 1] \in \mathscr{U}$ as is $(1/2, 3/4)$. Is $\{1\} \in \mathscr{U}$?

(4) Let Y be the plane $\mathscr{R} \times \mathscr{R}$. Let X be a fixed subset of Y and let $\mathscr{U}$ be the set of all possible intersections with X of the interiors of non-degenerate circles in Y. That is, we take the topology for the plane given in Example (b) and "relativize" it to X by intersecting

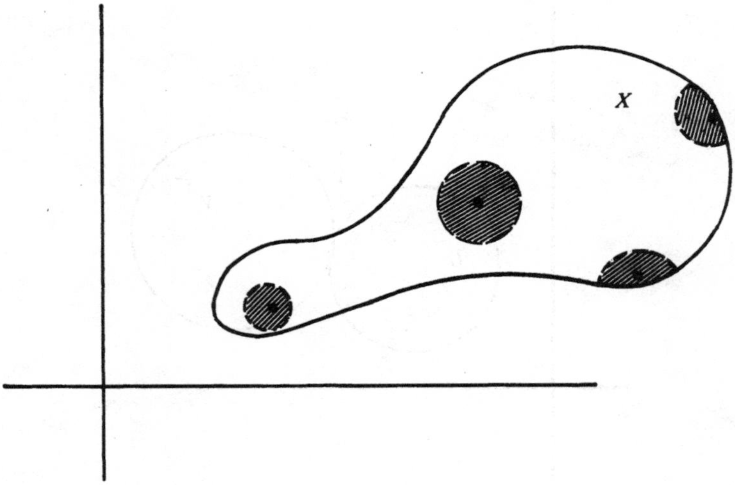

FIG. 4.8

all of the sets in this topology for the whole plane with just the set X. If X is as in Figure 4.8, some typical neighborhoods are indicated.

(5) Let X be the coodinate plane and let $\mathscr{U}$ be the set of all vertical lines in X.

(6) Let X be the coordinate plane and let $\mathscr{U}$ be the set of all interiors of non-degenerate squares in X.

(7) Let X be Euclidian 3-dimensional space as used for solid analytic geometry and let $\mathscr{U}$ be the set of all interiors of non-degenerate spheres in X.

(8) Let ω be a fixed object not in $\{1, 2, 3, \ldots\}$. Let $X = \{1, 2, 3, \ldots, \omega\}$. Let $\mathscr{U}$ consist of all singleton sets $\{k\}$ for $k \in \mathscr{N}$ and all sets of the form $\{\omega\} \cup \{k: k \geq n\}$ for *any* choice of n.

(9) Let X be a set, A, B and C subsets of X with $A \cap B$, $C \cap B$ and $A \cap C$ all non-empty. Let $\mathscr{U}$ contain A, B and C. What other sets must be in $\mathscr{U}$ to make $(X, \mathscr{U})$ a topological space?

(10) Prove that in any topological space $(X, \mathscr{U})$, $X = \bigcup_{U \in \mathscr{U}} U$.

(11) Let X consist of ten elements $a_1, a_2, \ldots, a_{10}$. Suppose that $\mathscr{U}$ is a topology on X with the following property: If $a, b \in X$ then $\exists U \in \mathscr{U} \ni a \in U$ and $b \notin U$. Prove that $\{a_n\}$ is a neighborhood of the point a_n—i.e. $\{a_n\} \in \mathscr{U}$ for $n = 1, 2, \ldots, 10$. *Hint*: let $K = \{k: k \neq n$ and $\exists U \in \mathscr{U} \ni a_n \in U$ but $a_j \notin U$ for $j = k, k+1, \ldots, 10$, and $j \neq n\}$. If $1 \in K$, this implies the desired conclusion. Why? If $1 \notin K$, obtain a contradiction. But first, before going through all this fancy but necessary stuff, convince yourself that the conclusion is indeed true and try to state of yourself why it is apparently true.

2. BASIC CONCEPTS: OPEN SETS AND CLOSED SETS

Like the first example we considered of an axiomatic structure, the axioms of topology don't assume very much. Before we can prove many meaningful theorems we need to introduce some basic notions related to a topology $\mathscr{U}$ on a set X. These are basic to the study of topology no matter what the particular aim in mind is. We can however motivate these definitions in relation to the problem of "connectedness."

Consider Figure 4.9. We want to think of X as being the set of points in the plane and $\mathscr{U}$ being the "usual" topology on $X = \mathscr{R} \times \mathscr{R}$. We are looking for an abstract condition which allows us to distinguish (in the context of the topology) the different relationships of the points u, v, w, x, y and z to the sets A and B. Notice that v is the only point "connecting" A to B so that this is related some how to our overall problem.

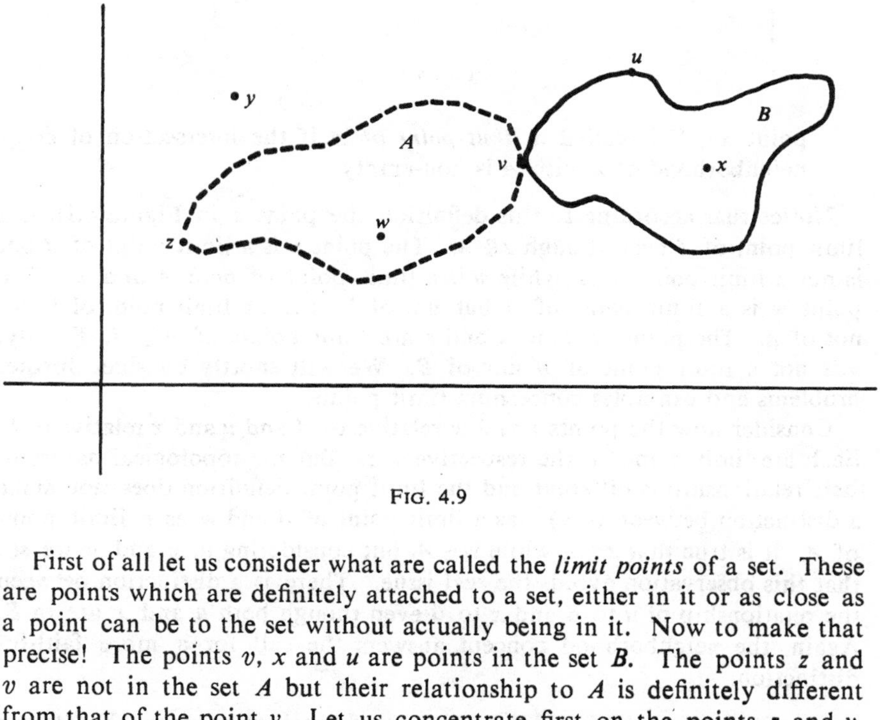

FIG. 4.9

First of all let us consider what are called the *limit points* of a set. These are points which are definitely attached to a set, either in it or as close as a point can be to the set without actually being in it. Now to make that precise! The points v, x and u are points in the set B. The points z and v are not in the set A but their relationship to A is definitely different from that of the point y. Let us concentrate first on the points z and y. The point z has the property that *every* neighborhood of z hits A. The point y has just the opposite property: there is a neighborhood of y which misses A. At the same time, neither point is actually in A. The following definition makes a distinction between the relationship of z and A as opposed to y and A, though neither is actually in A.

DEFINITION 4.2.1. If $(X, \mathscr{U})$ is a topological space and $A \subset X$, the

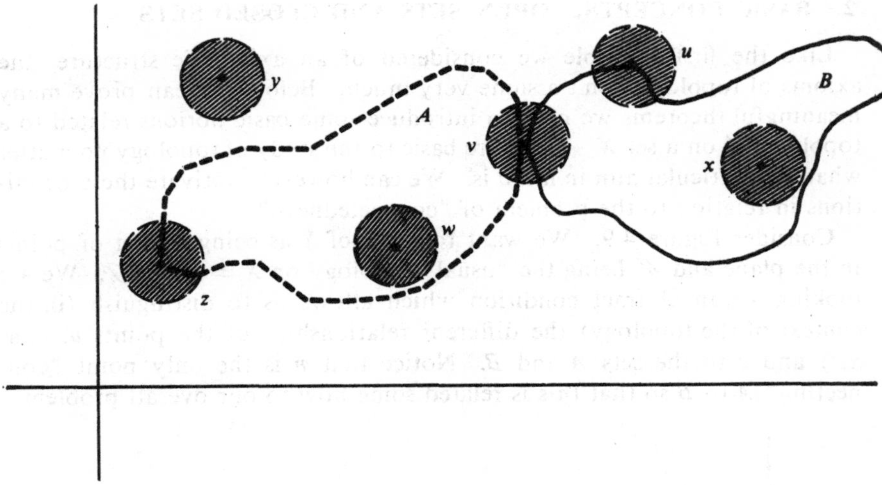

FIG. 4.10

point $x \in X$ is called a ***limit point of*** A if the intersection of *every* neighborhood of x with A is non-empty.

Notice that according to this definition the point z in Figure 4.10 is a limit point of A even though $z \notin A$. The point u is a limit point of B but is not a limit point of A, while v is a limit point of *both* A and B. The point w is a limit point of A but not of B as is x a limit point of B but not of A. The points u, v, w, x and z are limit points of $A \cup B$. Finally, y is not a limit point of A nor of B. We will shortly consider further problems and examples concerning limit points.

Consider now the points z and w relative to A and u and x relative to B. Each are limit points of the respective sets. But the topological nature of their relationship is different and the limit point definition does not make a distinction between (say) z as a limit point of A and w as a limit point of A. It is true that $z \notin A$ while $w \in A$, but considering u, x and B we see that this observation avoids the real issue. There is a distinction between the relationship of u to B and x to B even though both u and x are in B. Again, the neighborhood concept answers the call for a more faithful distinction.

DEFINITION 4.2.2. If A is a subset of topological space $(X, \mathscr{U})$, the point $x \in A$ is called an ***interior point*** of A if at least one neighborhood of x is contained entirely within A.

According to this definition w and x are limit points of A and B, respectively, which are also interior points of A and B, respectively. The points z, y, v, and u are not interior points of either set, though z, v, and u are limit points of one set or the other. No neighborhood of v, for example, lies entirely within A or within B or even $A \cup B$, even though v is a limit point of both.

There may be a feeling here that things are not quite complete. The definitions of limit point and interior point definitely characterize the relationship of y to A and w to A. But, again, what of z? It is a point on what we might naturally call the "boundary" of A. The point z, as opposed to the point w, which are both limit points of A, is also a limit point of $X \backslash A$. Here is the general

DEFINITION 4.2.3. If A is a subset of a topological space $(X, \mathscr{U})$ and x is a point in X which is a limit point of both A and of $X \backslash A$, we will call x a *boundary point* of A. The boundary of A is the set of all boundary points of A.

These three definitions well characterize the topological relations of the points in a topological space X to a given set A. The interior points of A are those (topologically) inside A. The non-limit points of A are those (topologically) well outside A. The boundary points, are those in between, possibly in A, possibly not, but limit points of both A and $X \backslash A$.

As it turns out, the notion of a boundary point is not really essential to our discussion of connectedness, despite appearances. But to better see that comment, think about it this way: no new ideas are introduced to define a boundary point. The definition is in terms of limit point and this is the really crucial notion. Notice again that the "connecting" point v is a limit point of both sets A and B and on the boundary of both.

EXAMPLES. (a) Let $(X, \mathscr{U})$ be the plane with the usual topology and let $A = \{(x, y): x^2 + y^2 < 1\}$. A is the inside of the unit circle. Every point in A is an interior point in A. No point outside of A is an interior point of A. Each point (x, y) such that $x^2 + y^2 = 1$ is a limit point of A though not in A. These statements are illustrated by Figure 4.11.

Here x is a typical interior point of A, y is a limit point of A not already in A and z is neither a limit point nor an interior point. The boundary of A is $\{(x, y): x^2 + y^2 = 1\}$.

(b) Let $(X, \mathscr{U})$ be as in (a) and let $A = \{(x, y): x = 1\}$. Then A has no interior points and all limit points of A are already in A.

(c) Let $(X, \mathscr{U})$ be the real line with the usual topology. Let $A = [0, 2)$. Then each $x \in A$ *greater than* 0 is an interior point of A. The point 2 is a limit point of A not in A. Each point of A is of course a limit point of A and there are no other limit points of A. The boundary of A is the set $\{0, 2\}$, the "end points" of the interval A. The next two examples should convince you of the folly of *proof* by pictures. Things are not at all as they appear!

(d) Now that you think you know what's going on let X be any set and $\mathscr{U}$ the discrete topology on X. If $A \subset X$, then the only interior points of A are those in A and every point of A is an interior point of A. The

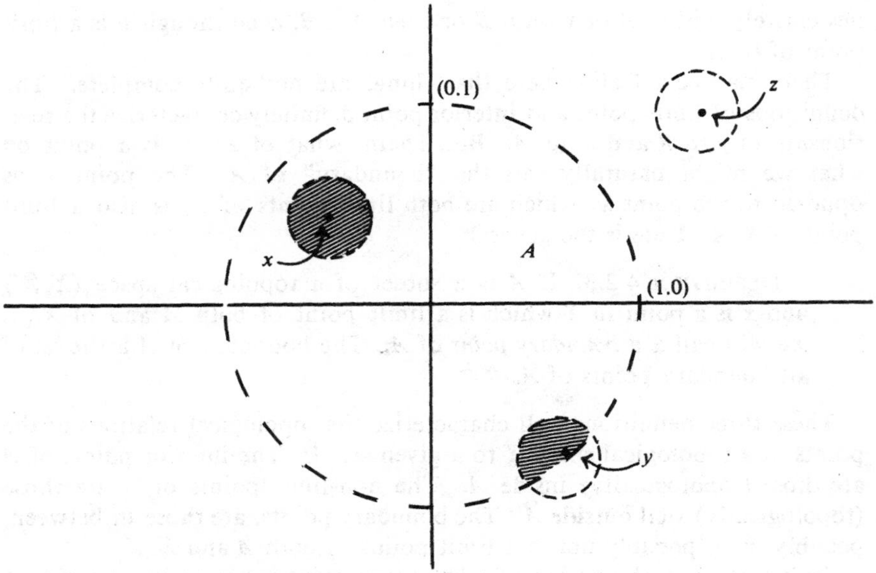

FIG. 4.11

same holds with "interior point" replaced by "limit point." Consequently, the boundary of any set A is empty.

(e) Let a set X have its indiscrete topology. If $\square \subsetneq A \subsetneq X$ then A has no interior points but every point of X is a limit point of A. The boundary of A is X!

With the aid of these two ideas, limit point and interior point, we will now define the most fundametal kinds of sets in a topological space.

Consider again Figure 4.10. We now want to make a topological distinction between the sets A and B each thought of as a whole. The set A has the property that *every* point in A is an interior point of A, since one is to think of $v \notin A$. The set B fails to have this property because of (say) the point u. On the other hand, the set B does have the property that it contains all of its limit points. Said in apparently another way: Every point of $X \backslash B$ is an interior point of $X \backslash B$. Again, a formal definition.

DEFINITION 4.2.4. A set A in a topological space $(X, \mathscr{U})$ is called ***open*** if every point of A is an interior point of A. A set B in X is called ***closed*** if its complement, $X \backslash B$, is open.

A connection with limit points is implicit in this definition as we shall soon see. The set A of Figure 4.10 is open; the set B is closed. Here are some further examples of such sets. These two kinds of sets are fundamental in any topological discussion, particularly one concerned with connectness.

EXAMPLES. (a) The set $A = \{(x, y): x^2 + y^2 < 1\}$ in the plane with the usual topology is open. The set $A' = \{(x, y): x^2 + y^2 \leq 1\}$ is closed and not open. The set $B = \{(x, y\}: x^2 + y^2 = 1\}$ is closed. The set $\{(1, 1)\}$ is closed. The set $X\backslash\{(1, 1)\}$ is open.

(b) The set $A = (0, 1)$ is open in the set of real numbers with the usual topology, for if $x \in (0, 1)$, then $(x/2, (x + 1)/2)$ is a neighborhood of x and $x \in (x/2, (x + 1)/2) \subset (0, 1)$. The set $B = [0, 1]$ is closed. The set $C = [0, 1)$ is not open, because $\mathscr{A} \in C$, and not closed, because $1 \notin C$.

(c) Every set in a space with its discrete topology is open. Hence every set is closed. The notion of a closed set is *not* (repeat: *not*) the negation of the notion of open set, and not even in less esoteric examples! Here are some problems.

Problems. In each problem determine the interior points and the limit points of each set that is given. Classify the set as open or closed, both or neither if this is possible.

(1) Let X be the plane with the usual topology. Answer the above question for each of the following sets.

(a) $A = \{(x, y): y = x \pm 1\}$.

(b) $B = \{(x, y): x + y \leq 2\}$.

(c) Let C be a square with interior but with one vertex removed.

(d) $D = \{(x, y): x^2 + y^2 = 1 \text{ and } (x, y) \neq (1, 0)\}$.

(e) $E = \{(x, y): (x, y) \neq (0, 0)\}$.

These sets are illustrated in Figure 4.12. Think in terms of pictures!

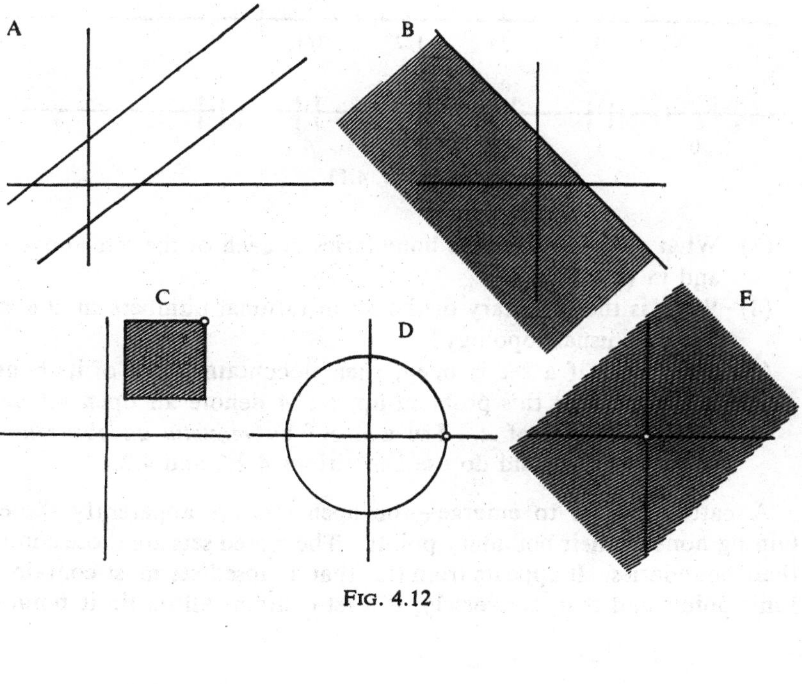

FIG. 4.12

(2) Let X be the real line with the usual topology. Answer the above questions for
   (a) $A = (0, 1]$.
   (b) $B = [2, \infty)$.
   (c) $C = (2, \infty)$.
   (d) $D = [a, b]$ for any choice of $a \leq b$.
   (e) $E = \{x: 0 \leq x \leq 1$ and x is rational$\}$. Here you will need to make use of some fundamental real number facts proven in Sections 2.7 and 2.8.
   (f) $F = \{1, 2, 3, \ldots\}$.
   (g) $G = \{x: x$ is a real number$\}$.
   These are illustrated in Figure 4.13.

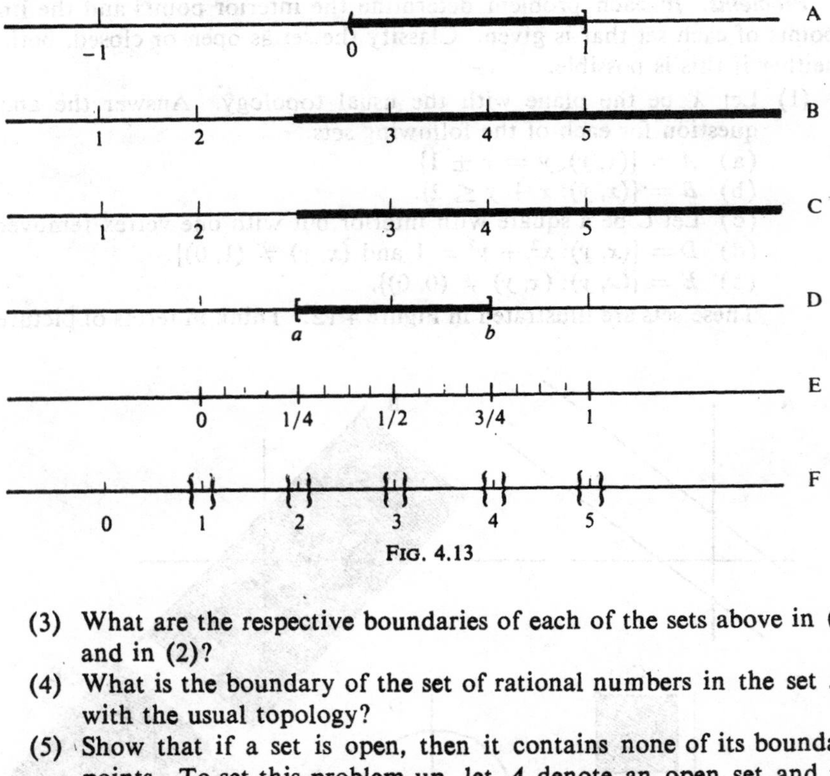

FIG. 4.13

(3) What are the respective boundaries of each of the sets above in (1) and in (2)?
(4) What is the boundary of the set of rational numbers in the set $\mathscr{R}$ with the usual topology?
(5) Show that if a set is open, then it contains none of its boundary points. To set this problem up, let A denote an open set and x a boundary point of A. Try a proof by *reductio ad absurdum* (i.e., suppose $x \in A$) and do use Definitions 4.2.3 and 4.2.4.

A pattern begins to emerge—the open sets are apparently those containing none of their boundary points. The closed sets are those containing their boundaries. It appears from this that a closed set must contain all its limit points and that, conversely, if a set contains all its limit points then it is closed. Let us try to prove this and its converse in an absolutely general setting and see how topological proofs can be made.

THEOREM 4.2.5. A set A in a topological space $(X, \mathscr{U})$ is closed iff every limit point of A belongs to A. Hence, if $\exists$ a limit point of A not in A, then A is not closed. Conversely, if A is not closed, then $\exists$ a limit point of A not in A.

Proof. Suppose A is closed. We will show any limit point of A must be in A. Let x be a limit point of A and suppose $x \notin A$. We will reach a contradiction.

Since $x \notin A$, then $x \in X \backslash A$. Since A is closed, $X \backslash A$ is open. By Definition 4.2.4, every point of an open set is an interior point. That is, $\exists$ a neighborhood $U \in \mathscr{U}$ of x such that $U \subset X \backslash A$. But then $U \cap A = \square$, contradicting the fact that x is limit point of A.

Now we will prove the converse: if A contains all its limit points, then A is closed. To show A is closed we must argue that $X \backslash A$ is open. That is, that any point of $X \backslash A$ is an interior point of $X \backslash A$. Suppose $x \in X \backslash A$. Then $x \notin A$. Since A contains all its limit points, then x is not a limit point of A. By Definition 4.2.1, this means there is a neighborhood $U \in \mathscr{U}$ of x whose intersection with A is empty. In other words, $U \subset X \backslash A$. But this means $X \backslash A$ is open by Definition 4.2.4. This is what we wished to prove.

This is a standard topological argument. Try to see it as a whole. Make careful note of the uses of definitions to prove it. It's nice, clean, pure mathematics, free of any tint of clumsiness or disorder. The work leading up to this theorem is also a good example of how mathematics can be done. Given ideas (open set, closed set) that may have a relationship, look at examples until a pattern emerges. Try to state the apparent rule this pattern implies. If you believe you've hit it right, you can call the rule a theorem by proving it. Something definitive, no matter how major or minor its importance, has been, with certainty, accomplished, This satisfying ability to *definitely* clear up complicated, if idealized, matters, has long attracted and held some of the most gifted minds of man's long history, despite the fact that only an almost invisibly small collection of fellow souls could possibly appreciate what they had done. One could hope for more, perhaps! Here are a couple of "easy" corollaries left for your (hopefully expanding!) mathematical mind.

COROLLARY 4.2.6. A subset A of a topological space is closed iff A contains its boundary.

COROLLARY 4.2.7. In any topological space $(X, \mathscr{U})$, both X and $\square$ are closed sets and both are open sets.

3. THE CLOSURE OF A SET

In the ever present hope of imposing as much structure on thought as is possible, Theorem 4.2.5 implies a further idea. Given *any* set A, Theorem 4.2.5 seems to say that it may be possible to obtain a closed set from A by adjoining to the set A all of its limit points. For example, the interval $(0, 1)$ is not closed in the usual topology of the real line but if we throw in the missing limit points 0 and 1, the resulting set $[0, 1] = (0, 1) \cup \{0, 1\}$ is closed. Let us try to decide if this is a general property by giving a name to the set of all limit points of A and then relating this set to A as best we can.

DEFINITION 4.3.1. If A is a subset of the topological space $(X, \mathscr{U})$, we define the *closure* of A, denoted by $\bar{A}$, to be the set $\bar{A} = \{x: x \text{ is a limit point of } A\}$.

We would be inclined to hope, on the basis of Theorem 4.2.5 and our previous remarks, that $\bar{A}$ is indeed a closed set. Considering the weird examples of toplogical spaces we have seen, we should think it necessary to prove $\bar{A}$ is closed before we can be certain of this. First an example. In $\mathscr{R} \times \mathscr{R}$ with the usual topology let $A = \bigcup_{k=1}^{\infty} A_k$, where $A_k = \{(1/k, y): 0 \leq y \leq 1\}$. The set A appears as the sequence of lines indicated in Figure 4.14. We want to compute $\bar{A}$. Incidentally, this example should give you an idea as to the genesis of the term limit point and you will notice an aspect of infinity about this problem. (You will also notice the absence

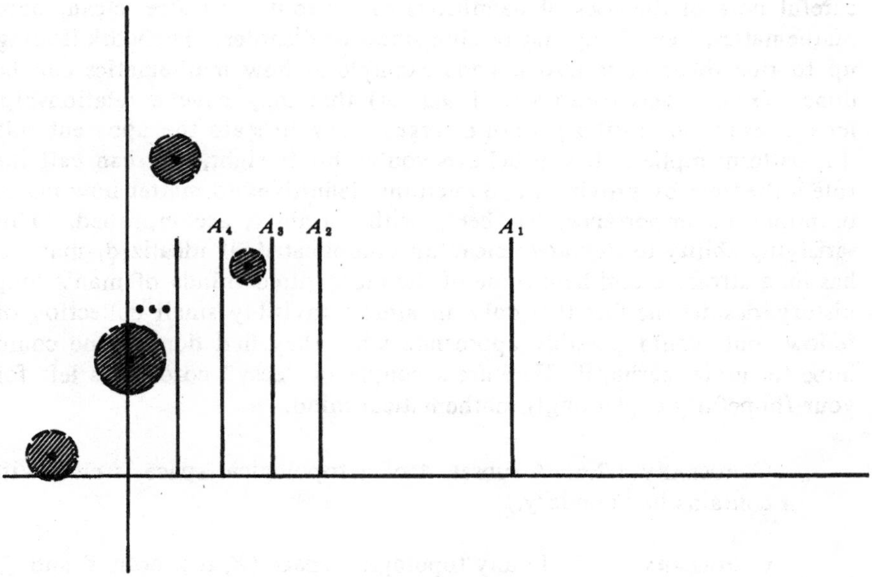

FIG. 4.14

and any mention of infinity, infinite closeness, etc. Such are nicely avoided in this topological context.) The claim, as the drawing suggests, is that $\bar{A} = A \cup \{(0, y): 0 \leq y \leq 1\}$, the latter set in this expression being the "limiting line" of the sequence of lines making up A.

Because every $z \in A$ is a limit point of A, one has $A \subset \bar{A}$. Let us show that any point $(0, \bar{y})$ is in $\bar{A}$ where $\bar{y}$ is a real number between 0 and 1. Suppose $\varepsilon > 0$ is a number and let $U = \{(x, y): \sqrt{(x-0)^2 + (y-\bar{y})^2} < \varepsilon\}$ be a arbitrary, but typical, neighborhood of $(0, \bar{y})$: that is, the interior of a circle of radius ε. By the Archimedean property of $\mathscr{R}$ $\exists$ an integer n such that $n > 1/\varepsilon$, or $1/n < \varepsilon$. The point $(1/n, \bar{y}) \in A_n$ and hence $(1/n, \bar{y}) \in A$. We claim $(1/n, \bar{y}) \in U$ also. Since $\sqrt{(1/n - 0)^2 + (\bar{y} - \bar{y})^2} = 1/n < \varepsilon$, then by definition U, $(1/n, \bar{y}) \in U$. Thus $U \cap A \neq \square$ and since U was arbitrary, $(0, \bar{y})$ is a limit point of A and hence $(0, \bar{y}) \in \bar{A}$. We can also draw a rather nice picture of this argument, as shown in Figure 4.15.

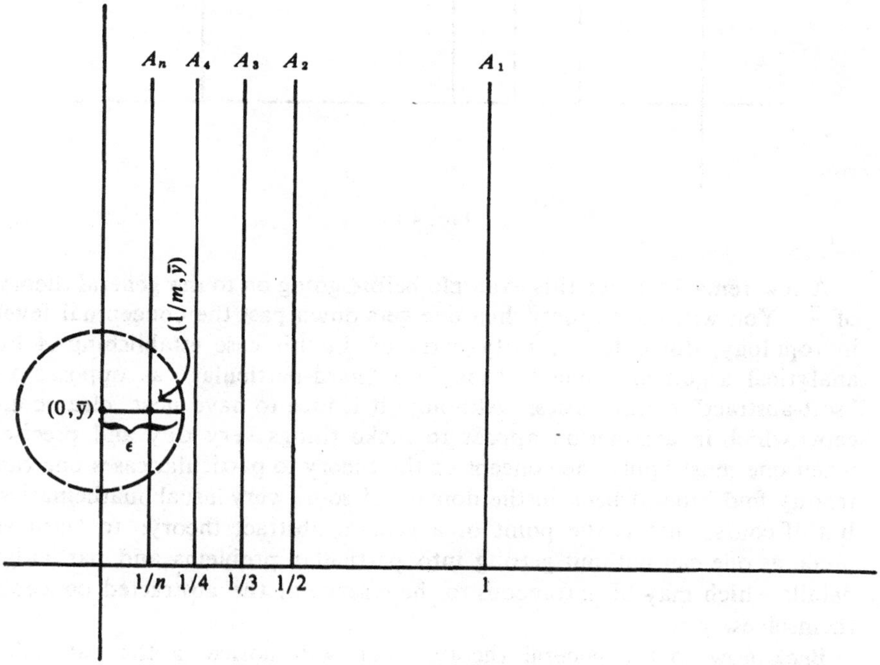

FIG. 4.15

We have only shown that $\bar{A} \supset A \cup \{(0, y): 0 \leq y \leq 1\}$! To complete the proof that $\bar{A} = A \cup \{(0, y): 0 \leq y \leq 1\}$ one must show that if $z \in \bar{A}$, then $z \in A$ or $z = (0, y)$ where $0 \leq y \leq 1$. We indicate by another drawing (Fig. 4.16) the intuitive veracity of this assertion and leave to the reader the analytical arguments. Note that for $z \notin A \cup \{(0, y): 0 \leq y \leq 1\}$ it *appears* that one can construct a neighborhood U of z such that $U \cap A = \square$.

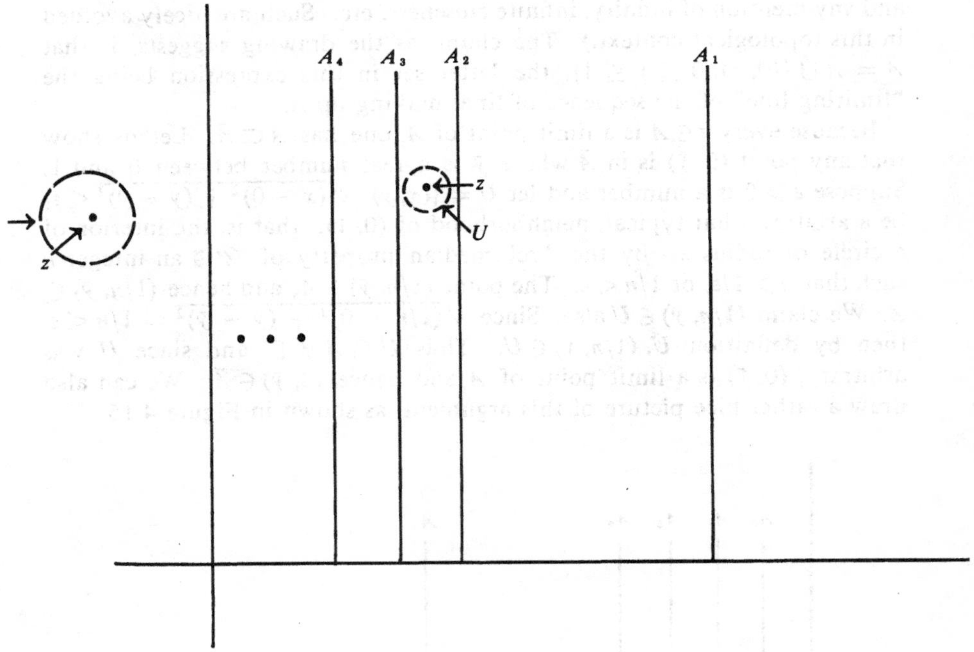

FIG. 4.16

A few remarks about this example before going on to the general theory of $\bar{A}$. You will notice that when one gets down past the conceptual level in topology, down to the nitty-gritty of, in this case establishing $\bar{A}$ by analytical arguments, one is back into "hard-particular" as opposed to "soft-abstract" mathematics. Although it is nice to have neat, clear concepts which in abstraction appear to make things very tidy and precise, when one must apply the concept or the theory to particular cases one can readily find himself back in the domain of some very usual mathematics. But of course that is the point of a general abstract theory: to learn as much as one can without getting into particular problems and particular details which may be extraneous to the essence of the abstracted concepts themselves.

Back now to the general theory. You will notice in the particular example above that $\bar{A}$, is indeed a closed set. We now claim that this is always true for any set A in any topological space, with a hint of proof.

THEOREM 4.3.2. For any set A in a topological space X, $\bar{A}$ is closed. Furthermore one always has $A \subset \bar{A}$.

To prove this, show that $X \backslash \bar{A}$ is open. If $x \in X \backslash \bar{A}$, then x is not a limit point of A. Hence $\exists$ a neighborhood U of x such that $U \cap A = \Box$. If you can conclude that $U \cap \bar{A} = \Box$, then $U \subset X \backslash \bar{A}$ and you are through.

The essential, and missing, point is just a bit subtle! Clearly $A \subset \bar{A}$.

Before trying to prove the next two corollaries draw a picture in $\mathscr{R} \times \mathscr{R}$ of the hypothesis and the conclusion.

COROLLARY 4.3.3. If A is closed, then $A = \bar{A}$ and conversely.

Here it suffices to show $\bar{A} \subset A$. An application of Theorem 4.2.5 should do it. The converse is manifest by Theorem 4.3.2.

COROLLARY 4.3.4. $\bar{A}$ is the smallest closed set containing A. That is, if B is closed and B contains A, then $B \supset \bar{A}$. Consequently, if $A \subset B$, then $\bar{A} \subset \bar{B}$.

Problems.

(1) Describe the closure of each set given in the problems in Section 4.2.

(2) This problem and the next two get at a subtlety necessary to understand "connectedness" as a topological property. Suppose $X = (0, 1)$ and $\mathscr{U}$ consists of all open intervals in $(0, 1)$. What is the closure of $(0, 1/2)$ *in* X? Note that $0 \notin X$ and hence is not even a part of the universe of discussion. What is the closure of $(0, 1)$ in X?

(3) Let $X = (0, 1) \cup (1, 2)$, $\mathscr{U}$ be the collection of all open subintervals of X. Let $A = (0, 1)$, $B = (1, 2)$. What is $\bar{A}$ and $\bar{B}$ *in* X? Would you want to call X "connected?" Is X the union of two disjoint sets each of which is *both open and closed* in X?

(4) Let $X = (0, 1) \cup [1, 2)$. Let $A = (0, 1)$, $B = [1, 2)$. What is $\bar{A}$ in X? Would you call X "connected?" Is X the union of disjoint sets? What is an apparent answer to the question: is X the union of two disjoint sets, both open and closed?

(5) What is $\bar{A}$ for $A = \{(x, y)\colon x$ and y are rational numbers$\}$ in the plane with the usual topology?

(6) In $\mathscr{R} \times \mathscr{R}$ with the usual topology, what is $\bar{A}$ for the "infinite staircase"

$$A = \bigcup_{k=1}^{\infty} A_k$$

where

$$A_1 = \{(x, y)\colon\ y = 1, 0 \leq x \leq 1\}$$

and

$$A_{2k+1} = \left\{(x, y)\colon\ y = \frac{1}{2^k}, \frac{3}{2} - \frac{1}{2^k} \leq x \leq \frac{3}{2} - \frac{1}{2^{k+1}}\right\}$$

and

$$A_{2k} = \left\{(x, y)\colon\ x = \frac{3}{2} - \frac{1}{2^k}, \frac{1}{2^k} \leq y \leq \frac{1}{2^{k-1}}\right\}$$

for $k = 1, 2, 3, \ldots$? You might well begin by drawing a picture.

(7) In $\mathscr{R} \times \mathscr{R}$ with the usual topology what is $\bar{A}$ for

$$A = \left\{(x_n, y_n)\colon\ x_n = 1 + \frac{1}{9} + \frac{1}{9} + \ldots + \frac{1}{9^n}\right.$$

and

$$\left. y_n = 1 - \frac{1}{2} + \frac{1}{2^2} - \frac{1}{2^3} + \ldots + \left(-\frac{1}{2}\right)^n \text{ for } n = 1, 2, \ldots\right\}?$$

(8) Here is one of the great counter examples to many things one might think should be true in an arbitrary topological space. We will have cause to look at it again. At this point it presents no more difficulty than the example of Figure 4.14. Let

$$A = \left\{\left(x, \sin\frac{1}{x}\right)\colon\ 0 < x \le \frac{1}{\pi}\right\}$$

be a subset of $\mathscr{R} \times \mathscr{R}$ with the usual topology. One can view A as the graph of the equation

$$y = \sin\frac{1}{x} \quad \text{for } x \in \left(0, \frac{1}{\pi}\right]$$

and this is show in Figure 4.17. What is $\bar{A}$?

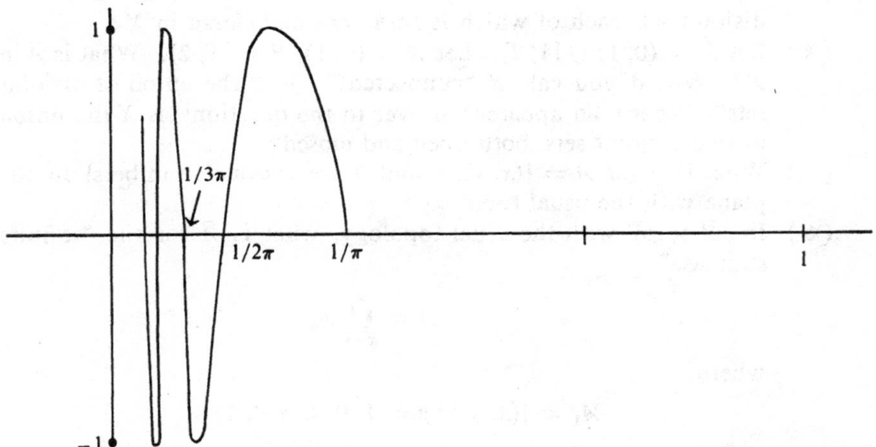

FIG. 4.17

(9) Show that $(1/10^3, \sqrt{2}) \notin \bar{A}$ where A is set indicated in Figure 4.14. To do this, describe a neighborhood U, in analytical terms, which misses A. Can you repeat a similar argument for a point of the form (x, y) where $y > 1$?

(10) Prove that the boundary of a set A in a topological space $(X, \mathscr{U})$ is just what any nice picture tells you it is, namely $\overline{A} \cap \overline{X \backslash A}$. To do this, first do draw a picture, let B denote the boundary of A as defined in Definition 4.2.3, and show that $B = \overline{A} \cap \overline{X \backslash A}$ beginning with the use of Axiom 3 of set theory. That is, show that $x \in B \rightarrow x \in \overline{A} \cap \overline{X \backslash A}$ and conversely.

(11) Prove Corollary 4.2.7 using only Corollary 4.3.3.

4. TOPOLOGY MEETS SET THEORY

In this section we are interested in the topological properties of openess, closedness and closure as they are gained, lost or preserved in the process of set theoretic operations.

There are two kinds of theorems that occur in mathematics. One kind, that which one ultimately seeks, gives a specific answer to a specific question and can be *usefully* applied to a specific example. Examples would be the answers to Questions I and II.

But before one gets such a result, particularly where one is dealing in structural results, one may need to prove a number of theorems whose use is purely theorectical. These results provide a crucial tool to be used in theory to eventually obtain a more practically useful result. Applied to a particular example, these may provide little new knowledge that is not already apparent. Such is the next result. It is crucial to the abstract development of topology, but applied to a particular concrete instance, it seldom gives knowledge not already apparent. It does, however, answer an obvious, interesting, and obviously important, question: If one applies the set theoretic operations of union, intersection and complimentation to two or more open (closed) sets, is the resulting set open (closed)? First a problem to set bounds on the general theory.

Problems. In the usual topology of the real line, show that:

(1) The intersection of the open sets $A_n = (-1/n, 1/n)$ for $n = 1, 2, \ldots$ is *not* open.

(2) The intersection of the sets $A_n = (-1/n, 1/n)$ for $n = 1, 2, \ldots 10$ is open.

(3) The intersection of the collection $\mathscr{C}$ of all sets $[-x, x]$ for $x > 0$, each of which is closed, is closed.

(4) The union of the closed sets $B_n = [0, 1 - 1/n]$ for $n = 1, 2, \ldots$ is *not* closed.

(5) The union of the closed sets $B_n = [0, 1 - 1/n]$ for $n = 1, 2, \ldots, 10$ is closed.

(6) The union of the collection $\mathscr{D}$ of all sets $(-x, x)$ for $x > 0$, each of which is open, is also open.

The results of this problem suggest the following general result as the best conclusion one can hope for.

THEOREM 4.4.1. Let $(X, \mathscr{U})$ be a topological space.

(1) A finite intersection of open sets in X is open.
(2) A finite union of closed sets in X is closed.
(3) The union of any collection of open sets is open.
(4) The intersection of any collection of closed sets is closed.

You will find De Morgan's Theorem 3.6.3 along with Definition 4.2.4 useful in obtaining (2) from (1) and (4) from (3). The proof of (1) will require the use of induction and the second axiom for a topological space. Here is a proof for the intersection of two open sets A_1 and A_2. If $x \in A_1 \cap A_2$, then Definition 4.2.4 says $\exists$ neighborhoods U_1 and U_2 of x such that $x \in U_1 \subset A_1$ and $x \in U_2 \subset A_2$. Since $x \in U_1 \cap U_2$ then by the second axiom for a topological space $\exists$ a neighborhood U of x such that $x \in U \subset U_1 \cap U_2$. But then $U \subset A_1 \cap A_2$ (why?) and hence $A_1 \cap A_2$ is open since x was arbitrary. To complete the proof for *any* finite collection proceed by induction. Incidentally, this proof shows why induction is an important theoretical tool. Suppose one knows that an intersection of any n open sets is open. One must show that the intersection of any $n + 1$ open sets is open. Thus suppose $A_1, A_2, \ldots, A_{n+1}$ are all open sets. One must show $\bigcap_{k=1}^{n+1} A_k$ is open, *given that* $\bigcap_{k=1}^{n} A_k$ is open. The remainder of the argument is left to you.

Now for (3). It is by far easier! Let $\mathscr{C}$ be any collection of open sets and let $A = \bigcup_{C \in \mathscr{C}} C$. Argue that any point $x \in A$ is an interior point of A.

There remains one final concept in a topological space to be related to set theoretic operations. This is the concept of the closure of a set. There is no reason to believe that, for example, $\overline{A \cup B} = \bar{A} \cup \bar{B}$, except that it would be a nice rule if valid. In fact, the closure operation is an extremely complicated one. It turns out that some of the outstanding problems of mathematical analysis can be stated in terms of determining the closure of certain sets in certain topological spaces. We can get a little insight into the wherewithal of that statement by considering the closure of the set $A = \{x: x \text{ is rational and } x^2 < 2\}$ in $\mathscr{R}$. One of its missing limit points is of course $\sqrt{2}$!

If the reader will pause to solve the

Problem.

(1) Give an example of two sets A and B such that $\overline{A \cap B} \subsetneq \bar{A} \cap \bar{B}$.
and
(2) Give an example of a sequence of sets $A_1, A_2, \ldots$ such that $\overline{\bigcup_{k=1}^{\infty} A_k} \neq \bigcup_{k=1}^{\infty} \bar{A}_k$.

he will see that the next theorem is the best one can hope to prove.

THEOREM 4.4.2. If A and B are subsets of a topological space $(X, \mathscr{U})$, then

$$(1)\quad \overline{A \cup B} = \bar{A} \cup \bar{B}.$$

$$(2)\ \overline{A \cap B} = \bar{A} \cap \bar{B}.$$

Let us go back to the nitty-gritty for a moment for a look at an old friend (?) and a thought the reader may have had. This question and its answer concerns a property of a set in a particular topological space, the set $\mathscr{R}$ with the usual topology and which will be of use in the sequel. Consider a bounded set $A \neq \square$ of real numbers. The set A has supremum by the "completeness axiom." If you think about it a minute, the number sup A is the *smallest* number greater than or equal to every number in A. Since a neighborhood U of the number sup A must be an open interval containing it, and hence contain numbers less than sup A itself, one gets the notion that sup A should be a limit point of A. In other words, one suspects that sup $A \in \bar{A}$ is a true statement. For example, if $A = \{x \in \mathscr{R}$: $0 < x < 1$ and x is rational$\}$ then $1 =$ sup A and indeed, $1 \in \bar{A}$. Here now is a proof of the general claim. It is suggested that you develop and draw a picture of the proof as it progresses.

THEOREM 4.4.3. If $A \neq \square$ is a bounded set of real numbers in $\mathscr{R}$, and $\mathscr{R}$ is given the usual topology, then sup $A \in \bar{A}$. Hence, if A is closed, then sup $A \in A$.

Proof. Let $c =$ sup A. We claim c is a limit point of A, hence in $\bar{A}$. Let U be a neighborhood of c. We must show $U \cap A \neq \square$. Since a neighborhood in the usual topology on $\mathscr{R}$ is an open interval, U is of the form (a, b). Since $c \in U$, then $a < c < b$. We claim $\exists x \in A \ni a < x < c$. If not, then $\forall x \in A$, $x \leq a$ since we know $x \leq c\ \forall\, x \in A$. But then, a is an upper bound of A less than c, which is a contradiction. Hence, there must exist $x \in A$ such that $a < x < c$ or what is the same thing, $x \in U \cap A$. Thus $c =$ sup $A \in \bar{A}$. Finally, if A is closed, then by Corollary 4.3.3, $A = \bar{A}$ and hence sup $A \in A$. This completes the proof.

This completes our preliminary study of general topology. In this and previous sections, we have introduced the notion of a topological space and developed the most important concepts of topology, open and closed sets, limit points and interior points in it. Moreover, it being clear that topology is but another, but vastly different and certainly special kind of set theory, we have related these fundamental topological concepts to the fundamental set theoretical concepts. This is the essence of the structural approach. Break down a problem into small pieces and obtain generalized structural knowledge about it. We now have some general theorems relating openess, closedness and closure to the principle ways of forming new sets from old, with no clear direction to these results save that these do appear intuitively related to connectedness in some sense. We also have a special result relating closure to familiar real number concepts. In a sense we are classifying our peripheral knowledge of these topological concepts much as a chemist or biologist classifies elements or organisms. Hopefully we are building to a whole theory which will provide ready directions and answers to pertinent questions.

Problems.

(1) Here is your turn to do some structural mathematics. Consider two sets A and B which are alternatively both open, both closed, A open, B closed and finally, vice versa. What is the story on $A \backslash B$? Look at examples first. Use examples to determine what is not true, and state and prove whatever general result you do believe is true.

(2) Prove that the boundary of a set A is itself alone a closed set using Problem 4.3(10) and Theorem 4.4.1. Then do a proof from definitions alone using no previous theorems.

(3) Prove that $\overline{\bigcup_{k=1}^{n} A_k} = \bigcup_{k=1}^{n} \overline{A}_k$, for any n sets $A_1, A_2, \ldots, A_n$, using Theorem 4.4.2(1) and induction on n.

(4) What is the story on $\overline{A \backslash B}$?

(5) Prove Theorem 4.4.2, using Theorem 4.3.2 and Corollary 4.3.4, then prove it again using only Definition 4.3.1.

(6) Prove that sup A is the largest limit point of a bounded set A of real numbers. To begin, let $c = \sup A$, c' be any limit point of A. One must argue that $c' \leq c$. Try supposing that $c' > c$.

(7) Can you characterize the *smallest* limit point of a set A of real numbers which is bounded below? (A set A is bounded below if $\exists$ a number m such that $m \leq x \ \forall \, x \in A$.)

(8) *The Cantor Set.* Here is one hard problem! The Cantor, or middle third set, is one of the outstanding examples of mathematics because so many conjectures have been demolished upon contact with the facts of its existence. It is described as follows:

From the unit interval $[0, 1]$ remove the middle third, $(1/2, 2/3)$ and let $C_1 = [0, 1/3] \cap [2/3, 1]$ denote the remainder. From each interval in C_1 remove the middle third, $(1/9, 2/9)$ and $(7/9, 8/9)$. Let $C_2 = [0, 1/9] \cup [2/9, 1/3] \cup [2/3, 7/9] \cup [8/9, 1]$ denote what is left. Construct C_3 similarly by removing middle thirds of length $1/3^2$. Assuming that $C_1, C_2, \ldots, C_n$ have been defined, with $C_n = \bigcup_{k=1}^{2^n} [a_k, b_k]$, define C_{n+1} by

$$C_{n+1} = \bigcup_{k=1}^{2^n} \left(\left[a_k, b_k - \frac{2}{3^{n+1}} \right] \cup \left[a_k + \frac{2}{3^{n+1}}, b_k \right] \right)$$

and note that

$$C_{n+1} = \bigcup_{k=1}^{2^n} \left(\left[a_k, a_k + \frac{1}{3^{n+1}} \right] \cup \left[b_k - \frac{1}{3^{n+1}}, b_k \right] \right)$$

as well. Now define the Cantor set C by $C = \bigcap_{n=1}^{\infty} C_n$. A drawing, very rough of course, of C_3 is

It is really hard to get a hold on this set. Its elements are rather wildly located on the interval [0, 1]. But there are two ideas that put what passes for, in mathematics, a locking grip on it. These two ideas are the notion of base 3 decimal representation and the notion of a left-end point. All decimals below are in base 3. Note that in describing this *set* one makes much use of topological ideas as well.

(a) Show the C is a closed set, using Theorem 4.4.1.

(b) Let $x = .a_1a_2a_3\ldots$ where $a_i = 0$ or 2. Letting $x_n = .a_1a_2\ldots a_n$ show by induction on n that $x_n \in C$ for all n. Using (a) and the inequality $|x - x_n| \dot{<} 1/3^n$ conclude that $x \in C$ as well.

(c) Show that once a number gets to be a left-end point of some set C_n then that's all it'll ever be in the process of defining the Cantor set. That is, if x is one of the left-end points a_k of

$$C_n = \bigcup_{k=1}^{2^n} [a_k, b_k],$$

then x is a left-end point of C_{n+j} for $j = 1, 2, 3, \ldots$.

(d) Show that if $x \in C$, then $x = .a_1a_2a_3\ldots$ where $a_i = 0$ or 2. *Hint*: If $x = .a_1a_2\ldots a_n1000\ldots$, then $x = .a_1a_2\ldots a_n0222\ldots$. If $x = .a_1a_2\ldots a_n100\ldots0b_nb_{n+1}\ldots$ with $b_n \neq 0$, show that $x \notin C$, a contradiction.

(e) Prove that $1/4 \in C$ but $1/4$ is not a left-end point or right-end point of any set C_n.

In summary, $x \in C$ iff $x = .a_1a_2a_3\ldots$ (base 3) where a_i is either 0 or 2.

5. CONNECTEDNESS IN A TOPOLOGICAL SPACE

In this section we begin with a precise and useful definition of a *connected object* and close with an initial study of connectedness in $\mathscr{R}$. For our purposes, we will take object simply to mean a set with a topology—a topological space—and we will attempt to define a connected topological space. We have already seen that topological concepts do apparently bear well on our intuitive notions of connectedness. We now want to lend precision to intuition and as before we begin with a motivating example of "connectedness" in the plane (see Fig. 4.18).

Let $X = A \cup B$, $Y = C \cup D$. Notice first that $x \in X$ but $y \notin Y$. One would presumably call X "connected" but would not call Y "connected." If we can abstractly characterize the difference between X and Y as topological spaces, then we will have a definition for *connectedness* which meets with our intuition, our sense of how things should be drawn from examples such as these.

Now there are two ways to proceed, one possibly more initially intuitive than the other. The initially intuitive one is the more difficult to

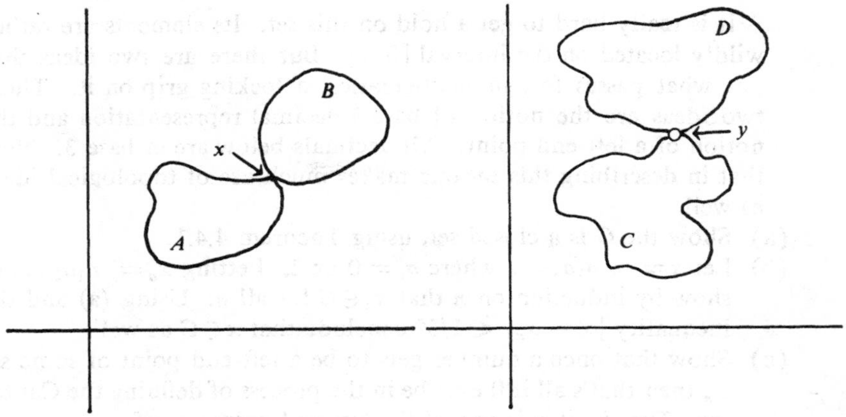

FIG. 4.18

handle in the end so we will instead consider the less intuitive but more topological one. An intuitive method would be to notice that any two points in X can be joined by a "path" or "curve" lying entirely in X, while in Y, no "path" could join a point in C with a point in D without going outside of Y, since $y \notin Y$. What makes this manner of viewing connectedness difficult is the catch word "path"; it would require a definition of "path" to make this idea precise. It turns out that such a definition, while nearly within our grasp, brings in unintended possibilities (look up the term "space-filling curve" in almost any book on topology). We will later consider this alternative view of connectedness in detail.

Another approach relies only on the basic undefined word *neighborhood* and ideas readily derived from it. Consider the set Y. You will notice the following: $Y = C \cup D$, $C \cap D = \square$, $C \neq \square$ and $D \neq \square$, and most importantly, $\bar{C} \cap D = \square$ and $\bar{D} \cap C = \square$. Note also a partial negation of this statement: $Y = C \cup D$, $C \cap D = \square$, $C \neq \square$, $D \neq \square$ and $\bar{C} \cap D \neq \square$ or $C \cap \bar{D} \neq \square$. Now consider X and let us try to imitate this description in X. Let $A_0 = A \backslash \{x\}$, $B_0 = B$, so that $X = A_0 \cup B_0$, $A_0 \cap B_0 = \square$, $A_0 \neq \square$ and $B_0 \neq \square$. But in the set X, which we think of as connected we find $\bar{A}_0 \cap B_0 \neq \square$; this description is the logical opposite of the situation in the "disconnected" space Y. And in what is truly to the point, the abstracted difference in these two descriptions of Y and X is due solely to the existence of X of the "connecting" point x and the non-existence of a corresponding point in Y, the lack of which gives Y the appearance of diconnectedness. This is our strongest case for the general definition.

DEFINITION 4.5.1. Let $(U, \mathscr{U})$ be a topological space. Two non-empty subsets A and B of X are said to be *separated* in X if $\bar{A} \cap B = \bar{B} \cap A = \square$. The set X is *disconnected* if there is a pair of non-empty separated sets A and B such that $X = A \cup B$; the sets A and B are called a separation of X. If X is not disconnected, then X is said to be *connected.*

In the expectation that you might not be too sure as to the need for considering intersections of the type $\bar{A} \cap B$, here is another example. Let $X = \mathscr{R}$ and $Y = (-\infty, 1) \cup (1, \infty)$. The claim is that Y is disconnected and X is apparently not. We will shortly prove that X is connected, but as you will see it is generally more difficult to prove connectedness than disconnectedness. Let $\mathscr{U}$ be the topology for X consisting of all open subintervals of X and let $\mathscr{V}$ be the topology for Y likewise consisting of all open subintervals of Y.

Let $A = (-\infty, 1)$, $B = (1, \infty)$. Then $Y = A \cup B$, $A \neq \Box$ and $B \neq \Box$. Since $1 \notin Y$ and hence is not in the universe of discussion, we also have $\bar{A} = A$ and $\bar{B} = B$ and consequently $\bar{A} \cap B = \bar{B} \cap A = \Box$. Hence A and B are separated and by Definition 4.5.1, Y is disconnected, as appearances would demand in accord with Figure 4.19. On the other hand, if we try a

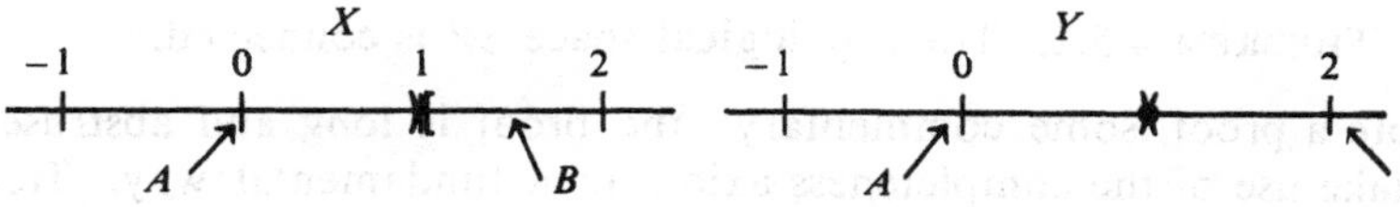

FIG. 4.19

similar thing in X, say $A = (-\infty, 1)$ and, since $1 \in X$ must go somewhere $B = [1, \infty)$ we find that $\bar{A} \cap B \neq \Box$ even though $\bar{B} \cap A = \Box$. This does *not* prove that X is connected by Definition 4.5.1, but only that *this* choise of A and B is not a separation of X.

Problems. For each pair of sets A and B decide whether or not A is separated from B in the indicated space with the usual topology. Draw a picture in each case where this is possible.

(1) A is the set of rational numbers and B is the set of irrational numbers in $\mathscr{R}$.

(2) In $\mathscr{R} \times \mathscr{R}$, $A = \{(x, 0)\colon\ x \geq 0\}$ and $B = \{(t, 1/t)\colon\ t > 0\}$.

(3) In $\mathscr{R} \times \mathscr{R}$, $A = \{(0, y)\colon -1 \leq y \leq 1\}$, $B = \bigcup_{k=1}^{\infty}\{(1/k, y)\colon -1 \leq y \leq 1\}$.

(4) In $\mathscr{R} \times \mathscr{R}$, $A = \{(1, y)\colon\ -1 \leq y \leq 1\}$, $B = \{(0, y)\colon\ -1 \leq y \leq 1\} \cup \bigcup_{k=2}^{\infty}\{(1/k, y)\colon\ -1 \leq y \leq 1\}$.

(5) In $\mathscr{R} \times \mathscr{R}$, $A = \{(0, y)\colon -1 \leq y \leq 1\}$, $B = \{(x, \sin 1/x)\colon 0 < x \leq 1/\pi\}$.

(6) In $\mathscr{R}$, $A = \{1 - 1/n\colon\ n \in \mathscr{N}\}$, $B = \{1\}$.

(7) In the three-dimensional space $\mathscr{R} \times \mathscr{R} \times \mathscr{R}$ with the topology of interiors of non-degenerate spheres, let

$$A = \{(x, y, z)\colon\ x^2 + y^2 + z^2 < 1\}$$

and

$$B = \{(x, y, z)\colon\ z = -1\}.$$

(8) In Problem 8, Section 4.1, let $A = \{\omega\}$, $B = \{1, 2, 3, \ldots\}$.

(9) In a space X with the discrete topology, show that, for any set $A \subset X$, A and $X \backslash A$ are separated.

(10) In a space, with the indiscrete topology, let $a, b \in X$, $a \neq b$. Show that $A = \{a\}$ is *not* separated from $B = \{b\}$.

(11) We will shortly prove that $\mathscr{R}$ with the usual topology is connected. The purpose of this problem is to show that changing the topology of a set ever so slightly can change its properties as a topological space. In $\mathscr{R}$, let $\mathscr{V} = \{(a, b]: a, b \in \mathscr{R}, a \leq b\}$. Show that $(\mathscr{R}, \mathscr{V})$ is a topological space which is *not* connected. Note in the proof of Theorem 4.5.2 (which follows) how crucial is the structure of the usual neighborhoods (a, b) in its proof.

At this point, with the idea of separated sets well in mind, it seems appropriate to prove the existence of at least one connected topological space. For our candidate we choose the set of real numbers, $\mathscr{R}$, with the usual topology $\mathscr{U}$ of open subintervals of X.

THEOREM 4.5.2. The topological space $\mathscr{R}$ is connected.

Before a proof, some commentary: the proof is long and abstruse. It will make use of the completeness axiom in a fundamental way. Here is an indication of why. Suppose the completeness axiom did not hold. For instance, suppose the set $A = \{x: x^2 < 2\}$ did not have the supremum we regularly greet as $\sqrt{2}$. Let $B = \{x: x^2 > 2\}$. Then without the existence in $\mathscr{R}$ of the number $\sqrt{2}$ we would find $\bar{A} \cap B = \bar{B} \cap A = A \cap B = \square$. The picture of $\mathscr{R}$ we would have in mind, without the number $\sqrt{2}$, would appear as shown in Figure 4.20. In other words, the completeness

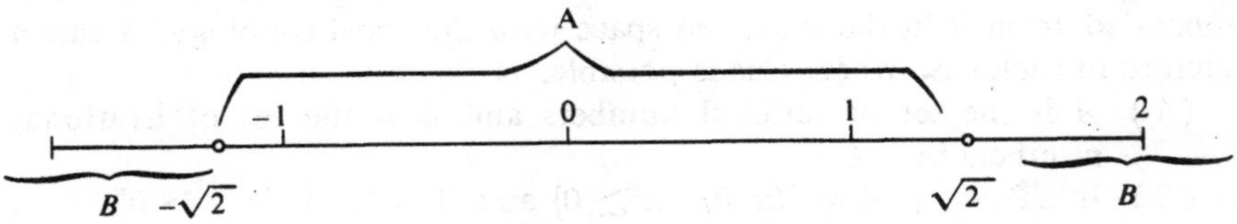

FIG. 4.20

axiom is apparently fundamental to the existence of the connectedness of $\mathscr{R}$. The reader was no doubt long ago to taught to think of $\mathscr{R}$, as a continuum of points—that is, as a "connected" line. Because of the repeated use and constant reiteration of this idea, he is no doubt now firmly convinced that the set $\mathscr{R}$ can be viewed as a connected line and perhaps feel no need to prove this "obvious" fact. It is obvious because he long ago ceased to question it without perhaps ever being convinced in the first place. But somebody did first have to definitely decide that such a view is correct. Here is a proof.

Proof of Theorem 4.5.2. Suppose $\mathscr{R}$ is not connected. Then $\mathscr{R}$ is disconnected and there must be non-empty sets A and B such that $\mathscr{R} = A \cup B$ and $\bar{A} \cap B = \bar{B} \cap A = \square$. Thus if we can produce a point $x \in X$ such that $x \in \bar{A} \cap B$ or $x \in \bar{B} \cap A$, then a contradiction will be obtained and the proof will be complete.

Since $B \neq \square$ $\exists$ an element $b_0 \in B$. From this there are two cases: $a \leq b_0$ $\forall$ $a \in A$ or $\exists$ a number $a_0 \in A$ such that $a_0 > b_0$.

In the first case, A is bounded above by b_0 and hence has a supremum, call it x. We claim $x \in \bar{A} \cap B$ or $x \in \bar{B} \cap A$. Since $x \in \mathscr{R} = A \cup B$ then $x \in A$ or $x \in B$.

Suppose $x \in A$. We claim $x \in \bar{B}$. Let U be a neighborhood of x. According to Definition 4.3.1, we must argue that $U \cap B \neq \square$. Since $x \in U$ and U is a open interval $\exists u \in U$ such that $u > x$. Since x is the supremum for A, $u \notin A$. Since $u \in \mathscr{R} = A \cup B$ then $u \in B$. Hence $U \cup B \neq \square$. Therefore $x \in \bar{B}$ and hence $x \in A \cap \bar{B}$.

Suppose on the other hand that $x \in B$. We claim $x \in \bar{A}$. But this has already been established in Theorem 4.4.3. This completes the proof in the first case.

The second case is that $\exists$ an $a_0 \in A$ such that $a_0 > b_0$. Let $B_0 = \{y \in B\colon y < a_0\}$. Then B_0 is bounded by a_0 and $B_0 \neq \square$ since, at least, $b_0 \in B_0$. Hence B_0 has a supremum, call it x, which is less than or equal to a_0. If we can show that $x \in \bar{B} \cap A$ or $x \in \bar{A} \cap B$ we are through. Again there are two possibilities: $x \in A$ or $x \in B$. If $x \in A$, then, again by Theorem 4.4.3, $x \in \bar{B}$ because x is the supremum of B_0 and hence $x \in \bar{B} \cap A$.

If $x \in B$, we must show $x \in \bar{A}$ to obtain the final contradiction. Let W be a neighborhood of x and let $w \in W$ be such that $w > x$. We must show $W \cap A \neq \square$. If $w \geq a_0$ then $a_0 \in [x, w] \subset W$ since $x \leq a_0$ by choice of x. Hence $W \cap A \neq \square$ since $a_0 \in A$. Suppose then that $w < a_0$. Hence, $x < w < a_0$. If $w \in B$ then $w \in B_0$ by definition of B_0. But since x is the supremum of B_0 this cannot be; hence $w \notin B$. Since $w \in \mathscr{R} = A \cup B$, then $w \in A$ and, hence, $w \in W \cap A$. Therefore, in either of these two cases, $W \cap A \neq \square$. Hence, $x \in \bar{A}$. Since $x \in B$, then $x \in \bar{A} \cap B$, and again a contradiction results.

In all possibilities then, the supposition that $\mathscr{R}$ is not connected leads to a contradiction and hence $\mathscr{R}$ must be connected. This completes this proof.

Now for a digression from the main theory which will hopefully prove interesting. The completeness axiom generally causes difficulty at first encounter. We have just used it to prove that the real number set $\mathscr{R}$ is connected in the usual topology defined by open intervals. Since an axiom that causes one difficulty would conceivably by an axiom one might wish to avoid, we ask the question: is the completeness axiom really *needed* to obtain the real numbers $\mathscr{R}$ as a connected topological space? The discussion preceding the proof of Theorem 4.5.2 suggests, but does not prove, that the answer is "yes."

To definitely answer this, suppose that $\mathscr{R}$ is axiomatized using only the Axioms A.1 through A.5, M.1 through M.5, D.1, O.1 and O.2 of Section 2.7. From Definition O.1 the notion of $a < b$ can be defined and the usual properties of order proven. Hence this set $\mathscr{R}$, thus axiomatized *without* the completeness Axiom C.1, can be given a topology $\mathscr{U}$ consisting of all open intervals $(a, b) = \{x \in \mathscr{R}: a < x < b\}$. The claim is that the following implication holds.

THEOREM 4.5.3. If the set $\mathscr{R}$ with the topology $\mathscr{U}$ just defined is a connected topological space, then Axiom C.1 must hold in $\mathscr{R}$.

In other words, the answer to the question above is: "yes," the completeness Axiom C.1 is *needed* to obtain the connectedness of $\mathscr{R}$, for, by Theorem 4.5.2, C.1 implies that $\mathscr{R}$ is connected and, by Theorem 4.5.3, if $\mathscr{R}$ is connected, then C.1 holds. In yet other words, if you'd like to just forget about the completeness axiom as an assumption in $\mathscr{R}$, then forget about viewing the set of real numbers as a connected set! Indeed forget about finding any connected sets in a familiar setting. Now for a proof of Theorem 4.5.3, which is really quite straightforward. For intuition keep the following example in mind:

Suppose M is the set of the form $(-\infty, b)$, where $b \in \mathscr{R}$. Let $A = (b, \infty)$, $B = \mathscr{R} \backslash A \supset M$. Notice that the one point $b \in \bar{A} \cap B$ is what we do call sup M.

Proof of Theorem 4.5.3. Suppose M is *any* non-empty bounded set in $\mathscr{R}$ whatsoever. We claim that M has a supremum in $\mathscr{R}$, *given* that $\mathscr{R}$ is connected. Since M is bounded, then M has at least one upper bound and hence $A = \{x \in \mathscr{R}: x \text{ is an upper bound for } M\}$ is a non-empty set. (To keep these ideas straight think of A as the collection of all numbers lying to the right of M—greater than everything in M.)

Let $B = \mathscr{R} \backslash A$. Now $B \neq \square$, for otherwise every number (being in A) would be an upper bound for M contradicting the fact $M \neq \square$. Moreover, $\mathscr{R} = A \cup B$ and, since $A \neq \square$ and $B \neq \square$ and we are assuming that $\mathscr{R}$ is connected, either $\bar{A} \cap B \neq \square$ or $\bar{B} \cap A \neq \square$. (Think again of a picture: A is everything in $\mathscr{R}$ to the right of M, B is everything else. There seems to be only one possible point that can connect them, and it appears to be less than every upper bound for M and at the same time greater than or equal to everything in M—in other words the number we would call sup M. Now for a *proof* of that.) There are thus two cases to consider.

Suppose $b \in A \cap \bar{B}$. We claim $b = \sup M$. Since $b \in A$, then b is an upper bound for M. According to Definition O.3, Section 2.7, all we need show is that if (say) a is any upper bound for M, then $b \leq a$. Suppose not, suppose $b > a$ for some $a \in A$. Then the interval $(a, b + 1)$ is a neighborhood of b which misses B. For suppose $t \in (a, b + 1)$. Then

$t > a$ and hence t is an upper bound for M because a is. Hence $t \in A$ therefore $t \notin B$. Hence $(a, b + 1) \cap B = \square$ and hence $b \notin \bar{B}$, a contradiction. Thus $b \leq a$ and $b = \sup M$ exists in $\mathscr{R}$.

In the other possible case, $\bar{A} \cap B \neq \square$, let $b \in \bar{A} \cap B$. Again we claim $b = \sup M$. First of all, b is an upper bound for M. For suppose $x \in M$ and, on the contrary, $b < x$. Then $b \in (b - 1, x)$, a neighborhood of b which must miss A. Because, if $a \in A$, then $x \leq a$. Since $b \in \bar{A}$ this cannot be and hence we must have $x \leq b$. This holds for all $x \in M$ and hence b is an upper bound for M. It remains to be shown that b is smaller than any upper bound for M. Let a be an upper bound for M. If $b > a$ then b would also be an upper bound for M. Hence one would have $b \in A$. But, we already know $b \in B$ and $A \cap B = \square$. Hence we cannot have $b > a$. Thus $b \leq a$ and by definition, $b = \sup M$.

Thus in either case, $\bar{A} \cap B \neq \square$ or $\bar{B} \cap A \neq \square$ we obtain the existence of the number defined as $\sup M$ according to Definition O.3. This completes the proof.

Again some brief commentary, part historical and part technical, though very imprecise. Briefly, Theorems 4.5.2 and 4.5.3 say that connectedness and completeness are equivalent properties in $\mathscr{R}$. We have alluded to these notions from time-to-time as being related. It was pointed out for example that the completeness axiom implies the existence of a number whose square is 2 and thus a number corresponding to the end point of the rotated hypotenuse in Figure 2.9, Section 2.7, thus "connecting" the sets $A = \{x: x^2 < 2\}$ and $B = \{x: x^2 > 2\}$. Richard Dedekind in 1872 firmly established the real numbers on the foundation of the rational numbers using just these ideas. Topology did not exist then, but Dedekind introduced the notion of what is now called a "Dedekind cut" on the set Q of rational numbers. A Dedekind cut is effectively a separation of Q as a topological space in the open interval topology; e.g., the two sets A and B above. Dedekind then used the existence of such "cuts" to assert the existence of a new number *connecting* the two sets A and B which make up the Dedekind cut or separation of Q.[2]

To close this section we prove something else that seems reasonable, a characterization of the possible connected subsets of $\mathscr{R}$, namely that any such set be an interval of numbers. A subset S of $\mathscr{R}$ is called an interval if $\forall x, y \in S$ with $x < y$ the entire set $[x, y] \subset S$. Here is the "expected" result.

THEOREM 4.5.4. If $S \subset \mathscr{R}$ and S is not the union of two separated sets, then S must be an interval.

A hint of proof: Suppose S is not an interval. Then $\exists x, y \in S$ such that

[2] H. Eves and G. V. Newsom: *Introduction to the Fundations and Fundamental Concepts of Mathematics*, p. 222 (New York: Holt, Rinehart and Winston Inc, 1965), contains a concise but good description of the method.

$[x, y] \not\subset S$. Hence $\exists a \in [x, y] \ni a \notin S$. Let $A = S \cap (-\infty, a)$ and $B = S \cap (a, \infty)$. Show that $S = A \cap B$ and A is separated from B, thus obtaining a contradiction.

6. THE GENERAL THEORY OF CONNECTED SETS

Knowing there is at least one connected topological space, and hence that this discussion is not vacuous we turn to the general theory of connectedness in topological spaces. You might care to speculate at this point, in view of the proof of Theorem 4.5.2, as to the probable difficulties in proving that anything more complicated than $\mathscr{R}$ is connected. Using the general theory along with Theorem 4.5.2, we will eventually, in the final chapter, produce many more connected spaces with ease. The plan of this section is to obtain a characterization of connected topological spaces in terms of topological concepts of minimum complexity. It turns out that neighborhoods are a bit too simple, but the next level of topological complexity, the notions of open and closed sets, is quite enough. We will also have cause to prove some general results on the construction of connected sets by way of theoretic and topological operations, for example those of union and closure. To do these things we must begin with an investigation of the connectedness of subsets of a topological space. This concept was not defined by Definition 4.5.1 since that definition applies only to a space as a whole.

We begin with the concept of the relative topology on a subset of a topological space. This is an important but essentially technical device for minimizing the conceptual clutter inherent in the discussion of connectedness of subsets. Consider Figure 4.21 where $Y = A \cup B \subset \mathscr{R} \times \mathscr{R}$. As noted, we have not given any meaning to connectedness for such a subset of a topological space. In the example of Figure 4.21, the relative topology on Y consists of intersections with Y of neighborhoods in the *usual* topology as $\mathscr{R} \times \mathscr{R}$. In other words, the neighborhoods in $\mathscr{R} \times \mathscr{R}$ are cut down to Y to produce the neighborhoods in the relative topology on Y. Some typical neighborhoods in the relative topology are indicated in the shaded portions of Figure 4.21. We will not find it necessary to make a general investigation of the relative topology on a subset. It will suffice as a concept to be used to define the connectedness of subsets and in notivating examples. There is one point though that should be cleared up: the ***relative topology*** is a topology! Think in terms of the above example to interpret the

THEOREM 4.6.1. Let $(X, \mathscr{U})$ be a topological space and let $Y \subset X$. Let $\mathscr{V} = \{Y \cap U: U \in \mathscr{U}\}$. Then $(Y, \mathscr{V})$ is a topological space and $\mathscr{V}$ is called the relative topology on Y.

Proof. According to the axioms for a topological space we must show (1) if $x \in Y$, then $\exists V \in \mathscr{V} \ni x \in V$, and (2) if $V, V' \in \mathscr{V}$ and $x \in V \cap V'$,

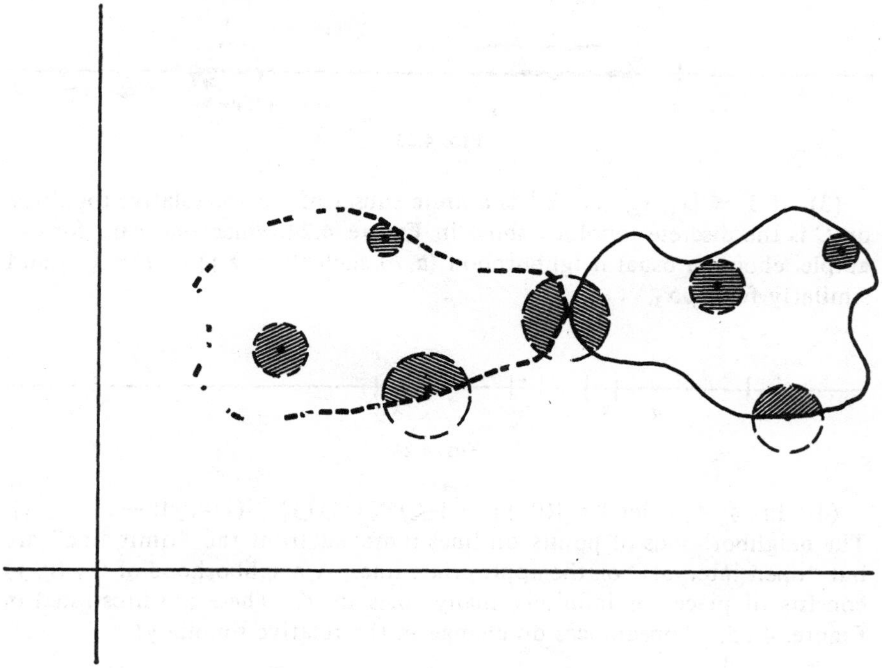

FIG. 4.21

then $\exists V'' \in \mathscr{V} \ni x \in V'' \subset V \cap V'$.

The first is easy. If $x \in Y$, then $x \in X$ and hence $\exists U \in \mathscr{U}$ such that $x \in U$. Let $V = Y \cap U$. Then $x \in V \in \mathscr{V}$. For the second, suppose $x \in V = Y \cap U$ and $x \in V' = Y \cap U'$ where $U, U' \in \mathscr{U}$. Then $\exists W \in \mathscr{U} \ni x \in W \subset U \cap U'$. Let $V'' = Y \cap W$. Then $x \in V'' = Y \cap W$ and $Y \cap W \subset Y \cap (U \cap U') = (Y \cap U) \cap (Y \cap U') = V \cap V'$. Hence the proof.

Here are some graphical examples of the relative topology on a subspace.

EXAMPLES. (1) In the space $\mathscr{R}$ with the usual topology, typical neighborhoods in the relative topology on an open interval (a, b) appear as in Figure 4.22—just as in $\mathscr{R}$ itself.

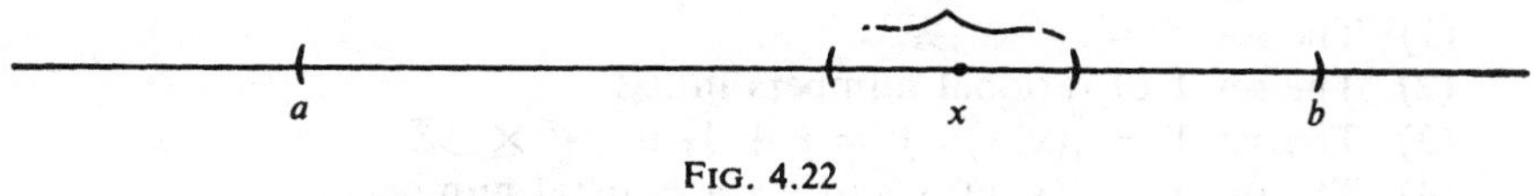

FIG. 4.22

(2) For a closed interval $[a, b]$ one has the neighborhoods of the end points being helf-open intervals (Fig. 4.23).

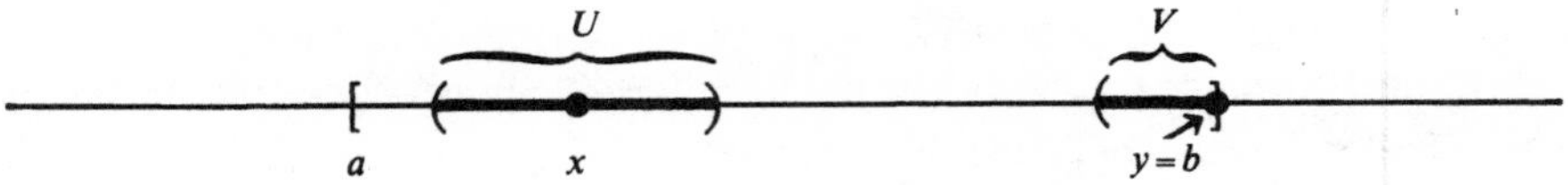

FIG. 4.23

(3) If $Y = \{x_1, x_2, \ldots, x_n\}$ is a finite subset of $\mathscr{R}$ the relative topology on C is the discrete topology show in Figure 4.24, since one can, for example, choose a usual neighborhood (a, b) such that $Y \cap (a, b) = \{x_1\}$ and similarly for $x_2, x_3, \ldots, x_n$.

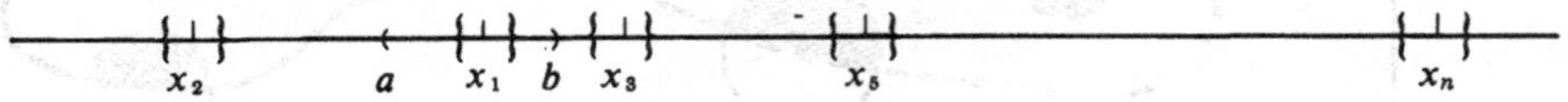

FIG. 4.24

(4) In $\mathscr{R} \times \mathscr{R}$ let $Y = \{(0, y)\colon -1 \leq y \leq 1\} \cup \bigcup_{k=1}^{\infty} \{(1/k, y)\colon -1 \leq y \leq 1\}$. The neighborhoods of points on lines removed from the "limit line" are but "open intervals" on the appropriate line. A neighborhood of say $(0, \bar{y})$ consists of pieces of infinitely many lines in Y. These are illustrated in Figure. 4.25. Appearances do change in the relative topology!

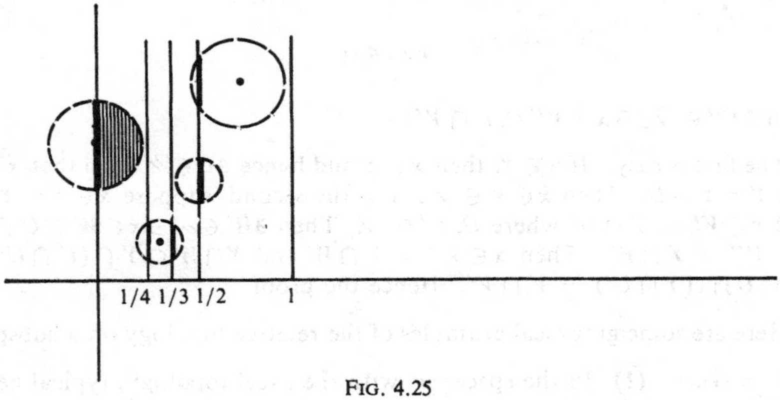

FIG. 4.25

Problems. Describe, graphically, neighborhoods in the relative topology on the indicated subsets of $\mathscr{R}$ or $\mathscr{R} \times \mathscr{R}$, respectively, both with the usual topology.

(1) The set $Y = \mathscr{N} \subset \mathscr{R}$.
(2) The set Y of rational numbers in $\mathscr{R}$.
(3) The set $Y = \{(x, y)\colon\ y = x + 1\}$ in $\mathscr{R} \times \mathscr{R}$.
(4) The set $Y = \{(x, y)\colon x$ and y are rational numbers$\}$.
(5) The set $Y = A$ in Problem 8, Section 4.3.
(6) The set $Y = \{(x, y)\colon\ x^2 + y^2 \leq 1\} \cup \{(x, y)\colon\ (x - 2)^2 + y^2 < 1\}$.
(7) The set $y = (0, 1]$ in $\mathscr{R}$.

With the concept of relative topology we can make a definition of the connectedness of a subset of a topological space which is certain to be in accord with the definition for the whole space. The idea is to cut the universe of discussion down to only the subset considered and use Definition 4.5.1.

DEFINITION 4.6.2. A subset Y of a topological space $(X, \mathscr{U})$ will be called connected if $(Y, \mathscr{V})$ connected (by Definition 4.5.1) in its relative topology $\mathscr{V}$.

For an example, consider the set $Y \subset \mathscr{R} \times \mathscr{R}$ indicated in Figure 4.26,

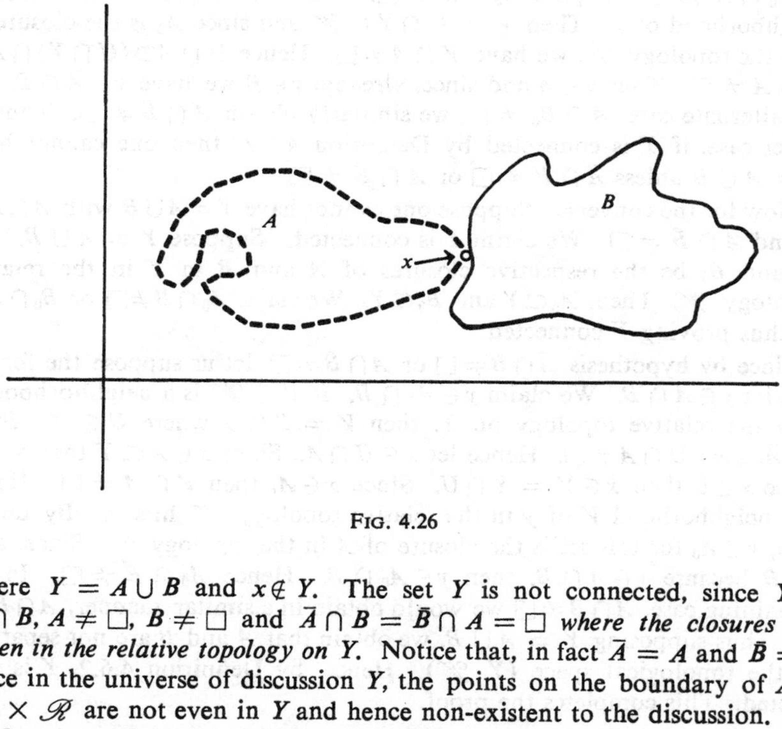

FIG. 4.26

where $Y = A \cup B$ and $x \notin Y$. The set Y is not connected, since $Y = A \cap B$, $A \neq \square$, $B \neq \square$ and $\bar{A} \cap B = \bar{B} \cap A = \square$ *where the closures are taken in the relative topology on* Y. Notice that, in fact $\bar{A} = A$ and $\bar{B} = B$ since in the universe of discussion Y, the points on the boundary of A in $\mathscr{R} \times \mathscr{R}$ are not even in Y and hence non-existent to the discussion.

Now you might get the idea from this example that the non-connectedness of Y should be decidable in $\mathscr{R} \times \mathscr{R}$ as a whole without regard to relative topologies and a restricted universe of discussion. You would be right in that conjecture. The point is that the relative topology gives one a *consistent* and straightforward method of *defining* the connectedness of a subset. We will now characterize connectedness of a subset in terms of the space as a whole without reference to relative topologies. This theorem will greatly simplify subsequent discussion.

THEOREM 4.6.3. The subset Y or the topological space $(X, \mathscr{U})$ is connected iff there do not exist non-empty subsets A and B of X such that $Y = A \cup B$ and $\bar{A} \cap B = \bar{B} \cap A = \square$, where the closures are taken *in X*.

Proof. Suppose Y is connected and $Y = A \cup B$, $A, B \neq \square$. We claim $\bar{A} \cap B \neq \square$ or $\bar{B} \cap A \neq \square$. Let $\mathscr{V}$ denote the relative topology on Y and let A_0 and B_0, respectively, denote the closures of A and B in the topological space $(Y, \mathscr{V})$. (Recall the example just above; there $A = A_0 \subset \bar{A}$, $x \in \bar{A} \backslash A_0$; $B = B_0 \subset \bar{B}$, $x \in \bar{B} \backslash B_0$. Thus closure in the relative topology yields a smaller set and may not even enlarge A.) Both A_0 and B_0 are subsets of Y. Since Y is connected, then by Definition 4.6.2, either $A_0 \cap B \neq \square$ or $B_0 \cap A \neq \square$. Suppose $A_0 \cap B \neq \square$. Then $\exists y \in A_0 \cap B$. Let $U \in \mathscr{U}$, a neighborhood of y. Then $V = U \cap Y \in \mathscr{V}$ and since A_0 is the closure of A in the topology $\mathscr{V}$, we have $V \cap A \neq \square$. Hence $U \cap A \supset (U \cap Y) \cap A = V \cap A \neq \square$. Thus $y \in \bar{A}$ and since, already, $y \in B$ we have $y \in \bar{A} \cap B$. In the alternate case, $A \cap B_0 \neq \square$, we similarly obtain $A \cap \bar{B} \neq \square$. Thus in either case, if Y is connected by Definition 4.6.2, then one cannot have $Y = A \cup B$ unless $\bar{A} \cap B \neq \square$ or $A \cap \bar{B} \neq \square$.

Now for the converse. Suppose one cannot have $Y = A \cup B$ with $\bar{A} \cap B = \square$ and $A \cap \bar{B} = \square$. We claim Y is connected. Suppose $Y = A \cup B$. Let A_0 and B_0 be the respective closures of A and B in Y in the relative topology $\mathscr{V}$. Then, $A_0 \subset Y$ and $B_0 \subset Y$. We claim $A_0 \cap B \neq \square$ or $B_0 \cap A = \square$, thus proving Y connected.

Since by hypothesis $\bar{A} \cap B \neq \square$ or $A \cap \bar{B} \neq \square$, let us suppose the former and let $y \in \bar{A} \cap B$. We claim $y \in A_0 \cap B$. If $V \in \mathscr{V}$ is a neighborhood of y in the relative topology on Y, then $V = Y \cap U$ where $U \in \mathscr{U}$. Since $y \ni \bar{A}$, then $U \cap A \neq \square$. Hence let $x \in U \cap A$. Since $x \in A \subset Y$ then $x \in Y$. Since $x \in U$ then $x \in V = Y \cap U$. Since $x \in A$, then $V \cap A \neq \square$. Hence any neighborhood V of y in the relative topology $\mathscr{V}$ hits A. By definition, $y \in A_0$ for this set is the closure of A in the topology $\mathscr{V}$. Since, also, $y \in B$ because $y \in \bar{A} \cap B$, then $y \in A_0 \cap B$. Hence $A_0 \cap B \neq \square$. In the remaining case, $\bar{A} \cap B \neq \square$ we would obtain in a similar manner, $A \cap B_0 \neq \square$. Thus supposing $Y = A \cup B$, we obtain that A and B are not separated in the topological space $(Y, \mathscr{V})$. Hence, by Definition 4.6.2, Y is connected. This completes the proof.

Problems.

(1) Use Theorem 4.6.3 to determine whether or not the subsets Y in Problems 1, 2 and 6 above are connected. Here you are allowed to assume that sets which "appear" to be connected are connected. For example in Problem 6 assume that the two sets defining Y are connected and determine whether or not they determine a separation of Y in the sense of the conditions in Theorem 4.6.3.

(2) Use Theorem 4.6.3 to show that every subset containing at least two points of a space X with its discrete topology is disconnected

and that every non-empty subset of a space X with its indiscrete topology is connected.

(3) Use Theorem 4.6.3 to further the discussion of connected sets in $\mathscr{R}$. In Theorem 4.5.4 it was shown that a set S which is not interval could be separated in $\mathscr{R}$. Show the converse, that any interval $(a, b) \subset \mathscr{R}$ is a connected subset of $\mathscr{R}$. To begin, suppose $(a, b) = A \cup B$ with $\bar{A} \cap B = \bar{B} \cap A = \square$. Since $[a, b] = \overline{(a, b)} = \overline{A \cup B} = \bar{A} \cup \bar{B}$ there are two possibilities: $a, b \in \bar{A}$ (or equivalently: $a, b \in \bar{B}$) or $a \in \bar{A}, b \in \bar{B}$. In the latter show that $A_0 = (-\infty, a) \cup A$ and $B_0 = B \cup (b, \infty)$ is a separation of $\mathscr{R}$, a contradiction. The former case is slightly different.

The Theorem 4.6.3 furnishes a test for the connectedness of subspaces of a topological space. Let us use it to prove another result, this one quite useful in the sequel, which one should expect to be true. Consider again $\mathscr{R} \times \mathscr{R}$ and suppose A and B are two separated subsets of $\mathscr{R} \times \mathscr{R}$, as in Figure 4.27. Can you draw in this picture a *connected* subset C of $\mathscr{R} \times \mathscr{R}$ which lies in $A \cup B$ but does not lie *entirely* in A or *entirely* in B? The picture strongly suggests you cannot! Here now is the general statement to this effect, valid in *any* topological space.

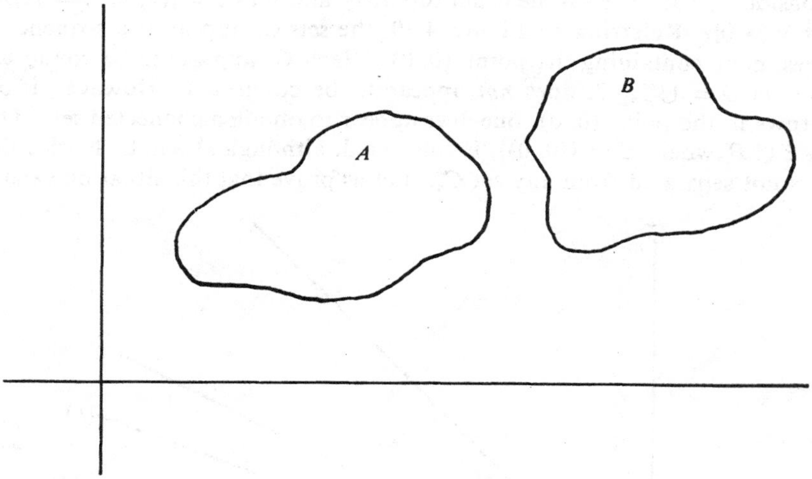

FIG. 4.27

THEOREM 4.6.4. Let $(X, \mathscr{U})$ be a topological space and suppose A and B are separated subsets of X. Suppose C is a connected subset of X and $C \subset A \cup B$. Then $C \subset A$ or $C \subset B$.

This is left for reader to prove using Theorem 4.6.3. A picture drawn strongly suggests a proof. Suppose $C \not\subset A$ and $C \not\subset B$. Let $D = A \cap C$ and $E = B \cap C$. Because $C \not\subset B$ and $C \subset A \cup B$, we know $D \neq \square$. Similarly

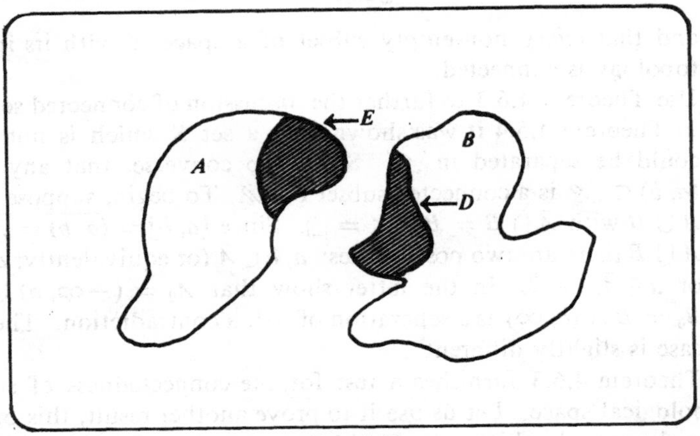

FIG. 4.28

$E \neq \square$ and we have a picture to the effect of Figure 4.28. Show that this is a faithful picture of these conditions. That is, show that $C = D \cup E$ and $\bar{D} \cap E = \square$ and $D \cap \bar{E} = \square$ which, by Theorem 4.6.3, is a contradiction to the assumption that C is connected. Hence a proof.

Now another general result well motivated by a particular example: Consider $\mathscr{R} \times \mathscr{R}$ with the usual topology and let $C_n = \{(x, y): y = (1/n)x$ for $x > 0\}$. Referring to Figure 4.29, the sets C_n appear as a sequence of lines, none containing the point $(0, 0)$. Each C_n appears to be connected. The set $D = \bigcup_{n=1}^{\infty} C_n$ does *not* appear to be connected. However, if one throws in the point $(0, 0)$, one does appear to obtain a connected set. That is, $E \cup D$, where $E = \{(0, 0)\}$, is connected, although D is not. Notice that E is not separated from any set C_n. Let us prove that this situation extends

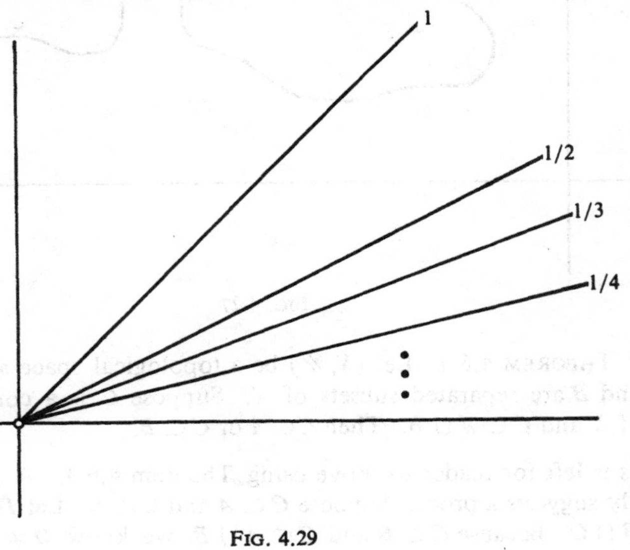

FIG. 4.29

to the ultimate generality.

THEOREM 4.6.5. Let $(X, \mathscr{U})$ be a topological space and let $\mathscr{C}$ be a completely arbitrary collection of subsets C of X, each of which is connected. Let $E \subset X$ be a fixed connected subset of X. Suppose that $\forall C \in \mathscr{C}$, C is not separated from E. Let $D = \bigcup_{C \in \mathscr{C}} C$. Then $E \cup D$ is connected.

To prove this, make use of Theorem 4.6.4. Let $Y = E \cup D$ and suppose $Y = A \cup B$ with $A \neq \square$ and $B \neq \square$. The claim is that Y is connected; by Theorem 4.6.3 it suffices to show that $\bar{A} \cap B \neq \square$ or $\bar{B} \cap A \neq \square$. Since the set $E \subset Y = A \cup B$ is a connected set then Theorem 4.6.4 says $E \subset A$ or $E \subset B$. Suppose $E \subset A$. Conclude that $\exists C \in \mathscr{C}$ such that $C \subset B$ and from this and the hypothesis of the theorem, that $\bar{A} \cap B \neq \square$ or $\bar{B} \cap A \neq \square$ thus obtaining a proof.

The previous two theorems are quite powerful theoretical tools. We will now use these to give two different proofs of yet another expected result. Consider a connected subset A a topological space $(X, \mathscr{U})$. The limit points of A are closely related to A! And, in fact, in the usual diagram, one tends to believe that the limit points of A are connected to A and that $\bar{A}$ itself is connected. The claim is that this holds in a completely general situation. This is really a remarkable result and will be an extremely useful for obtaining connected sets, for, as previous examples have shown, $\bar{A}$ can add quite a lot to the set A. After the general theorem, some neat applications will follow.

THEOREM 4.6.6. If A is a connected subset of topological space $(X, \mathscr{U})$, then $\bar{A}$ is also connected.

Here is a hint of proof dependent on Theorem 4.6.5. In the notion of Theorem 4.6.5 let E denote the connected set A. Let $\mathscr{C}$ be the collection of all sets $C = \{x\}$ such that x is a limit point of A. Then if $D = \bigcup_{C \in \mathscr{C}} C$ we have $\bar{A} = D \cup E$. Hence it suffices to use Definition 4.6.5 to obtain that $\bar{A} = D \cup E$ is connected. For this it suffices to show that if $C = \{x\} \in \mathscr{C}$ then C is not separated from $E = A$. The details remain.

Problems.

(1) Prove Theorem 4.6.4.

(2) Prove Theorem 4.6.5.

(3) Prove Theorem 4.6.6.

These problems are a bit of the harvest from the general theory.

(4) Give that $A = \{(x, \sin 1/x): 0 < x \leq 1/\pi\}$ is connected, conclude that the set $B = A \cup \{(0, y): -1 \leq y \leq 1\}$ is connected, which ought to shake your intuition up a bit!

(5) Let $C = \bigcup_{n=1}^{\infty} \{(x, y): y = 1/nx, x > 0\}$ and let $D = \{(x, y): y = 0, x > 0\}$. Given that all the apparently connected sets are connected, is $C \cup D$ connected? Is $C \cup D \cup \{(0, 0)\}$ connected?

(6) Show that the set Y of Figure 4.25 is not connected.

(7) Consider $\mathscr{R} \times \mathscr{R}$ and let ∞ be a symbol for something *not in* $\mathscr{R} \times \mathscr{R}$. (For example, let ∞ denote the set $\mathscr{R} \times \mathscr{R}$, if you wonder wherefrom such things might come in a rigorous set theory!) Let $X = \mathscr{R} \times \mathscr{R} \cup \{\infty\}$ and define a topology $\mathscr{U}$ on X by letting the neighborhoods at points (x, y) be the usual neighborhoods of interiors of circles. Let a neighborhood of ∞ be any set of the form $U = \{\infty\} \cup \{(x, y): \sqrt{x^2 + y^2} > n\}$ where $n \in \mathscr{N}$. Now here are the questions. Let $A = \{(x, y): x > 0, y = 0\}$. $B = \{(t, 1/t): t > 0\}$. What are $\bar{A}$ and $\bar{B}$ in X? In $\mathscr{R} \times \mathscr{R}$? Is A separated from B in X? Let $C = A \cup B \cup \{\infty\}$. Is C connected in X? Note that C is *not* a subset of $\mathscr{R} \times \mathscr{R}$.

(8) Use Theorem 4.6.4 to give an alternate proof of Theorem 4.6.6 Suppose $\bar{A} = C \cup D$, C separated from D. Since $A \subset C \cup D$ and A is connected, Theorem 4.6.4 says . . . .

(9) Suppose A is connected and $A \subset B \subset \bar{A}$. Prove that B is connected. Apply this to prove again that $\bar{A}$ is connected.

(10) Give an example of a set A in a topological space $X \ni \bar{A}$ is connected and A is not connected.

(11) Let $A = \{(x, \sin 1/x): 0 < x < 1/\pi\}$, $B = \{(x, \sin 1/x): -1/\pi < x < 0\}$. Let $Y = A \cup B$. Is Y connected, supposing A and B are? Is $Y \cup \{(0, y): -1 \leq y \leq 1\}$ connected? Is $Y \cup \{(0, 0)\}$ connected? Is $Y \cup \{(0, 1)\}$ connected? For these latter questions you may wish to consider the space Y as the universe of discussion and use the relative topology on Y.

(12) Show that any interval $[a, b] \subset \mathscr{R}$ is connected.

You can see that the general theory is beginning to bear substantive results. As will be seen, we have in this section established a firm basis for the study of connectedness in $\mathscr{R}$, in $\mathscr{R} \times \mathscr{R}$, and in even more abstract spaces. The final result of this section is a characterization of connectedness in terms of the concepts of open and closed sets, a quite exquisite abstract result.

Consider the set $X = A \cup B \subset \mathscr{R} \times \mathscr{R}$ with its relative topology as a subset of $\mathscr{R} \times \mathscr{R}$, as in Figure 4.30.

Again $x \notin X$, and let X be the universe of discussion. Consider now the set A. Since $x \notin X$, $\bar{A} = A$ and hence by Corollary 4.3.3, A is closed in X. For the same reason $B = B$ and B is closed. Since $A = X \backslash B$ then A is open. In other words, A is both open *and* closed and $\square \subsetneq A \subsetneq X$. Hence the disconnected space X contains a non-empty, proper, open and closed subset.

Consider now the space X of Figure 4.31, with x included. Let us try the same thing. If $A_0 = A \backslash \{x\}$ then A_0 is *open* in the topological space $X = A \cup B$. But A_0 is not closed for $\bar{A}_0 \neq A \neq A_0$. Try as you might you would be unable to find a non-empty proper open and closed subset of

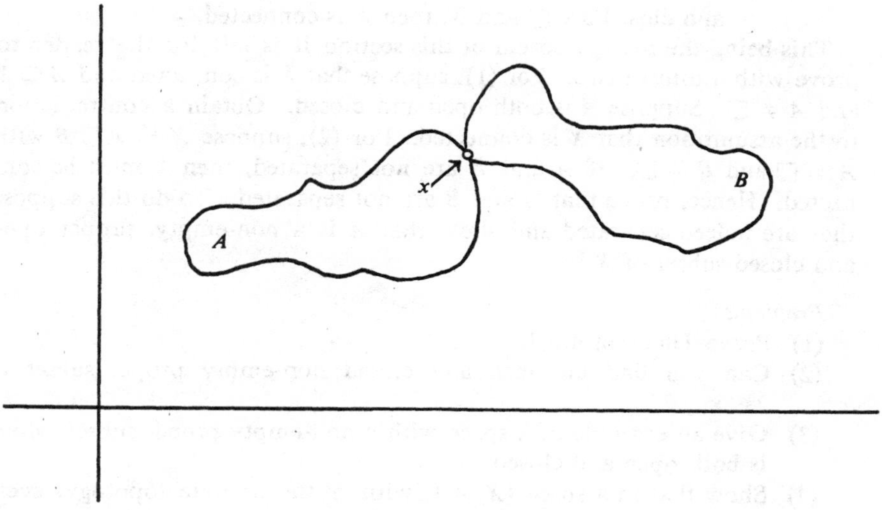

FIG. 4.30

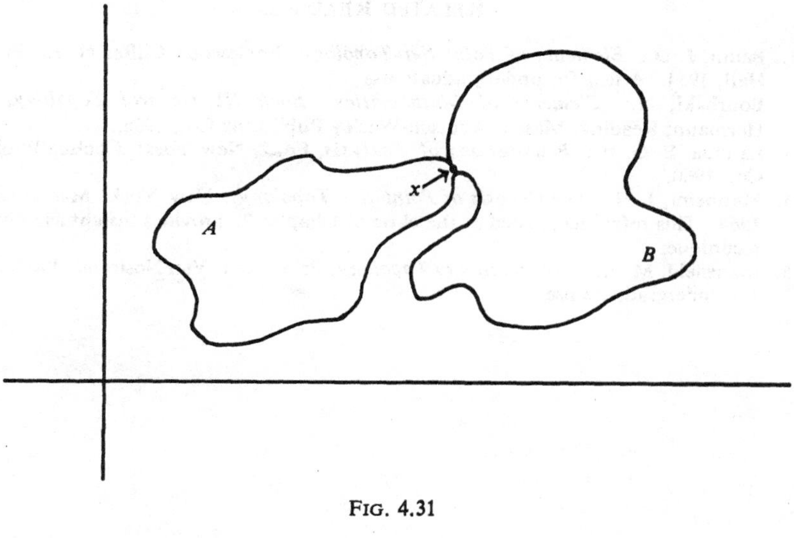

FIG. 4.31

this *connected* space X. That is the claim of this last and completely general theorem on connected topological spaces. First, recall Corollary 4.2.7, which bears directly on it.

THEOREM 4.6.7. Let $(X, \mathscr{U})$ be a topological space.

(1) If $(X, \mathscr{U})$ is connected, then the only subsets of X which are simultaneously both open and closed are X and $\square$.

(2) If the only subsets of X which are simultaneously both open

and closed are $\square$ and X, then X is connected.

This being the main theorem of this section it is left for the reader to prove with a hint or two. For (1), suppose that X is connected and $A \subseteq X$ and $A \neq \square$. Suppose A is both open and closed. Obtain a contradiction to the assumption that X is connected. For (2), suppose $X = A \cup B$ with $A \neq \square$ and $B \neq \square$. If A and B are not separated, then X must be connected. Hence, prove that A and B are not separated. To do this suppose they are indeed separated and show that A is a non-empty, proper open and closed subset of X.

Problems.

(1) Prove Theorem 4.6.7.

(2) Can you find an open and closed non-empty proper subset of $\mathscr{R} \times \mathscr{R}$?

(3) Give an example of a space with a non-empty proper subset which is both open and closed.

(4) Show that in a space $(X, \mathscr{U})$, with $\mathscr{U}$ the discrete topology, every set is both open and closed.

Related Readings

1. Baum, J. D.: *Elements of Point Set Topology*, Englewood Cliffs, N. J.: Prentice-Hall, 1964. A text for undergraduate use.
2. Bourbaki, N.: *Elements of Mathematics*: *Book III, General Topology*, Paris: Hermann; Reading, Mass.: Addison-Wesley Publishing Co., 1966.
3. Landua, E. G. H.: *Foundations of Analysis*, Ed. 2, New York: Chelsea Publishing Co., 1960.
4. Manheim, J. H., *The Genesis of Point Set Topology*. New York: Macmillan Co., 1964. This reference, listed at the close of Chapter 2, provides insight as opposed to technique.
5. Mansfield, M. J.: *Introduction to Topology*, Princeton: Van Nostrand, 1963. A text for undergraduate use.

chapter 5

"The best ideas are simple."

FUNCTIONS

1. INTRODUCTION

In the halls and coffee rooms of any university one can hear groups of mathematicians discussing some non-mathematical matters. Ever so often there will creep into these conversations some very mathematical phrases —"It's one-to-one," "That's not invertible," "Modulo the usual obfuscation," etc.—and everyone better understands what is meant. This happens, not so much because mathematicians do believe in the manner of mathematical expression, but more because the most basic of mathematical concepts are but abstractions of the elementary ways by which men structure their thoughts, when they want to structure them. Said another way, the most fundamental concepts of mathematics are naively simple and occur widely outside of mathematics as will as within.

For example, take the notion of a function that we will shortly define. At its root is the notion of correspondence. We are forever being confronted by the institutions and their graphs and charts illustrating this or that correspondence. On a personal level we find ourselves habitually thinking in terms of relations and correspondences and, in turn, existing correspondences affect the way we think act or feel. We are concerned with the correspondence between the population density of automobiles and the quality of the air we breathe. We are concerned with the correspondence between population growth and the number of persons now on this earth. This corrospondence is not so simple, being a function of the educational level and cultural outlook of each individual as well. We are confronted, too, with the correspondence between the sheer number of existing people and the degree of privacy and peace a person can enjoy, and this in turn influences the way we think, act and feel. If one's concern is the quality of life, then one in these United States may think in terms of the lack or presence of a correspondence between quantity and quality. One can obviously go on and on, but the idea is that correspondences and relative thinking abound. And, clearly, the essential idea in every case is one of associating or relating each element of a given set of possibilities with another element of a given set of possibilities and thinking of this *relation itself as a whole.* Mathematicians have abstracted this altogether common idea and find it to be the most pervasive of all mathematical concepts, even more so than the idea of a set. But while they have carried the concept to the ultimate abstraction it does not generally offer utility in the very general examples of correspondences just mentioned, they are too vague. It may, on the other hand, have great utility in studying in detail a specific aspect of any of the above which is subject to some kind of qualification. The point for the reader is that correspondences (and hence functions) are nothing new to him, though their abstract use may be.

When one begins to think about a given concern in terms of correspondences between its various aspects, one is imposing an order and structure on one's thoughts. This is one reason why mathematicians use the idea—ever the quest for order and structure. For example, consider Question I of the introduction to Part 2. As will be seen, ***the key*** to its solution is that in the removal of points from the interval we do this one at a time, in a sequential order. We remove a first point, p_1, then a second, p_2, and so on. In doing so we realize a *correspondence* between natural numbers and points removed and it is this which will yield an answer to Question I. Consider now Question II where we suppose we have a connected object X and distort or transform it into another object Y. How can one relate Y to X in a useful way ? One simple way which will prove useful is as follows. If we think of X and Y, respectively, as being the set of "points" making up the original object X and the resulting object Y, then one can keep track of things in at least one sense by establishing a definite describable correspondence between each point $x \in X$ and the point $y \in Y$ it is moved to in the distortion.

For example, take a rubber strip, X, designate it to be of unit length, and stretch it "uniformly" to twice its length and call the resulting strip Y. Now imbed the matter on the real number line and describe the distortion thus:

$$x \in X \text{ corresponds to the number } y \in Y \text{ such that } y = 2x.$$

But more yet: mathematicians being always prone to concise expressive notation, this correspondence is sometimes expressed in a more abstruce fashion, in the end subject to better discussion, in certain respects. To wit, this correspondence is described as the set $C \subset X \times Y$ defined by

$$C = \{(x, y): y = 2x\} \quad \text{or}$$
$$C = \{(x, 2x): x \in X\}.$$

Note that this does get the idea of the same correspondence over, using less words, though with some increase in abstraction. In return, one obtains expression in terms of previous concepts.

Such correspondences and their uses are the subject matter of this chapter. We will, as indicated, call those correspondences of use in mathematics *functions*, a designation rooted in mathematical history, so first a bit again of history so that you might better appreciate the essentials. The notion of a function, in this case a correspondence between numbers, was implicit in early developments of calculus and in Descartes' analytic geometry, but a definition of sorts did not appear until 1748 in Euler's *Introduction in Analysis Infinitorum.* Euler defined a function as any analytic expression made up of variables and constants. Thus an equation

$$y = x^2 + 2x - 1$$

described a function. This definition established the long needed *concept* of a correspondence *per se* petween numbers, but it had two defects. What does "analytic" mean? For example, would an equation

$$y = \int_0^x (t^3 - 1)^{1/2} dt$$

establish a function to Euler's satisfaction? Such expressions are now fundamental in many applications of mathematics. Another defect we can appreciate is that such a concept of function simply leaves out too many interesting correspondences. For example, the correspondence between the triple variable of engine timing, air-fuel ratio and exhaust system design and the volume of pollutants set free by an automobile is not clearly a function by Euler's definition. Moreover, mathematicians and statisticians constantly deal with more general phenomena than this. The weakness of pre-20th century concepts of function and functional correspondence lay with the insistence on having some equation, such as

$$y = x + \sin x,$$

before one would admit the existence of a function or correspondence between a number x and another number y. Something more general is now known to be needed for most mathematics. With the advent of set theory, such became possible and the attitude taken was grossly simple: admit *any correspondence between the elements of two sets* as a function (with but one essential restriction we shall shortly mention). Such a function concept, because of its generality as nothing more than a definite existing correspondence, occurs naturally theoughout mathematics.

There is one more important and sometimes overlooked point which we will later emphasize in a more precise way. If one has a correspondence in his thoughts then that correspondence is a thing in itself and is most usually thought of as a whole. As such the correspondence itself is different from any one element or relation it expresses. Thus a function as we will define it will be a correspondence associating each element of one set with exactly one element of the other, this *correspondence to be thought of as a whole,* just as a set is the idea of a collection of distinct things thought of as a whole.

2. RELATIONS AND FUNCTIONS

In fact, a function is but a special kind of set. In order to see that and to develop a definition common to almost all mathematics which is amenable to as thorough and certain a discussion as we will need, we take a somewhat abstract approach. We will begin with the more general idea of a *relation*, which because of its additional generality occurs more often but demands, and yields, less.

Consider the set P of all people in this world and the relationship of father-to-son. If one wishes to, and in other instances mathematicians do wish to, one *can view* this relationship as a certain subset of $P \times P$. In particular, the set

$$S = \{(x, y)\colon y \text{ is the son of } x \text{ and } x \text{ is a male}\}.$$

This does not add one whit to the content of the designated relationship; it does provide an alternate view of the idea of a relationship. But now look at things this way: *any* subset R of $P \times P$ describes some relationship between pairs of people in this world. Two people x and y are related by the relationship R iff $(x, y) \in R$. If it happens that $R = \{(x, y)\colon$ y is the daughter of x and x is a male$\}$, then R is but a previously recognized relationship ordinarily described in more personal terms—the father-daughter relation. The point is that any subset R of $P \times P$ defines some relationship between pairs of people, even though it may not be a useful or interesting relationship, and conversely, any designated relationship between people establishes a subset of $P \times P$. Mathematicians carry the idea one small step further.

DEFINITION 5.2.1. Let A and B be any two sets what so ever. A *relation between A and B* is a fixed subset R of $A \times B$.

Now for a few examples to show how inclusive an idea this is.

EXAMPLE (a) Let A and B be sets. The set $R = \Box \subset A \times B$ is the relation of ***chaos***! Nothing is related to anything. The set $R = A \times B$ is the relation of ***boredom***! Everything is related to everything else, and (on this abstract level) there is nothing more to be said.

(b) Let A be a set. The equality relationship $a = a$ between elements of A can be thought of as the set

$$R = \{(a, a)\colon a \in A\} \subset A \times A.$$

(c) The inequality relationship $a < b$ on the set of real numbers $\mathscr{R}$ can be thought of as the set

$$R = \{(a, b)\colon b = a + c \text{ where } c \text{ is positive}\}.$$

(d) Let $s(t)$ denote the vertical displacement at time t of an object thrown directly upward with initial velocity v_0. As is well known, $s(t) = (-a/2)t^2 + v_0 t$ where a is acceleration due to gravity. The relation between time in flight and vertical displacement is the set $R = \{(t, s)\colon s = -a/2t^2 + v_0 t\} \subset \mathscr{R} \times \mathscr{R}$.

(e) Let A be the set of books in the library, B the set of cards in the card catalog. The set $R = \{(a, b)$: b directs one to book $a\}$ is the relationship established by the card catalog.

(f) Let $X = \{1, 2, 3, 4\}$ and interpret 1 as a grade-school educational level, 2 as high school, 3 as college, 4 as postgraduate. Let P denote the set of all people in the United States. The set $Y = \{(x, p)$: p has completed educational level $x\} \subset X \times P$ is the relationship between each individual and his educational level, if it falls into one of the above categories: if p denotes you, then $(2, p) \in Y$ while $(1, p) \notin Y$. At some later date you may be able to say that $(3, p) \in Y$ and $(2, p) \notin Y$. If a certain person p admits to no educational level, then $(x, p) \notin Y \; \forall \; x \in X$. The set $I = \{(y, a)$: $y \in Y$ has income $a\} \subset Y \times \mathscr{R}$ is the relationship between a given (educational level, person) *pair* y and his income a. The set $C = \{(z, n)$: $z \in I$ has n children$\} \subset I \times \mathscr{N}$ is the relationship between a given ((educational level, person), income) *triple* and its propensity to populate this earth. Notice that $C \subset [(X \times p) \times \mathscr{R}] \times \mathscr{N}$. These things do build up notation wise.

Now for the abstract concept of a function. Any relationship R on a set A to a set B does establish a correspondence between certain elements of A and certain elements of B in this sense: $a \in A$ corresponds to $b \in B$ iff $(a, b) \in R$. In short the most general notion of a correspondence is that of a relation as defined above. For example, the correspondence between a person, his educational level and his income is the relation (I) above. Such a general notion of correspondence is—would you believe—*too general* for much mathematics! In the presence of so much abstraction, mathematicians commonly place one constraint on the correspondences they must often make use of—these must be "single valued." For example, (I) would be such an admissible correspondence, because a given (person, educational level) pair has but *one* income. But R of Example (c) would not be admissible because, for example, $(1, 2) \in R$, $(1, 3) \in R$, $(1, 3/2) \in R$, and so on—too many second elements for a given first element. Only such admissible correspondences as (I) and, for example, those of (b), (d) or (e) are relations to be called *functions*.

DEFINITION 5.2.2. Let A and B be any two sets whatsoever. A *function* between A and B is a nonempty relation $f \subset A \times B$ such that if $(a, b) \in f$ and $(a, b') \in f$, then $b = b'$.

This is a precise statement to the effect that a function on A to B is a correspondence between a's in A and b's in B such that at most one $b \in B$ corresponds to a given $a \in A$—the correspondence is single-valued. For a mildly facetious point, notice that if B has more than one element, then any function f cannot be the relation of boredom (Example (a) above) and hence there must be something interesting about almost any function!

You are about to be inundated with notation, so a word of advice. Don't lose the essential idea amid all the notation. And more, view the notation as but a better way to communicate the idea, that is its sole function, no pun intended. If you view subsequent notation as anything more than a precise way of keeping track of things in a given correspondence, it will become but a series of meaningless symbols. On the other hand, if you do not learn well to *use* the notation to handle certain aspects of thinking in terms of correspondences, then the door of much useful mathematics will be closed to you. The concept and the notation that goes with it are the product of centuries of mathematical thought, beyond the imagination of the most gifted minds of previous epochs; in this day they are as fundamental as the proverbial $1 + 1 = 2$ once was. One further point: you must come to see the place of every bit of abstraction that goes into this development or you will not appreciate what is to be done. We are, for the first time, going to prove some rather non-intuitive, unexpected results that can be handled only in rather abstract terms. The abstract function concept is the only thing that will allow us to do so.

So much for advice and caution, here is some basic notation to be used henceforth. The idea is to avoid the unwieldly aspect of discussing functions as subsets of $A \times B$, now that a definition exists. This notation is the standard one you are most likely familiar with from previous courses.

DEFINITION 5.2.3. Let A and B be sets and f a function between A and B. The *domain* of f is the subset D_f of A defined by $D_f = \{a \in A\colon \exists\, b \in B \ni (a, b) \in f\}$. The *range* of f is the subset R_f of B defined by

$$R_f = \{b \in B\colon \exists\, a \in A \ni (a, b) \in f\}.$$

If $(a, b) \in f$ we will instead write $b = f(a)$, read "b equals f of a" or "b is the function f of a." The element $f(a) \in B$ is also called the "value of f at a." Hence $\forall\, a \in A$, $(a, f(a)) \in f$ when f is thought of as a subset of $A \times B$.

With this definition we begin thinking in terms of correspondences between sets. A function f corresponds the element $a \in D_f \subset A$ to the *one* element denoted by $f(a) \in B$. That is, in essence, the worth of requiring a single-value to the correspondence—so that one knows that $f(a)$ denotes *one and only one* element in the abstract set B. But don't let the notation lose significance—$f(a)$ is but the name applied to whatever corresponds to a; it is a way of keeping track of a and what it corresponds to in B.

Another point, this one, too, of conceptual importance. It is missing the point to speak of "the function $f(a)$"; it is to the point to say "the function f." Why? Conceptually, the function f is, as mentioned earlier, the *whole* correspondence, whereas $f(a)$ is but one element that is a *part* of the correspondence. Would you call a single raindrop "rainfall," as it dampens your brow ? Here is an example of a function and the accompanying notation.

Let $A = [-1, 1] \subset \mathscr{R}$ and $B = [-2, 3] \in \mathscr{R}$, Let $f = \{(a, b)$: b $= +\sqrt{a} + 1$ when $\sqrt{a}$ exists as a real number$\}$. Then $D_f = \{a \in A: \exists\, b \in B \ni (a, b) \in f\} = [0, 1] \subset A$ because $\sqrt{a}$ exists as a real number only for $a \in [0, 1]$. Furthermore, $R_f = \{b \in B: \exists\, a \in A \ni (a, b) \in f\} = \{b \in B: \exists\, a \in A \ni b = +\sqrt{a} + 1\} = [1, 2] \subset B$. Notice that D_f and R_f are in this case both proper subsets of A and B. In the "$f(a)$" notation we would express this function as follows: since $(a, b) \in f$ iff $b = +\sqrt{a} + 1$ and in the notational concept introduced above $b = f(a)$, we have $f(a) = +\sqrt{a} + 1$. Hence, f is the function defined by the equation $f(a) = +\sqrt{a} + 1$ for all $a \in A$ such that $\sqrt{a}$ is a real number.

Here is a good way to think about D_f and R_f: D_f is the collection of all things in A that enter into the correspondence *at all*. More importantly, R_f is the collection of all things in B that *result* from the correspondence—all things in B that are *values* or *results* of applying the correspondence to things in A.

You will notice in the above example that there are lots of "extra" numbers in A and in B which really do not enter into the correspondence at all. We will devote the remainder of this section and the next to refining the notation and concept of a function toward a progressively idealized concept which classifies just such unwanted or undesirable properties. We begin with the idealization that D_f consist of the entire set A.

> DEFINITION 5.2.4. Let f be a function between a set A and to a set B. If $D_f = A$ we will say that f is a function *from* A to B. Moreover, to denote this property we will always write $f: A \to B$, read "f is a function from A to B."

The notation $f: A \to B$ is simply a short way of specifying that $D_f = A$—no more. Thus in the example just above it would be *incorrect* to write

$$f: [-1, 1] \to [-2, 3]$$

but it *is correct* tc write

$$f: [0, 1] \to [-2, 3].$$

Moreover, since in this example it is also true that f is a function between $[-1, 1]$ and $[1, 3]$ and *from* $[0, 1]$ to $[1, 3]$ it is equally correct to write

$$f: [0, 1] \to [1, 3].$$

Finally, one would describe this function thus:

Let $f: [0, 1] \to [1, 3]$ be the function defined by

$$f(x) = +\sqrt{x} + 1,$$

where from this we are to understand that $x \in [0, 1]$ and $f(x) \in [1, 3]$.

EXAMPLES. The following are examples of functions and the uses of notation.

(a) Let A be the set of library cards in the card catalog of the library, B the set of books in the library. Let $f: A \to B$ be the function defined by the relation

$$f = \{(a, b): \text{card } a \text{ directs one to book } b\}.$$

Here, if you wish, $f(a)$ is the book card a directs one to. In this case, $D_f = A$ and $R_f = B$ (in a well run library!).

(b) Given an equation, say $y = \sin 1/x$, one relates to this in a natural way a function. Let $A = \mathscr{R}/\{0\}$, $B = \mathscr{R}$. Let $f: A \to \mathscr{R}$ be defined by $f(x) = \sin 1/x$. Here $D_f = A$ and $R_f = [-1, 1] \subsetneq \mathscr{R}$.

(c) Let A be any set. Fix a subset $E \subset A$ and define a function χ_E: $A \to \mathscr{R}$ by

$$\chi_E(a) = \begin{cases} 1 & \text{if } a \in E \\ 0 & \text{if } a \notin E. \end{cases}$$

If that's not clearly a function, consider it in terms of Definition 5.2.2. Let $\chi_E = \{(a, b): b = 1 \text{ if } a \in E, b = 0 \text{ is } a \notin E\} \subset A \times \mathscr{R}$. This function is called the ***characteristic function*** or ***indicator*** of E and is extremely useful in probability and statistical theory. Here, $D_{\chi_E} = A$ and $R_{\chi_E} = \{0, 1\}$. This is an example of a function definable on any abstract set

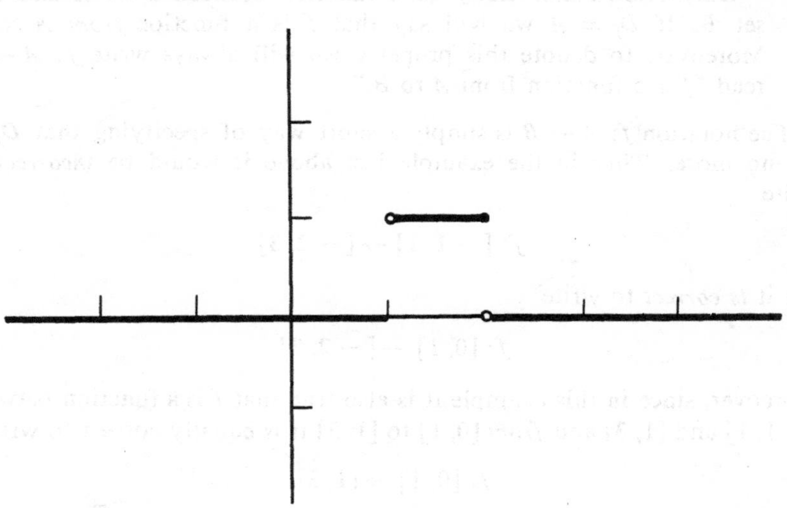

FIG. 5.1

whatsoever.

If we abandon abstraction for the moment and let (say) $A = \mathscr{R}$ and $E = (1, 2]$, then χ_E is a function whose value at any number $a \in (1, 2]$ is 1 and whose value at any point $a \notin (1, 2]$ is 0. Its "graph," in the usual sense, thought of as a subset or $\mathscr{R} \times \mathscr{R}$ is as in Figure 5.1. Notice that its graph is precisely the set

$$\{(a, b): b = 1 \text{ if } a \in E = (1, 2], 0 \text{ if } a \notin E\}.$$

In other words, its "graph" is the set defining the function itself when thought of as a relation $\chi_E \subset \mathscr{R} \times \mathscr{R}$.

(d) Let $A = B = \mathscr{N}$. Define f: $A \to B$ by $f(n) = 2^n$. Here $D_f = A$ and $R_f = \{2, 4, 8, \ldots\} = \{k \in \mathscr{N}: \text{k} = 2^n \text{ for some } n\}$.

(e) Let us distort the half-open unit interval to an interval of infinite length. Let f: $[0, 1) \to \mathscr{R}$ be defined by $f(x) = 1/(1 - x)$. Here $D_f = [0, 1)$ and $R_f = [1, \infty)$.

In calculus and in analytical geometry one habitually identifies functions, graphs and equations as nearly identical things. This is fine in many cases. Being familiar we will not dwell on it, but on another view of more use to us. Given two sets, A and B, and a function, f, between A and B, we can think of a diagram as in Figure 5.2. The idea is that each $a \in D_f \subset A$ corresponds through the function f to an element $b \in R_f \subset B$.

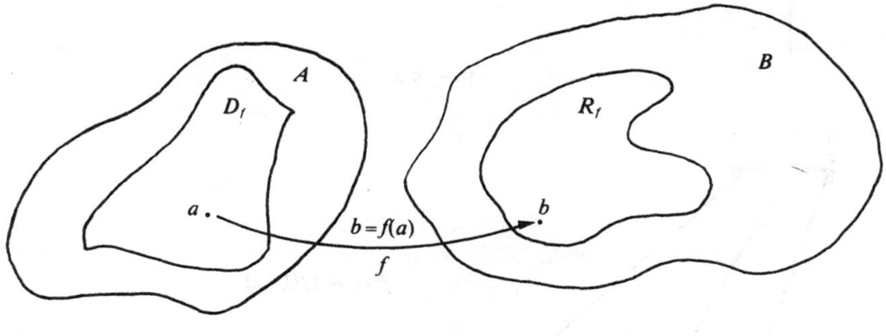

FIG. 5.2

As the sets and functions become more particular we can draw more particular pictures. For Example (d) above we can think in terms of

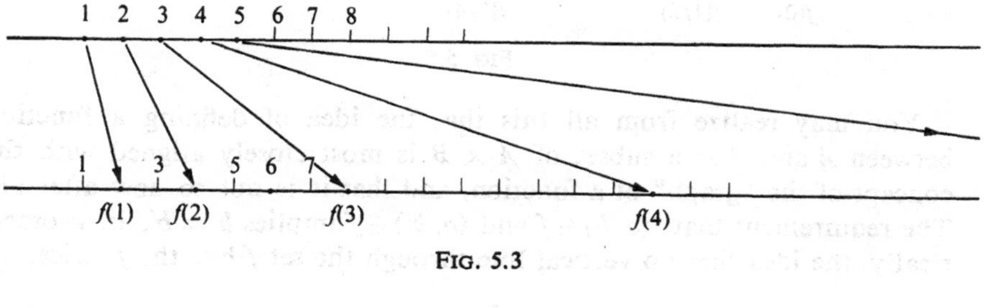

FIG. 5.3

Figure 5.3. If, on the other hand, we wish to view things in the usual way, we have a graph in $\mathscr{N} \times \mathscr{N}$ as in Figure 5.4. For the function f of (*e*), we have Figure 5.5. In the usual graph in $\mathscr{R} \times \mathscr{R}$, one instead has Figure 5.6.

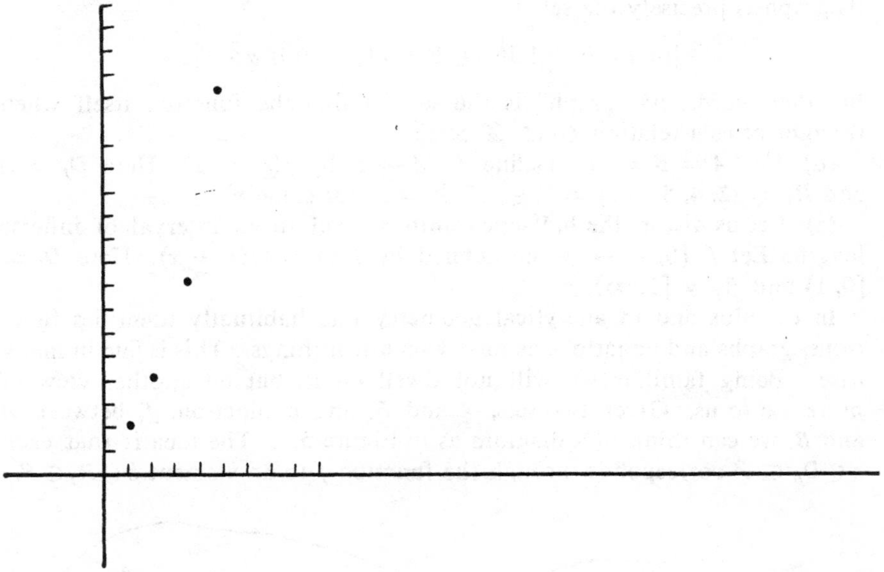

FIG. 5.4

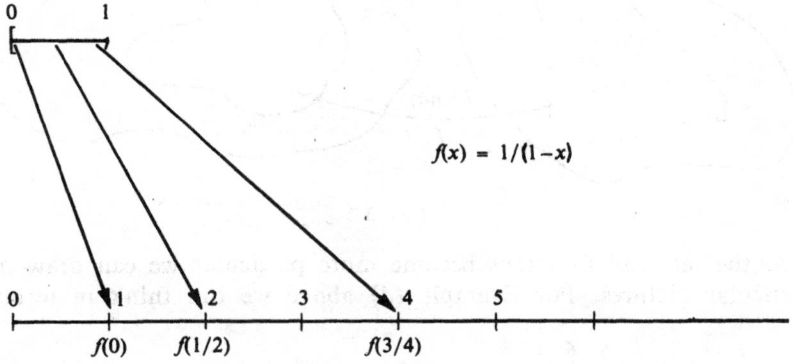

FIG. 5.5

You may realize from all this that the idea of defining a function between A and B as a subset of $A \times B$ is most closely aligned with the concept of the "graph" of a function, and that it is not so new after all. The requirement that $(a, b) \in f$ and $(a, b') \in f$ implies $b = b'$, is, geometrically, the idea that no vertical line through the set f hits the f twice.

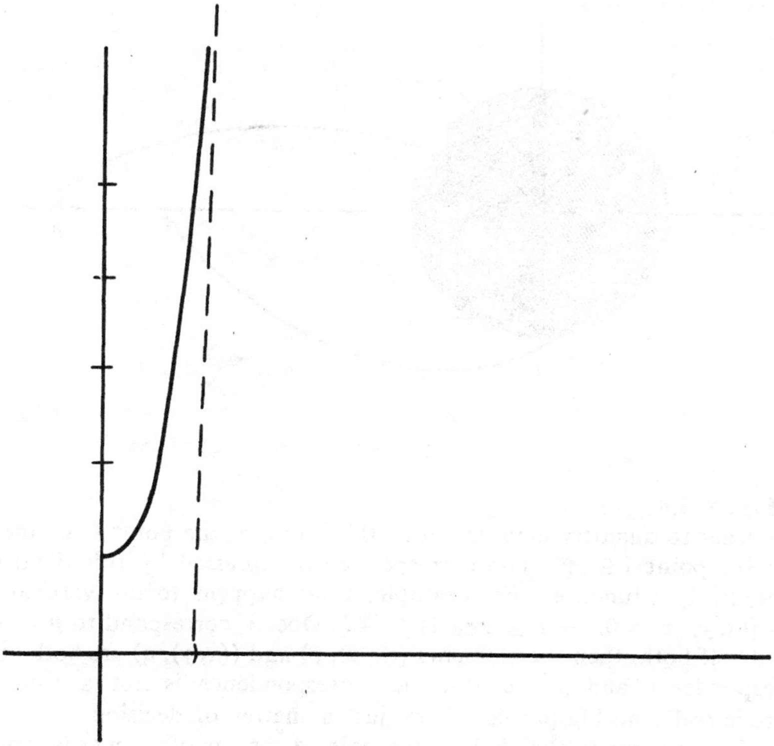

FIG. 5.6

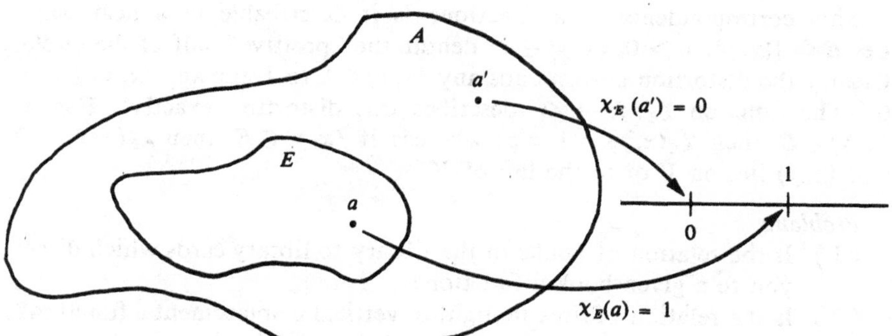

FIG. 5.7

Figure 5.7 is a representation of the function in Example (c).

Our final example will be related to the problem of distorting a connected set. Consider the unit circle in $\mathscr{R} \times \mathscr{R}$ and its interior and call this set X. Let us distort X by taking that half of X on the positive side of the vertical axis, tearing it off and then shrinking it to a point p. Do the same to the remaining piece of X, shrinking it to another point q.

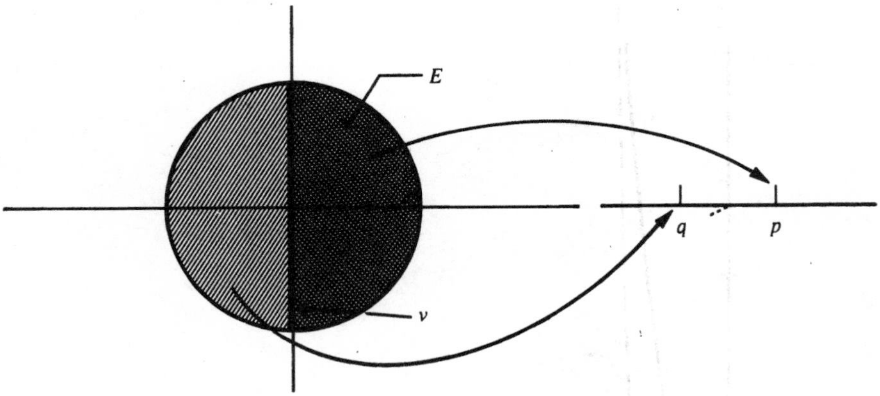

FIG. 5.8

See Figure 5.8.

In order to quantify matters, let us think of q as the point 0 in the set $\mathscr{R}$, p the point $1 \in \mathscr{R}$. The correspondence expressed by this distortion is *not* clearly a function. For example, what happens to the vertical line $V = \{(x, y): x = 0, \ -1 \leq y \leq 1\} \subset X$? Does V correspond to $p = 1$ or $q = 0$? If both, then the elements $((x, y), \text{p})$ and $((x, y), \text{q})$ are both in the correspondence and $p \neq q$ and the correspondence is not a function. There is really no big problem here, just a matter of decision and precision. Let us agree to shrink V to the point q, or, in other words, correspond any $(x, y) \in V$ to the point q.

This correspondence is a function. Is it describable in a neat way? Let $E = \{(x, y): x > 0, (x, y) \in X\}$ denote the "positive" half of the circle. Clearly the distortion corresponds any $(x, y) \in E$ to 1 and any $(x, y) \notin E$ to 0. The function $\chi_E: X \to \mathscr{R}$ describes this distortion exactly! For, if $(x, y) \in E$, then $\chi_E(x, y) = 1 = p$; whereas if $(x, y) \notin E$, then $\chi_E(x, y) = 0$ and (x, y) lies on V or to the left of V.

Problems

(1) Is the relation of books in the library to library cards which direct you to a given book a function?

(2) Is the relation of time in flight to vertical displacement a function? If an object is thrown upward and falls freely, is the relation of vertical displacement to time in flight a function?

(3) Suppose that three aspects of automotive engine design influence the emission of pollutants. Is the relation of each possible triple (of these design aspects) to the volume of pollutants a function? Would you expect that the reverse relation is a function?

*(4) What is the "natural" domain of the function indicated by the formula $y = 1/(x^2 - 1)$ for x a real number? Define this function and specify its range.

(5) Let $A = \{1, 2\}$, $B = \{a, b, c\}$. Describe all possible functions with domain A and range a subset of B. For example, $f(1) = a$, $f(2) = c$ is one such function. How many functions are there with domain B and range a subset of A? Which of the two notations A^B or B^A is more natural for the set of all functions from A to B?

(6) Show that the definition of a function $f: A \to B$ comes down to this: f maintains equations in A as equations in B. That is, a *relation* f is a *function* iff $a = a'$ in A implies $f(a) = f(a')$ in B.

(7) Show that a relation f is a function iff $(a, b) \in f$ and $(a', b') \in f$ and $b \neq b'$ implies $a \neq a'$. That is, f is a function iff $f(a) \neq f(a')$ implies $a \neq a'$.

(8) Define a function f which "tears" a unit interval in two at its midpoint and "stretches" each part "uniformly" to twice its length.

(9) Take the real line R and roll it up into a circle of radius 2 without a "north pole" and centered at (0, 1) in the plane $\mathscr{R} + \mathscr{R}$. That is, define a function which describes the correspondence indicated below

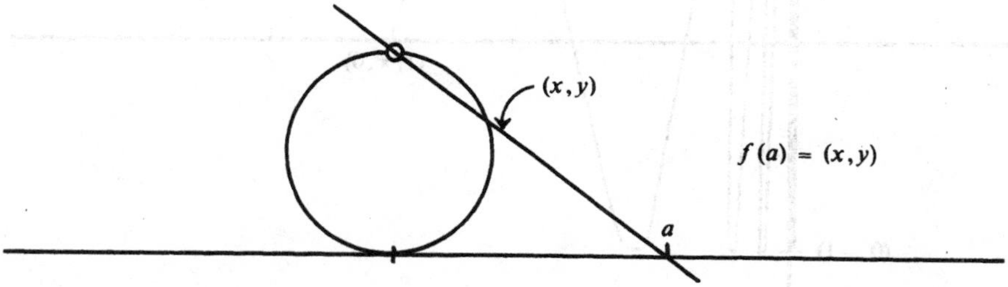

Hint: Find the intersection of the line $y = (-2/a)x + 2$ and the circle $x^2 + (y - 1)^2 = 1$.

(10) Which of the following two functions best describes the trans-

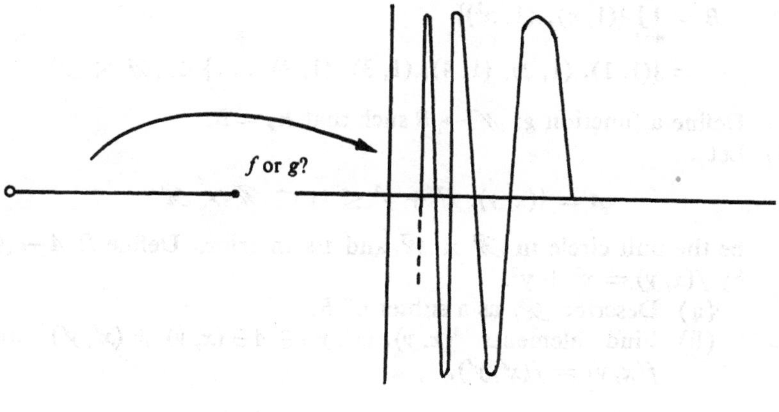

formation or distortion indicated by the figure, where a straight line open on one end is stretched and warped infinitely often.

(a) The function f: $(0, 1/\pi] \to \mathscr{R}$ given by $f(x) = \sin 1/x$ or

(b) The function g: $(0, 1/\pi] \to \mathscr{R} \times \mathscr{R}$ given by

$$g(x) = \left(x, \sin \frac{1}{x}\right)?$$

Hint: What are R_f and R_g subsets of respectively?

(11) Consider a straight line L represented by the interval $[-1, 1/\pi]$ and distort L so that it looks like

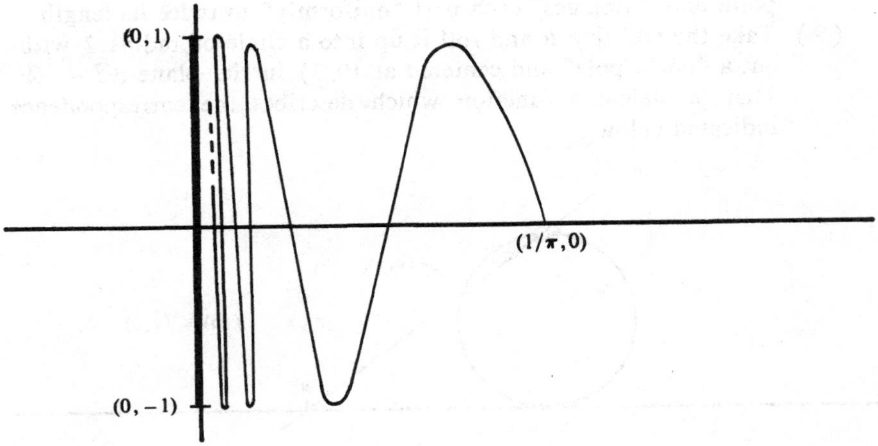

Define a function describing this distortion, such that $f(-1) = (0, 1)$, $f(0) = (0, -1)$ and $f(1/\pi) = (1/\pi, 0)$

(12) Let S be a set. Define a function ϕ: $S \to 2^S$.

(13) Let

$$B = \bigcup_{n=1}^{\infty} \{(1, n), (1, n^2)\}$$
$$= \{(1, 1), (1, 2), (1, 4), (1, 3), (1, 9), \ldots\} \subset \mathscr{R} \times \mathscr{R}.$$

Define a function g: $\mathscr{N} \to B$ such that $R_g = B$.

(14) Let

$$A = \{(x, y): x^2 + y^2 \leq 1\} \subset \mathscr{R} \times \mathscr{R}$$

be the unit circle in $\mathscr{R} \times \mathscr{R}$ and its interior. Define f: $A \to \mathscr{R}$ by $f(x, y) = x^2 + y^2$.

(a) Describe $\mathscr{R}_f$ as a subset of R.

(b) Find elements $(x, y), (x', y') \in A \ni (x, y) \neq (x', y')$ and $f(x, y) = f(x', y')$.

(c) Draw a figure analagous to those of Figures 5. 2, 5.3 and 5.5 to illustrate f.

(15) Define a function with domain the set X of Problem 8, Section 4.1, and range a subset of [0, 1].

(16) Take an ordinary coin and flip it 3 times. Define a function describing the outcome; e.g., $f(1) = H, f(2) = T, f(3) = H$. Be sure to specify domains and ranges. Define functions such that each of the possible outcomes is described by exactly one of these functions.

(17) Most quantitative correspondences you have ever observed are either between numbers, or, at most, points in the plane $\mathscr{R} \times \mathscr{R}$. Perhaps you've wondered why mathematicians go in for such generality, considering functions between abstract sets. To help allay some suspicions you may have, let us consider an example from probability theory. Take two coins, say two pennies. Flip them simultaneously. The possible outcomes are (T, H), (T, T), (H, T), (H, H) where H represents "heads," T, "tails," and the outcome of the coin from the left hand is listed first.

Now probability is, at its base, the assignment (correspondence) of a *numerical value to a logical thought process.* Seeing there are four possible outcomes to flipping two pennies simultaneously, one assigns a probability of 1/4 to any one of them. Consider now the possibility of the outcome (T, T) *or* (H, H). Here either of two possibilities of four is satisfactory and theprob ability of (T, T) or (H, H) is said to be 1/2. Note that 1/2 is the sum of the separate probabilities. Let us consider how to represent all possible outcomes. Let $U = \{(T, T), (T, H), (H, H), (H, T)\}$. Clearly, the set of all possible outcomes is represented by the set $\mathscr{E} = 2^U$ of all possible subsets of U. For example, the set $\{(T, T), (H, H)\} = \{(T, T)\} \cup \{(H, H)\}$ represents the outcome of both tails or both heads on both coins. For an outcome or event $E \in \mathscr{E}$ let $P(E)$ represent the probability of the outcome E. Clearly, we should desire $P(\square) = 0$ since there will never be any empty outcome ! Clearly, $P(U) = 1$ since there will always be some outcome! Let $E_1 = \{(T, T)\}$, $E_2 = \{(H, H)\}$, $E = \{(T, T), (H, H)\}$. We have already observed that $P(E) = 1/2 = 1/4 + 1/4 = P(E_1) + P(E_2)$. Since $E = E_1 \cup E_2$, this becomes $P(E_1 \cup E_2) = P(E_1) + P(E_2)$.

Here now is the abstract definition of a probability on a finite set. (Incidentally, this definition, in a slightly more general form, dates back to only the year 1931 and the work of the Russian mathematician Kolmogorov.) A probability on a collection $\mathscr{E}$ of subsets (events) E of a finite set U is a function $P: \mathscr{E} \to [0, 1]$ such that

(a) $P(\square) = 0$, $P(U) = 1$
(b) For any $A, B \in \mathscr{E}$ with $A \cap B = \square$,
$P(A \cup B) = P(A) + P(B)$.

Recall the correspondence between the logical quantifier v, the set operation $\cup$, and the observation that probability is a measure of logic, in interpreting (b). The assumptions (a) and (b) are the axioms for a finite probability space, and are, as noted, of quite recent origin. Notice, too, that they depend on the existence of a set theory and an abstract concept of function.

Now for some problems.

(a) Work out the values of $P(E)$ for all 16 subsets (event) E in the example above where $\mathscr{U} = \{(T, T), (H, H), (H, T), (T, H)\}$. You may and should use (b) to do so. For example,

$$\begin{aligned} P(\{(T, T), (T, H)\}) &= P(\{(T, T)\} \cup \{(T, H)\}) \\ &= P(\{T, T\}) + P(\{T, H\}) \\ &= \frac{1}{4} + \frac{1}{4} = \frac{1}{2} \end{aligned}$$

For each event, do "see" that the answers agree with your "logical intuition."

(b) Find events A and B such that

$$P(A \cup B) \neq P(A) + P(B).$$

(c) Prove that $P(A \cup B) = P(A) + P(B) - P(A \cap B)$ for any two events $A, B \in \mathscr{E}$. Does this abstract equation measure well your good (probability) sense?

(d) Does the function P assign proper measure to the logical connective "and"? That is, does P assign proper values to sets of the form $A \cap B$ which represent of course events of the form: event A and event B?

(18) Here is an extremely important kind of function for mathematics and its applications, including, especially, a sophisticated theory of probability. This one is defined on quite an abstract set. We assume a knowledge of integral calculus and at least the idea of a continuous function. Here we want to consider a set whose elements are themselves functions! Let A denote the set of all continuous functions on the interval $[0, 1] \subset \mathscr{R}$. For example, the function f defined by $f(x) = x^2$ is an *element*, a point in A. One cannot draw a picture of the set A. Since any f in A is continuous on $[0, 1]$ then $\int_0^1 f(x)dx$ exists and is a real number. Define $I: A \to \mathscr{R}$ by by $I(f) = \int_0^1 f(x)dx$. For example, if $f(x) = x^2$, then $I(f) = \int_0^1 x^2 dx = x/3]_0^1 = 1/3$.

(a) Find elements $f, g, h \in A$ such that $I(f) = 1$, $I(g) = 0$, $I(h) = \sqrt{2}$.
(b) Find elements f and g in A such that $f \neq g$ and $I(f) = I(g)$.
(c) Express the integration property

$$\int_0^1 f(x)dx + \int_0^1 g(x)dx = \int_0^1 (f(x) + g(x)dx$$

in terms of the function I.

(19) Imagine removing the points 1/2, 1/4, 3/8, 5/16, 7/16, 9/32, 11/32, 13/32, 15/32, ... from the interval [0, 1]. Recall Question I (introduction to Part 2). Define a function $f: \mathscr{N} \to [0, 1]$ such that R_f is the set of *all* points so removed. *Hint*: note that $\mathscr{N} = \{1, 2, 2 + 1, 2^2, 2^2 + 1, 2^2 + 2, 2^2 + 3, 2^3, 2^3 + 1, 2^3 + 2, \ldots 2^3 + 7, 2^4, 2^4 + 1, \ldots\}$.

(20) In the article "Mathematics of Aesthetics" in Volume IV of *The World of Mathematics*, the mathematician G. D. Birkhoff defines the aesthetic measure $M(A)$ of a work of art A as the ratio 0(A)/ $C(A)$ where $0(A)$ is the "order, harmony, or symmetry" of A and $C(A)$ is the "complexity or demand for perceptive attention" of A. Let $\mathscr{A}$ represent some collection of artwork (music, painting, etc.) and discuss Birkhoff's aesthetic measure m for its possibilities as a function on $\mathscr{A}$.

3. IDEALIZING THE FUNCTION CONCEPT

The previous exercises indicate that functions can arise in some variety, and in rather abstract circumstances. Moreover, special problems of definition and range can arise for a particular function even between sets of real numbers. On top of all this there is the concept of function itself —the idea of a single valued correspondence. The aim of this section is to find out as much as one can about *any* function, merely due to its nature as a correspondence between sets, and nothing more. This is the structure game again. The harvest is a list of true facts about *any* correspondence, to be used in studying any *particular* correspondence. (Here we use the word "*correspondence*" to mean a *function* as defined in Definition 5.2.2.) The approach is to identifify those properties of a function which, as abtsract properties, would make it a nicer, more tidy, correspondence. In other words, we are going to idealize the function concept.

The first idealization is a somewhat technical one that is nice to have around at times. It is the concept of an *onto* or *surjective* function, and concerns how large the range of a function is. First of all, a small fact, merely worth listing.

LEMMA 5.3.1. Let A and B be sets, $f: A \to B$. Then $R_f = \{f(a) \in B: a \in A\}$.

Proof. The two sets R_f and $\{f(a) \in B: a \in A\}$ must be shown to be equal. If $b \in R_f$, then by Definition 5.2.3 $\exists a \in A \ni (a, b) \in f$. According again to Definition 5.2.3, the symbol $f(a)$ denotes the element $b \in B$; i.e., $b = f(a)$. Hence $b \in \{f(a) \in B: a \in A\}$. If, on the other hand, $b \in \{f(a) \in B: a \in A\}$, this means $\exists a \in A \ni b = f(a)$. Since $(a, f(a)) \in f$, then $(a, b) \in f$ and hence, by Definition 5.2.3, $b \in R_f$. This completes the proof.

We have already noted that $R_f \subset B$; Lemma 5.3.1 merely identifies R_f in terms of the more familiar and natural notation $f(a)$, and repeats more formally our earlier observation that R_f is the set of all "function values" $f(a)$ for $a \in A$.

Now there is nothing in the definition of a function that says that one must have $R_f = B$. Indeed for the function $f: [1, 2] \to \mathscr{R}$ defined by

$$f(x) = \frac{1}{x}$$

we have $R_f = [1/2, 1] \subsetneqq \mathscr{R}$. It is not always easy to determine whether or not the set $R_f = B$ when $f: A \to B$. For example, if $f: [0, 1] \to \mathscr{R}$ is defined by

$$\int_0^x [1/\log (1/t)^{1/2}]dt$$

it is difficult to determine just what the values of f are.

On the other hand, if say,

$$f(x) = \frac{1}{x} \text{ for } x \in \left[\frac{1}{2}, 1\right],$$

then it is natural to describe f as a function from $A = [1, 2]$ to $B = [1/2, 1]$; i.e., $f: A \to B$. Moreover, with this choice of B, one has $R_f = B$.

Here then is an initial idealization of the function concept.

DEFINITION 5.3.2. Let $f: A \to B$. If $R_f = B$, then f is said to be a function from A *onto* B.[1] If $R_f \subset B$, f is said to be (merely) *into*.

Here is a formal restatement of this property in terms of function values. The definition, you will note, is in terms of set equality. The property stated below is the one that you should make use of in most cases where an hypothesis of *onto* is given.

THEOREM 5.3.3. Let $f: A \to B$. Then f is onto B iff $\forall b \in B \exists a \in A \ni b = f(a)$. Hence f is not onto B iff $\exists b \in B \ni b \neq f(a) \, \forall a \in A$.

Proof. According to Definition 5.3.2 and Lemma 5.3.1, f is ontc iff $B = R_f = \{f(a) \in B: a \in A\}$. Hence f is onto iff $b \in B$ implies $b \in \{f(a) \in B: a \in A\}$ iff $\exists a \in A \ni b = f(a)$. The remaining part of the theorem follows

[1] The name *surjective* rather than *onto* is coming into vogue of late.

by negation of what has been proven. Hence the proof.

Thus given *any* function which, for one reason or another, is known to be onto, one has the following fact: no matter how arbitrary or abstract an element $b \in B$ that is given, it must equal $f(a)$ for some $a \in A$—it must be *involved* in the correspondence f. Said yet another way, any element of B is a function value of f on A—everything in B corresponds to something in A, when f is onto.

Let us complete the idealization of function concept. In a perfect world, presumably all functions would be of the type we are about to delineate. The fact is that most functions are not.

Consider two sets A and B and a function $f: A \to B$. In the best of all possible mathematical worlds the equation $f(a) = f(a')$ would imply the equation $a = a'$. Said another way, the elements $a \in A$ would *not lose their identity* by way of the correspondence f. But we need not look far to find this lack of perfection in the function concept. If $f: [-1, 1] \to \mathscr{R}$ is given by $f(x) = x^2$, then $f(1) = f(-1)$ and $1 \neq -1$. Thus -1 loses its identity in R_f through the correspondence f. Nevertheless, since a fair and important class of functions do not share this kind of defect, we single these out and dub them appropriately.

> DEFINITION 5.3.4. Let A and B be sets and $f: A \to B$. If $\forall a, a' \in A$ the equation $f(a) = f(a')$ implies the equation $a = a'$, we call f a 1-1 (one-to-one) function.[2]

The name is appropriate. If f is a function, then $f(a)$ represents exactly one element b of R_f. Moreover, if f is 1-1, then there is only one element $a \in D_f$ corresponding to this b. For if one also has $f(a') = b$, then $f(a) = f(a')$ and by Definition 5.3.4, $a = a'$. Thus a one-to-one function is a rather ideal correspondence. One thing in the domain corresponds to one thing in the range and *conversely*.

Considering some correspondences already mentioned, the correspondence between engine design and pollution output of an automobile is not likely to be 1-1. Nor would the correspondence between education and affluence be 1-1. Some economists conjecture a 1-1 correspondence between the levels of unemployment and inflation in the form of an inverse ratio. Then there is the conjecture that the correspondence between the quality of life and the quantity of goods and services available is becoming non one-to-one with time. On a mathematical level, folding a square into a right triangle with hypotenuse on its diagonal is not a one-to-one correspondence. Stretching an interval to twice its width in a uniform way is one-to-one.

Before further examples and problems, let us make an important point: Definition 5.3.4. is a very usable one. It is truly a *rule for identifying*. To

[2] A surjective (onto) 1-1 function is called ***bijective*** in the parallel terminology. The appellation ***injective*** is applied to a 1-1 function.

identify a function $f: A \to B$ as being 1-1, observe the following rule: suppose a and a' are in A; show that if $f(a) = f(a')$, then $a = a'$. This observation always gives one something to work with, the equation $f(a) = f(a')$, from which to try to draw the conclusion $a = a'$.

EXAMPLES. (a) $A = [-1, 1]$, $B = [-5, 5]$ and define $f: A \to B$ by $f(x) = 2x + 1$. The claim is that f is 1-1. Let us first consider a particular element, say $x = 1/2 \in A$. Then $f(x) = 1$. Is there an $x' \in A \ni f(x') = 1$ and $x' \neq 1/2$. There is not, for if $f(x') = 1$, then $2x' + 1 = 1$ and hence $x' = 1/2$ by ordinary algebra.

Now the general case: suppose $x, x' \in A$ and $f(x) = f(x')$. Then $2x + 1 = 2x' + 1$ and hence $2x = 2x'$ or $x = x'$. Thus f is 1-1 by Definition 5.3.4.

(b) Let $A = \{k \in \mathscr{N}: k \text{ is even}\}$, $B = \mathscr{N}$. Define $f: A \to B$ by $f(k) = k/2$. Then f is 1-1 for $f(k) = f(k') \to k/2 = k'/2 \to k = k'$. Note also that f is onto, for $n \in B \to f(2n) = n$.

(c) Define $f: \mathscr{R} \times \mathscr{R} \to \mathscr{R}$ by $f(x, y) = x$. Then f is not 1-1, for $f(0, 1) = f(0, 2) = 0$.

(d) Let D denote the set of differentiable functions on the real line interval $(0, 1)$ and let E denote the set of all real-valued functions on $(0, 1)$. Define a function δ: D $\to E$ by $\delta(f) = f'$, the derivative of f. For example, if $f(x) = 3x^2 + 1$, then $\delta(f)$ is the function $f'(x) = 6x$. That is, $\delta(f)(x) = 6x$. The function δ is not 1-1. For, if $g(x) = 3x^2 + 3$ then $g \neq f$ and yet $\delta(f)(x) = 6x = \delta(g)(x)$. Hence $g \neq f$ and $\delta(g) = \delta(f)$ and by Definition 5.3.4 δ is not 1-1.

Problems.

(1) Is $f(x) = \sin x$ a function from $\mathscr{R}$ onto $\mathscr{R}$?

(2) Is $f(x) = x^3$ a function from $\mathscr{R}$ onto $\mathscr{R}$?

(3) If $f(x) = 1/\sin x$ a function from $(0, \pi)$ onto $[1, \infty)$?

(4) Define a function from $\mathscr{N}$ onto $B = \{2, 4, 6, 8, \ldots\}$.

(5) Define a function from $A = \{2, 4, 9, \ldots\}$ onto $\mathscr{N}$.

(6) Flatten a square onto one of its sides by means of a function defined on a subset of $\mathscr{R} \times \mathscr{R}$.

(7) Fold a square onto a triangle by similar means. Can you unfold it by a function?

(8) Let $\mathscr{A} = 2^{\mathscr{N}}$. For each $E \in \mathscr{A}$, $E \subset \mathscr{N}$, define a real number $f(E)$ as follows. Using base 2 for the digital representation of a real number (see Section 2.9), let the n-th digit of $f(E)$ be 1 if $n \in E$, 0 if $n \notin E$. Thus if $E = \{1, 2\}$, then $f(E) = .11000\ldots = 1/2 + 1/4 = 3/4$. If $E' = \{2, 4\}$ then $f(E') = .010100\ldots = 1/4 + 1/16 = 5/16$. If $E'' = \{2, 4, 6, \ldots\}$, then $f(E'') = .010101\ldots = 1/2^2 + 1/2^4 + 1/2^6 + \ldots = 1/2^2 + (1/2^2)^2 + (1/2^2)^3 + \ldots = 1/[1 - (1/2)^2] - 1 = 1/3$. Let $f: \mathscr{A} \to [0, 1]$ be the correspondence defined in this way. Using the fact that any real number

has a base 2 representation, prove that f is *onto* [0, 1]. It is not necessary to *really* know the values of f. Let $b \in [0, 1]$. According to Theorem 5.3.3, one must show $\exists E \in \mathscr{A}$—i.e., $E \subset \mathscr{N}$—$\ni f(E) = b$. We know that $b = .a_1a_2a_3\ldots$ where $a_i = 0$ or 1. Use this to define a set E and show $f(E) = b$.

(9) Which of the functions in Problems 1, 9, 10, 12, 13, 14, 17 and 18, Section 5.2, are into?

(10) Let $f: A \to B$. When is it proper in a proof to assert the following: "Let $b \in B$. Then $b = f(a)$ for some $a \in A$?"

(11) Show that the function $f: \mathscr{N} \to \mathscr{N}$ defined by $f(n) = n(n + 1)/2$ is 1-1.

(12) Let $A = [0, 1]$, $B = [0, 1]$, $f(x) = x^2$ for $x \in A$. Is f 1-1?

(13) Let $A = [0, 1]$, $B = [-1, 2]$. Define a function $f: A \to B$ that is 1-1 and onto. *Hint*: Define f so that (say) $f(0) = -1, f(1) = 2$ and generalize.

(14) Define a 1-1 function from $A = \{2, 4, 6, \ldots\}$ onto $B = \{1, 2, 3, \ldots\}$.

(15) Let $A = B = \{1, 2, 3, \ldots\}$. Define functions $f: A \to B$ such that
  (a) f is 1-1 but not onto.
  (b) f is onto but not 1-1.
  (c) f is neither onto nor 1-1.
  (d) f is onto and 1-1.

(16) Would you expect a probability function to be 1-1? Is P of Problem 17, Section 5.2, 1-1?

(17) Which of the functions in Problems 1, 3, 9, 10, 11 and 18, Section 5.2, are 1-1?

(18) Show that if $f: A \to B$ is 1-1, then $a \neq a' \to f(a) \neq f(a')$.

(19) Which of the functions in Problems 1, 2, 3 and 6 above are 1-1?

(20) Let $f: A \to B$. Under what conditions on f is it correct to assert: "Let $b \in B$. Then $\exists$ exactly one $a \in A \ni f(a) = b$"?

(21) Let f: (a, b) $\to R$ be a differentiable function. The "mean value theorem" of calculus says that if $x, y \in$ (a, b) then $\exists z$ between x and $y \ni f(x) + f(y) = f'(z)(x\text{-}y)$, where $f'(z)$ is the derivative of f at z. Suppose $f'(t) \neq 0 \forall t \in (a, b)$. Show that f is 1-1. Use Definition 5.3.4 as a rule of procedure for doing this.

(22) Let $f: \mathscr{R} \to \mathscr{R}$ such that for *any* two numbers a and a, $f(a) + f(a') = f(a + a')$ and $f(aa') = af(a')$. Show that f is 1-1. An example of a function with these two properties is $f(x) = 2x$, or in fact, any linear function, $f(x) = mx$, where m has a fixed value.

(23) (a) Show that the function f of problem 8 above is not 1-1, considering the sets $E = \{1\}$ and $E' = \{2, 3, 4, \ldots\}$ and the material of Section 2.9.
  (b) Let $2^{\mathscr{N}}_f = \{E \subset \mathscr{N}: E \text{ is finite}\}$ and define $g: 2^{\mathscr{N}}_f \to [0, 1]$ by $g(E) = .a_1a_2a_3\ldots$ where $a_i = 1$ if $i \in E$, 0 if $i \notin E$. Show

that g is 1-1 but not onto.

(c) Let $X = 2^{\mathcal{N}}/2_f^{\mathcal{N}} = \{E \subset N: E \text{ is not finite}\}$. Let $h: X \to [0, 1]$ be defined by $h(E) = .a_1a_2a_3\ldots$ where $a_i = 1$ if $i \in E$, $a_i = 0$ if $i \notin E$. Regarding $h(E) = .a_1a_2a_3\ldots$ in base 2 show that h is 1-1 and onto $[0, 1]$.

4. FUNCTIONS ACTING ON SETS

This is a most natural topic though its abstract manifestation seems to be very troublesome at first. Let us begin by examining why an individual might care to think in terms of correspondences at all.

Consider the correspondence p between the pair (engine timing, air-fuel ratio) and the volume of pollutants expelled by a given internal combustion engine per second of operation at idle. It is known that too high a value T of engine timing will ruin the engine, hence there are practical limits, $-T$ and T on the value one can assign to engine timing. Also, an air-fuel ratio of more than (say) 20 is impractical, and hence one can think of the pollution function p as a function from $[-T, T] \times [0, 20]$ into the real numbers $\mathcal{R}$ which corresponds each pair (t, r) of a timing value and air-fuel ratio to a particular volume of pollutants $p(t, r)$.

Given that one is thinking in terms of a correspondence, there are two broad and natural questions that can be asked, each the converse of other. These are:

(a) Given a collection C of feasible (timing, air-fuel ratio) pairs S, what is the range of pollution volumes that correspond to setting $(t, r) \in C$?

(b) Given a fixed range D of pollution volumes acceptable to public health (or whatever), what class of engine settings $(t, r) \in [-T, T] \times [0, 20]$ yield pollutant volumes within the range D?

It does seem that if one finds any value at all in having a correspondence to think in terms of, it is to be able to first *state* and then answer such questions. If there were no correspondence present, the questions would be meaningless.

Now the mathematician provides a formal conceptual answer to such questions, but that is all at this level where he is interested only in the abstract idea of a correspondence and in what is true about *any* function. At this level, an answer to question a is that the range of pollution volumes corresponding to settings $(t, r) \in C$ is the set of numbers

$$\{p(t, r): (t, r) \in C\} \subset \mathcal{R}$$

which is ordinarily denoted by $p(C)$. In answer to question b, one can say that the range of engine settings yielding pollutant values in the set D is the set of pairs

$$\{(t, r): p(t, r) \in D\} \subset [-T, T] \times [0, 20],$$

which is ordinarily denoted by $p^{-1}(D)$. These descriptions don't really answer the respective questions, of course; more particular information about the function p is needed, but they do pick out a concept naturally related to the function concept. As we are dealing herein only on the conceptual level, the sets $p(C)$ and $p^{-1}(D)$ are examples of the kinds of sets and function-related operations on sets to be considered in this section. The above example merely serves to support the claim that sets of this type arise in a very natural way and are an integral part of thinking in terms of correspondences. Here is a formal, completely abstract, definition.

DEFINITION 5.4.1. Let A and B be sets and let $f: A \to B$.

(a) If $C \subset A$, we use the notation $f(C)$ to represent the subset $\{f(x): x \in C\}$ of B. We call $f(C)$ the image of C in B.

(b) If $D \subset B$, we use the notation $f^{-1}(D)$ to represent the subset $\{x: f(x) \in D\}$ of A. We call $f^{-1}(D)$ the pre-image of D in A, or simply "f inverse of D."

We will shortly be manipulating such sets and it is worthwhile to again point out out that Definition 5.4.1 serves as a rule for identifying certain sets related to a particular function. Sets are identified by what's in them. By Definition 5.4.1, an element $b \in B$ is in the set $f(C)$ iff $b = f(x)$ for some $x \in C$—i.e., iff $\exists\, x \in C \ni b = f(x)$. Similarly, an element $a \in A$ is in the set $f^{-1}(D)$ iff $f(a) \in D$—iff $\exists\, x \ni x = a$ and $f(x) \in D$. In Figure 5.9, we indicate a way to think about such sets.

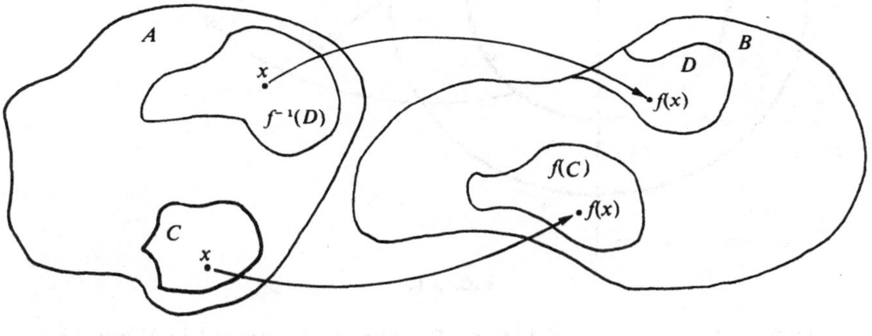

FIG. 5.9

EXAMPLES. (a) Let $f: \mathscr{R} \to \mathscr{R}$ be defined by $f(x) = x^2$.
If $C_1 = [-2, 2]$, then $f(C_1) = [0, 4]$.
If $C_2 = [-2, 1]$, then $f(C_2) = [0, 4]$.
If $C_3 = [-2, 3]$, then $f(C_3) = [0, 9]$.
If $C_4 = [-2, -1]$, then $f(C_4) = [1, 4]$.
If $D_1 = [0, 1]$ then $f^{-1}(D_1) = [-1, 1]$.
If $D_2 = [-2, 1]$ then $f^{-1}(D_2) = [-1, 1]$.
If $D_3 = [1, 4]$ then $f^{-1}(D_3) = [-2, -1] \cup [1, 2]$.
If $D_4 = [1/4, 2]$ then $f^{-1}(D_4) = [-\sqrt{2}, 1/2) \cup (1/2, \sqrt{2}]$.

(b) Here is an example worth keeping in mind throughout this section as a guide to what's true and what's not true in general. Let

$$A = \{(x, y): x^2 + y^2 < 1\} \subset \mathscr{R} \times \mathscr{R}$$

and $B = \mathscr{R}$ and define $f: A \to B$ by

$$f(x, y) = x^2 + y^2 .$$

Thinking in terms of Figure 5.10, we see that f corresponds *all* points on a circle C_r of radius r centered at (0, 0) in A to the *single* real number r^2, for $(x, y) \in C_r$ iff $x^2 + y^2 = r^2$.

Let

$$C = \{(x, y) \in A: 1/4 \leq \sqrt{x^2 + y^2} < 1/2\} \subset A.$$

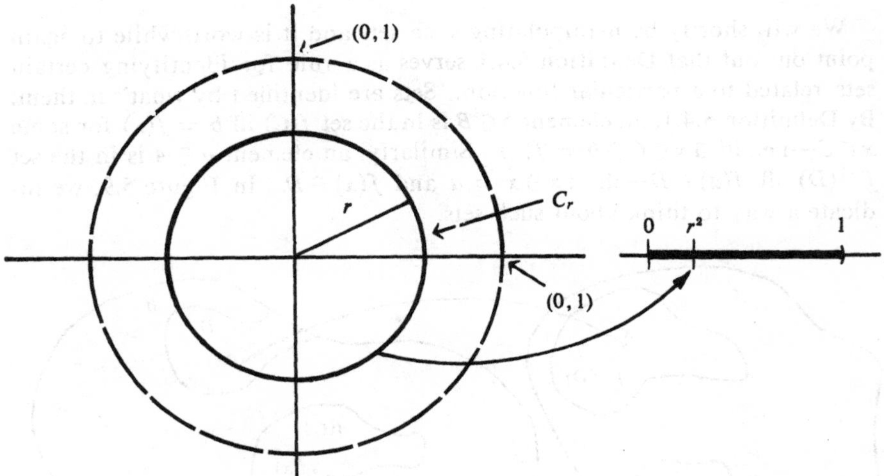

FIG. 5.10

Then $f(C) = \{r: 1/16 \leq r < 1/4\} \subset B$. This is illustrated in Figure 5.11.

Consider the set

$$C = \{(x, y): \sqrt{x^2 + y^2} = 1/2\} ,$$

the circle of radius 1/2, and the set $C' = \{(0, 1/2)\}$. Then $f(C) = \{1/2\}$ and $f(C') = \{1/2\}$! Hence, one can have sets $C \neq C'$ such that $f(C) = f(C')$. Notice also these distinctions in notation: $f(C') = \{f(0, 1/2)\} = \{1/2\}$ while $f(0, 1/2) \in f(C')$ or $f(0, 1/2) \in \{1/2\}$. Such distinctions are again significant and useful at an abstract conceptual level.

Now we take in the other direction. Let $D = [0, 1/4] \subset \mathscr{R}$. By Definition 5.4.1,

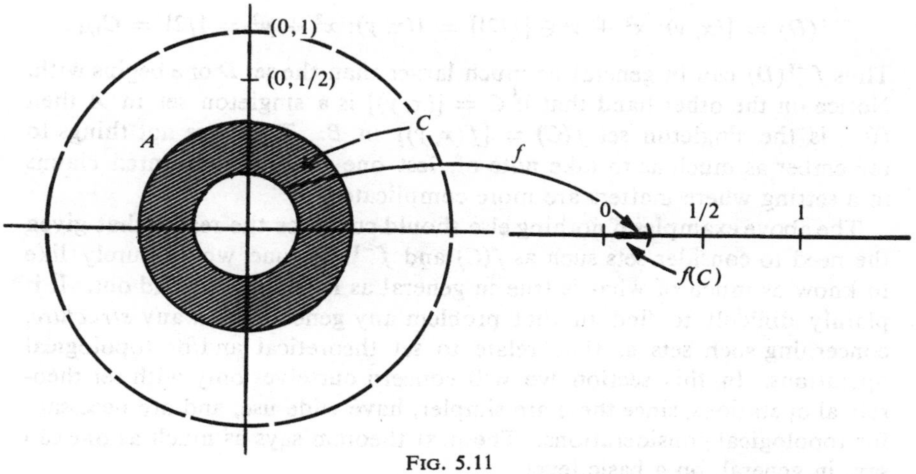

FIG. 5.11

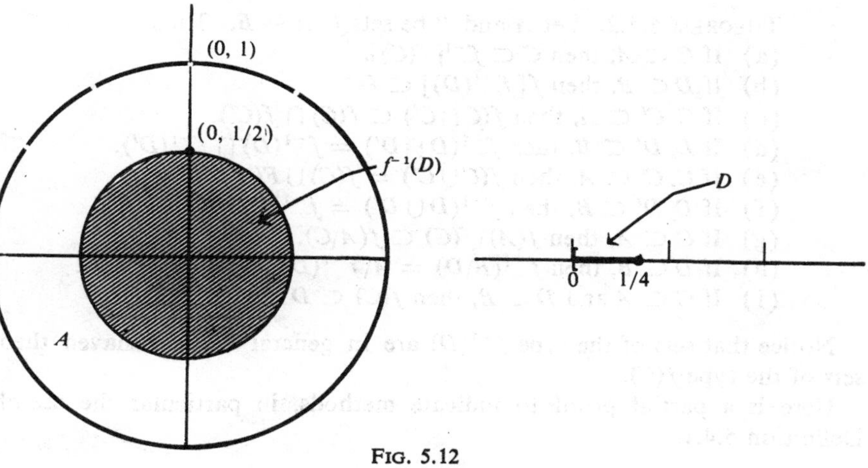

FIG. 5.12

$$\begin{aligned} f^{-1}(D) &= \{(x, y)\colon x^2 + y^2 \in [0, 1/4]\} \\ &= \{(x, y)\colon x^2 + y^2 \leq 1/4\}\,. \end{aligned}$$

Hence $f^{-1}(D) = \bigcup_{0 \leq r \leq 1/2} C_r$, where C_r is the circle of radius r centered at (0, 0). This is illustrated in Figure 5.12.

Let $D' = [-1, 1/4] \subset \mathscr{R}$. Again by Definition 5.4.1,

$$f^{-1}(D') = \{(x, y)\colon x^2 + y^2 \in [-1, 1/4]\} = \{(x, y)\colon x^2 + y^2 \in [0, 1/4]\}$$

since $x^2 + y^2 \geq 0 \ \forall \ (x, y)$. Hence, although $D \neq D'$, we get $f^{-1}(D) = f^{-1}(D')$. One thing this means is that in an abstract setting an implication: $f^{-1}(D) = f^{-1}(D') \to D = D'$, need not be true.

Consider the set $D'' = \{1/2\}$, a singleton set. Here

$$f^{-1}(D) = \{(x, y): x^2 + y^2 \in \{1/2\}\} = \{(x, y): x^2 + y^2 = 1/2\} = C_{1/4}.$$

Thus $f^{-1}(D)$ can in general be much larger than the set D one begins with. Notice on the other hand that if $C = \{(x, y)\}$ is a singleton set in A then $f(C)$ is the singleton set $f(C) = \{f(x, y)\}$ in B. These are not things to remember as much as to take note of, lest one make unwarranted claims in a setting where matters are more complicated.

The above examples if nothing else should convince the reader that given the need to consider sets such as $f(C)$ and $f^{-1}(D)$, one would surely like to know as much of what is true in general as is possible to find out. It is plainly difficult to find in that problem any general rules, any *structure*, concerning such sets as they relate to set theoretical and/or topological operations. In this section we will concern ourselves only with set theoretical operations, since these are simpler, have wide use, and are necessary for topological considerations. The next theorem says as much as one can say, in general, on a basic level.

THEOREM 5.4.2. Let A and B be sets $f: A \to B$. Then

(a) If $C \subset A$, then $C \subset f^{-1}[f(C)]$.

(b) If $D \subset B$, then $f[f^{-1}(D)] \subset D$.

(c) If $C, C' \subset A$, then $f(C \cap C') \subset f(C) \cap f(C')$.

(d) If $D, D' \subset B$, then $f^{-1}(D \cap D') = f^{-1}(D) \cap f^{-1}(D')$.

(e) If $C, C' \subset A$, then $f(C \cup C') = f(C) \cup F(C')$.

(f) If $D, D' \subset B$, then $f^{-1}(D \cup D') = f^{-1}(D) \cup f^{-1}(D')$.

(g) If $C \subset A$, then $f(A) \setminus f(C) \subset f(A \setminus C)$.

(h) If $D \subset B$, then $f^{-1}(B \setminus D) = A \setminus f^{-1}(D) = f^{-1}(B) \setminus f^{1}(D)$.

(i) If $C \subset A$ and $D \subset B$, then $f(C) \subset D$ iff $C \subset f^{-1}(D)$.

Notice that sets of the type $f^{-1}(D)$ are in general better behaved then sets of the type $f(C)$.

Here is a partial proof to indicate methods, in particular the use of Definition 5.4.1.

Proof of (a). We must show $C \subset f^{-1}[f(C)]$. Let $x \in C$. It must be argued that $x \in f^{-1}[f(C)]$. Let $D = f(C)$. According to Definition 5.4.1, $x \in f^{-1}(D)$ iff $f(x) \in D = f(C)$. Now, since $x \in C$, then $f(x) \in f(C)$ by Definition 5.4.1. Hence $x \in f^{-1}(D) = f^{-1}(f(C))$. Thus the proof.

Proof of (b). Let $y \in f[f^{-1}(D)]$. It must be shown that $y \in D$. Since $y \in f[f^{-1}(D)]$, then according to Definition 5.4.1, $\exists\, x \in f^{-1}(D)$ such that $y = f(x)$. Since $x \in f^{-1}(D)$, then Definition 5.4.1 again says that $f(x) \in D$. Hence $y = f(x) \in D$. Thus the proof.

Proof of (d). Let $x \in f^{-1}(D \cap D')$. Then $f(x) \in D \cap D'$ and hence $f(x) \in D$ and $f(x) \in D'$. Thus $x \in f^{-1}(D)$ and $x \in f^{-1}(D')$. Hence, $x \in f^{-1}(D) \cap f^{-1}(D')$. Therefore $f^{-1}(D \cap D') \subset f^{-1}(D) \cap f^{-1}(D')$. It remains to show that $f^{-1}(D) \cap f^{-1}(D') \subset f^{-1}(D \cap D')$. Let $x \in f^{-1}(D) \cap f^{-1}(D')$. Then

$x \in f^{-1}(D)$ and $x \in f^{-1}(D')$ and hence $f(x) \in D$ and $f(x) \in D'$. Thus $f(x) \in D \cap D'$ and consequently, $x \in f^{-1}(D \cap D')$. Thus the proof.

Proof of (*g*). Let $y \in f(A)\backslash f(C)$. Then $y \in f(A)$ and hence $\exists\, x \in A$ such that $y = f(x)$. Moreover, $x \notin C$. For, if x where in C, then $f(x) \in f(C)$ and hence $y = f(x) \notin f(A)\backslash f(C)$, a contradiction. Hence $x \in A$ and $x \notin C$ and therefore $x \in A\backslash C$. Thus $y = f(x) \in f(A\backslash C)$ by Definition 5.4.1. Thus the proof.

Proof of (*i*). If $f(C) \subset D$, then $x \in C$ implies $f(x) \in f(C)$ and hence $f(x) \in D$. Thus, $x \in f^{-1}(D)$ by Definition 5.4.1. This shows that $C \subset f^{-1}(D)$. Conversely, suppose $C \subset f^{-1}(D)$ and $y \in f(C)$. Then $\exists\, x \in C \ni y = f(x)$ by Definition 5.4.1. Hence, since $x \in C$, then $x \in f^{-1}(D)$. Thus $y = f(x) \in D$. Hence $f(C)$. Thus the proof.

Problems.

(1) Prove the remaining parts of Theorem 5.4.2.

(2) Show by example that the results (a), (b), (d) and (h) in Theorem 5.4.2 are the best result one can have in general. For example, for part (a) find sets A, B, a set $C \subset A$ and a function f such that $C \subsetneq f^{-1}f(C)$. Previous problems may provide examples.

(3) Let $\mathscr{C}$ be a collection of subsets of a set A and $f\colon A \to B$. Prove that $f^{-1}(\bigcup_{C\in\mathscr{C}} C) = \bigcup_{C\in\mathscr{C}} f^{-1}(C)$ and $f^{-1}(\bigcap_{C\in\mathscr{C}} C) = \bigcap_{C\in\mathscr{C}} f^{-1}(C)$.

(4) Let $(X, \mathscr{U})$ and $(Y, \mathscr{V})$ be topological spaces and let $f\colon X \to Y$. Suppose f has this property: if $x \in X$ and $V \in \mathscr{V}$ is a neighborhood of $f(x)$, then $\exists$ a neighborhood U of $x \ni f(U) \subset V$. This is illustrated in the figure below.

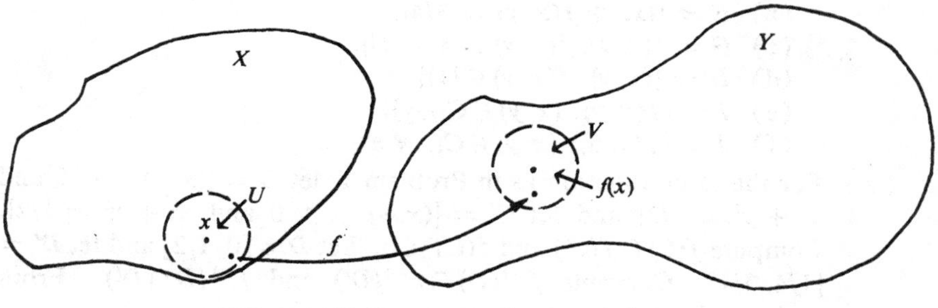

Conclude that f has the following property: if Q is an open set in Y, then $f^{-1}(Q)$ is an open set in X. It must be shown that if $x \in f^{-1}(Q)$, then $\exists$ a neighborhood U of $x \ni U \subset f^{-1}(Q)$. Then conclude that if P is closed in Y, then $f^{-1}(P)$ is closed in X.

Before closing this section we will refine Theorem 5.4.2 further by placing additional conditions on the function f—again a structuring of abstract knowledge. These results are "predictable" on the basis of previous examples and problems.

THEOREM 5.4.3. Let A and B be sets and $f: A \to B$. Then
(a) If f is onto, then $f(f^{-1}(D)) = D \forall D \subset B$.
(b) If f is 1-1, then $f^{-1}(f(C)) = C \forall C \subset A$.
(c) If f is 1-1, then $f(C \cap C') = f(C) \cap f(C')$ for any pair of sets $C, C' \subset A$.
(d) If f is 1-1 and $C \subset A$, then $f(A \backslash C) = f(A) \backslash f(C)$. Hence if f is 1-1 and onto, then $f(A \backslash C) = B \backslash f(C)$.
(e) If f is 1-1 and onto, then for $C \subset A$ and $D \subset B, f(C) = D$ iff $C = f^{-1}(D)$.

Proof of (b). By Theorem 5.4.2, we have $f^{-1}f(C) \supset C$. We wish to show that $x \in f^{-1}f(C) \to x \in C$. If $x \in f^{-1}(f(C))$ then by Definition 5.4.1, $f(x) \in f(C)$. Hence by Definition 5.4.1, $\exists\, x' \in C \ni f(x) = f(x')$. Since f is 1-1, then $x = x'$. Since $x' \in C$, then $x \in C$ and the proof is complete.

Proof of (e). Let $D = f(C)$. Then by (b), $f^{-1}(f(C)) = C$. But $f^{-1}(D) = f^{-1}(f(C))$ and hence $f^{-1}(D) = C$. Conversely, suppose $C = f^{-1}(D)$. Then by (a), $f(C) = f(f^{-1}(D)) = D$. Hence the proof.

Problems.

(1) Let f be the function in Example (b). Describe each of the sets below as subsets of A or B, as appropriate, in terms of Definition 5.4.1 and simply as sets themselves. For example, $E = f(C)$ where $C = \bigcup_{r \in [1/3, 2/3]} C_r$ and $E = [1/9, 4/9]$.
(a) $E = \{f(x, y): (x, y) \in C_r \forall r \in [1/3, 2/3]\}$.
(b) $F = \{(x, y): f(x, y) = 3/4\}$.
(c) $G = \{(x, y): f(x, y) \in (1/4, 2]\}$.
(d) $H = \{(x, y): f(x, y) \in \{1\}\}$.
(e) $I = \{f(x, y): (x, y) \in C_{1/\sqrt{2}}\}$.
(f) $J = \{f(x, y): (x, y) \in C_{1/n} \forall n \in \mathcal{N}\}$.

(2) For the same function as in Problem 1, let $C = \{(x, y): x < 0$ and $x^2 + y^2 = 1/2\}$ and let $C' = \{(x, y): x > 0$ and $x^2 + y^2 = 1/2\}$. Compute $f(C) \cap f(C')$ and $f(C \cap C')$. Let $D = [0, 1/2]$ and let $D' = [1/4, 3/4]$. Compute $f^{-1}(D) \cap f^{-1}(D')$ and $f^{-1}(D \cap D')$. From their results, what can you say is *not* true in general.

(3) For the same function, f, let $D = [1/4, 2]$. What $f(f^{-1}(D))$? To do this, first figure out $C = f^{-1}(D)$ and then work on $f(C)$. How does $f(f^{-1}(D)$ relate to D? Let $C = \{(0, 1/2)\}$. How does $f^{-1}(f(C))$ relate to C? What do these two examples indicate as being *not* true in general?

(4) Let f be the correspondence of a library card to the book that it directs you to. What does f^{-1} (*The Anatomy of Mathematics*) consist of? Let C denote the set of library cards listed under the letter

a. Describe $f(C)$ in similar language.

(5) Let $f(t)$ denote the vertical displacement of an object thrown directly upward at time, t. Suppose the total time in motion (both up and down) is 10 seconds. What is $f^{-1}\{1\}$ and $f^{-1}([2, 6])$ in approximate terms (i.e., describe in words)? What is $f([0, 10])$?

(6) Let A be the coordinate plane, B the set of real numbers. Let $P(x, y) = x$; $P(x, y)$ is called the *projection* of the point (x, y) onto the x-axis.

(a) Let $C = \{(x, x): x \in \mathrm{R}\}$. What is $P(C)$?
(b) Let $E = \{(1, y): 0 \leq y\}$. What is $P(E)$?
(c) Let $F = \{(x, y): -1 \leq x \leq 3/2, y \text{ arbitrary}\}$. What is $P(F)$?
(d) Let $D = [0, 1)$. What is $p^{-1}(D)$?
(e) Let $G = [0, \infty)$. What is $p^{-1}(G)$?
(f) What is $P^{-1}\{2\}$?
(g) What is $P\{(x, y): x - y \text{ is a positive integer}\}$.
(h) Let $H = \{(x, 1): 2 \leq x \leq 3\}$. What is $P^{-1}(P(H))$? Let $I = [3, 5]$. What is $P(P^{-1}(I))$?

(7) Let us tear the real line $\mathscr{R}$ in half at 0 and shrink both parts to a point. This can be described, for example, by the function f: $\mathscr{R} \times \mathscr{R}$ given by

$$f(x) = \begin{cases} 1 & x \geq 0 \\ -1 & x < 0 \end{cases}$$

for the range of f consists of only two points, which we choose to be 1 and -1 for convenience. Consider $\mathscr{R}$ with the usual topology. Classify the following sets as open or closed in $\mathscr{R}$.

(a) $C = (-\infty, 0)$ and $f(C)$.
(b) $D = \{-1\}$ and $f^{-1}(D)$.
(c) $C' = (-2, 3]$ and $f(C')$.
(d) $D' = (0, 2)$ and $f^{-1}(D')$.
(e) $D'' = \{-1, 1\}$ and $f^{-1}(D'')$.

(8) Take a rubber band of unit length and stretch it "uniformly" to twice its length and then slide it over one unit length to the right. This distortion can be described by the function f: $[0, 1] \rightarrow \mathscr{R}$ given by $f(x) = 2x + 1$.

(a) Let $V = (2 - 1/4, 2 + 1/4)$ be a neighborhood of $f(1/2) = 2$. Find a neighborhood U of 1 (in the relative topology on $[0, 1]$) such that $f(U) \subset V$.
(b) Let $V = (2 - \varepsilon, 2 + \varepsilon)$ where $\varepsilon > 0$ be a neighborhood of $f(1/2)$. Find a neighborhood U of 1 (in the relative topology on $[0, 1]$) such that $f(U) \subset V$.
(c) Let $V = (3 - \varepsilon, 3 + \varepsilon)$ be a neighborhood of $f(1)$. Find a neighborhood U in the relative topology on $[0, 1]$ such that $U \subset f^{-1}(V)$.

(d) Give [0, 1] its relative topology as a subset of $\mathscr{R}$ with the relative topology. Let Q be a open set in $\mathscr{R}$. Then $f^{-1}(Q)$ is a set in [0, 1]. Show that $f^{-1}(Q)$ is open. For a start, let $x \in f^{-1}(Q)$. Then $f(x) \in Q$ and Q is open. Hence $\exists$ an interval $V = (f(x) - \varepsilon, f(x) + \varepsilon) \subset Q$. If you can find an interval U containing x such that $f(U) \subset V$, then you will also have $x \in U \subset f^{-1}(Q)$ and hence a proof that $f^{-1}(Q)$ is open.

(9) For the function f of Problem 8, prove that if P is a closed set in $\mathscr{R}$ then $f^{-1}(P)$ is closed in [0, 1] with its relative topology. *Hint*: Use Problem 8 and show that $f^{-1}(\mathscr{R}\backslash P) = [0, 1]\backslash f^{-1}(P)$.

(10) For the function f of Problem 7 find an open set Q and a closed set P in $\mathscr{R}$ such that $f^{-1}(Q)$ is not open and $f^{-1}(P)$ is not closed. Notice that this is a distortion with tearing, whereas Problem 8 is not.

(11) Let X be the set in Problem 8, of Section 4.1. Define $g: X \to \mathscr{R}$ by $g(n) = 1/n$ and $g(\omega) = 0$. Let $V = (-1/4, 1/4) \subset \mathscr{R}$. What is $g^{-1}(V)$? Is there a neighborhood U of ω such that is $g(U) \subset V$? Let $V' = (-\varepsilon, \varepsilon)$ where $\varepsilon > 0$. Then $g(\omega) \in V'$. Find a neighborhood U' of ω such that $g(U') \subset V'$. Let $V'' = (1/3, 3/4)$. Then $g(2) \in V''$. Is there a neighborhood U'' of 2 such that $g(U'') \subset V''$?

(12) Find a function $f: \mathscr{R} \to \mathscr{R}$ and a set $C \subset \mathscr{R}$ such that $f(\bar{C}) \neq \overline{f(C)}$. Find a function $g: \mathscr{R} \to \mathscr{R}$ and a set $E \subset \mathscr{R}$ such that $g(\bar{A}) \subset \overline{g(A)}$. All closures are to be taken in the usual topology.

(13) Prove the remaining parts of Theorem 5.4.3.

(14) Let $f: A \to B$ and suppose f is 1-1. Let $b \in B$. Show that $f^{-1}\{b\}$ is either empty or a singleton set.

(15) Topology is pure mathematics. Only recently have there been attempts to use it in non-mathematical studies. This problem indicates one natural way by which topological spaces can result any time one has any numerically valued correspondence.

Typically in applications of mathematics one begins with some collection of objects (elements) he wishes to study. (Perhaps X is the collection of all students on this university or perhaps X is a culture of bacteria.) The investigator subjects, or observes the reactions of, these elements to some force or influence. (The university trustees forbid all visiting speakers under age 65, or the biologist starves the bacteria.) The investigator then constructs a correspondence, a function, which relates each $x \in X$ to some measure of x's reaction. He then has a function $f: X \to \mathscr{R}$.

Suppose that X is any set and $f: X \to \mathscr{R}$ is any function. We will make X into a topological space. Let $\mathscr{U} = \{f^{-1}(0): 0 \text{ is an open set in } \mathscr{R}\}$. Use Theorem 5.4.2(d) to show that $(X, \mathscr{U})$ is a topological space.

Moral: We are all elements in some (potential) topological space.

(16) There is a certain poetry of meaning in the above problem. Suppose X is the collection of all students in this university, the trustees have spoken as above, and Professor I. M. Outraged over in Sociology has noted the reaction. He lets $A = \{x: x$ goes into a state of intellectual shock and can no longer attend class$\}$. $B = \{x: x$ does nothing$\}$, $C = \{x: x$ applauds$\}$ and defines $f: X \to \mathscr{R}$ by

$$f(x) = \begin{cases} 1 & x \in A \\ 1/2 & x \in B \\ 0 & x \in C \end{cases}$$

(a) Show that the topology $\mathscr{U}$ which results has only the neighborhoods $\square$, X, A, B, C, $A \cup B$, $A \cup C$, $B \cup C$.

(b) Show that A is *separated* from B and from C.

(c) Show that $(X, \mathscr{U})$ is not a connected topological space.

(d) A topology $\mathscr{U}$ is said to distinguish the points $x, y \in X$ where $x \neq y$ if $\exists\, U, V \in \mathscr{U} \ni x \in U, y \in V$ and $U \cap V = \square$. Show that this topology $\mathscr{U}$ cannot distinguish points taken from B, nor points taken from A, nor from C. They are, in their reaction, a topologically amorphous mass. Show that $\mathscr{U}$ does distinguish any $x \in A$ from any $y \in B$.

As superficial and facetious as this problem is, it does illustrate a point: as soon as Professor Outraged decides upon no greater distinction between students than the groupings A, B and C, then the constructed topological space *faithfully absorbs* his viewpoint into its structure and certain thought processes are automatically and consistently take care of—such is the mathematical way of thinking.

(17) In practice, topologies such as the above would arise not from a single function, a single measurement or experiment, but by a collection of such. Suppose X is a set and $\mathscr{F}$ is a collection of real number valued functions defined on X. Let $\mathscr{W} = \{\bigcup_{i=1}^{n} f_i^{-1}(0_i)$: $n \in \mathscr{N}, f_i \in \mathscr{F}$ and 0_i is an open set in $\mathscr{R}\}$. Show that $(X, \mathscr{W})$ is a topological space; $\mathscr{W}$ is called the weak topology of $\mathscr{F}$ on X.

5. INVERSE AND COMPOSITE FUNCTIONS

The list of basic functional concepts is finite, and these two complete it! Both will prove to be crucial to an understanding of infinite sets. We take up the concept of an inverse function first. This is but the idea of thinking of a given function $f: A \to B$ as establishing a correspondence in the opposite direction, from B to A. Unfortunately for simplicity, but fortunately for intrinsic interest, not all functions have an inverse.

DEFINITION 5.5.1. Let A and B be sets, $f: A \to B$. Let f^{-1} be the relation in $B \times A$ defined by

$$f^{-1} = \{(b, a): (a, b) \in f\} = \{(b, a): b = f(a)\}.$$

The relation f^{-1} is in general a relation only, and not a function. If

$$f(x) = x^2, A = [-1, 1], B = [0, 1],$$

then, for example, $(-1, 1) \in f^{-1}$ and $(1, 1) \in f^{-1}$ and hence f^{-1} is not a function according to Definition 5.2.2. We, however, do have

THEOREM 5.5.2. Let $f: A \to B$. Then, f^{-1} is a function iff f is 1-1.

Proof. Suppose f is 1-1. We claim f^{-1} is a function. According to Definition 5.2.2 it must be shown that $(b, a) \in f^{-1}$ and $(b, a') \in f^{-1}$ imply $a = a'$. If $(b, a) \in f^{-1}$ and $(b, a') \in f^{-1}$ then $(a, b) \in f$ and $(a', b) \in f$. Hence $f(a) = b = f(a')$. Since f is 1-1, then $a = a'$. Hence f^{-1} is a function.

Conversely, suppose f^{-1} is a function. We claim f is 1-1. Suppose $a, a' \in A$ and $f(a) = f(a')$. It must be shown that $a = a'$. Let $b = f(a) = f(a')$. Then $(a, b) \in f$ and $(a', b) \in f$. Hence $(b, a) \in f^{-1}$ and $(b, a') \in f^{-1}$. Since f^{-1} is supposed to be a function, $a = a'$ by Definition 5.2.2. Hence the proof.

Thus f^{-1} is a function iff f is 1-1. We now have

THEOREM 5.5.3. Let $f: A \to B$ which is 1-1. Then $D_{f^{-1}} = B$ iff f is onto.

Proof. Suppose f is onto. We claim $b \in B \to b \in D_{f^{-1}}$. Since f is onto, $\exists\, a \in A \ni f(a) = b$. Hence $(a, b) \in f$ and hence $(b, a) \in f^{-1}$. According to Definition 5.2.3, $b \in D_{f^{-1}}$. Hence $B \subset D_{f^{-1}}$. Since $D_{f^{-1}} \subset B$ by definition of f^{-1}, we now have $B = D_{f^{-1}}$.

Conversely, suppose $D_{f^{-1}} = B$. We claim f is onto. Let $b \in B$. Then $b \in D_{f^{-1}}$. Hence $\exists\, a \in A \ni (b, a) \in f^{-1}$. But by Definition 5.5.1 this means $(a, b) \in f$. Therefore $b = f(a)$. By Theorem 5.3.3, f is onto. Hence the proof.

With these two theorems we now define the inverse of a function, f.

DEFINITION 5.5.4. Let $f: A \to B$ which is 1-1 and onto. The *inverse function* of f, or simply the *inverse* of f, is the function f^{-1} defined by Definition 5.5.1.

Now we collect all of the above into the important theorem on inverse functions, the idea being to relate these ideas to the usual functional notation.

THEOREM 5.5.5. Let $f: A \to B$ which is 1-1 and onto. Then
(a) $f^{-1}: B \to A$ is 1-1 and onto.
(b) For $b \in B, f^{-1}(b) = a$ iff $f(a) = b$.
(c) $f^{-1}(f(a)) = a \;\forall\, a \in A$.

(d) $f(f^{-1}(b)) = b \; \forall \, b \in B$.

Proof: According to Theorem 5.5.2 f^{-1} is a function. According to Theorem 5.5.3 $D_{f^{-1}} = B$. Hence $f^{-1}: B \to A$. Moreover,

$$(f^{-1})^{-1} = \{(a, b): (b, a) \in f^{-1}\} = \{(a, b): (a, b) \in f\} = f.$$

Since f is a function and $(f^{-1})^{-1} = f$, then $(f^{-1})^{-1}$ is a function. But according to Theorem 5.5.2 this means f^{-1} is 1-1. Moreover, the equation $f = (f^{-1})^{-1}$ also means, $D_{(f^{-1})^{-1}} = D_f = A$ and hence by Theorem 5.5.3 f^{-1} is onto. Hence (a) holds.

Let $b \in B$. For an $a \in A$, $(b, a) \in f^{-1}$ iff $(a, b) \in f$. *Translated*: according to Definition 5.2.3, $f^{-1}(b) = a$ iff $f(a) = b$. Hence (b) holds.

For (c) if $b = f(a)$ then by (b), $f^{-1}(b) = a$. But since f^{-1} is a function and $b = f(a)$, then $f^{-1}(b) = f^{-1}f(a)$.

Hence $a = f^{-1}(a)$ by substitution.

Finally, for (d), if $b \in B$, then $a = f^{-1}(b)$ iff $f(a) = b$ by (b). Hence, since f is a function and $a = f^{-1}(b)$ we have $b = f(a) = ff^{-}(b)$. Hence the proof.

Both the haphazard and the careful reader may be confused about this notation and that of Definition 5.4.1. He should not be. Notice, for example that from Definition 5.4.1 and Theorem 5.5.5 (2) $f^{-1}\{b\} = \{f^{-1}(b)\}$ denotes the singleton set consisting of the $a \in A$ such that $f(a) = b$. Furthermore, suppose for the moment that g denotes the inverse, f^{-1}, of f. Then by Definition 5.2.2, with $D \subset B$,

$$\begin{aligned} g(D) &= \{g(b): b \in D\} = \{f^{-1}(b): b \in D\} \\ &= \{a: f(a) = b \in D\} = f^{-1}(D) \end{aligned}$$

by Definition 5.2.2. Replacing g by f^{-1}, we obtain $f^{-1}(D) = f^{-1}(D)$ so that there is no inconsistency in notation. Up to this point the notation $f^{-1}(D)$, defined for *any* function f, could not be separated in the conceptual form: f^{-1} of D. For 1-1, onto functions, *only*, it now can be.

The above paragraphs are a rather direct and abstract development of the basic theory of the inverse function of a function. It is important to notice that we only speak of f^{-1} as a *function* when f is 1-1 and onto. This will be seen to be sufficient for all cases. Now for some examples. Part (b) of Theorem 5.5.5 tells us that we can think of f^{-1} as simply establishing the reverse correspondence of that of f. Figure 5.13 illustrates this idea.

Here is an example of how to obtain a formula for f^{-1} knowing a formula for f. Let $A = [1, 3]$, $B = [3, 7]$ and $f: A \to B$ be given by $f(x) = 2x + 1$. Then, f is 1-1 and onto and according to Theorem 5.5.5 (b), $x = f^{-1}(y)$ iff $y = f(x) = 2x + 1$. Hence $f^{-1}(y) = x = (y - 1)/2$ is a formula for f^{-1}. Notice that $f^{-1}(f(x)) = x \; \forall \, x \in A$ and $f(f^{-1}(y)) = y \; \forall \, y \in B$.

The specific determination of an inverse function is rarely this easy. If

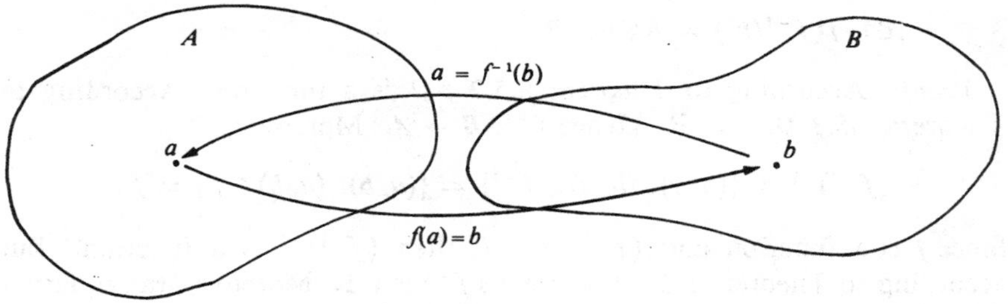

FIG. 5.13

a function f describes a real-world correspondence, then f^{-1} is a function iff the process described by f is reversible, and to find a formula for f^{-1} means that one can give a method for recovering x knowing only $f(x)$—for one must have: $f^{-1}f(x) = x$. A large part of subsequent use of inverse functions is not tied to the finding of a formula for f^{-1} but merely to the use of the *concept* of f^{-1}.

Consider the function, f: $[0, \infty) \to [1, \infty)$ given by $f(x) = e^x$. According to Problem 21, Section 5.3, f is 1-1. It is not so easy to establish the fact that f is onto, but f is and f^{-1} is well known: $f^{-1}(x) = \log_e x$. For, $\log_e e^x = x$ and $x = e^{\log_e x}$—that is, $f^{-1}f(x) = x$ and $ff^{-1}(x) = x$—according to well known formulas.

We now take up the idea of composition of functions. There are two ways to look at this one and here is the first.

Suppose that $f(x) = \sin x^2$ and $h(x) = x^2$. What is the nature of the relationship of f and g? It is this: let $g(x) = \sin x$, then $f(x) = g(h(x))$ for each x, by blind, unthinking substitution in the formulas. Thus "f is a function of h by g," or $f(x)$ can be obtained from g operating on $h(x)$.

Here is a second: suppose A, B and C are sets and h: $A \to B$, g: $B \to C$. If $x \in A$ then $h(x) \in B$. Since $h(x) \in B$, $g(h(x))$ is defined and $g(h(x)) \in C$. Thus a correspondence, x corresponding to $g(h(x))$, exists from A to C. Let f denote this correspondence. That is, let f denote the subset of $A \times C$ defined by $f = \{(x, h(g(x))): x \in A\}$. Is f a function? Let (a, c), $(a, c') \in f$. Then (a, c) is some pair $(x, h(g(x))$ and (a, c') is some pair $(x', h(g(x'))$. By Theorem 5.3.2(a), $a = x, c = h(g(x))$ and $a = x', c' = h(g(x'))$. Since $x = a = x'$, then $g(x) = g(x')$ because g is a function. Finally,

$$c = h(g(x)) = h(g(x')) = c'$$

because h is a function. Thus $(a, c) \in f$ and $(a, c') \in f$ imply $c = c'$ and f is a function from A to C by Definition 5.2.2.

It is this latter idea which is used for the following definition and the work in the paragraph above insures the consistency of the

DEFINITION 5.5.6. Let A, B and C be sets and suppose h: $A \to B$ and g: $B \to C$. The *composition of h by g* is the function f: $A \to C$ defined by the equation $f(x) = g(h(x))$ for each $x \in A$.

Ordinarily, the composition of h by g is denoted by $g \circ h$ rather than introducing the third symbol f; i.e., $g \circ h$ is the function defined by

$$(g \circ h)(x) = g(h(x)) .$$

We read this as "g of h," or "g op h." The value of the function $g \circ h$ at x is the element $g(h(x))$ in C. One can illustrate the function $g \circ h$, as in Figure 5.14.

This entire development of function concepts has been carried on without even defining the equality of two functions! This is because we had

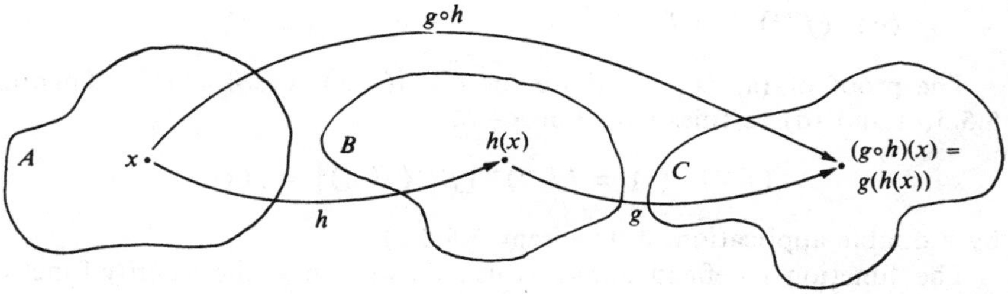

FIG. 5.14

no reason to compare two functions in any sense. Here is a definition of equality.

DEFINITIOM 5.5.7. Let g: $A \to B$ and h: $A \to C$. We say that $g = h$ iff $\forall\, x \in A$, $h(x) = g(x)$.

Clearly if two function h and g are equal they must have the same range, for,

$$R_g = \{g(x)\colon x \in A\} = \{h(x)\colon x \in A\} = R_h .$$

Notice though that equality of functions is only defined in the case where the functions are already *known* to have the *same domain*.

Here now is the principle result of this section. We will use it over and over again in the study of infinite sets.

THEOREM 5.5.8. If h: $A \to B$ and g: $B \to C$ and both g and h are 1-1 and onto, then

(a) $g \circ h$: $A \to C$ is 1-1 and onto.
(b) g^{-1}: $C \to B$, h^{-1}: $B \to A$ are 1-1 and onto and hence $h^{-1} \circ g^{-1}$: $C \to A$ is 1-1 and onto.
(c) $(g \circ h)^{-1}$ exists and $(g \circ h)^{-1} = h^{-1} \circ g^{-1}$.

There is a lot of information stored in Theorem 5.5.8. (For a particular example to keep in mind in attempting a proof, see Problems 8 and 9 below.) Use Definition 5.3.4 twice to obtain that $g \circ h$ is 1-1; it is straightforward that $g \circ h$ is onto. The first part of (b) follows from Theorem 5.5.5 and the latter part then follows from (a) applied to g^{-1} and h^{-1}. The first part of (c) then follows from (a) and Theorems 5.5.2 and 5.5.3. The last equation is proven by using Definition 5.5.6 on the rule for equality given in Definition 5.5.7.

THEOREM 5.5.9. Let S be a set and let $i\colon S \to S$ be defined by $i(x) = x \ \forall\, x \in S$ and let $f\colon S \to S$ be 1-1 and onto. Then
(a) i is 1-1 and onto and $i = i^{-1}$.
(b) $f \circ f^{-1} = i = f^{-1} \circ f$.
(c) $(f^{-1})^{-1} = f$.

The proof of (a) is immediate for $i = \{(x, x)\colon x \in S\} = i^{-1}$. Theorem 5.5.5(c) and (d) certifies (b). For $x \in S$,

$$(f^{-1})^{-1}(x) = (f^{-1})^{-1}[f^{-1}(f(x))] = f(x)$$

by a double application of Theorem 5.5.5(c).

The function i defined above is usually known as the *identity* function on S since it leaves each $x \in S$ *unchanged* in the correspondence.

To close this section we take up a theorem on the solution of certain kinds of functional equations. You will recall that if $a, b \in \mathscr{R}$ and $a \neq 0$ then $\exists x \in \mathscr{R} \ni xa = b$ and $\exists y \in \mathscr{R} \ni ay = b$. In fact, in $\mathscr{R}$, $x = y = ba^{-1}$. The operation of composition of functions has similar properties.

THEOREM 5.5.10. Let A, B and C be sets, $g : A \to C$ and $h : B \to C$. If h is 1-1 and onto, then $\exists$ a function $\chi\colon A \to B \ni g = h \circ \chi$. Moreover $\chi = h^{-1} \circ g$ and χ is unique.

This theorem is illustrated in Figures 5.15 and 5.16.

For a proof of Theorem 5.5.10, suppose h is 1-1 and onto. Then $h^{-1} : B \to C$ exists by Theorems 5.5.2. and 5.5.3. Let $\chi = h^{-1} \circ g$. We claim $h \circ \chi = g$. For $x \in A$,

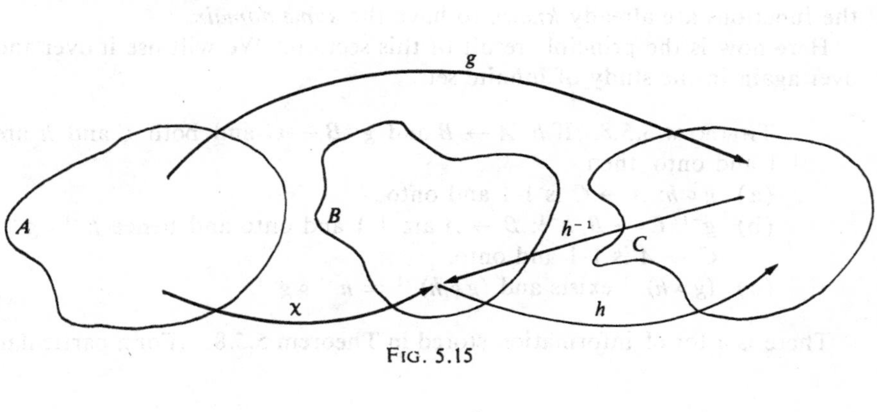

FIG. 5.15

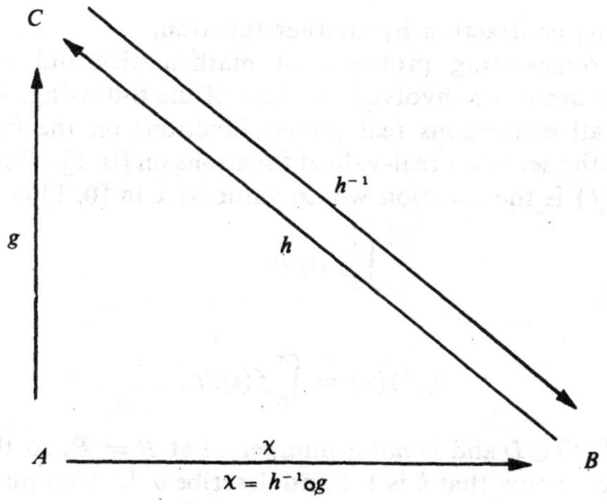

FIG. 5.16

$$(h \circ \chi)(x) = h(\chi(x)) = h((h^{-1} \circ g)(x)) = h(h^{-1}(g(x))) = g(x)$$

using Definition 5.5.6 and 5.5.5.(d). Hence $\forall x \in A \ (h \circ \chi)(x) = g(x)$ and therefore $h \circ \chi = g$.

Hence the function $\chi = h^{-1} \circ g$ solves the equation $h \circ \chi = g$. We claim this is the only solution. Suppose $h \circ \chi' = g$ with $\chi' : A \to B$. We claim $\chi = \chi'$. Let $x \in A$. Then,

$$(h \circ \chi)(x) = g(x) = (h \circ \chi')(x)$$

or

$$h(\chi(x)) = h(\chi'(x)) .$$

Since h^{-1} exists only when h is 1-1, we have $\chi(x) = \chi'(x)$ from Definition 5.3.4. Hence $\chi = \chi'$ and $\chi = h^{-1} \circ g$ is the only solution.

Another way of viewing this result is as follows. In an equation $h \circ f = g$ where h^{-1} exists we can assert: $f = h^{-1} \circ g$. Note that this same results from the following observation. Let x be fixed. Since h^{-1} is a function, $h(f(x)) = g(x)$ implies $h^{-1}(h(f(x))) = h^{-1}(g(x))$ and hence

$$f(x) = h^{-1}(g(x))$$

from Theorem 5.5.5(c). That is, $f = h^{-1} \circ g$.

Problems.

(1) Let $f: [0, 3] \to [-1, 8]$ be given by $f(x) = 3x - 1$. Describe f^{-1}.

(2) Let $f(x) = ax^2 + bx + c$ be a function defined on $\mathscr{R}$ into $\mathscr{R}$. Describe the relation f^{-1} using the quadratic formula.

(3) Describe a function f which stretches a rubber band of unit length uniformly to twice its length. Let it go, and describe the

resulting contraction by another function.

(4) Some outstanding problems of mathematics and mathematical physics are more involved versions of the following. Let C be the set of all continuous real-valued functions on the interval [0, 1] and D the set of all real-valued functions on [0, 1]. Define $\delta : C \to D$ by: $\delta(f)$ is the function whose value at x in [0, 1] is

$$\int_0^x f(t)dt .$$

Thus

$$\delta(f)(x) = \int_0^x f(t)dt .$$

Note: $\delta(f) \in D$ and is *not* a number. Let $B = R_\delta$ so that $\delta : C \to B$ is onto. Show that δ is 1-1 and describe δ^{-1}. You may recall from the "fundamental theorem of calculus" that

$$\frac{d}{dx}\left[\int_0^x f(t)\,dt\right] = f(x) \qquad \text{for } f \in C .$$

(5) Let A be the set of points in $R \times R$ indicated in the figure below.

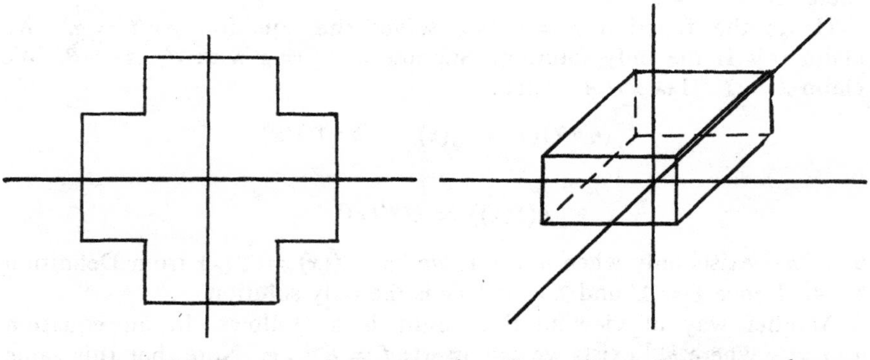

Fold this figure up into a box in $\mathscr{R} \times \mathscr{R} \times \mathscr{R}$ open at the top. Describe this process by a function f. Is f^{-1} a function?

(6) Let $f : A \to B$ which is 1-1 and onto. Let $g : B \to A$. Suppose $g(f(x)) = x \; \forall x \in A$. Prove that $g(y) = f^{-1}(y) \; \forall y \in B$. Hence f^{-1} is the only function that reverses the correspondence f.

(7) Give two examples of functions g and h such that $g \circ h \neq h \circ g$. Would you expect that for most applicable (i.e., $h \circ g$, $g \circ h$ are defined) pairs h and g, one would have $g \circ h = h \circ g$?

(8) Let $A = B = C$ be a circle of radius one in the plane. Let $h : A \to B$ be defined by: $h(x)$ is the point arrived at by x upon rotating A by $+30°$. Let $g(x)$ be the point arrived at by rotating x by $+45°$. What does $g \circ h$ represent?

(9) In Problem 8 notice that g and h are 1-1 and onto. What rotations

do g^{-1} and h^{-1} represent? What does $g \circ h^{-1}$ represent? What about $g^{-1} \circ h^{-1}$, $h^{-1} \circ g^{-1}$ and $(g \circ h)^{-1}$?

(10) Fold a square into a triangle, as indicated below and describe this by a composition of functions.

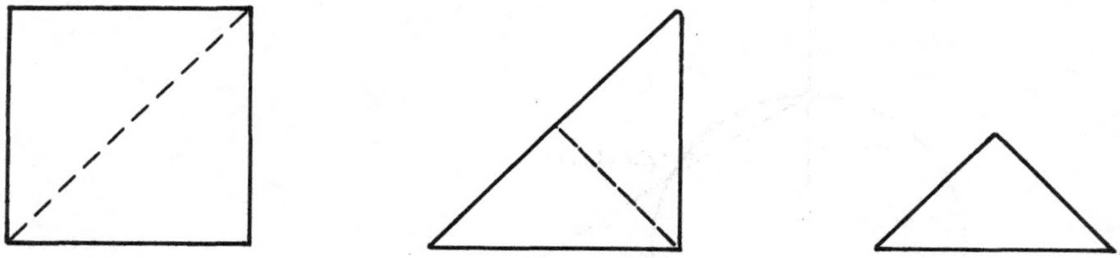

(11) Let $f: \{0, 1, 2\} \to \{0, 1, 2\} \ni f(0) = 0, f(1) = 2$ and $f(2) = 2$. Find all functions $g: \{0, 1, 2\} \to \{0, 1, 2\} \ni f \circ g = g \circ f$.

(12) One of the outstandingly useful ideas of contemporary mathematics is the technique of studying a space, an algebraic structure, or even a set by instead studying the set of all functions on it that incorporate some property of the space or structure. We will encounter one material example of this approach in the sequel. For the moment here is an example of how the functions on a set give information about a set. Let S be a non-empty set and suppose that $\forall f, g: S \to S$ one has

$$(f \circ g)(x) = (g \circ h)(x) \forall x \in S.$$

Prove that S has only one element.

(13) Let S be any set. Just as any set admits a topological structure (the discrete and the indiscrete topology) any set also gives rise to a group structure. Incidentally, the applications of group theory to quantum physics arise from a consideration of the kinds of groups we are about to consider. Furthermore, this problem is another aspect of structured abstract thought, and should provide a better perspective for viewing Theorem 5.5.10. This problem is another aspect of the concept of examining collections of functions on a set in order to learn things about the set itself, for given a set with no particular structure, we claim there naturally arises from it another set with the structure of a group. (See Section 2.5 for a refresher on groups.) Specifically let

$$G = \{f: S \to S: f \text{ is 1-1 and onto}\}.$$

(a) Show that G with the operation $\circ$ of composition of functions is a group. You will find Theorems 5.5.8 and 5.5.9 especially useful in this problem.

(b) Referring to Section 2.5, interpret Theorems 1 and 2 and Problems 2, 3 and 5 in the context of the group G above.

(c) Let g and h be functions. Place conditions on g and h such that the equation $\chi \circ h = g$ has a solution χ.

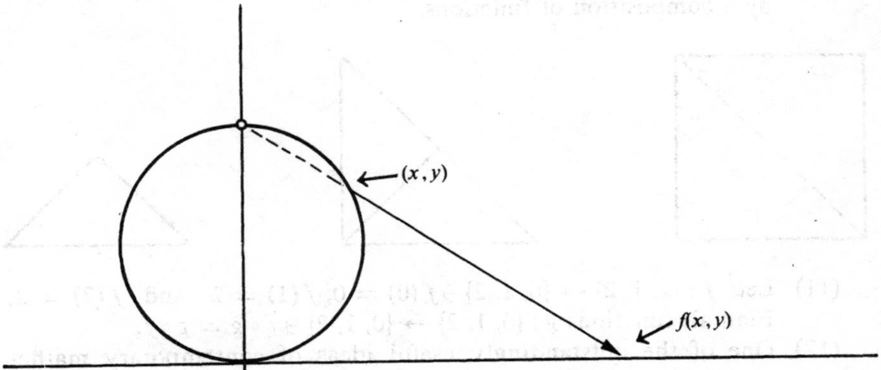

chapter 6

"Je le vois, mais je ne le crois pas!"—Cantor

COUNTING THE INFINITE

1. INTRODUCTION

Each vertical line of a rectangle has zero area. Put them altogether and get something positive. It's all the obscure insignificant nothings that yield something.

In this chapter we will reach some substantial and non-trivial conclusions. The intent is to present the basics of Cantor's theory of infinite sets, a subject which appears at first to be somewhat esoteric but in fact later emerges naturally in wide areas of mathematics, especially including significant questions in Fourier analysis and probability theory. In the process we will answer Question I (introduction to Part 2) and in doing so will have cause to use almost all the material so far developed. Additionally, because we will for the first time prove things that are not at all obvious, our previous concern for generality and precision will be vindicated.

There is a sub-theme running throughout this book that, it is thought, especially bears on this chapter. This theme is that the creative force that moves one man to do mathematics is not unlike that force that moves another to write music or poetry, to produce art or literature. Beethoven was stone deaf but heard music that no one had ever heard before and was impelled to share it. The often tormented artist sees and senses mood and appearance perhaps unique unto himself, perhaps common to all, and trys to fix and transmit it through his work. Cantor, with a mind that eventually led him to a mental breakdown, created in man's mathematical mind "a paradise from which no one shall drive us" as Hilbert was moved to say in defense of Cantor's theory of infinite sets. Despite its importance in even the most mundane applications, the reader might well derived maximum satisfaction from and insight into the mathematical infinite if he views the matter as but an instance of mathematical art, created by one man's mathematical mind, to salve one small corner of man's esthetic sense. Cantor's theory of infinite sets is not music, art, nor literature, but it springs from and serves the same spirit.

The reader has no doubt noticed that reaching out to infinity can have a certain leveling effect on one's confidence in his mental prowess. This is not a unique experience. The technical concepts, definitions, notations and so on of previous chapters were developed largely to aid the attempts of the best of mathematical (but finite) minds to reach and to hold a place in Cantor's paradise. These concepts and methods have proven to be virtually the only guide in an area of thought where intuition can lead one astray and the expected is verily uncommon. To travel from the finite to the infinite is to follow Alice through-the-looking-glass and to find common sense turned upside down. Time, absolute zero, and the quantum jump are the nearest comparisons that are not purely mathematical. The reader is duly warned then to take along a fully equipped pack of previous concepts before venturing into this domain.

As has been pointed out, problems of infinity have always generated a special interest in mathematical man. He can, relatively speaking, easily grasp the finite because it is all around him. His approach to infinity has in consequence been one of imitation and extension of finite methods to an infinite context. This is the plan of this section. We will define and delineate the essentials of counting and finite sets. We will then extend the same concepts and methods to the infinite and derive the important consequences. This approach leads to some interesting results that any mathematically literate one ought to be aware of.

A finite set S is, in the usual sense, one with only so many, say 5, 10, 20, or n elements for some $n \in \mathcal{N}$. To count the fingers on one hand one corresponds the number 1 with one finger, 2 with the next and so on. It is speculated that primitive man counted the number of stone cutting tools he had crafted some bleak winter day by pairing these, one-to-one, with his fingers, toes and maybe those of an acquaintance, if he was especially productive. It is further speculated that, to perhaps gain independence of an uncooperative neighbor he began instead to make marks in the dirt floor of his abode, one for each cutting tool. Such was presumably the beginning of mathematical abstraction. To this day we still count in essentially the same way! To count is to set up a one-to-one correspondence between a set $I_n = \{1, 2, \ldots, n\} \subset \mathcal{N}$ and those objects one is counting and conclude that one has n of these. Hence.

Definition 6.1.1. A set S is said to be *finite* if $\exists\, n \in \mathcal{N}$ and a function $f\colon I_n \to S$ which is 1-1 and onto. Such a set S is then said to have n elements. If $S = \square$, S is also said to be finite having no elements.

Notice that we only specify the existence of *a* function, for in determining totality one cares not in what order one counts whatever one has.

Here is an important technical point. Given a non-empty set S, a number n and a 1-1 function f from I_n onto S one can assign names to the elements in S, naming the element that corresponds to $k \in \mathcal{N}$, "$f(k)$." Thus one can write

$$S = \{f(1), f(2), \ldots, f(n)\} = \{f(k)\colon k \in I_n\} = R_f\,.$$

More often, one uses the following terminology. Let $s_k = f(k)$ for $k \in I_n$.

Then, in this notation, $S = \{s_1, s_2, \ldots, s_n\}$. Thus to say that a set S is finite and has n elements, is to say that one can subscript the elements of the set with the first n natural numbers, one and only one element of S for each number $k = 1, 2, \ldots, n$. We will have more to say on this later, when it will be given more significance.

Another consequence of Definition 6.1.1 is that one can always discuss the counting of the elements of S in terms of the familiar set I_n. For, $I_n = f^{-1}(S)$ and because f is 1-1, there is no loss of identity between the elements of S and those of I_n. This point will be well reinforced when we begin to develop some further consequences of Definition 6.1.1. From this last observation arises a natural question with which we can just as well begin the theory of counting the finite. Suppose two sets S and T are given, both having (say) ten elements. We then usually say S and T have the same number of elements and believe that there should be a one-to-one correspondence between S and T. Does this follow from Definition 6.1.1? An affirmative answer is the assertion of

THEOREM 6.1.2. Let S and T be sets both having n elements. Then $\exists f: S \to T$ which is 1-1 and onto.

Proof. According to Definition 6.1.1, $\exists$ functions $g: I_n \to S$ and $h: I_n \to T$ both 1-1 and onto. By Theorem 5.5.5(a), $g^{-1}: S \to I_n$ is 1-1 and onto. By Theorem 5.5.8(a), $h \circ g^{-1}: S \to T$ is 1-1 and onto. Let $f = h \circ g^{-1}$ to complete the proof.

This is of course not an astounding result. Its importance and meaning lies in the subsequent discussion of infinite sets. There we will use the idea expressed in Definition 6.2.1 to give an exact meaning to the concept of "same number of elements" when S and T are *not* finite.

Consider a room filled with 366 people. You, no doubt on a second thought, believe that 2 of these must have the same birth date. If you examine your reasoning for this you will see that it is roughly as follows. Let P be the set of all people in the room. Each person $p \in P$ has a unique birth date and thus there exists a correspondence $f: P \to I_{365}$ where $f(p)$ is the birth date of person p. But there are 366 people in the room and hence $\exists\, p, p' \in P$ such that $p \neq p'$ and $f(p) = f(p')$. Hence f cannot be 1-1, meaning that two persons have the same birth date.

This reasoning is generalized in the following fundamental theorem on finite sets. We prove it, despite its "obvious" character, because it is fundamental and because we will then know exactly what it depends on. (And, of course, because in mathematics one proves all things!) But a further point. In the above example you may be thinking that the essential point is that there are *more* persons in P than days of the year and hence at least 2 have the same birth date. Of course, that is the point. But what does "more" mean? If you try to be precise about it, to say that "S has more elements than T" means there cannot exist a function $f: S \to T$ which is 1-1. But then one has come full circle. At some point, a proof should be given that places the natural concept of "more" on a sure foundation. Here it is, and the foundation is the axiom of induction. Because all questions of counting the finite can, by virtue of Definition 6.1.1, be settled in terms of the sets I_n, we will find it sufficient to state and prove this next result in the context of sets $I_n \subset \mathcal{N}$ and need not consider arbitrary finite sets.

THEOREM 6.1.3. Let $n \in \mathcal{N}$. Then
(a) If $A \subsetneq I_n$, there cannot exist a 1-1 function $f: I_n \to A$.
(b) If $D \subset I_n$ and $g: D \to I_n$ is onto, then $D = I_n$.

Proof: We claim first that (a) follows from (b). We will then prove (b) instead. For suppose (b) holds and suppose there is a 1-1 function $f: I_n \to A$.[1] Then $f^{-1}: R_f \to I_n$ would be onto by Theorem 5.5.5(a). Since $R_f \subset A \subsetneq I_n$ this would contradict the conclusion of (b) that $R_f = I_n$. Hence (a) follows from (b) and it suffices to prove (b).

We prove (b) by induction. If $n = 1$, then $I_n = \{1\}$ and if $g: D \to I_n$ is onto where $D \subset I_n$, then $\exists\, x \in D \ni g(x) = 1$. But $x \in D \to x \in I_n \to x = 1 \to D = I_n$. Hence (b) holds for $n = 1$.

Suppose (b) holds for $n = k$. We claim (b) must hold for $n = k + 1$. Suppose then that $g: D \to I_{k+1}$ and that g is onto, where $D \subset I_{k+1}$. We must show $D = I_{k+1}$. Let $D_0 = \{x \in D: g(x) \in I_k\}$. Since g is onto I_{k+1} $\exists\, x \in D \ni g(x) = k + 1$ and hence $x \in D \backslash D_0$. From this we conclude that $D_0 \subsetneq D$.

Consider now only the set D_0. Either $k + 1 \in D_0$ or $k + 1 \notin D_0$. Suppose first that $k + 1 \notin D_0$. Then $D_0 \subset I_k$. We claim $D_0 = I_k$. To see this define $f: D_0 \to I_k$ by $f(x) = g(x)$ for each $x \in D_0$. Clearly f is onto I_k because g is onto. Since $D_0 \subset I_k$ and we are assuming the truth of (b) for $n = k$, this means $D_0 = I_k$. Since $I_k = D_0 \subsetneq D \subset I_{k+1}$ this leaves only the conclusion that $D = I_{k+1}$ since $I_{k+1} = I_k \cup \{k + 1\}$.

Consider now the contrary case $k + 1 \in D_0$. Then $g(k + 1) \in I_k$. Let $E = D_0 \backslash \{k + 1\} \cup D \backslash D_0$. Then $E \subset I_k$ because $D_0 \backslash \{k + 1\} \subset I_k$ and $D \backslash D_0 \subset I_k$ (because $k + 1 \in D_0$). Now define $f: E \to I_k$ by

$$f(x) = \begin{cases} g(x) & x \in D_0 \backslash \{k + 1\} \\ g(k + 1) & x \in D \backslash D_0 . \end{cases}$$

Because $D \backslash D_0 \cap D_0 \backslash \{k + 1\} = \square$ this correspondence is a function. Furthermore, since $g(D_0 \backslash \{k + 1\}) \subset I_k$ and $g(k + 1) \in I_k$, then $R_f \subset I_k$. That is, the correspondence f really is into I_k. We claim f is onto.

Let $m \in I_k$. If $m = g(k + 1)$, then $f(x) = m$ for $x \in D \backslash D_0$ by definition of f. Since $D \backslash D_0 \neq \square$ this puts $m \in R_f$. If $m \neq g(k + 1)$, then, because $m \in I_k$, and g is onto $\exists\, y \in D_0 \ni g(y) = m$. Since $m \neq g(k + 1)$ and g is

[1] Part (b) of this theorem is of course motivated by the idea that a function is always a *single-valued* correspondence.

a function we know $y \neq k + 1$ and hence that $y \in D_0\backslash\{k + 1\}$. But then $f(y) = g(y) = m$ by definition of f. Hence

$$\forall\, m \in I_k \,\exists\, x \in E \ni f(x) = m .$$

Thus f is onto and by the induction hypothesis we must have $E = I_k$. That is, $D_0\backslash\{k + 1\} \cup D\backslash D_0 = I_k$. Thus $D \supset I_k$ and because $k + 1 \in D_0 \subset D$, then D must equal I_{k+1}.

This completes the proof.

The reader may be wondering—was it worth it? Isn't there an easier way to prove such an intuitively obvious result? Not likely. The proof is of such difficulty because (surprisingly enough) it deals with such a fundamental concept, because it is based on such minimal knowledge of the natural numbers and finally because we demand a precise and completely abstract argument. We have only used the ideas of addition of natural numbers and the induction axiom.

But yet, what was the point of this? We have, since the age of four or five, noticed that one cannot correspond each of four apples to each of five oranges without having at least two oranges related to one apple; we have observed similar things like this in a large number of cases, usually without being aware of it in the function sense. But we have never, and never can, consider all possible cases. Yet we believe, by virtue of experience, that matters will continue this way in each and every case. Our experience or intuition here is really inductive reasoning, from a number of previous cases to the next case. In answer to the question then: Theorem 6.1.3 shows that these matters are as we expect them to be in *all* possible cases and one sees from its proof what this conclusion rests on—the assumption of the validity of inductive reasoning. This is more than one knew, with certainty, before. Note now how the next result faithfully follows one's intuitive concept of "more than" in the context of counting.

COROLLARY 6.1.4. **Let S and T be sets with m and n elements, respectively, and suppose $m < n$. Then**

(a) $\exists$ a 1-1 function $f\colon S \to T$

(b) No function $g\colon S \to T$ is onto.

Proof. By Definition 6.1.1, $\exists$ 1-1, onto functions $h\colon I_m \to S$ and $k\colon I_n \to T$. Since $m < n$ then $I_m \subset I_n$. Let $f = k \circ h^{-1}$. Then $f\colon S \to T$ is 1-1 by Theorems 5.5.5 and 5.5.8. This proves (a).

Now suppose $g\colon S \to T$. We claim g cannot be onto. Suppose g is onto. Let $f_0 = k^{-1} \circ [g \circ h]$. Then $f_0\colon I_m \to I_n$ is onto because h is onto S, g is onto T and k^{-1} is onto I_n by Theorem 5.5.5(a). But by Theorem 6.1.3(b) there cannot exist a function $f_0\colon I_m \to I_n$ since $I_m \subsetneq I_n$. Thus g cannot be onto. This completes the proof.

The following problem establishes the remaining basic facts about counting the finite.

Problems.

(1) Prove that if every student in this university is not acquainted with at least one other student, then at least 2 students have the same number of acquaintances. Can you conclude that more than 2 students have the same number of acquaintances?

(2) Suppose $f: I_n \to I_n$ is onto. Prove that f must be 1-1.

(3) Suppose $f: I_n \to I_n$ is 1-1. Prove that f must be onto.

(4) This problem should remove a bit of mystery as to the origin of some long-standing notation. Prove that I_n has exactly 2^n subsets. That is, 2^{I_n} has 2^n elemets.

(5) Prove that if S is a set having n elements, then 2^S has 2^n elements.

(6) Let $\mathscr{F}_n = \{f: f: I_n \to \{0, 1\}\}$. Prove that $\mathscr{F}_n$ has exactly 2^n elements.

(7) Problems 4 and 6 say that $\mathscr{F}_n$ and 2^{I_n} have the same number, 2^n, of elements. According to Definition 4.6.2, $\exists$ a 1-1 onto function $\phi: 2^{I_n} \to \mathscr{F}_n$. Describe such a function. This problem should tell you that proving existence (as in Theorem 6.1.2) does not necessarily yield a specific example or realization of that which is supposed to exist!

(8) Prove that a subset S of a finite set T is itself finite. You might try induction on the number of elements in T.

(9) Suppose S and T are sets and $\exists f: S \to T$ such that f is 1-1 and onto and suppose that S has n elements. Prove that T has n elements.

(10) Writing the real numbers in base 2, fix n and let $S = \{.a_1a_2 \ldots a_n: a_i = 0 \text{ or } 1\}$. How many elements does S have? Prove your answer in the sense of Definition 6.1.1.

(11) Put the answers to Problems 4, 5, 6, 7 and 10 all together into a coherent whole.

2. EXTENSION: COUNTING THE INFINITE

Previous results establish the essentials of counting the finite in a precise and abstracted manner. Definition 6.1.1 gives a definite meaning to *counting* and the "number of elements in a set." Theorem 6.1.2 characterizes finite sets with the ***same number*** of elements. Theorem 6.1.3 and Corollary 6.1.4 give definite criteria for establishing when one finite set has *less* (or *more*) elements than another finite set. Because these latter criteria are abstracted solely in terms of sets and functions we can extend them to infinite sets *without changing a single word.* This is what Cantor did, and this is what we will now do. The consequences are neither obvious nor predictable. We begin with

DEFINITION 6.2.1. A set S is said to be ***infinite*** if it is not finite.

Such a simple idea! Even a 4-year-old can count, can determine that a collection placed before him is finite. The infinite is but the negation of this idea yet the consequences are profound.

There is, as you no doubt already believe, an infinite set.

THEOREM 6.2.2. The set $\mathscr{N}$ is infinite and hence so is the set $\mathscr{R}$.

Proof. Suppose not. Then $\exists\ f\colon I_n \to \mathscr{N}$ which is 1-1 and onto for some choice of n. Let $A = \{x \in I_n : f(x) \in I_{n+1}\} \subset I_{n+1}$. If $g(x) = f(x)$ for $x \in A$, then $g\colon A \to I_{n+1}$ is 1-1 and onto, in contradiction to Theorem 6.1.3(a). Thus $\mathscr{N}$ is infinite. Because $\mathscr{N} \subset \mathscr{R}$, $\mathscr{R}$ is infinite by Problem 8, Section 6.1. Hence the proof.

Another proof of the obvious! But there is a small world of mathematical thought implicit in Definition 6.2.1 and there are strange consequences to be extracted concerning infinite sets. A strong belief in the correctness of the *methods* used to study the infinite, buttressed by an occasional application of these methods to the finite or the obvious, may make them more familiar and trustworthy when they are needed for the unobvious.

We now extend the ideas expressed in Theorem 6.1.2 and Corollary 6.1.4 to arbitrary sets.

DEFINITION 6.2.3. Let S and T be sets. We will say that:

(a) S and T have the *same-number of elements* if $\exists$ a 1-1, onto function $f\colon S \to T$.[2]

(b) T has *more elements* than S if $\exists\ f\colon S \to T$ which is 1-1, but no function $g\colon S \to T$ is onto.

The effort expended in making Definition 6.2.3 would be wasted if *all* infinite sets had the same-number of elements. Yet at first thought, such does appear reasonable, though after further investigation, not so at all!

Consider the set $\mathscr{N} = \{1, 2, 3, \ldots\}$. We tend to think of $\mathscr{N}$ as quite large and in fact non-ending. It is difficult to imagine a set with more elements than $\mathscr{N}$. On the other hand, we have noticed that for any finite set there is always a set with more elements. For example, if S has n elements then by Problem 5 above 2^S has 2^n elements. Here is the first remarkable result about infinite sets: the same conclusion, that 2^S has more elements than S, holds.

Before proving this result, we must make a special agreement regarding the words *set and collection.*

In Chapter 3 we made careful distinction between the term *set* and *collection.* For the sake of ease of definition and discussion, we will henceforth disregard the distinction between the two and in making further definitions it is to be understood that, in a given instance, the word *set* refers to either a *set* or *collection.* We will not, in a given discussion, consider a single entity to be both a set and a collection for, as we will soon see, the distinction

[2] The hyphenated "same-number" is intended, is no misprint, and is not to be disregarded, for the time being. We are not yet ready to define the "number of elements in a set" (see Section 7.3.17).

we have just agreed to disregard becomes more important than ever. This no doubt sounds somewhat ridiculous but be assured that all will be better understood in the end because of this agreement not to distinguish between the two words at every turn, but only when necessary for consistency. The strict basis for this lack of distinction between sets and collections is that the weight of our theorems and proofs rests on correspondences (functions) between elements and not upon equality of elements—as will be seen. Regarding then the collection 2^S of subsets of a set S as but another set T we have

THEOREM 6.2.4. Let S be a set. Then 2^S has more elements than S.

Proof. Define $f\colon S \to 2^S$ by $f(x) = \{x\}$. Clearly f is 1-1. To complete the proof we must show that there is no function $g\colon S \to 2^S$ which is onto.

Suppose there were. Let $g\colon S \to 2^S$ and suppose g is onto. For each $x \in S$, $g(x) \in 2^S$ and hence $g(x) \subset S$. For each $x \in S$, either $x \in g(x)$ or $x \notin g(x)$. Let $E = \{x \in S\colon x \notin g(x)\}$. Since $E \subset S$ then $E \in 2^S$. Since g is onto $\exists\, x_0 \in S \ni g(x_0) \in E$. We leave it to the reader to prove that this is nonsense, hence completing the proof.

In particular, $2^{\mathcal{N}}$ has more elements than $\mathcal{N}$, hard as that is to imagine. But on the other hand there are quite a lot of elements in $2^{\mathcal{N}}$! Theorem 6.2.4 immediately gives rise to a paradox, known to Cantor himself, and brings us to the aforementioned importance of the distinction between sets and collections. Suppose we admit the existence of the set of all sets. That is, suppose we try to treat the collection of all sets as a set. Surely, no set has more elements than this one! In other words the concept of "the set of all sets" must lead to nonsense. This realization was a plague on early set theorists and is one of the reasons for an extremely careful formulation of the axioms of set theory. If these axioms should admit the collection of all sets as a set, then they would be inconsistent. In this regard, Russell proposed a method of avoiding the difficulty known as the "*theory of types.*" Russell's solution, which, so far as is known does avoid the known paradoxes of set theory, was to admit a collection as a set only if that collection consisted of elements of the same "type" or "level." All of its elements could be elements (Type I), or all could be sets and subsets (Type II), or all could be collections of sets (Type III) and so on, but no entity of a given type could be considered as being of a previous type. These comments are as far as we will go into the matter, being content to deal in each theorem only within a given universal set, all of whose elements are never to be thought of as being both elements and sets simultaneously.[3]

Problems.

(1) Show that the sets S_∞ and T_∞ used to lead up to Question I are

[3] See H. Meschowski: *Evolution of Mathematical Thought*, pp. 47–54. (San Francisco: Holden-Day Inc., 1965). This brief discussion would be well worth reading at this point.

infinite sets.

(2) Complete the proof of Theorem 6.2.4.

(3) Prove that $S = [0, 1]$ and $T = [0, 2]$ have the same-number of elements. The whole is the sum of its parts?

(4) Prove that $S = (0, 1)$ and $T = \mathscr{R}$ have the same-number of elements.

(5) Prove that $S = (0, 1)$ and $T = [0, 1)$ have the same-number of elements.

(6) Prove that $\mathscr{N}$ and $I = \{\pm n: n \in \mathscr{N} \text{ or } n = 0\}$ have the same-number of elements.

(7) Let $S = (0, 1) \times (0, 1]$, $T = (0, 1]$. If $(x, y) \in S$, then $x = .x_1x_2x_3\ldots$ and $y = .y_1y_2y_3\ldots$ and this decimal representation is unique in the sense of Theorem 2.9.3. Define $f: S \to T$ by $f(x, y) = .x_1y_1x_2y_2x_3y_3\ldots$ and show that f is 1-1. Hence a square cannot have more points than a line! Do you think these ought at least have the same number? If so, why? If not, why not?

(8) Let S be an infinite set and let $2_i{}^S = \{T \subset S: T \text{ is infinite}\}$. Show that $2_i{}^S$ has more elements than S.

(9) Let S be an infinite set and let $2_f{}^S = \{T \subset S: T \text{ is finite}\}$. Can you extend the method of Problem 7 to show that $2_f{}^S$ has more elements than S?

(10) Suppose sets S and T have the same-number of elements. Show that 2^S and 2^T have the same number of elements.

(11) Do the axioms of set theory in Chapter 3 and the logic of Chapter 1 admit the collection of all sets as a set? *Hint*: if so, consider the Russell paradox and the possibilities for Axiom 2.

3. COUNTABLY INFINITE SETS AND UNCOUNTABLY INFINITE SETS

Theorem 6.2.4 forces the mathematician to conside infinite sets and kinds of infinity beyond that of the infinity of $\mathscr{N}$. Knowing this it is both advantageous and natural to single out those sets that are no more infinite than the set $\mathscr{N}$, especially since $\mathscr{N}$ possesses a kind of infinity one feels some acquaintance with.

DEFINITION 6.3.1. Let S be a set. If $\exists\, f: \mathscr{N} \to S$ which is 1-1 and onto, then S is said to be ***denumerable.*** If S is finite or denumerable, S is said to be ***countable.*** If S is not countable, then S is called ***uncountable.***

We have examples of finite sets (I_n for any n), a countable set ($\mathscr{N}$ itself), and an uncountable set ($2^{\mathscr{N}}$). Then of course there is the uncountable set $2^{2^{\mathscr{N}}}$ which has more elements than $2^{\mathscr{N}}$ by Theorem 6.2.4. One can go on, but we have no reason in this text to appreciate the worth of it. The overall goal now is to determine the status of the set $\mathscr{R}$ with respect to Definition

6.3.1. Note, too, that the sets S_∞ and T_∞ leading up to Question I are by Definition 6.3.1 both denumerable sets. This will be the key to answering Question I.

The origin of the term countable is of interest because of a common conceptual error. One would like to believe that given a set S, it is perfectly reasonable to subscript the elements of S with the natural numbers. For example, it is assumed that one can say: let $s_1, s_2, s_3, \ldots$ denote the elements of S; or simply: let $S = \{s_1, s_2, \ldots\}$. That is, that one can "count" the elements of S in the order of the natural numbers $\mathcal{N}$. But if this were so, one could then define a 1-1, onto function $f: \mathcal{N} \to S$, namely $f(n) = s_n$, and S would be countable. Since we have at least one example of an uncountable set, namely $S = 2^{\mathcal{N}}$, such an assertion about any arbitrary set S is false. Moreover, as we will soon see, many (indeed most) familiar sets are uncountable. Of course, it is conceptually "nice" to be able to subscript all the elements of a set S with the natural numbers. This is warranted only when it is ***first known*** that S is countable and for this reason we single out countable sets for special attention. We are now going to investigate the properties of countable sets in detail, and, in particular, look for familiar countable sets other than $\mathcal{N}$ itself.

In contrast to finite sets (Theorem 6.1.3(a)), and to display the peculiar nature of infinite sets we have,

THEOREM 6.3.2. If S is denumerable, then S has the same-number of elements as proper subset of itself.

Proof. Since S is denumerable, $\exists\ f: \mathcal{N} \to S$ which is 1-1 and onto. Define $g: \mathcal{N} \to \{2, 4, 6, \ldots\}$ by $g(n) = 2n$. Let $T = \{f(2k): k \in \mathcal{N}\} \subsetneq S$. Let $h = f \circ [g \circ f^{-1}]$. Then $h: S \to T$ and because f, g and f^{-1} are 1-1 and onto, h is 1-1 and onto by Theorem 5.5.8(a). Hence the proof.

We have, at this point, a basic denumerable set, the set $\mathcal{N}$. Any set in 1-1 correspondence with $\mathcal{N}$ is, by Definition 6.3.1, denumerable. In seeking other denumerable and countable sets our approach is to examine the property of countability as it is lost or preserved in the process of adding or deleting elements from a set, for this is the natural way to construct larger or smaller sets. The matter of adding elements to a set of course concerns sets formed by set theoretic union. The matter of deletion concerns sets formed by complimentation. The fundamental, most difficult to prove, and most necessary theorem, in the study of countability, is also the most believable. This is

THEOREM 6.3.3. Let $A \subset \mathcal{N}$. Then A is either finite or denumerable.

Proof. Suppose A is not finite. We will show that A is denumerable. Let $M = \{n \in \mathcal{N}: \exists$ exactly one element of A greater than exactly $n - 1$ elements of $A\}$.

Suppose we can show that $M = \mathcal{N}$. Then for each $n \in \mathcal{N}$ we can let (say) $f(n)$ denote the single element of A greater than the first $n - 1$ elements of A. Then $f: \mathcal{N} \to A$ and we have only to show that f is 1-1 and onto to complete the proof. This is the idea behind the definition of the set M.

To see that $M = \mathcal{N}$, first of all note that $1 \in M$ because A, being nonempty, has a least element and this least element is preceded by exactly $1 - 1 = 0$ elements of A.

Suppose $k \in M$. We will show $k + 1 \in M$. Since $k \in M$, then there is an element $a \in A$ preceded by exactly $k - 1$ elements of A. Let $B = A \backslash \{1, 2, 3, \ldots, a - 1, a\}$; thus B consists of A with all the $k - 1$ elements preceding a, and a itself, thrown out. Since A is denumerable, then $B \neq \square$ (Problem 8, Section 6.1). Since $B \subset A \subset \mathcal{N}$, B has a least element b and $b \in A$. Furthermore b is preceded by a and (the) exactly $k-1$ elements of A preceding a. Thus b is preceded by exactly $(k - 1) + 1 = k = (k + 1) - 1$ elements of A. Hence $k + 1 \in M$.

Thus, by the induction axiom $M = \mathcal{N}$ and we can define a function f as above. Since $f(n)$ and $f(m)$ denote the unique elements of A preceded by exactly $n - 1$ and $m - 1$ elements of A, respectively, then $f(n) = f(m)$ implies $n = m$ and f is 1-1.

To see that f is onto, observe that for all k, $k \leq f(k)$. This is clear for $k = 1$. For $k = n$, $f(n + 1)$ is the unique element of A greater than exactly n elements of A. Since $f(n)$ is greater than exactly $n - 1$ elements of A, then $f(n + 1) > f(n)$. Assuming the claim $f(k) \geq k$ is true for $k = n$, we then have $f(n + 1) > f(n) \geq n$ and hence $f(n + 1) \geq n + 1$.

Now suppose c is an arbitrary element of A. We will show that $c = f(n)$ for some $n \in \mathcal{N}$. Since $A \subset \mathcal{N}$, c is a natural number, call it k; hence $c = k \leq f(k)$. Suppose $c = f(n)$ for no n. Then $c > f(1)$, the least element in A. If $c > f(j)$, the least element in A preceded by $j - 1$ elements in A, then $c \geq f(j + 1)$. But $c = f(n)$ for no n implies $c > f(j + 1)$. Hence for any $j \in \mathcal{N}$, $c > f(j)$. But this contradicts the inequality $c = k \leq f(k)$. Thus $c = f(n)$ for some n and by definition f is onto. This completes the proof.

With this result the reader will find it easy to prove

THEOREM 6.3.4. If T is countable and $S \subset T$, then S is countable.

Using Theorem 6.3.4, we are ready to handle the matter of deletions from a countable or uncountable set.

THEOREM 6.3.5.

(a) If T is denumerable and $S \subset T$ is finite, then $T \backslash S$ is denumerable,

(b) If T is uncountable and $S \subset T$ is countable, then $T \backslash S$ is uncountable.

Proof of (a). Since $T \backslash S \subset T$ and T is countable then by Theorem 6.3.4,

$T\backslash S$ is countable. Hence $T\backslash S$ is either finite or denumerable. If denumerable, then we are through.

Hence suppose $T\backslash S$ is finite. Then $\exists$ an $n \in \mathscr{N}$ and a function g: $I_n \to T\backslash S$ which is 1-1 and onto. Since S is finite $\exists$ an $m \in \mathscr{N}$ and a function h: $I_m \to S$. Now define f: $I_{m+n} \to T$ by $f(k) = g(k)$ for $k = 1, 2, \ldots, n$ and $f(n + j) = h(j)$ for $n + j = n + 1, n + 2, \ldots, m$. Clearly f is 1-1 and onto T, a contradiction to the supposition that T is denumerable. Hence the proof.

We leave the proof of (b) to the reader. The problems below indicate that Theorem 6.3.5 is the best result one can get along these lines.

Let us turn to the matter of set theoretic union. We will prove that a countable union of countable sets must be countable and that this is the best possible result concerning the matter of set theoretic union and countability. In order to make the proof we must first find out more about countable sets. This is the concern of the next two theorems.

A 1-1 correspondence cannot produce "more or less than" it begins with and that is the point of the next result, a result which will prove most useful.

THEOREM 6.3.6. If S and P are sets and g: $P \to S$ is 1-1 and S is countable, then P is countable.

To prove this, apply Theorem 6.3.4 to $T = R_g$. Recall that when a set T is known to be countable, one is given a function f: $\mathscr{N} \to T$ (or f: $I_n \to T$) which is 1-1 and onto to work with.

Up to this point the theory of infinite sets has proceeded more-or-less reasonably with few really surprising results. We now come to the first of these. These next two theorems get to the core of countability vis-à-vis the enlarging operation of set theoretic union. The first and fundamental result is that $\mathscr{N}$ and $\mathscr{N} \times \mathscr{N}$ have the same-number of elements, despite their "appearance," an "appearance" displayed in Figure 6.1. Said another way, there are as many lattice points as vertical lattice lines—in an infinite lattice. The formal statement

THEOREM 6.3.7. $\mathscr{N} \times \mathscr{N}$ is denumerable.

Proof. Apply Theorem 6.3.6 to the function g: $\mathscr{N} \times \mathscr{N} \to \mathscr{N}$ given by $g(m, n) = 2^m 3^n$.

The cross-product operation on $\mathscr{N}$, forming the set $\mathscr{N} \times \mathscr{N}$, creates quite a number of new elements but yet does not produce a set having more elements than $\mathscr{N}$ itself. A closer look at cross-products and countability will follow in the up coming problems. Here is the principle result on set theoretic union.

THEOREM 6.3.8 For each $n \in \mathscr{N}$ let A_n be a countable set and suppose that $A_i \cap A_j = \square$ for any i and j with $i \neq j$. Then $\bigcup_{k=1}^{\infty} A_k$ is countable.

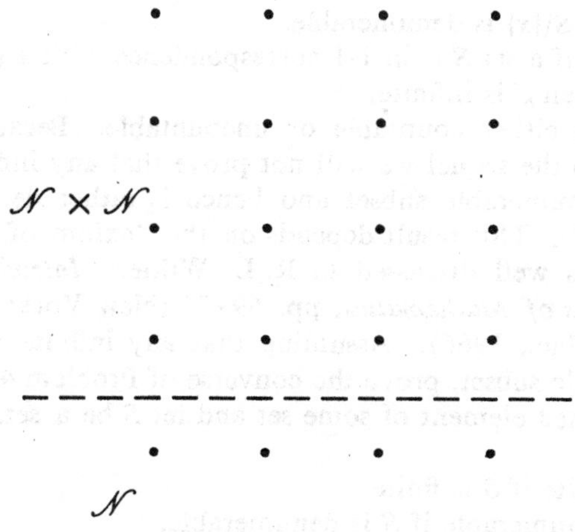

FIG. 6.1

Proof. Each set A_k is countable and hence there is an onto 1-1 correspondence (function) $f_k: N_k \to A_k$ where $N_k = \mathscr{N}$ if A_k is denumerable and $N_k = I_n$ for some n if A_k is finite. In either case, $N_k \subset \mathscr{N}$ and this is what will matter.

Let $S = \mathscr{N} \times \bigcup_{k=1}^{\infty} N_k \subset \mathscr{N} \times \mathscr{N}$ and define $g: S \to \bigcup_{k=1}^{\infty} A_k$ by $g(k, n) = f_k(n) \in A_k$. Think of $g(k, n)$ as the n-th element in the k-th set A_k.

We claim that the correspondence g, which is easily seen to be a function, is 1-1 and onto. First g, is onto for if $x \in \bigcup A_k$, then $x \in A_k$ for some k and since f_k is onto $A_k \ \exists\ n \in N_k$ such that $f_k(n) = x$. Hence $g(k, n) = x$.

To see that g is 1-1, suppose $g(k, n) = g(k', n')$. Then, $f_k(n) = f_k{}'(n')$. Now $f_k(n) \in A_k$ and $A_k \cap A_k{}' = \square$ if $k \neq k'$. Since $f_k{}'(n') = f_k(n) \in A_k \cap A_k{}'$ we must have $k = k'$. Thus, $f_k(n) = f_k(n')$. But f_k is 1-1 and hence $n = n'$. Thus $g(k, n) = g(k', n')$ implies $(k, n) = (k', n')$ and hence g is 1-1 by Definition 5.3.4.

Let $f = g^{-1}$. By Theorem 5.5.5(a), $f: \bigcup_{k=1}^{\infty} A_k \to S$ is 1-1 and onto and since $S \subset \mathscr{N} \times \mathscr{N}$, then $f: \bigcup_{k=1}^{\infty} A_k \to \mathscr{N} \times \mathscr{N}$ is 1-1. By Theorems 6.3.7 and 6.3.6, $\bigcup_{k=1}^{\infty} A_k$ is countable. Hence the proof.

Roughly, Theorem 6.3.8 says that one cannot transcend countability by countable means. We will remark on this again.

Problems.

(1) Show that the sets S_∞ and T_∞ leading up to Question I are denumerable sets.

(2) Is the following statement a faithful (i.e., equivalent) restatement of Question I: can a countable subset of [0, 1] be a subinterval?

(3) Using Definition 6.3.1 only, show that if S is denumerable and

$x \in S$, then $S \backslash \{x\}$ is denumerable.

(4) Prove that if a set S is in 1-1 correspondence with a proper subset of itself, then S is infinite.

(5) Any set is either countable or uncountable. Because it is not material to the sequel we will not prove that any infinite set contains a denumerable subset and hence is either denumerable or uncountable. This result depends on the "axiom of choice" and its proof is well discussed in R. L. Wilder: *Introduction to the Foundations of Mathematics*, pp. 69–74 (New York: John Wiley Co. Sons, Inc., 1966). Assuming that any infinite set contains a denumerable subset, prove the converse of Problem 4.

(6) Let a be fixed element of some set and let S be a set. Prove that $\{a\} \times S$ is
  (a) Finite if S is finite.
  (b) Denumerable if S is denumerable.
  (c) Uncountable if S is uncountable.

(7) Prove Theorem 6.3.4.

(8) Prove Theorem 6.3.5(b).

(9) Prove Theorem 6.3.6.

(10) Prove that $2^m 3^n = 2^{m_1} 3^{n_1}$ implies $m = m_1$ and $n = n_1$. From this prove Theorem 6.3.7.

(11) You will notice in the proof of Theorem 6.3.8 that one had to assume the sets $\{A_k\colon k = 1, 2, \ldots\}$ were mutually disjoint in order to prove that g is a 1-1. We can drop this assumption by an appeal to an earlier observation. If $\{A_k\colon k = 1, 2, \ldots\}$ is a collection of countable sets, then for each n, let $B_n = A_n \backslash \bigcup_{k=1}^{n-1} A_k$. Conclude that B_n is countable and that $B_i \cap B_j = \Box$ for $i \neq j$ and hence that $\bigcup_{k=1}^{\infty} A_k$ is countable whether or not the A_k are mutually disjoint.

(12) Give an example of a collection of countable sets whose union is not countable.

(13) Prove that if A and B are finite with m and n elements, respectively, then $A \times B$ is finite with mn elements.

(14) Prove that if A and B are countable, then $A \times B$ is countable.

(15) For each $n \in \mathscr{N}$ let A_n be a set and $g_n\colon \mathscr{N} \to A_n$ which is 1-1 and onto. Define $g\colon \mathscr{N} \times \mathscr{N} \to \bigcup_{n=1}^{\infty} A_n$ by $g(n, k) = g_n(k)$. Is g necessarily a function? If not, give an example. Under what conditions is g a function? When is g 1-1 and onto?

The next few problems are aimed at counting cross products and taking a more detailed look at $2^{\mathscr{N}}$.

(16) We have defined the cross product of two sets in a rigorous way. One can extend this to the cross product of 3, 4 or n sets in a simiiliar way. Disregarding the necessity for such details, given n sets $A_1, A_2, \ldots, A_n$ let us admit into existence all n-tuples $(a_1, a_2, \ldots, a_n)$ where $a_k \in A_k$. Two of these $(a_1, \ldots, a_n)$ and $(b_1, \ldots, b_n)$ are

called equal iff $a_1 = b_1, a_2 = b_2, \ldots, a_n = b_n$. Let

$$A = \{(a_1, a_2, \ldots, a_n): a_k \in A_k\} = A_1 \times A_2 \times \cdots \times A_n$$

denote the *cross-product* of the sets A_k. (One also uses the notation $A = \mathbf{P}_{k=1}^{n} A_k$). Prove by induction on n that $A_1 \times A_2 \times \cdots \times A_n$ is countable if each A_k is countable.

(17) Let $\mathscr{A}_n$ be the collection of all subsets of $\mathscr{N}$ having exactly n elements. Then $\mathscr{A}_n \subset 2^{\mathscr{N}}$ and we have seen (Theorem 6.2.4) that $2^{\mathscr{N}}$ is uncountable. We wish to prove that $\mathscr{A}_n$ is *countable*. To do this, recall that since any set $A \in \mathscr{A}_n$ has exactly n elements, $\exists$ a 1-1 onto function $f_A: I_n \to A$ (the notation f_A is used because this function depends on the set A one is given). Let $S = \mathbf{P}_{k=1}^{n} \mathscr{N}$ be the set of all n-tuples $(a_1, a_2, \ldots, a_n)$ where $a_k \in \mathscr{N}$. By Problem 16, S is countable. Define $g: \mathscr{A}_n \to S$ by $g(A) = (f_A(1), f_A(2), \ldots, f_A(n)) \in S$. Show that g is 1-1 and apply Theorem 6.3.6.

(18) Use Problem 12 to prove that the sub-collection of $2^{\mathscr{N}}$ consisting of all *finite* subsets of $\mathscr{N}$ is denumerable. Note Problem 8, Section 6.2.

(19) With all the other details to be arranged in Problem 12 it was not pointed out that: (a) there are many 1-1, onto functions $f: I_n \to A$ for a given $A \in \mathscr{A}$ because one can count the elements of A in any order and, hence, (b) picking one, f_A, to be used in the proof is to make use of the "axiom of choice." This can be avoided if we make use of an idea used in the proof of Theorem 6.3.3. Prove that among all the 1-1, onto functions $f: I_n \to A$ there is one, and there is only one, function $f_A: I_n \to A \ni f_A(k)$ is the least element in the set $A \backslash \{f_A(1), \ldots, f_A(k-1)\}$. With this, the "axiom of choice" can be avoided in analogy to forming the set of all left shoes (see Section 3.7) from all pairs of shoes, or all left socks if these could *first* be marked. The mark here is that the function chosen have the property: $f_A(k)$ is the least element in

$$A \backslash \{f_A(1), \ldots, f_A(n-1)\}.$$

(20) Now for "infinite" products and the set $2^{\mathscr{N}}$. Let for each $k \in \mathscr{N}$, $A_k = A = \{0, 1\}$, a very finite set. According to Problem 15, $A \times A \times \cdots \times A = \mathbf{P}_{k=1}^{n} A_k = \{(x_1, \ldots, x_n): x_i \in A_i = \{0, 1\}\}$. Hence $\mathbf{P}_{k=1}^{n} A_k$ consists of all n-tuples $(x_1, \ldots, x_n)$ such that x_i is either 0 or 1 and is in fact a finite set. Let $\mathbf{P}A = \mathbf{P}_{k=1}^{\infty} A_k$ denote the set of all infinite-tuples $(x_1, x_2, \ldots) \ni x_i \in A_k = \{0, 1\}$. Thus $\mathbf{P}_{k=1}^{\infty} A_k$ consists of all infinite sequences of 0's and 1's. How many elements does $\mathbf{P}_{k=1}^{n} A_k$ have? Prove that $2^{\mathscr{N}}$ and $\mathbf{P}_{k=1}^{\infty} A_k$ are in 1-1 correspondence and hence that $\mathbf{P}_{k=1}^{\infty} A_k$ is uncountable. From this draw at least two conclusions: (a) Problem 15 is the best possible result on preserving countability in cross products, and

(b) countability can be lost by an operation (forming cross products) of countable "extent," even though a simple union (Theorem 6.3.8) will not destroy countability.

(21) The set $\mathbf{P}_{k=1}^{\infty} A_k$ is thought of as "infinite-dimensional" in opposition to the "two-dimensional" set $A_1 \times A_2 = \{(x_1, x_2): x_i \in A_i\}$. Such "infinite-dimensional" objects have proven to be of basic use in modern physics. The set $\mathbf{P}_{k=1}^{\infty} A_k$ itself is of much use in probability theory; here is an indication. Suppose a coin is to be tossed a denumerably infinite number of times. Let 0 represent "tails," 1 represent "heads." Whatever the outcome of this infinite sequence of coin tosses it would correspond to exactly one element of $\mathbf{P}_{k=1}^{\infty} A_k$. What do you think is the probability of tossing "head" only after the first 100 tosses? In practice any such infinite coin tossing sequence is impossible, but in studying what must happen in the "long run," the set $\mathbf{P}_{k=1}^{\infty} A_k$ is indispensible.

(22) With A_k as in (20), let $F_n = \{f: f \text{ is a function from } I_n \text{ into } \{0, 1\}\}$. Define a 1-1 correspondence between F_n and $\mathbf{P}_{k=1}^{n} A_k$. Define a 1-1 correspondence between 2^{I_n} and $\mathbf{P}_{k=1}^{n} A_k$.

(23) Let $F = \{f: f \text{ is a function with domain } \mathscr{N} \text{ and range in } \{0, 1\}\}$.
   (a) Prove that $2^{\mathscr{N}}$ and F are in 1-1 correspondence and compare this with Problem 6, Section 6.1.
   (b) Define a 1-1 correspondence between F and $\mathbf{P}_{k=1}^{\infty} A_k$ after first proving, using Problem 19 and (a), that there must be such a correspondence.

(24) One can challenge the existence of the set $\mathbf{P}_{k=1}^{\infty} A_k$. This problem yields one way to meet that challenge. Let $\mathscr{N}$ be the universe of discussion. Each function $f: \mathscr{N} \to \{0, 1\}$ is a relation, hence a subset of $\mathscr{N} \times \{0, 1\}$. Why does the set F exist as a consequence of the axioms of set theory? Given that F exists, and with Problem 22a in mind, the mathematician simply defines $\mathbf{P}_{k=1}^{\infty} A_k$ to be, in fact, the set F and thinks of the point $(x_1, x_2, \ldots, x_n, \ldots) \in \mathbf{P}_{k=1}^{\infty} A_k$ as the function $f: \mathscr{N} \to \{0, 1\}$ defined by $f(n) = x_n$.

Such is a justification for the existence of "infinite-dimensional" spaces, a concept of some importance in modern physics as well as mathematics.

(25) Let $A \subset \mathscr{R}$ be a countable set. Prove that the set of all solutions to any quadratic equation $ax^2 + bx + c = 0$ where $a, b, c \in A$ is countable. *Hint*: identify the quadratic $ax^2 + bx + c$ with the triple $(a, b, c) \in A \times A \times A$, this last set being countable by Problem 11.

4. BEYOND THE COUNTABLY INFINITE: $\mathscr{R}$ AND THE ANSWER TO QUESTION I

Hopefully the reader now has some feeling, no doubt mixed and uncertain, for what may be countable. We will now develop the most important conclusion of this Chapter, the proof that the familiar set $\mathscr{R}$ is uncountable. At the same time we will answer Question I. We have of course an example of an uncountable set, $2^{\mathscr{N}}$, but it is hard to regard $2^{\mathscr{N}}$ as familiar or even substantive in a certain sense. Not so for $\mathscr{R}$, for we do regard $\mathscr{R}$ as known and useful, having operated with real numbers since learning advanced arithmetic. But the reader may object: might there be other sets, familiar and less complicated than $\mathscr{R}$, which are uncountable? For instance, the set $\mathscr{Q}$ of all positive fractions

$$1, 1/2, 2, 1/3, 2/3, 3, \ldots$$

appears to contain a lot more than $\mathscr{N}$ alone. Fortunately, or unfortunately, however you chose to look at it, $\mathscr{Q}$ doesn't make it, $\mathscr{Q}$ is not large enough for $\mathscr{Q}$ is denumerable.

THEOREM 6.4.1. $\mathscr{Q}$ and $\mathscr{N}$ are in 1-1 correspondence and hence $\mathscr{Q}$ is denumerable.

For a proof, correspond $\mathscr{Q} = \{p/q \colon p, q \in \mathscr{N}, q \neq 0\}$ with a natural subset of $\mathscr{N} \times \mathscr{N}$. Then use Theorems 6.3.7, 6.3.4 and 6.3.6. There is one subtlety to be dealt with: $1/2 = 2/4 = 3/6 = \cdots = n/2n$ and the implied non-uniqueness of fractional representation must be dealt with.

The discovery, by Cantor, that $\mathscr{R}$ is uncountable was one of his most startling results. We will provide two proofs, one our own and the other, Cantor's classic proof. They are radically different. Canter's proof was subject to a certain kind of doubt, especially noted by Kronecker, a matter we will look into. Our approach is through the topological concept of connectedness (!), a tribute to the organic unity of mathematics and a material support to the earlier comment that questions concerning infinite sets arise naturally throughout mathematics. Here we find connectedness, defined with no regard for infinite sets, in fundamental relationship with the principle distinction between infinite sets: countability. Cantor, it should be mentioned, worked at a time when topology was non-existent. Incidentally, our approach will partially answer another question that may have occurred to the reader: How large must a subset of $\mathscr{R}$ be in order to be connected? We are going to prove that a countable subset of $\mathscr{R}$ cannot be connected and hence that $\mathscr{R}$, being connected, cannot be countable.

This matter goes deep. Time, order, number, topology and a view of past and future evolved through the eons of time that man has spent trying to comprehend this universe all relate in part to the concepts that yield the next theorem. It is true that the proof, as it always must, boils down to cold hard mathematics in which none of its evolution appears. It is true too that it is difficult to make a precise case for the impingement of other ideas on its realization, but if the reader will grant to this mathematical matter a little poetic license we will look further into this well of thought, searching for a glimmer of intuition.

Perhaps the best way to get some feeling for the relationship of countability and connectedness is to keep in mind the fact that any non-empty subset of $\mathcal{N}$ has a first element and that one thinks of $\mathcal{N}$ as an arrayed sequence

$$1, 2, 3, \ldots$$

containing a first, a second, and exactly one item *immediately* preceding and *immediately* following each other item. This, let us say, is the nature of a *discrete sequence*. In contrast consider the set $\mathcal{R}$. The set $\mathcal{R}$ has no first element, open subsets have no first element, no number $x \in \mathcal{R}$ immediately precedes nor immediately follows any other number in $\mathcal{R}$.

Let us compare this with the intuitive concept of time. We find it impossible to imagine an initial moment of time. Time appears to have no beginning and no end. Given a moment in time we usually find it impossible to speak of an *immediately* preceding or *immediately* following moment. Time is like $\mathcal{R}$ with the usual ordering, and from that ordering, the usual topology. An event occurring at some particular time can usually be associated with a neighborhood of events following and preceding it, somewhat in correspondence with neighborhoods in the usual topology of $\mathcal{R}$. Moreover, we think of time as connected, continuous, flowing on-and-on with no disruption. We have already proven than $\mathcal{R}$ with the usual topology is connected, a result that, in the sense alluded to here, agrees with our concept of time itself.

Consider now specific events in time occurring in a definite sequence. Here we imagine a first or initial event, and each event immediately precedes and immediately follows another. For example, the bouncing of an infinitely elastic rubber ball, each successive bounce being followed and preceded by exactly one bounce. Here we can *conceptually separate* each event (bounce) from any other, for each event follows or precedes any other. That is, using our intuition as a guide, the set of bounces of the ball is not "topologically connected." One can, for example, speak of the set A of all events (bounces) preceding the tenth one and the set B of this event (bounce) and all those following it and these form a definite mental, and, with the appropriate (discrete) topology, topological separation of the whole sequence of bounces. In contrast one cannot apparently do this with non-discrete events in time such as the infinity of positions of an arrow in flight. It seems impossible to separate its motion into (topologically) separated parts.

Thus, some matters appear to us to occur in sequence, countable, each event separated in some definite sense from all that immediately precede and all that immediately follow it, such as the bounces of a ball. Others appear to be like the flight of an arrow, like time itself, no moment of flight or time completely and definitely separated from those moments that follow and precede it.

What then about this matter of sequences and sequential events? The idea of a sequence or a sequential ordering is very much connected with countability, for to have a set S and a 1-1 and onto function $f: \mathcal{N} \to S$ is to assign an order

$$f(1), f(2), f(3), \ldots, f(n), \ldots$$

to the elements of S in the sense that one is able to display the elements of S in sequence: f: $f(1), f(2), \ldots$. Roughly speaking, a set S is countable only when it can be arranged in a sequence like the natural numbers. For example, consider the set $\mathcal{N} \times \mathcal{N}$ arrayed in Figure 6.2.

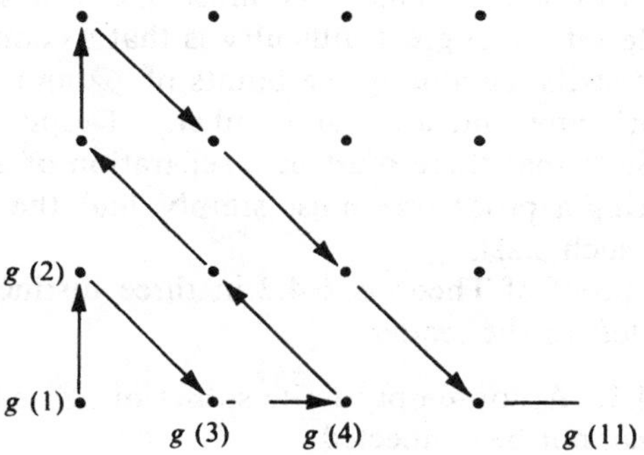

FIG. 6.2

While in one sense, the countability of $\mathcal{N} \times \mathcal{N}$ is surprising, if on the other hand one views this diagram and perceives the elements of $\mathcal{N} \times \mathcal{N}$ as occurring in the indicated sequence, then it is not so hard to imagine a natural one-to-one correspondence g between $\mathcal{N}$ and $\mathcal{N} \times \mathcal{N}$ as indicated above.

The point is that countability is very much associated with discreteness and a definite order, with a first element and exactly one immediate successor and predecessor for each element (save for the first element) in whatever is thought of as countable. In contrast, non-discrete, flowing, continuous, *connected* phenomena do not appear to have such a property.[4] That is the point of our next theorem, which we might call the "theorem of intermediate events," if we interpret it to say that whatever occurs in a definite sequential order cannot be connected in time. This is something you may intuitively believe already, no doubt because your concept of time and of sequential occurrence is very much imbedded in the usual topological structure of $\mathcal{N}$ and $\mathcal{R}$. Here then is

THEOREM 6.4.2. A countable subset of $\mathcal{R}$ cannot be connected.

[4] P. W. Bridgman: *The Nature of Physical Theory*, pp. 29–32. Princeton: Princeton University Press, 1963.

Before beginning a proof there is an additional purely mathematical point to be made. If one rather dislikes the idea of an uncountable set, of a set having a "greater degree" of infinity than the set $\mathcal{N}$, Theorem 6.4.2 says that one is then left with no familiar connected sets. In other words, the acceptance of connectedness as a desirable property demands the existence of uncountable sets as it demanded the completeness axiom!

Let us begin with an example that indicates the nature of the difficulty of a proof for Theorem 6.4.2. We have just noted that the set $\mathcal{Q}$ of rational numbers is countable. If Theorem 6.4.2 is to be true, then $\mathcal{Q}$ cannot be connected and there must be a separation of $\mathcal{Q}$ in $\mathcal{R}$. Because $\sqrt{2} \notin \mathcal{Q}$, we have that $A = (-\infty, \sqrt{2}) \cap Q$ and $B = (\sqrt{2}, \infty) \cap \mathcal{Q}$ is a separation of $\mathcal{Q}$. In the proof of Theorem 6.4.2 we must obtain a separation for an *arbitrary* countable set. The great difficulty is that a countable set, like $\mathcal{Q}$, can be quite well distributed among the points of $\mathcal{R}$ and, in the sense that $\overline{\mathcal{Q}} = \mathcal{R}$, be "near" any and all real numbers. Despite this possibility, Theorem 6.4.2 asserts that there must be a separation of any countable set in $\mathcal{R}$, and in making a proof one must simply find the "right way" by which to separate such a set.

We present the proof of Theorem 6.4.2 in three distinct parts, the first two of which are left to the reader.

LEMMA 6.4.3. A non-empty finite subset of $\mathcal{R}$ having more than one element cannot be connected.

LEMMA 6.4.4. If $a, b, c \in \mathcal{R}$ and $a < b$, then there exists a closed interval $[a', b'] \subset [a, b]$ such that $c \notin [a', b']$ and $a' < b'$.

We now turn to a proof of Theorem 6.4.2, where we will make use of Lemma 6.4.4 infinitely often, in the sense of inductive reasoning.

Proof. Suppose $T \subset \mathcal{R}$ and T is countable. We will prove that T cannot be connected. If T is finite this follows from Lemma 6.4.3. Hence let us suppose that T is denumerable and that $f\colon \mathcal{N} \to T$ which is 1-1 and onto.

The idea of the proof is to use Lemma 6.4.4 to obtain a sequence of closed intervals $[a_k, b_k]$, one for each $k \in \mathcal{N}$, such that $f(k) \notin [a_k, b_k]$, and from these obtain a separation of T of the form

$$A = \bigcup_{k=1}^{\infty} (-\infty, a_k) \cap T, \quad B = \bigcup_{k=1}^{\infty} (b_k, \infty) \cap T.$$

Specifically, we claim that $\forall\, k \in \mathcal{N}\ \exists$ a closed interval $J_k = [a_k, b_k]$ such that $a_k < b_k$ and

(1) $T \cap (-\infty, a_1) \neq \square$ and $T \cap (b_1, \infty) \neq \square$.

(2) $J_{k+1} \subset J_k$ for $k = 1, 2, \ldots$.

and

(3) $f(k) \notin J_k$ for $k = 1, 2, \ldots$.

We will prove this by induction on $\mathscr{N}$. The conditions (1), (2) and (3) may be visualized as in Figure 6.3.

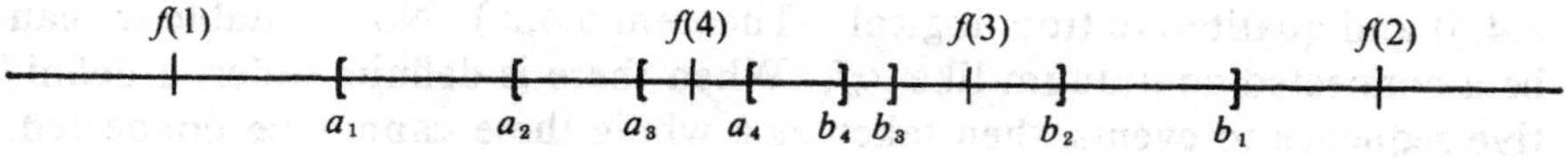

FIG. 6.3

For $k = 1$ we obtain $J_k = J_1$ as follows: either $f(1) < f(2)$ or $f(2) < f(1)$. If (say) $f(1) < f(2)$, then there exist numbers a_1, b_1 such that $f(1) < a_1 < b_1 < f(2)$. Letting $J_1 = [a_1, b_1]$ we see that (1) and (3) above hold, while (2) is not yet in question. We proceed similarly in the case $f(2) < f(1)$. Hence J_1 exists and (1) and (3) hold.

Given $J_1 = [a_1, b_1]$ we can use Lemma 6.4.4 to obtain J_2. With $a = a_1$, $b = b_1$ and $c = f(2)$, Lemma 6.4.4 asserts that there is an interval $J_2 = [a_2, b_2] \subset [a_1, b_1] = J_1$ such that $c = f(2) \notin J_2$. Hence (1), (2) and (3) hold for $k = 1, 2$.

Suppose that $J_1, J_2, \ldots J_n$ have been chosen, with (1), (2) and (3) satisfied for $k = 1, 2, \ldots, n$. Then, using Lemma 6.4.4 with $a = a_n$, $b = b_n$ and $c = f(n + 1)$, we obtain an interval $J_{n+1} = [a_{n+1}, b_{n+1}] \subset [a_n, b_n] = J_n$ such that $c = f(n + 1) \notin J_{n+1}$.

Hence by the axiom of induction the intervals J_k, one for each $k \in \mathscr{N}$, exist and satisfy (1), (2) and (3). Now let $A = T \cap \bigcup_{n=1}^{\infty} (-\infty, a_n)$ and $B = T \cap \bigcap_{n=1}^{\infty} (b_n, \infty)$. We claim $A \neq \square$, $B \neq \square$, $T = A \cup B$ and $\bar{A} \cap B = \bar{B} \cap A = \square$, thus yielding a separation of T.

First of all, (1) implies that $f(1) \in T \cap (-\infty, a_1)$ and hence $f(1) \in A$. Similarly, $f(2) \in T \cap (b_1, \infty)$ and hence $f(2) \in B$. Thus $A \neq \square$ and $B \neq \square$.

Suppose that $x = f(k) \in T$. We claim $x \in A \cup B$. Since $f(k) \notin J_k = [a_k, b_k]$, either $f(k) < a_k$ or $f(k) > b_k$. Thus $f(k) \in T \cap (-\infty, a_k)$ or $f(k) \in T \cap (b_k, \infty)$. Hence $x = f(k) \in A \cup B$. Thus $T = A \cup B$.

To see that $\bar{A} \cap B = \square$ let $B_0 = \bigcup_{k=1}^{\infty} (b_k, \infty)$. We claim $A \subset \mathscr{R} \backslash B_0$. Now, $x \in A$ implies $x < a_n$ for some n which in turn implies that $x < a_n < b_n < b_j$ for $j \leq n$ since $J_n = [a_n, b_n] \subset J_j = [a_j, b_j]$ by (2). Hence $x \leq b_j$ for all $j \leq n$. If $j > n$, then $x < a_n < a_j \leq b_j$ since $J_j = [a_j, b_j] \subset J_n = [a_n, b_n]$ by (2). Hence $x \leq b_j$ for all $j > n$ as well and therefore $x \notin B_0 = \bigcup_{k=1}^{\infty} (b_k, \infty)$. Thus $A \subset \mathscr{R} \backslash B_0$.

Since B_0 is an open set (Theorem 4.4.1), then $\mathscr{R} \backslash B_0$ is closed and hence by Corollary 4.3.4, $\bar{A} \subset \mathscr{R} \backslash B_0$. Thus, $\bar{A} \cap B \subset \bar{A} \cap B_0 = \square$ since $B \subset B_0$. Hence $A \cap B = \square$. Similarly we can prove that $A \cap \bar{B} = \square$. Thus T is a union of non-empty sets A and B such that $\bar{A} \cap B = \bar{B} \cap A = \square$. By Theorem 4.6.3, T is not connected. Hence the proof.

And now from Theorems 4.5.2 and 6.4.2,

THEOREM 6.4.5. $\mathscr{R}$ is uncountable for $\mathscr{R}$ is connected.

Theorem 6.4.5 was one of the great breakthroughs of modern mathematics.

It, especially in conjunction with Theorem 6.4.2, shows the existence of a fundamental distinction between the two most important number sets of mathematics $\mathscr{N}$ and $\mathscr{R}$—the distinction is both quantitive (size—Theorem 6.4.5) and qualitative (topological—Theorem 6.4.2.)′ No countable set can be a connected continuum like $\mathscr{R}$. When there is definite order, a definitive sequence to events, then taken as a whole these cannot be connected. Only when one has something more complicated than a sequential phenomenon can he hope for connectedness in the usual sense. That sounds much like an answer to Question I, so let us formalize and be sure of it.

Question I asks: can one remove points $p_1, p_2, p_3, \ldots, p_n, \ldots$, one at a time and, putting these altogether, remove an entire interval from $\mathscr{R}$? Since the function $f\colon \mathscr{N} \to \mathscr{R}$ given by $f(n) = p_n$ makes $T_\infty = \{p_1, p_2, \ldots, p_n, \ldots\} = \mathscr{R}_f$ a countable subset of $\mathscr{R}$, then by Theorem 6.4.2, T cannot be connected. To conclude that T_∞ cannot be an interval we need know only that an interval must be connected. This is a consequence of the connectedness of $\mathscr{R}$ and the material of Section 4.6 and for the reader who has not proven this already we interject a proof.

THEOREM 6.4.6. Any interval (a, b) or $[a, b]$ in $\mathscr{R}$ is connected.

Proof. Suppose (a, b) is not connected. Then $(a, b) = A \cup B$ with A separated from B in $\mathscr{R}$ (Theorem 4.6.3).

There are two possibilities. If $a, b \in \bar{A}$, let $B_0 = B$ and let $A_0 = A \cup (-\infty, a] \cup [b, \infty)$. Since $(a, b) = A \cup B$ it follows that $A_0 \cup B_0 = \mathscr{R}$ and since $\bar{A} \cap B = \bar{B} \cap A = \square$ it follows that $\bar{A}_0 \cap B_0 = \bar{B}_0 \cap A_0 = \square$ because $a, b \in \bar{A}$ and $a, b \notin B$. But this means $\mathscr{R}$ is not connected, a contradiction.

The other possibility is that $a \in \bar{A}$ and $b \in \bar{B}$ for $[a, b] = \overline{(a, b)} = \overline{A \cup B} = \bar{A} \cup \bar{B}$ by Theorem 4.4.2. In this case let $A_0 = (-\infty, a] \cup A$, $B_0 = B \cup [b, \infty)$. It again follows that $\mathscr{R} = A_0 \cup B_0$ but A_0 is separated from B_0, a contradiction.

Thus (a, b) must be connected. By Theorem 4.6.6, $[a, b]$ is also connected.

Putting Theorem 4.6.4 together with Theorem 6.4.2, we have the answer to Question I: no, one cannot remove an interval from $\mathscr{R}$ one point at a time.

Without an interest in certain deeper problems of mathematics and an acquaintance with some of the brilliant ideas of 19th century analysis, it is difficult to convey the importance of Question I in mathematics. Among other things, the reader is invited to consider Problem 5, Section 7.4, where the matter of the length or measure of a countable set in $\mathscr{R}$ is considered. This problem resolves the question of the length of the sets $T_1 = S\backslash\{p_1\}$, $T_2 = S\backslash\{p_1, p_2\}, \ldots, T_n = S\backslash\{p_1, p_2, \ldots, p_n\}$ and particularly the particular the length of T_∞. The reader will recall that the accompanying observations of the length of the sets T_n is what initially leads one to believe (expect, hope, wonder) that T_∞ could not be connected, thus anticipating the final proven answer to Question I. Though the reader presumably has little knowledge of either Fourier series, probability on an infinite space of events, integration theory, the theory of differentiable functions or several other advanced fundamental concepts, we will still try to relate Question I and its answer to fundamental problems related to all of these.

It will be recalled that mathematical man has tried to understand some one single complex phenomenon by approximation to it by infinitely many simpler, related phenomena. Such is the approach of Fourier's beautiful theory of approximation of functions describing complex phenomena by means of infinitely many *finite* sums of sine and cosine functions (recall Section 2.9). It happens in some of these approximation theories that the end result of the approximations, the limit of the approximations, ***does not agree at every point*** with the complex end one was aiming for. But it often is the case that the failure to agree only happens at ***countably many*** points and so by Theorem 6.4.2 the failure of agreement, the failure to approximate exactly, cannot occur over an entire interval, over an entire connected set, but must occur only on a separated point set. This fact allows one to make good use of an approximation theory that is not exact, because it fails only at separated, and in the proper neighborhood sense, sufficiently isolated points. For a full understanding of this matter one must make a study of the theory of measure and show that a countable set, known not to be an interval, also has at most zero measure—the failure of the approximation technique then fails only or a set that has "no measure worth considering" to put it roughly, but as closely as we can.

Problems.

(1) Prove Theorem 6.4.1.

(2) Prove Lemma 6.4.3.

(3) Prove Lemma 6.4.4.

(4) Let $a = .a_1a_2a_3\ldots$ be a real number. For each n, let

$$I_n = [.a_1a_2\ldots.a_n0, .a_1a_2\ldots a_n9].$$

Show that

(a) $a \in I_n$ for all n.

(b) $I_{n+1} \subset I_n$.

(c) $\bigcap_{k=1}^{\infty} I_k = \{a\}$. (*Hint*: if $b \in \bigcap_{k=1}^{\infty} I_k$, show that $|b - a| < 1/n$ for all $n \in \mathscr{N}$ and hence that $b = a$.)

(d) Let $A = \bigcup_{n=1}^{\infty} (-\infty, .a_1a_2\ldots a_n0)$, $B = \bigcup_{n=1}^{\infty} (.a_1a_2\ldots a_n9, \infty)$. Show that $A \cup B = \mathscr{R} \backslash \bigcap_{k=1}^{\infty} I_k$ and that $\bar{A} \cap B = \bar{B} \cap A = \square$.

(e) Let $A_0 = A \cup \{a\}$, $B_0 = B$. Show that $A_0 \cap \bar{B}_0 = \{a\}$.

(5) For each n, let $I_n = [a_n, b_n]$ be a closed interval with $I_{n+1} \subset I_n$ for all n. Using the ideas in the proof of Theorem 6.4.2, show that $\bigcap_{n=1}^{\infty} I_n \neq \square$.

(6) Remove the points $1/2$, $3/4 = 1/2 + 1/2^2$, $5/8 = 1/2 + 0/2^2 + 1/2^3$, $9/16 = 1/2 + 0/2^2 + 0/2^3 + 1/2^4$, $11/16 = 1/2 + 0/2^2 + 1/2^3 +$

$1/2^4, 17/32, \ldots$ and so on from $(0, 1)$ and described by means of a base 2 decimal representation one number lying between $1/2$ and $3/4$ that is not removed.

(7) In base 2, let $C_1 = \{.1\}$, $C_2 = \{.1, .11\}$, $C_3 = \{.1, .101, .11\}$, $C_4 = \{.1, .1001, .101, .1011, .11\}$, and in general if $C_n = \{.1, a_1, a_2, \ldots, a_2n - 2_{-1}, .11\}$ let $C_{n+1} = \{.1, a_1, a_1 + 1/2^{n+1}, a_2, a_2 + 1/2^{n+1}, \ldots, a_2n - 2_{-1}, +1/2^{n+1}, .11\}$. Describe $[1/2, 3/4]\backslash \bigcup_{k=1}^{\infty} C_k$.

(8) Define a 1-1 correspondence between $\mathscr{R}$ and any interval (a, b). Conclude that (a, b) and $[a, b]$ are uncountable.

(9) In base 10 any decimal $.a_1a_2 \ldots a_n$ for any n represents the rational number $a_1/10 + a_2/10^2 + \cdots + a_n/10^n$. Let $2_f{}^{\mathscr{N}}$ denote the collection of finite subsets of $\mathscr{N}$. Using the correspondence $\phi(E) = .a_1a_2 \ldots a_n$ where $a_k = 1$ if $k \in E$, $a_k = 0$ if $k \notin E$ show that $2_f{}^{\mathscr{N}}$ is denumerable.

(10) Suppose one removes points $p_1, p_2, \ldots, p_n, \ldots$ from the interval $(0, 1)$, with $p_1 < p_2$. According to Theorem 6.4.2 $\exists$ a point $p \in (0, 1)\backslash\{p_1, p_2, p_3, \ldots\} \cap [p_1, p_2]$. Show that
  (a) There are in fact countably many more points $q \in (0, 1)\backslash \{p_1, p_2, p_3, \ldots\} \cap [p_1, p_2]$ using mathematical induction, Theorems 6.4.2 and 6.3.8.
  (b) Extend (a) by showing that, in fact, $(0, 1)\backslash\{p_1, p_2, p_3 \ldots\} \cap [p_1, p_2]$ is uncountable.

Thus not only can one not remove all the points from an interval by picking these one at a time but in fact one can remove very few—what remains between the points removed must be uncountable! "I see it but I do not believe it" said Cantor.

The next two results say that any bounded denumerable set must have a limit point distinct from all but possibly one element in the set.

(11) Let $M \in \mathscr{R}$ and let $f: \mathscr{N} \to [-M, M]$ such that $f(n) \leq f(n + 1) \forall n \in \mathscr{N}$. Prove that $\exists\, x \in [-M, M]$ that is close to infinitely many of the number $f(n)$: precisely, $\exists\, x \in [-M, M]$ such that every neighborhood of x contains infinitely many numbers $f(n)$. To have an example in mind consider the function $f(n) = 1 - 1/n$, $M = 1$. What number x will serve here? You can use Problem 5 above if you find it sensible, letting $I_n = [f(n), M]$.

(12) Generalize Problem 11. Let $g: \mathscr{N} \to [-M, M]$ and let $f(n) = \sup \{g(k)\colon k \geq n\} \leq M$. Show that $f(n + 1) \leq f(n)$ and hence that $\exists\, x \in [-M, M]$ such that for any neighborhood U of x, U contains infinitely many numbers $g(k)$. Prove that if g is 1-1 (hence $g(\mathscr{N})$ is denumerable), then x is distinct from all but one element $g(n)$ for some n. The idea here is that the numbers $g(k)$ must tend to "cluster or gravitate" around a single number x as k "approaches infinity." For example consider the function $g(n) = 1 + (-1)^n/n$, $M = 2$. What is $f(n)$ in this case? What number x can serve?

5. CANTOR'S PROOF THAT $\mathscr{R}$ IS UNCOUNTABLE

When first proven by Cantor, Theorem 6.4.5 was considered to be both surprising and quite remarkable. One does tend to think of the infinity of $\mathscr{N}$ as being the only kind of infinity there is. But Theorem 6.4.5 says there is another kind and it occurs in a familiar setting. For this reason we want to look into the matter a bit further. We begin with Cantor's original proof that $\mathscr{R}$ is uncountable, wherein he showed that the interval (0, 1), and hence $\mathscr{R}$ itself, must be uncountable. Here is Cantor's proof.

Suppose the set (0, 1) were countable. Clearly (0, 1) is not finite so suppose $\exists$ a 1-1, onto function $f: \mathscr{N} \to (0, 1)$. According to Section 2.8, each $f(n)$ has a unique decimal representation in the sense of Theorem 2.8.3. Let

$$f(n) = .a_{n1}a_{n2}a_{n3}\ldots$$

be the decimal representation for $f(n)$. We thus have a correspondence:

$$1 \longleftrightarrow .a_{11}a_{12}a_{13}\ldots$$

$$2 \longleftrightarrow .a_{21}a_{22}a_{23}\ldots$$

$$3 \longleftrightarrow .a_{31}a_{32}a_{33}\ldots.$$

Because f is onto, all numbers in (0, 1) appear at least once on the right in this array, and because f is 1-1, only once.

Now define the infinite decimal $r \in (0, 1)$ as follows:

$$r = .r_1r_2r_3\ldots$$

where

$$r_i = \begin{cases} 1 & \text{if} \quad a_{ii} \neq 1 \\ 2 & \text{if} \quad a_{ii} = 1 \end{cases}$$

Thus, $r_1 \neq a_{11}, r_2 \neq a_{22}, \ldots, r_n \neq a_{nn}$, for *every* n. Hence $r \neq f(n)\ \forall\, n$. For suppose that $r = f(n)$ for some n. But then $r_n = a_{nn}$ a contradiction. However, $r \in (0, 1)$ and f was supposed to be onto. This is a contradiction. Hence (0, 1) cannot be countable.

We now have two proofs that $\mathscr{R}$ cannot be countable, radically different in appearance. But both the connectedness of $\mathscr{R}$ and the infinite decimal expansion of every real number in the end fall back on the same thing: the completeness axiom. These two proofs use this same axiom in but different ways.

Cantor's proof above is subject to a certain kind of doubt. To illustrate the point let us consider two examples, the first a consequence of his proof that $\mathscr{R}$ is uncountable.

A number $r \in \mathscr{R}$ is called *algebraic* if r is a root of a polynomial equation

$$a_0 + a_1x + a_2x^2 + \cdots + a_nx^n = 0$$

with rational coefficients a_i. Such a polynomial has at most n roots. Since the coefficients are rational and the rational numbers are countable one can prove, using the techniques of this chapter, that there are *only countably many* such polynomials. Hence, since each has only a finite number of roots, Problem 11, Section 6.3, can be used to assert that there are only countably many algebraic numbers. But $\mathscr{R}$ is uncountable and hence there exist non-algebraic (called "transcendental numbers," of which π and e are examples) numbers. But this proof does *not exhibit* or *construct* a non-algebraic number, it only asserts their existence as a matter of absence. Kronecker could not accept such proofs of existence, and other notable mathematicians have also objected. Still, Cantor's reasoning is clever, but keep that word, *existence*, in mind, for we are not yet finished with the matter.

We will now construct a number, in a manner exactly like that in Cantor's proof above, which we cannot distinguish from the number 0 without proving or disproving a centuries old conjecture that has never been settled. Pierre Fermat, in the margin of one of his works, asserted that

$$x^n + y^n = z^n$$

has no solution for $x, y, z \in \mathscr{N}$ for $n > 2$. For $n = 2$, the numbers $x = 3$, $y = 4$ and $z = 5$ give a solution since $4^2 + 3^2 = 5^2$.

Fermat's assertion is known as "Fermat's last theorem." Fermat claimed to prove it, but no one has yet found his or any other proof. Let us call $n > 2$ an F-number iff

$$x^n + y^n = z^n$$

has *no* solution for $x, y, z \in \mathscr{N}$. Thus $n > 2$ is an F-number iff Fermat's last theorem is true about n. No one has ever determined whether or not all $n > 2$ are F-numbers. No one has yet found an n that is not an F-number.[5] Fermat passed away in 1665 and both professional mathematicians and amateurs have been trying ever since to settle the question.

Now define a real number r as follows: let $r = .r_1r_2r_3, \ldots$ where $r = 0$, $r_2 = 0$ and for $n > 2$

$$r_n = \begin{cases} 0 & \text{if } n \text{ is an } F\text{-number.} \\ 1 & \text{if } n \text{ is not an } F\text{-number.} \end{cases}$$

Clearly, $r = 0$ iff every $n > 2$ is an F-number, iff Fermat's last theorem is true! Recall now Godel's result that there must exist an undecidable theorem in any system. If "Fermat's last theorem" is of this sort, then the status of r is undecidable.

The implication then is that Cantor's proof of the uncountability of (0, 1) uses a technique for defining a number r which may *not* allow one to decide just what number r is. The number r is constructed but only in a limited

[5] Computer and other techniques have to date shown that $n = 3, 4, \ldots, 2{,}000$ or so are all F-numbers.

sense. Its definition does not give a workable "rule for identifying." This is the objection to Cantor's proof.

Now one does not need to be a learned or unlearned mathematician in order to question the existence of the number r, as a real number, as defined in Cantor's proof above. Indeed this proof is invariably questioned by at least one student each time this author has presented it. Allowing that the student is not sure of the precise nature of his doubt, it appears that he is unsure whether one should allow the definition of any number r in such a way. If one takes the position that every definition should yield a "rule for identifying," then one should obstinately refuse to accept Cantor's proof. Granted this refusal, is it still a fact that $\mathscr{R}$ is uncountable? What of our proof, which ultimately relies on Theorems 6.4.2 and 4.5.2? There appears to be a very slight distinction.

The proof of Theorem 6.4.2 does involve a subtlety of inductive reasoning that we have glossed over, though the method of proof is quite common and acceptable without further detail by most mathematicians.[6] Our proof does possess one property that distinguishes it from that of Cantor: we at no point have need to assert the *existence* of any particular real number, defined by specifying its digits. On the other hand, we do in Theorem 6.4.2 assert the existence of infinitely many closed intervals having certain properties, and this, in the presence of the completeness axioms is equivalent to asserting the existence of certain real numbers with certain properties. And therein lies the key to distinguishing between the two proofs: our proof of Theorem 6.4.2 does not make use of the completeness axiom in any way. It is only when we assert that $\mathscr{R}$ is connected that the completeness axiom comes into play. But now, if the reader will turn all the way back to the proof (Theorem 4.5.2) that $\mathscr{R}$ is connected, he will find that though the completeness axiom is used, it is not used to assert the existence and uniqueness of any particular number. A very small point, that!

Problems.

(1) Prove that if T is a finite nonempty set and $S = \{f: f: \mathscr{N} \to T\}$, then S is uncountable, modeling your proof after that of Cantor's proof for the uncountability of $(0, 1)$.

(2) Give a complete proof that there are only countably many *algebraic numbers.*

(3) Let $S \subset (0, \infty)$ be a set and suppose that for every finite subset $T = \{x_1, \ldots, x_k\}$ of S, one has $x_1 + x_2 + \cdots + x_k = \sum_{i=1}^{k} x_i \leq 10$. (For ease of notation in discussion, simply denote $\sum_{i=1}^{k} x_i$ by $\sum_{x \in T} x$.) Prove that S is countable. *Hint*: consider the set $S_n = \{x \in S: x \geq 10/n\}$. Can S be denumerable? Can you generalize a

[6] See R. R. Stoll: *Set Theory and Logic* (San Francisco: W. H. Freeman and Co., 1963). See Section 2.3 for a thorough discussion of the inductive construction referred to here.

bit?

(4) Given that a closed interval $[a, b]$ with $a < b$ must be uncountable, give a short proof that a countable subset of $\mathscr{R}$ cannot be connected. *Hint*: if $f: \mathscr{N} \to \mathscr{R}$, consider the closed interval with end points $f(1)$ and $f(2)$, which by hypothesis, cannot be countable. (Hence Question I and Theorem 6.4.2 both follow from Cantor's proof of the uncountability of $(0, 1)$ and can be settled without resort to "connectedness" concepts.)

6. THE SCHRODER-BERNSTEIN THEOREM

The preceding discussions and problems indicate that there are many countable sets and many uncountable sets and that determining when a set is countable, having the same-number of elements as $\mathscr{N}$, or simply determining when two given sets have the same-number of elements, can be difficult. There is one general result which in particular enables one to settle such questions in a fairly easy way. This is the Schroder-Bernstein theorem, a result which asserts just what one would like to believe is true, and in the context of infinite sets must be called surprising! Its proof is not easy.

THEOREM 6.6.1. *Let A and B be sets. If there is a 1-1 function f: $A \to B$ and a 1-1 function g: $B \to A$, then there is a 1-1 function ϕ from A onto B. Thus A and B have the same-number of elements.*

Proof. We must construct a 1-1 function ϕ: $A \to B$ which is *onto*.

Let $h = f^{-1} \circ g^{-1}$. The function h is defined on some subset of A with range in A. (Specifically $D_h = \{x \in A\colon x \in R_g \text{ and } g^{-1}(x) \in R_f\}$ but this specific fact is not important.) Define ϕ_1: $R_g \to B$ by $\phi_1(x) = g^{-1}(x)$ and for each n, define ϕ_n by $\phi_n(x) = g^{-1} \circ h_n(x)$ where $h_n = h \circ h \circ \cdots \circ h$ to n factors. The domain of ϕ_n is some subset of A and the range is some subset of B. Let $A_0 = \{x \in A\colon \exists\, n \text{ such that } \phi_n(x) \text{ is defined and } \phi_n(x) \notin R_f\}$.

A given $x \in A$ is either in A_0 or not in A_0. Because of this we can define a function ϕ: $A \to B$ by

$$\phi(x) = \begin{cases} f(x) & x \notin A_0 \\ g^{-1}(x) & x \in A_0 \end{cases}$$

We claim that ϕ is 1-1 and onto.

First ϕ is 1-1. If $\phi(x) = \phi(x')$ we claim $x = x'$. If both $x, x' \in A_0$ or both $x, x' \notin A_0$, then $x = x'$ because g and f are, respectively, 1-1. Hence suppose $x \notin A_0$ and $x' \in A_0$. We claim this is impossible. For if so, then $f(x) = \phi(x) = \phi(x') = g^{-1}(x')$. Since $x' \in A_0$ there is an n such that $\phi_n(x')$ is defined, but $\phi_n(x') \notin R_f$. First, $n > 1$. For if $n = 1$, then $\phi_1(x') = g^{-1}(x')$ is defined and $g^{-1}(x') = f(x) \in R_f$. Knowing $n > 1$ we then have, $\phi_n(x') = g^{-1} \circ h_n(x') = g^{-1} \circ h_n \circ (g \circ f)(x) = g^{-1} \circ h_{n-1}(x) = \phi_{n-1}(x)$. Since

$\phi_n(x_l) \notin R_f$, then $\phi_{n-1}(x) \notin R_f$. But this puts $x \in A_0$ a contradiction. Hence ϕ is 1-1.

To show that ϕ is onto, suppose $b \in B$. We must find an $a \in A$ such that $\phi(a) = b$. Let $a = g(b)$. If $a \in A_0$, then $\phi(a) = g^{-1}(a) = b$ and we are through. Hence, consider the other possibility: $a \notin A_0$. Then for *all n*, $\phi_n(a)$ is not defined or $\phi_n(a) \notin R_f$. But for $n = 1, \phi_1(a) = g^{-1}(a)$ is defined. Hence, $\phi_1(a)$ must be in R_f. Let $a' = f^{-1}(\phi_1(a)) = (f^{-1} \circ g^{-1})(a)$. We claim $a' \notin A_0$. For if $a' \in A_0$ then there is an n such that $\phi_n(a')$ is defined and $\phi_n(a') \notin R_f$. But, $\phi_n(a') = \phi_n(f^{-1} \circ g^{-1})(a) = \phi_{n+1}(a)$. This contradicts the earlier fact that for all n either $\phi_{n+1}(a)$ is not defined or $\phi_{n+1}(a) \in R_f$. Hence $a' \notin A_0$. Then, by definition,

$$\phi(a') = f(a') = f(f^{-1} \circ g^{-1})(a) = g^{-1}(a) = b\,.$$

Thus ϕ is onto and the proof is complete.

We will use the Schroder-Bernstein theorem to settle a question whose answer and difficulty has been hinted at throughout this chapter. We have seen that $\mathscr{R}$ and $2^{\mathscr{N}}$ both have more elements than $\mathscr{N}$. We will show that in fact they have the same-number of elements.

THEOREM 6.6.2. The set $\mathscr{R}$, the open interval (0, 1) and the collection $2^{\mathscr{N}}$ all have the same-number of elements.

Proof. First, $\mathscr{R}$ and (0, 1) have the same number of elements. For the function f: $(0, 1) \to \mathscr{R}$ given by $f(x) = x$ is 1-1 and the function g: $\mathscr{R} \to (0, 1)$ given by $g(x) = 1/(1 + e^x)$ is 1-1 since the exponential function is 1-1. Thus $\mathscr{R}$ and (0, 1) have the same-number of elements.

To complete the proof we show that (0, 1) and $2^{\mathscr{N}}$ have the same number of elements. First define f: $[0, 1] \to 2^{\mathscr{N}}$ as follows: if $x \in (0, 1)$ is written $x = .x_1x_2x_3\ldots$ in base 2 where this decimal representation is unique in the sense of Section 2.9, let $f(x) = \{n \in \mathscr{N}: x_n = 1\}$. Hence f: $(0, 1) \to 2^{\mathscr{N}}$. If $f(x) = f(x')$, then with $x = .x_1x_2x_3\ldots$ and $x' = .x_1'x_2'x_3'\ldots$ we have $x_n = x_n' = 1$ for all $n \in f(x) = f(x')$ and $x_k = x_k' = 0$ for all $k \notin f(k) = f(x')$. Thus $x = x'$ and f is 1-1.

We now seek a 1-1 function g: $2^{\mathscr{N}} \to (0, 1)$. Here we use base 10 decimal representations. For $E \in 2^{\mathscr{N}}$ let $g(E) = .a_1a_2a_3\ldots$ where $a_n = 1$ if $n \in E, a_n = 2$ if $n \notin E$. The fact that g is 1-1 now easily follows from Theorem 2.8.3.

According to the Schroder-Bernstein theorem, $2^{\mathscr{N}}$ and (0, 2) must have the same-number of elements. This completes the proof.

The Schroder-Bernstein theorem is one of those powerful theorems of mathematics. It can be used to settle quite difficult problems in a relatively easy way.

Problems. Use the Schroder-Bernstein theorem to show that the following sets have the same-number of elements.

(1) $[0, 1)$ and $(0, 1)$.
(2) $(0, 1)$ and $[0, 1]$.
(3) $(0, 1)$ and $2_i^{\mathcal{N}} = \{E \in 2^{\mathcal{N}}: E \text{ is infinite}\}$
(4) $(0, 1)$ and $\mathbf{P}_{n=1}^{\infty} X_n$ where $X_n = \{0, 1\}$ for all n (see Problem 21, Section 6.3).
(5) A line and a square.
(6) $\mathcal{R}$ and $\mathcal{R} \times \mathcal{R}$.
(7) $2^{\mathcal{R}}$ and $2^{\mathcal{R} \times \mathcal{R}}$.
(8) $\mathcal{R}$ and $\{(x_1, x_2, x_3): 0 \leq x_i < 1\} = \mathcal{R} \times \mathcal{R} \times \mathcal{R}$ (a line and a cube have the same-number of elements.)
(9) $2^{\mathcal{R}}$ and $F = \{f: f \text{ is a function from } \mathcal{R} \text{ into } \{0, 1\}\}$.
(10) Show that the Cantor set C of Problem 8, Section 4.4, is uncountable, having the same number of elements as $\mathcal{R}$.

7. THE CONTINUUM HYPOTHESIS

Closely related to the uncountability of $\mathcal{R}$ is an outstanding and only recently settled problem of contemporary mathematics, the problem of the *continuum hypothesis*. We have seen that $\mathcal{N}$ is countable and that $\mathcal{R}$ is uncountable. We have seen that the set $\mathcal{Q}$, of rational numbers with the property that $\mathcal{N} \subset \mathcal{Q} \subset \mathcal{R}$ has the same-number of elements as $\mathcal{N}$. Cantor conjectured, and the following statement is the continuum hypothesis, that there is *no* set S having more elements than $\mathcal{N}$ but less element than $\mathcal{R}$. In other words, the continuum hypothesis asserts that if $\mathcal{N} \subset S \subset \mathcal{R}$, then either $\mathcal{N}$ and S are in 1-1 correspondence or S and $\mathcal{R}$ are in 1-1 correspondence. For most of this century the assertion remained unsettled, but before stating its resolution, consider some of its aspects. In particular, let us try to pick out a set S "between" $\mathcal{N}$ and $\mathcal{R}$, in the sense of having more elements than $\mathcal{N}$ but less elements than $\mathcal{R}$.

If one begins with a set I_n with n elements, then 2^{I_n} has 2^n, and hence more elements than I_n, and there are $2^n - (n + 1)$ sets having more elements than I_n but less than 2^{I_n}. Try this with $\mathcal{N}$. One finds that $2^{\mathcal{N}}$ has more elements than $\mathcal{N}$ but by Theorem 6.6.2 exactly as many elements as $\mathcal{R}$. Hence $2^{\mathcal{N}}$ furnishes no set S between $\mathcal{N}$ and $\mathcal{R}$. Knowing then that $2^{\mathcal{N}}$ and $\mathcal{R}$ have the same-number of elements let us recast the continuum hypothesis in terms of $\mathcal{N}$ and $2^{\mathcal{N}}$.

If there is a set S with more elements than $\mathcal{N}$ and less elements than $\mathcal{R}$, then S should have more elements than $\mathcal{N}$ and less elements than $2^{\mathcal{N}}$. So let us try for a set S between $\mathcal{N}$ and $2^{\mathcal{N}}$. (It might be helpful here to think of $\mathcal{N}$ as a subset of $2^{\mathcal{N}}$ under the 1-1 correspondence $n \rightarrow \{n\}$). Let us try to distinguish some set "between" $\mathcal{N}$ and $2^{\mathcal{N}}$, by specifying, as we must, its elements by some rule applicable to the elements of $2^{\mathcal{N}}$. If $T \in 2^{\mathcal{N}}$ then $T \subset \mathcal{N}$ and T is finite or denumerable and that is *all* we know about T. We might consider then the set

$$2_f^{\mathcal{N}} = \{T \in 2^{\mathcal{N}}: T \text{ is finite}\}.$$

But by Problem 1 below, $2_f^{\mathcal{N}}$ is denumerable and hence is in 1-1 correspondence with $\mathcal{N}$ itself; in other words, $2_f^{\mathcal{N}}$ fails as choice for S. What then about

$$2^{\mathcal{N}} \backslash 2_f^{\mathcal{N}} = \{T \in 2^{\mathcal{N}}: T \text{ is denumerable}\}\ ?$$

This set cannot be countable for then $2^{\mathcal{N}} = 2^{\mathcal{N}} \backslash 2_f^{\mathcal{N}} \cup 2_f^{\mathcal{N}}$ would be countable. Hence it must be uncountable. In fact, according to Problem 2 below, $2^{\mathcal{N}} \backslash 2_f^{\mathcal{N}}$ and $2^{\mathcal{N}}$ have the same-number of elements.

Now consider what this means. Using the at least more obvious applicable words (concepts) of the theory—finite and denumerable—we are unable to pick out a set "between" $\mathcal{N}$ and $2^{\mathcal{N}}$ in the sense of having more elements than $\mathcal{N}$ and less elements than $2^{\mathcal{N}}$. Recall Godel's result that within a theory there is a statement which cannot be prove or disproven within the theory. We, in our remarks have of course not settled the matter, but the remarks along with the awareness of Godel's results, make the couclusion of Paul Cohen, arrived at in 1963, seem reasonable:

> Set theory with the assumption of the continuum hypothesis is consistent; set theory with the denial of the continuum hypothesis is consistent.

It is fortunate that the continuum hypothesis, unlike the axiom of choice, does not lie at the base of important contemporary theories. But it does appear from time-to-time at certain stages of research, and, when it does, there is little chance that its unmentioned use would be overlooked, for, unlike the axiom of choice, the continuum hypothesis is extremely technical, rather far out—for now, in this century.

Problems.

(1) Show that $\mathcal{N}$ and $2_f^{\mathcal{N}}$ have the same-number of elements, using the techniques of Section 6.

(2) As in Problem 1, use Section 6 to show that $2^{\mathcal{N}} \backslash 2_f$ and $2^{\mathcal{N}}$ have the same-number of elements.

Related Reading

1. Bell, E. T.: *Men of Mathematics*, New York: Simon and Schuster, 1937. See the chapter "Paradise Lost" on Cantor's life and work.
2. Lieber, L. R.: *Infinity*, Holt, Rinehart and Winston Inc., 1953. Here the theory of infinite sets is presented vividly and intuitively.
3. Manheim, J. H.: *The Genesis of Point Set Topology*, New York: Macmillan Co., 1964. This text provides more insight into the matter and significance of infinite sets, particularly as they touch on other areas of mathematics.
4. Steen, L. A.: New models of the real number line, *Scientific American*, August 1971. This is a fascinating article that relates and touches upon almost all the ideas of this and previous chapters in as non-technical a way as possible, while introducing new and surprising models of the real number line that were only uncovered in the 1960's.

One model admits "infinitesimals," and in another, the continuum hypothesis fails to be true.
5. Stoll, A. E.: *Set Theory and Logic*, San Francisco: W. H. Freeman and Co., 1963. This text goes to much greater lengths to prove everything in full detail.
6. Vilenkin, N. Y.: *Stories about Sets*, New York: Academic Press Inc., 1968. If you would like more intuition, ...
7. Wilder, R. L.: *Introduction to the Foundations of Mathematics*, Ed. 2, New York: John Wiley and Sons, Inc., 1965. For a more sophisticated approach see this text, particularly Chapters 3, 4, 5, 6 and 10.

chapter 7

"I'm quite sure that . . . I have no . . . prejudices. All I care to know is that a man is a human being . . . ; he can't be any worse."—Mark Twain

EQUIVALENCE RELATIONS

1. INTRODUCTION

In this chapter we introduce a neat device that helps one to order and handle certain ideas in a nice way: again the structure game, the structure game again. There is one basic theorem and it will be used to fix three exceptionally diverse problems related to previous work. Specifically, using this one result we will: (1) show that any function can be "factored" into a 1-1 function and an onto function, (2) motivate the Frege-Russell definition of cardinal number, tying it in with abstract concept of number and (3) characterize the open sets in $\mathscr{R}$.

The reader no doubt suspects that the concept to be discussed, *equivalence relations*, is a purely technical one foreign to the minds of most men. The mathematician's name for this concept is no doubt quite foregin, but the concept of an equivalence relation does seem to be in very deliberate use all around us. The essence of the mathematical concept is one of viewing all things of a certain collection that share some property as a single class, as equivalent, and to view any one member of the class as representative of all in the class. Sometimes one finds that useful, sometimes not. But the viewpoint is a natural one. Take, for instance, prejudice. What is prejudice but the concept (habit) of considering all of a certain class as being one, and that what is believed or observed about one member of this class is true about all. Few men, if any, have not at some time thought in this way. Sometimes it's useful, helping one to concentrate on a key point, for without some ignorance of all the variety there is, one would probably reach inanity rather than new thought when considering most experiences. Sometimes such thinking is not very useful or likely to do much good, and no doubt the reader can provide his own examples.

The situation is much the same in mathematics. This is the use of equivalence relations in mathematics: to consider some things the same when (because of the context or aim) their differences don't matter, of course in an orderly and well defined way.

As we develop this topic it will not be initially clear that it does ultimately involve the separation of objects into distinct classes, each pair in a given class to be thought of as equivalent. That is the point of the main theorem of this section. We begin with an abstracted approach which initially appears as but a generalized concept of equality.

In Section 5.1 we introduced the concept of a relation R between A and B, thought of as a subset of $A \times B$. We noticed that the equality relation on a set A could be identified with the set $E = \{(a, a): a \in A\} \subset A \times A$. Hence if $a, a' \in A$ then $(a, a') \in E$ iff $a = a'$. With this in mind we introduce new notation.

DEFINITION 7.1.1. Let $R \subset A \times B$ be a relation. We will write aRb to mean that $(a, b) \in R$ and read

as "a is related to b by the relation R."

The initial use of the relation concept was to define a function, by putting a condition on the relation. Here we take the same tack, putting conditions on a relation R which will make it more interesting. After all, an arbitrary relation R can be almost anything and thus cannot be too useful. Specifically, we generalize the three properties of the equality relation E:

(a) aEa holds for all $a \in A$.
(b) aEb implies bEa for all $a, b \in A$
(c) aEb and bEc implies aEc for all $a, b, c \in A$.

For,

(a) $a = a$ for all $a \in A$
(b) $a = b \rightarrow b = a$ for all $a, b \in A$
(c) $a = b$ and $b = c \rightarrow a = c$ for all $a, b, c \in A$

are three fundamental properties of equality. It goes without saying, so let us say it, that one would not generalize unless it were known that other relations have these three properties of equality. Here are the formal definitions.

DEFINITION 7.1.2. Let A be a set and let $R \subset A \times A$ be a relation.
(a) R is called *reflexive* if xRx for all $x \in A$.
(b) R is called *symmetric* if $xRy \rightarrow yRx$ for all $x, y \in A$.
(c) R is called *transitive* if xRy and yRz implies xRz for all $x, y, z \in A$.

DEFINITION 7.1.3. A relation R on a set A is called an *equivalence relation* iff it is reflexive, symmetric and transitive.

We will be concerned throughout the remainder of this section only with equivalence relations. Notice that an equivalence relation is only defined as a relation between the elements of a single set. That is, only a relation $R \subset A \times A$ (not $A \times B$) is a candidate for an equivalence relation.

EXAMPLES.

(a) As noted above the equality relation $E = \{(a, a): a \in A\}$ is reflexive, symmetric and transitive and hence an equivalence relation.

(b) In $\mathscr{R} \times \mathscr{R}$ the relation $I = \{(a, b): a < b\}$ is neither reflexive nor symmetric, but is transitive.

(c) Let P be the set of all people, $R = \{(p, q): p$ and q have same general skin color$\}$. Then R is reflexive, symmetric and transitive and hence is an equivalence relation.

(d) Let A be the collection of all atoms in the universe, and let $R = \{(a, a'): a$ and a' have the same number of electrons$\}$. Clearly R is reflexive, symmetric and transitive. Hence an equivalence relation.

There is a little geometry (apologies to all geometers for the loose use of the word) that gets into all of this. Consider a relation $R \subset A \times A$ and draw $A \times A$ as though it were a plane, as in Figure 7.1.

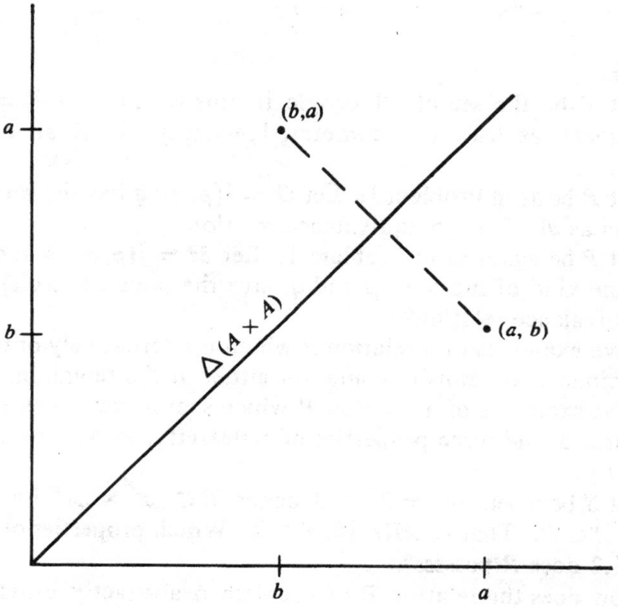

FIG. 7.1

To say that R is reflexive is to say that R contains the "diagonal" $\Delta(A \times A)$ of $A \times A$. To say that R is symmetric is to say that the graph of R must be geometrically symmetric about the diagonal $\Delta(A \times A)$. That is, if aRb, i.e., $(a, b) \in R$, then its "reflection" about $\Delta(A \times A)$, the point (b, a), is in R iff R is symmetric. Transitivity is a little more complicated and is indicated in Figure 7.2, where $(a, b) \in R$ and $(b, c) \in R$. Here R is transitive iff the missing vertex of the appropriate triangle is in R.

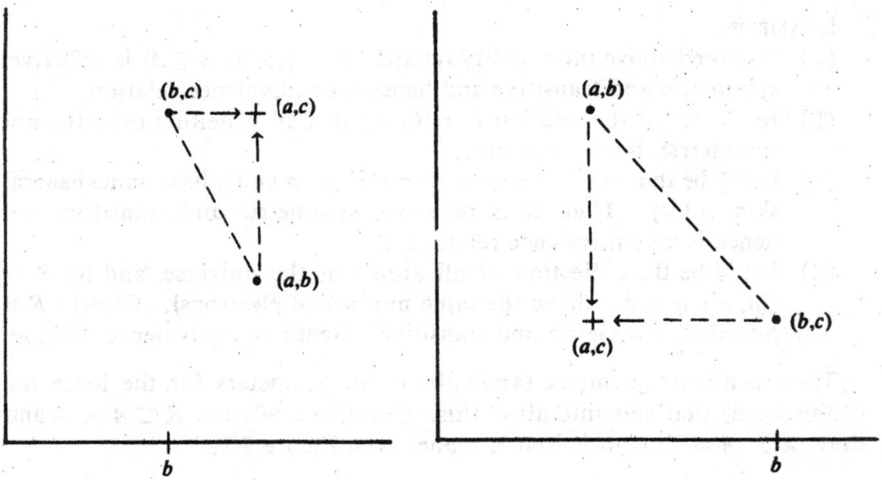

FIG. 7.2

Problems.

(1) Let P be the set of all people in this world. Let $R = \{(p, q): p$ q speak at least one common language$\}$. Is R an equivalence relation?

(2) Let P be as in Problem 1. Let $O = \{(p, q): q$ has the same occupation as $q\}$. Is O an equivalence relation?

(3) Let P be again as in Problem 1. Let $M = \{(p, q): p$ and q like the same kind of music or p and q have the same life style$\}$. Is M an equivalence relation?

(4) Give examples of a relation R which is alternatively only reflexive, symmetric or transitive and not either of the remaining.

(5) Give examples of a relation R which shares every two, but not the third, of the three properties of reflexivity, symmetry and transitivity.

(6) Let S be a set, $\mathscr{A} = 2^S$ and define $R \subset \mathscr{A} \times \mathscr{A}$ by $(A, B) \in R$ iff $A \subset B$. That is, ARB iff $A \subset B$. Which properties of Definition 7.1.2 does R possess?

(7) How does the relation R of Problem 6 abstractly differ from the relation $J = \{(a, b): a \leq b\} \subset \mathscr{R} \times \mathscr{R}$? That is, is there a property that J has and R does not have, or vice versa? More precisely, find a propositional function $p(x)$ which is true for x replaced by J but false for x replaced by R—or vice versa.

(8) Let S and T be sets, $f: S \to T$. Define $R \subset S \times S$ by the condition: $(x, y) \in R$ iff $f(x) = f(y)$. That is xRy iff $f(x) = f(y)$. Show that R is an equivalence relation.

(9) In $\mathscr{N} \times \mathscr{N}$ let $R = \{(m, n): 2$ divides $m - n\}$. Show that R is an equivalence relation. *Note*: 2 divides $m - n$ means

$$\exists q \in \{0, \pm 1, \pm 2, \ldots\} \ni m - n = 2q.$$

(10) In the set C of all continuous functions on the interval $[0, 1]$ define $R \subset C \times C$ by fRg iff $\int_0^1 f(t)dt = \int_0^1 g(t)dt$. Let $f(x) = x$. Find a function $g \neq f$ such that fRg. Show that R is an equivalence relation,

(11) Here is an interesting example of a non-transitive relation. Let A consist of four dice numbered as indicated below.

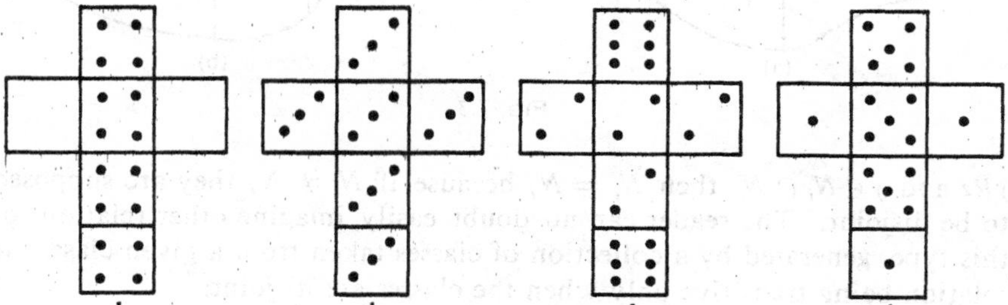

Notice in comparing d_1 and d_2 that d_1 will beat d_2 in 24 of 36 tries, hence d_1 beats d_2 with probability 2/3. Let $R = \{(d_i, d_j): d_i$ beats d_j with probability $\geq 2/3\} \subset A \times A$. Show that R is not transitive. Is R symmetric or reflexive?[1]

2. EQUIVALENCE RELATIONS AND EQUIVALENCE CLASSES

How do our earlier comments on the separation of individuals or things into classes, and considering any one of the class as representative, relate in some sense to the notion of an equivalence relation?

Almost any individual as he grew up probably identified himself in some sense with some neighborhood of his locale which was more-or-less well defined in his mind. Suppose a city population C is given and imagined to be partitioned by some rule into neighborhoods $N_1, N_2, \ldots, N_n$. Define a relation R in $C \times C$ by xRy iff x and y belong to the same neighborhood. Clearly xRx because every person lives in some neighborhood. Further more, if xRy then yRx. Hence R is reflexive and symmetric.

Now suppose xRy and yRz. Is xRz? Not necessarily. For suppose there are two different neighborhoods N_i and N_j such that $y \in N_i \cap N_j$ and that $x \in N_i$ only, and $z \in N_j$ only. Then xRy for both x and y are in N_i and yRz because both are in N_j. But xRz is false because x and z are not both in any single neighbohood N_R (see Figure 7.3(a)). Hence R need not be transitive. However if we suppose this city to be broken up into *mutually disjoint* neighborhoods, Figure 7.3(b), this cannot happen. For if xRy and

[1] The example is taken from the December 1970 issue of *Scientific American*.

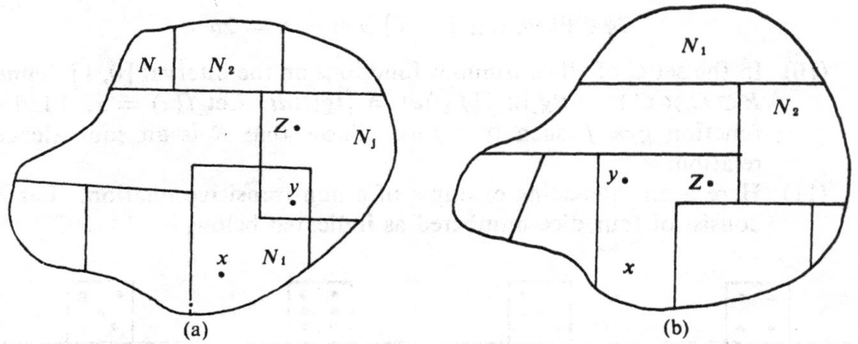

FIG. 7.3

yRz and $y \in N_i \cap N_j$ then $N_i = N_j$ because if $N_i \neq N_j$ they are supposed to be disjoint. The reader can no doubt easily imagine other relations of this type, generated by a collection of classes taken from a given class, the relation being transitive only when the classes are disjoint.

If one abstracts the reasoning above the structure seems to be as follows.

THEOREM 7.2.1. Let S be any set and let $\mathscr{C}$ be a fixed collection of subsets of S. ($\mathscr{C}$ may be an infinite collection.) Define R in $S \times S$ by xRy iff x any y belong to some set $C \in \mathscr{C}$. Then

(a) R is symmetric.
(b) If $S = \bigcup_{C \in \mathscr{C}} C$, then R is reflexive.
(c) If $\mathscr{C}$ is a mutually disjoint collection, then R is transitive.

We leave the proof to the reader and point out the

COROLLARY 7.2.2. Let $\mathscr{C}$ be a mutually disjoint collection of subsets of a set S such that $S = \bigcup_{C \in \mathscr{C}} C$. If the relation R in $S \times S$ is defined by

$$xRy \text{ iff } \exists C \in \mathscr{C} \ni x \in C \text{ and } y \in C,$$

then R is an equivalence relation on S.

Clearly then, equivalence relations can arise in many a context. All one needs is a rule for dividing a set into a collection of mutually disjoint subsets whose union is the entire set.

For discussions sake we make the

DEFINITION 7.2.3. If S is a set and $\mathscr{C}$ is a collection of mutully disjoint subsets of S such that $S = \bigcup_{C \in \mathscr{C}} C$, then $\mathscr{C}$ is called a ***partition*** of S.

We have just seen (Corollary 7.2.2) that any partition $\mathscr{C}$ on a set gives rise to a natural equivalence relation R on S. Consider now the set P and the relation R of Example (c), Section 7.1. This biological relationship determines a partition of the people of this earth that has been well-noted. In one class there is all those people with white skin, in another all those with black, in a third all those with yellow, as the main divisions, forged in the evolutionary furnace, go.

Let x be a person of white skin and let $R_x = \{y: yRx\}$. Clearly R_x is the collection of all white-skinned people. Moreover, let x' be the another white-skinned person and $R_{x'} = \{y: yRx'\}$. Clearly $R_x = R_{x'}$. In general, if $p \in P$ let $R_p = \{y: yRp\}$. The set R_p is an example of what we will soon call an *equivalence class*. We are finally getting to the essence of equivalence relation. The two people, x and x' with the same skin color, and skin color is the *only* criteria that matters in *this* abstract relation, determine the *same equivalence class*. More generally, if pRp' then $R_p = \{y: yRp\} = \{y: yRp'\} = R_{p'}$ because R is transitive. Less abstractly, if two people p and p' have the same skin color, then the class of all people having the same skin color as p and the class of all those of the same as p' are identical classes. This kind of thing is of course too vague for rigorous mathematics, but it appears that an equivalence relation R on a set S naturally partitions the set S into classes wherein all those elements "equivalent" by the relation R belong to the same class. This is the worth of the concept in mathematics and this is the subject of the basic theorem of this chapter

THEOREM 7.2.4. Let S be a set, R an equivalence relation on S. For $x \in S$, let $R_x = \{y \in S: yRx\}$. Then

(a) $x \in R_x$ and hence $S = \bigcup_{x \in S} R_x$

(b) xRx' iff $R_x = R_{x'}$.

(c) x is not related to x' iff $R_x \cap R_{x'} = \square$.

(c) $\mathscr{C} = \{R_x: x \in S\}$ is a partition of S.

Notice (d), it is the converse of Corollary 7.2.2. Not only does every partition of a set determine a natural equivalence relation, but, conversely, every equivalence relation on a set determines a partition on the set. These concepts, partitions and equivalence relations are one and the same.

Proof of Theorem 7.2.4.

(a) Since R is an equivalence relation, then xRx holds for any x. Hence $x \in R_x$. Thus $S \subset \bigcup_{x \in S} R_x$. Since $R_x \subset S$, clearly $\bigcup_{x \in S} R_x = S$.

(b) If xRx', we claim $R_x = R_{x'}$. Let $y \in R_x$. Then yRx and by transitivity yRx'. Hence $y \in R_{x'}$. Thus $R_x \subset R_{x'}$. Similarly $R_{x'} \subset R_x$ and hence $R_x = R_{x'}$.

Conversely, if $R_x = R_{x'}$ then since $x \in R_x$ by (a) then $x \in R_{x'}$. By definition of $R_{x'}$, xRx'.

(c) If x is not related to x', then $(x, x') \notin R$. Suppose $y \in R_x \cap R_{x'}$. Then yRx and yRx'. By symmetry, xRy and hence by transitivity, xRx', a contradiction. Thus $R_x \cap R_{x'} = \square$.

Conversely, if $R_x \cap R_{x'} = \square$, then since $x \in R_x$ by (a) one has

$x \notin R_{x'}$ and hence x is not related to x'.

(d) From (a) $\bigcup_{C \in \mathscr{C}} C = \bigcup_{x \in S} R_x = S$. From (b) and (c), if $C \in \mathscr{C}$, $C' \in \mathscr{C}$ and $\mathscr{C} = R_x$, $C' = R_{x'}$ then $C \cap C' = \square$ or $C = C'$. Hence $\mathscr{C}$ is a mutually disjoint collection and by Definition 7.2.3 is a partition of S.

There is really only one thing to remember regarding Theorem 7.2.4: any equivalence relation on a set naturally breaks up the set into mutually disjoint pieces each piece consisting of all those elements considered "equivalent" by the relation. For ease of discussion we make the

DEFINITION 7.2.5. If R is an equivalence relation on a set S, the set $R_x = \{y: yRx\}$ is called the *equivalence class determined by x*. If $y \in R_x$ we will call x and y *equivalent*. The collection of all equivalence classes is denoted by S/R, read "S mod R." That is, $S/R = \{R_x: x \in S\}$.

In case the reader missed it note that

$$\text{(a) } R \subset S \times S \qquad \text{(b) } R_x \subset S \qquad \text{(c) } S/R \subset 2^S .$$

To complete this discussion, we have a corollary to the effect that any individual element in an equivalence class represents the whole class as far as the equivalence relation is concerned.

COROLLARY 7.2.6. If R is an equivalence relation and $y \in R_x$, then $R_y = R_x$.

Proof. If $y \in R_x$ then yRx. By Theorem 7.2.4(b), $R_yR = R_x$.

Problems.

(1) Can you partition whatever area you call home into a collection $\mathscr{C}$ of the type hypothesized in Corollary 7.2.2? How about a partition which at least gives a reflexive and symmetric relation R?

(2) In $\mathscr{N}$, let $\mathscr{C} = \{C \subset \mathscr{N}: C$ contains all even numbers or C contains all odd numbers$\}$. Describe the relation R of Corollary 7.2.2.

(3) Let $a_1 \leq a_2 \leq a_3 \leq \ldots \leq a_n \leq \ldots$ be a collection of real numbers, one for each $n \in \mathscr{N}$. Let $\mathscr{C} = \{[a_n, a_{n+1}]: n = 1, 2, \ldots\}$. Is the relation R of Theorem 7.2.1 an equivalence relation? If not, place conditions on the numbers a_n which make it an equivalence relation. What about $R' = \{(a_n, a_{n+1}): n = 1, 2, 3, \ldots\}$.

(4) Prove Theorem 7.2.1.

(5) Prove Corollary 7.2.2.

(6) Describe the eqivalence classes for Example (a) of 7.1.

(7) In Example (d) describe the typical equivalence class. How many equivalence classes are there known to be? Let a be an atom with one electron. What is an appropriate name for R_a? Equivalence relations classify, categorize, put into "little boxes" as the song

goes. They are not beloved in human relations, though perhaps "For all things there is a season."

(8) Describe the typical equivalence class for each relation in Problems 1 through 9, Section 7.1, which are in fact equivalence relations.

(9) Let $f(x) = x^2, x \in [-1, 1]$. Define R on $[-1, 1]$ by xRy iff $f(x) = f(y)$. Show that, R is an equivalence relation. Describe the typical equivalence class for $\mathscr{R}$.

(10) Let $f(x) = \sin 1/x, 0 < x \leq 1/\pi$. Define R by xRy iff $f(x) = f(y)$ for $x, y \in X = (0, 1/\pi]$. Let $T = [-1, 1]$. Define a 1-1, onto function $g: X/R \to T$.

(11) In $X = \mathscr{R} \times \mathscr{R}$ let

$$R = \{((x, y), (x', z)): x = x'\} \subset (\mathscr{R} \times \mathscr{R}) \times (\mathscr{R} \times \mathscr{R}).$$

Show that R is an equivalence relation and that X/R is in 1-1 correspondence with $\mathscr{R}$ itself. In other words view X/R as $\mathscr{R}$. It is suggested that you draw a picture of the typical equivalence class in $\mathscr{R} \times \mathscr{R}$.

(12) Let X be a set, R an equivalence relation on X and define $\eta: X \to X/R$ by $\eta(x) = R_x$. Does this make sense? Is η a function? Is η an onto function? Try an example first, say the relation R of Problem 2, Section 7.1. Let x represent you. What is $\eta(x)$?

(13) This problem wraps up the theory of partitions and equivalence classes. Let $\mathscr{C}$ be a partition on a set S and let R be the relation of Corollary 7.2.2. Show that R_x is exactly the set $C \in \mathscr{C}$ to which x belongs. That is, $x \in C$ iff $C = R_x$.

3. CARDINAL NUMBER

Just what does "3" represent to you? Nothing really concrete. You can give definite meaning to "3 friends" or "3 record albums" or "3 books" and except for one detail those examples have nothing in common—the one detail they possess is the idea or property of "3" or "three-ness." This is the motivation for the Frege-Russell definition of cardinal number.

We have previously (Definition 6.2.3(a)) given a definite meaning to the *same-number* of elements for sets S and T. We have not given a meaning to the number of elements in a set, or more properly, the *cardinal number* of a set or collection. A good way to approach this is through Theorem 7.2.4.

Let S be a set containing at least all material things in this world and let 2^S be, as usual, all of its subsets. Define a relation R on 2^S by ARB iff A and B have the same-number of elements. Hence ARB iff there is a 1-1 onto function $f: A \to B$.

Clearly ARA, for $i(x) = x$ for $x \in A$ is 1-1 and onto from A to A. If $f: A \to B$ is 1-1 and onto, then, Theorem 5.5.5 implies $f^{-1}: B \to A$ is 1-1 and onto. Hence ARB implies BRA. If ARB and BRC, then $\exists$ 1-1, onto function $f: A \to B$ and $g: B \to C$. By Theorem 5.5.8 $g \circ f: A \to C$ is 1–1 and onto, and hence ARC. Thus R is an equivalence relation and by Theorem 7.2.4, 2^S is the union of the mutually disjoint classes R_A.

Consider "3" again, and let A be the set consisting of your three friends (hoping you have more). The class R_A then contains the set consisting of your three record albums and also the set consisting of your three books. (Again, hoping you have more of each.) Carry it on: R_A consists of *all* sets having exactly three elements. Moreover if B is any set having three elements, then BRA and hence by Corollary 7.2.6, $R_A = R_B$. What could be a better representative of 3 than the class R_A defined by R and your three friends?

At least that is the way Russell and Frege saw the matter of cardinal number. They made the

DEFINITION 7.3.1. If $A \subset S$, then the *cardinal number* of A is the equivalence class R_A.

So you see, a cardinal number, the number that expresses *how many*, can very well be viewed as but an abstract set. Take a set A with n elements. Since I_nRA then $I_n \in R_A$, in fact $R_{I_n} = R_A$. To associate n with R_{I_n} is to form a 1–1 correspondence between the number n and the cardinal number R_{I_n} and in the sense of this correspondence (only) to identify them as the same. In other words let n represent R_{I_n}, just as you have always done. For example you commonly let 3 represent that one property all sets of three things have in common, the property being that they all one in 1–1 correspondence, that they all belong to $R_{\{1,2,3\}}$.

Really this discussion has not been too precise, the original set S not being well fixed. We can take S to be the set of all sets relevant to our purpose in given discussion and let it go, vaguely, at that. We cannot use "the set of all sets." The aim here is but to get across a point on the essential nature of the abstract number concept.[2] We will not use it in the sequel. The concept of same-number suffices herein. But an important point must be made. This definition does allow one to associate with *any* set A a well defined object called a cardinal number, the collection R_A.

From this one can construct an *arithmetric of infinite cardinal numbers* and this was one of Cantor's main interests. We can also now write same-number as we might prefer: *same-number* means *same cardinal number*, for sets A and B have the same-number of elements iff the cardinal number of A, R_A equals R_B, the cardinal number of B.

Two more points to wrap up the matter of cardinal number. The cardinal number of the set $\mathscr{N}$, that is, the equivalence class $R_{\mathscr{N}}$ is universally denoted by $\aleph_0$ (alpha-null). The cardinal number of the set $\mathscr{R}$

[2] For a more complete discussion see R. R. Stoll: *Set Theory and Logic* (San Francisco: W. H. Freeman and Co., 1963) or R. L. Wilder: *Introduction to the Foundations of Mathematics* (New York: John Wiley and Sons, Inc., Ed. 2, 1965).

is denoted by c and is called the cardinal of the continum. If we define the cardinal R_A to be less than R_B, denoted by $R_A < R_B$, when B has more elements than A (Definition 6.2.3), then $R_{\mathcal{N}} = \aleph_0 < c = R_{\mathcal{R}}$ by Theorem 6.4.5. Thecontinuum hypothesis then says that there is no cardinal number R_A suchthat $\aleph_0 < R_A < c$. Finally, there is the matter of determining when two sets have the same cardinal number. The main result here is of course the Schroder-Bernstein theorem of Section 6.6.

Problems.

(1) Show that if cardinals R_A and R_B are given and we define $R_A \leq R_B$ to mean $R_A < R_B$ or $R_A = R_B$, then, $R_A \leq R_B$ and $R_B \leq R_A$ implies $R_A = R_B$.

(2) Show that given any cardinal number R_A there is a cardinal number R_B such that $R_A < R_B$.

(3) Define the sum $R_A + R_B$ of cardinal numbers A and B to be the cardinal R_C where $C = A \cup B$. Show that $\aleph_0 + \aleph_0 = \aleph_0$, $\aleph_0$ being $R_{\mathcal{N}}$ of course.

(4) Let k denote the cardinal number of R_{I_k}. Let $C = \{f\colon f\colon I_n \to I_m\}$ for $m, n \in \mathcal{N}$. Show that the cardinal number of C is m^n. That is, that C and $R_{I_{m^n}}$ have the same cardinal number.

(5) Let S be a set. Let $C = \{f\colon f\colon S \to \{0, 1\}\}$. Show that 2^S has the same cardinal number as the set C.

In general, if A and B are sets and α and β denote the cardinal numbers R_A and R_B, then the β-th power of α, denoted α^β, is defined to be the cardinal number of the set $C = \{f\colon f\colon A \to B\}$.

4. A CHACTERIZATION OF OPEN SETS IN $\mathcal{R}$

We proceed now to a second application of Theorem 7.2.4, this one to measure and length. For centuries it was known how to assign a length to an open interval $(a, b) \subset \mathcal{R}$ in a useful way, related to most existing mathematics. Specifically, the length of the interval (a, b) is defined to be the number $b - a$. In the early years of this century Henri Lebesgue developed his theory of integration, a theory much better than that of Riemann or Darboux, but requiring a definition of length for virtually *any* kind of set—open, closed, half open, discrete, etc. At the base of this theory is the result we will prove, which allows one to assign a meaningful definition of length to any open set in $\mathcal{R}$. Having a concept of length for any open set in $\mathcal{R}$ one can extend this concept to any set in $\mathcal{R}$ from the observation that any set is contained in an open set and in some sense approximated by the open sets containing it. The theorem we will prove is then a landmark in the theory of measure and length.

Before we state it, first try to imagine a non-trivial open set in $\mathcal{R}$. Around any of its points it must contain a neighborhood, an open interval. One can easily imagine an open set which is not an open interval, say

$(0, 1) \cup (2, 3)$. One can imagine an open set like $(0, 1/2) \cup (3/4, 7/8) \cup (15/16, 31/32) \cup \ldots$ this one a countable union of open intervals. Past this point things get rather hazy, hidden in the mist at infinity, to repeat a phrase, and with good reason as our theorem asserts.

THEOREM 7.4.1. Any open set in $\mathscr{R}$ is a countable union of mutually disjoint non-empty open intervals. That is, if $O \subset \mathscr{R}$ is open then

$$O = \bigcup_{n=1}^{\infty} (a_n, b_n)$$

where $(a_i, b_i) \cap (a_j, b_j) = \square$ if $i \neq j$ and we allow the possibility that some a_i represents $-\infty$ and some b_j represents $+\infty$.

Notice that the conclusion of the theorem is precisely that there is a partition of 0 by open intervals. Hence, the method of proof below is a natural one.

Proof. Define a relation R in $O \times O$ by xRy iff $\exists$ a closed interval $I \ni x, y \in I$ and $I \subset O$. (Thus, if $x, y \in O$ and $x \leq y$ then xRy iff $[x, y] \subset O$, in the special case $x \leq y$.) We leave it to the reader to prove that R is an equivalence relation.

Since R is an equivalence relation, $O = \bigcup_{x \in O} R_x$. We claim two things:

(1) Each set R_x is a non-empty open interval.

(2) There are only countably many sets R_x.

For (1) we first show that R_x is an interval (Theorem 4.5.4) and then that R_x is open. Suppose $y < z$ and $y, z \in R_x$. We claim $[y, z] \subset R_x$. Since $y, z \in R_x$ and R is an equivalence relation we have $y \in R_y = R_z = R_x$. Thus yRz and hence, by definition of R, $[y, z] \subset O$. If $t \in [y, z]$ then $[t, z] \subset [y, z] \subset O$ and hence tRz or $t \in R_z = R_x$. Thus $[y, z] \subset R_x$. Thus R_x is an interval.

We claim R_x is open. Since $x \in O$, $\exists$ a neighborhood, an open interval (a, b), $\ni x \in (a, b) \subset O$. If $t \in (a, b)$, then $[t, x] \subset (a, b) \subset O$ in the case $t \leq x$. In the case $x < t$ $[x, t] \subset (a, b) \subset O$. In either event, $t \in (a, b)$ implies tRx or $t \in R_x$. Hence $x \in (a, b) \subset R_x$. Thus R_x is open about x.

Now let $y \in R_x$ be arbitrary. From Corollary 7.2.6, $R_y = R_x$ and hence by the above paragraph $\exists$ a neighborhood $(c, d) \ni y \in (c, d) \subset R_y = R_x$. Thus $\forall y \in R_x \exists$ a neighborhood of y contained in R_x and hence R_x is open. Thus (1) is proven.

For (2), observe that from Section 2.7 and the fact that R_x is an open interval, R_x must contain at least one rational number. Using the "axiom of choice," choose one rational number from each distinct class R_x and call it $f(R_x)$. We claim that the *relation* $\{(R_x, f(R_x)): R_x \in O/R\}$ is a 1-1 *function*. It is a function because only one number $f(R_x)$ is chosen from R_x and the sets R_x are mutually disjoint. It is 1-1 because if $R_x \neq R_y$ then $R_x \cap R_y = \square$ and $f(R_x)$, being in R_x, cannot be $f(R_y)$ since $f(R_y) \in R_y$. That is, $R_x \neq$

$R_y \to f(R_x) \neq f(R_y)$. Call this function (naturally) f.

Hence $f: O/R \to Q$, where Q is the set of rational numbers. By Theorem 6.4.1, Q is countable. By Theorem 6.3.6, O/R is countable. Hence there are only countably many equivalence classes R_x. Thus (2) is proven.

Finally (1) and (2) combined with the equality $O = \bigcup_{R_x \in O/R} R_x$ is exactly the statement of the theorem. This completes the proof.

Such is one good example of the use of the concept of an equivalence relation in an abstract setting. Notice how it handles, very formally and very neatly, certain expected properties of O, through the conclusions of Theorem 7.2.4 and Definition 7.2.5, that would otherwise be difficult to state and manipulate. Equivalence relations and Theorem 7.2.4 are used throughout modern algebra, analysis and topology though we will not go further into the specifics of these matters here. If one encounters a situation where one wishes to think of disparate things as equivalent in a formal manipulative way, then more than likely the equivalence relation concept will serve well.

Let us use Theorem 7.4.1 to *define* the length of any subset of $\mathscr{R}$. To begin we go all the way back to the completeness axiom for $\mathscr{R}$ (Section 2.8) and note the following.

LEMMA 7.4.2. If $\square \neq M \subset R$ and there is a number b such that $b \leq x$ for all $x \in M$, then $\exists$ a number c such that

(1) $c \leq x$ for all $x \in M$

and

(2) $a \leq c$ for all a such that $a \leq x$ for all $x \in M$.

Proof. Let $-M = \{-x: x \in M\}$. By hypothesis, $-M$ is bounded above by $-b$. Hence $-M$ has a supremum, call it $-c$. It now easily follows that $c = -(-c)$ satisfies (1) and (2).

DEFINITION 7.4.3. A set $M \subset \mathscr{R}$ satisfying (1) above is said to be *bounded below* and a number c satisfying (1) is called a *lower bound* for M. A number c satisfying both (1) and (2) is called the *infimum* of M, denoted by inf M.

The number inf M, like sup M, when M is bounded above, is unique. It is also called the greatest lower bound for M. We can now define the length, first of any open set, and then of any set, in $\mathscr{R}$.

DEFINITION 7.4.4. If O is an open set in $\mathscr{R}$ with $O = \bigcup_{n=1}^{\infty} (a_n, b_n)$ with $a_n, b_n \in \mathscr{R}$ and $(a_i, b_i) \cap (a_j, b_j) = \square$ for all $i \neq j$, we define the length of O, $\ell(O)$, to be the number $\ell(O) = \sup \sum_{k=1}^{n} (b_k - a_k): n \in \mathscr{N}\}$. If any $a_n = -\infty$ or $b_n = +\infty$ or if the supremum does not exist, we set $\ell(O) = +\infty$.

For example, if $O = \bigcup_{n=1}^{\infty} (n-1, (n-1) + 1/2^{n-1}) = (0, 1) \cup (1, 3/2) \cup (2, 9/4) \cup \ldots$ then $\ell(O) = 1 + 1/2 + 1/4 + \ldots = 2$, while $\ell((2, +\infty)) =$

$+\infty$, $\ell((-\infty, -1)) = +\infty$ and $\ell(-1, 3) = 4$. Notice that it is Theorem 7.4.1 that allows us to make this definition for any set O.

Now, if $E \subset \mathscr{R}$, then there are many open sets O containing E and for any of theses we should wish the length of E to be smaller than $\ell(O)$ no matter which $O \supset E$. Hence we define $\ell(E)$ to be the largest number smaller than every length $\ell(O)$. That is,

DEFINITION 7.4.5. If $E \subset \mathscr{R}$ define $\ell(E) = \inf\{\ell(O)\colon O \supset E$ and O is open$\}$.

For example, if $E = \{1/2\}$, then any interval $O = (1/2 - t, 1/2 + \varepsilon) \supset E$ and hence $\ell(E) \leq 2\varepsilon$ for any $\varepsilon > 0$. Thus $\ell(E) = 0$. If $E = [0, 1]$, then any set $O = (-\varepsilon, 1 + \varepsilon) \supset E$ and hence $\ell(E) \leq 1 + 2\varepsilon$ for all $\varepsilon > 0$. Hence $\ell(E) \leq 1$. In fact, $\ell(E) = 1$ for if $\ell(E) < 1$, then

$$a = \ell(E) + \frac{[1 - \ell(E)]}{2} \leq \ell(O)$$

for all $O \supset E$ since $\ell(O) \geq 1$. But $\ell(E) = \inf\{\ell(O)\colon O \supset E, O \text{ is open}\}$ and this would make a a lower bound greater than the largest lower bound, a contradiction.

Problems.

(1) Compute $\ell(O)$ for

$$O = (0, 9) \cup (9, .99) \cup (.99, .999) \cup \ldots \cup (.99\ldots9, .99\ldots99) \cup \ldots.$$

(2) Compute $\ell([0, 1))$.

(3) Let $p_1, p_2, \ldots, p_n$ be n points in $\mathscr{R}$ with $p_1 < p_2 < \ldots < p_n$. Show that $\ell(\{p_1, p_2, \ldots, p_n\}) = 0$.

(4) Let $E = \{1, 1/2, 1/4, 1/8, \ldots\}$. Compute $\ell(E)$.

(5) Let $E \subset \mathscr{R}$ be a countable set, say $E = \{p_1, p_2, \ldots\}$ with

$$p_1 < p_2 < \ldots < p_n < p_{n+1} < \ldots.$$

Can you show that $\ell(E) = 0$? Compare this with the problems posed by Question I.

(6) Let $I = \{x \in [0, 1]\colon x$ is irrational$\}$. Prove that if $O \supset I$ and O is open then $O \supset [0, 1]$ and hence that $\ell(I) = 1$.

(7) Let $E = \{x \in [0, 1]\colon x$ is rational$\}$. Then E is countable so write $E = \{x_1, x_2, x_3 \cdots\}$. Let $\varepsilon > 0$ and let

$$O_\varepsilon = \bigcup_{n=1}^{\infty} \left(x_n - \frac{\varepsilon}{2^n}, x_n + \frac{\varepsilon}{2^n}\right).$$

Then $O_\varepsilon \supset E$ and $\ell(O_\varepsilon) \leq 2\varepsilon$. Conclude that $\ell(E) = 0$. Compare with Problems 5 and 6.

5. A FACTORIZATION THEOREM

Not all functions $f: S \to T$, where S and T are sets need be 1-1 nor need be onto. Certainly 1-1 functions and onto functions are much easier to deal with, particularly in an abstract setting. We wish to show that any function f can be factored in the form $f = g \circ \eta$ where g is 1-1 and η is onto. The key idea is based on the fundamental idea of an equivalence relation—when one wishes to consider different things as equivalent, equivalence relations naturally arise. Suppose one considers the view of f from the set T for a moment. As far as the set T is concerned, if $s, s' \in S$ $f(s) = f(s')$ then s and s' have no difference, for they produce the same element $t = f(s)$ of T. That is, from the viewpoint of T, elements s and s' such that $f(s) = f(s')$, are "equivalent" elements of S.

Consider than the relation R is $S \times S$ given by sRs' iff $f(s) = f(s')$. The relation R is an equivalence relation and this is left for the reader to verify.

Consider now the collection S/R of equivalence classes and for $R_x \in S/R$ define $g(R_x) = f(x) \in T$ so that g is a correspondence from S/R into T. Indeed g is a function, for if $R_x = R_y$ then by Theorem 7.2.4(b), xRy and hence $g(R_x) = f(x) = f(y) = g(R_y)$. As far as T is concerned g is a more sensible function than f because g is in fact 1-1, treating all elements $s, s' \in S$ such that $f(s) = f(s')$ alike—as one! For, suppose $g(R_x) = g(R_y)$. Then $f(x) = f(y)$, hence xRy, hence $R_x = R_y$. Thus, putting all this together we have $R_x = R_y$ iff $g(R_x) = g(R_y)$ and hence g is 1-1 function.

At this point a diagram of sorts (Figure 7.4) appears and we will proceed to fill it in below.

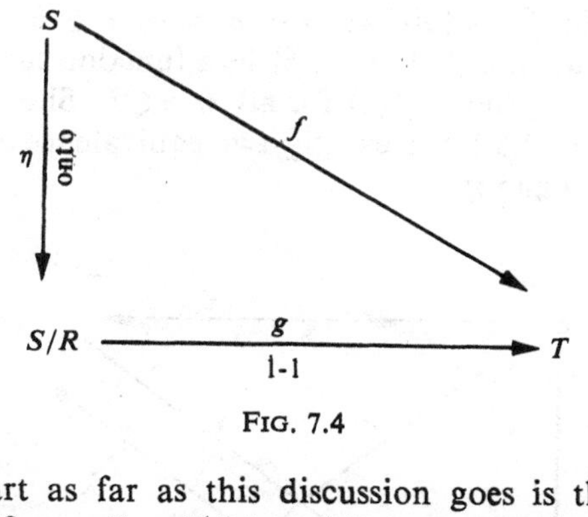

FIG. 7.4

The missing part as far as this discussion goes is the correspondence denoted by η. Define $\eta: S \to S/R$ by $\eta(x) = R_x$. Then η is a function, for if $R_x = \eta(x) \neq \eta(y) = R_y$ then by Theorem 7.2.4(c), $R_x \cap R_y = \square$ and since $x \in R_x$ and $y \in R_y$ we have $x \neq y$. That is, $\eta(x) \neq \eta(y) \to x \neq y$, so that η is a function. Furthermore, η is onto S/R. For if $R_x \in S/R$ then $x \in S$ exists and $\eta(x) = R_x$ by definition.

One more fact remains and it is stated in

THEOREM 7.5.1. Let S and T be sets, $f: S \to T$. For the equivalence relation R defined on S by xRy iff $f(x) = f(y)$,

(1) $\exists$ an onto function $\eta: S \to S/R$ given by $\eta(x) = R_x$

(2) $\exists$ a 1-1 function $g: S/R \to T$ given by $g(R_x) = f(x)$ such that

(3) $f = g \circ \eta$.

That is, $f(x) = g(\eta(x)) \forall x \in S$.

The function η is usually referred to as the *natural map* from S onto S/R. To complete the proof we observe that (1) and (2) are proven above, whereas for (3), if $x \in S$ is given, then $\eta(x) = R_x$, $g(R_x) = f(x)$ and hence $f(x) = g(\eta(x))$.

Hence, any function can be *factored under composition* into a 1-1 and an onto function with a natural set S/R as an intervening domain. The following problems will substantiate the use of the word "natural."

Problems.

(1) Let E denote the set of all the employed people in this country and define $f: E \to \mathscr{R}$ by $f(p)$ is the income of p. In the contex of Theorem 7.5.1 describe R, E/R, g and η in ordinary descriptive language. Does Theorem 7.5.1 make sense in this context?

(2) Set up a "real-world" realization of Theorem 7.5.1 different from that in Problem 1.

(3) Let $D = \{1, 2, 3, 4, 5, 6\}$. Suppose one is given two die. Each roll of the pair produces an element of $D \times D$, a pair of numbers between 1 and 6. Define $f: D \times D \to I_{12}$ by $f(x, y) = x + y$. Discuss this function in the context of Theorem 7.5.1

(4) Let $I = \{0, \pm 1, \pm 2, \ldots\}$. Define an operation $+$ on a two element set $\{\alpha, \beta\}$ as follows, $\alpha + \alpha = \alpha$, $\alpha + \beta = \beta$, $\beta + \alpha = \beta$ and $\beta + \beta = \alpha$. Let $f: I \to \{\alpha, \beta\}$ be a function such that f is onto and $f(m + n) = f(m) + f(n)$ for all $m, n \in I$. Show that the relation R or Theorem 7.5.1 has exactly two equivalence classes. Describe the functions g and η.

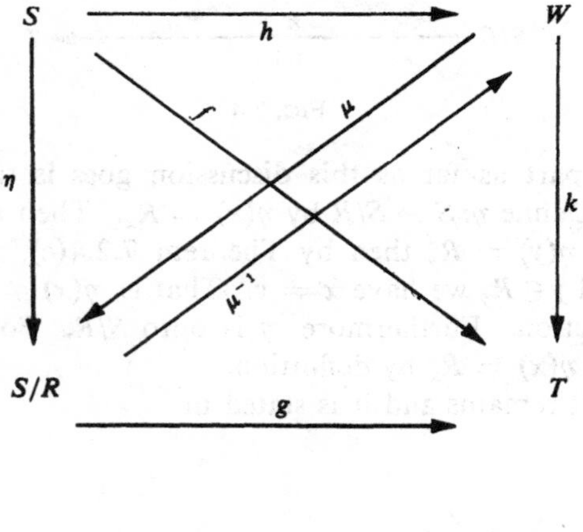

(5) Suppose $f: S \to T$, W is a set and $h: S \to W$ is onto, $k: W \to T$ is 1–1 and $f = k \circ h$. Define a 1–1 onto function $\mu: W \to S/R$ such that $\eta = \mu \circ h$ and $g = k \circ \mu^{-1}$, where R is as in Theorem 7.5.1. Note that $g \circ \eta = k \circ h$ and one has the diagram above.

Related Readings

1. Dinkines, F.: *Elementary Theory of Sets*. New York: Appleton-Century-Crofts, 1964. One can find here additional explanations and examples of relations, functions and infinite sets.
2. Eves, H., and Newsom, C. V.: *Introduction to the Functions and Fundamental Concepts of Mathematics*, New York: Holt, Rinehart and Winston, 1965. See Chapters VIII and IX.
3. Foulis, D. J.: *Fundamental Concepts of Mathematics*, Boston: Prindle, Weber and Schmidt, Inc., 1969. See this one for further explanations of relations and functions and a slightly more sophisticated development.
4. Halmos, P. R.: *Naive Set Theory*, Princeton: Van Nostrand, 1960. The matter of ordinal as well as cardinal numbers is considered and related herein.

chapter 8

"I regret ... to administer such a ... dose of four-dimensional geometry. I do not apologize ... that nature ... is four dimensional. Things are what they are ...."—A. N. Whitehead

CONTINUITY, CONNECTEDNESS AND COMPACTNESS

1. INTRODUCTION

In this chapter we move beyond elementary modern mathematics. Much of our work has been done in the context of the real number system, though much of it has at the same time been more generally stated and developed. Nearly all previous work is at the foundation of most advanced mathematics and of universal use in mathematics. In this chapter we turn to the fundamental concept of "continuous mathematics"—topology and analysis—the concept of a *continuous function*—a continuous relation between sets. Just that description tells us that the sets will be more than simply abstract sets.

The concept of a continuous transformation or function is well motivated by a consideration of Question II of this text. We will settle that question completely in the most general sense, draw some consequences and then attack certain more conceptually difficult but related problems. In particular we will take another, different look at connectedness, and then venture into n-dimensional spaces, guided by intuition developed in one, two and three dimensional spaces and the abstract concepts of topology. By their set-theoretic nature, basic topological concepts are independent of dimension and allow one to operate just as surely in spaces he cannot visualize as in those he can, the logic of his senses being replaced by the logic of his mind.

This chapter beings with a completely general notion of continuity, or non-tearing, in change, a concept fundamental to many, if not all investigations of change and transformation. In a very broad sense, a change or transformation is continuous if it does not at some point sharply break with its past, or perhaps better, with its trend. Implicit in this view of continuous transformation is the belief that knowledge of the nature or quality of the transformation *near* (but not necessarily at) a single point in transition, *implies* knowledge of the nature or quality of the transforma tion *at* the point, since what is true at the point cannot differ sharply from what is true near the point.

One might sense from these comments that topological concepts naturally enter into a discussion of continuity, if it is to be idealized and made precise. As stated earlier, topology is the study of the qualitative properties of abstract space. The neighborhood concept has been used before to assert conclusions about a point from knowledge gained near the point. Continuity is related too to the essential point of Section 2.9, that approximate knowledge to any *arbitrarily good* degree of approximation can yield a certain conclusion. These general remarks may be kept in mind to some advantage in the sequel, particularly if the reader is to get past the special details of each problem and obtain a general view of continuity in transformation to be used wherever he finds it appropriate.

2. CONTINUITY, OR DISTORTION WITHOUT TEARING

The answer we conjecture to Question II is that; if a connected object X is given and if X is distorted *without tearing* to an object Y, then Y is itself connected. But what can we mean, in a suitably precise fashion, by "without tearing"?

Let us suppose an object X is given along with a distortion Y of it defined by some function $f: X \to Y$ which we can suppose is onto for the sake of discussion. Suppose X is given a topology $\mathscr{U}$ and Y is given a topology $\mathscr{V}$. Using what is at hand, namely $(X, \mathscr{U})$ $(Y, \mathscr{V})$ and the function f, we want to develop a good definition of "tearing" or "nontearing"—take your choice. For motivation we again go to figures in the plane. Suppose the situation is as Figure 8.1 where we suppose a line is cut in X through the point x and f simply pulls the figure X apart along this line. Let us try to abstractly contrast what happens near x as opposed to what happens near x', where no "tearing" takes place.

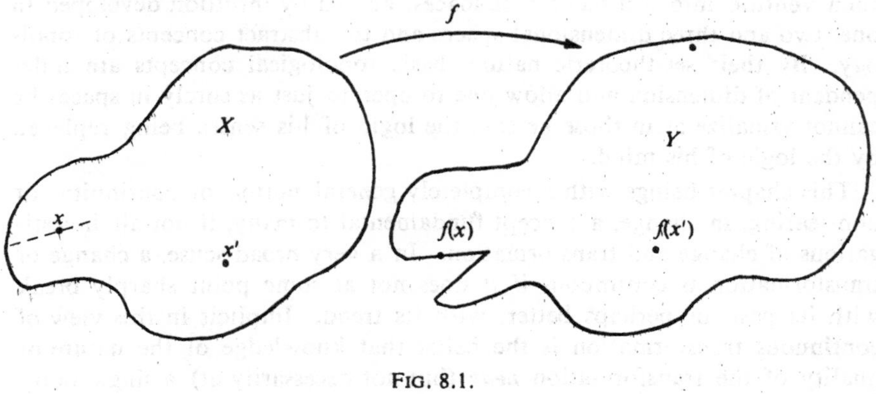

FIG. 8.1.

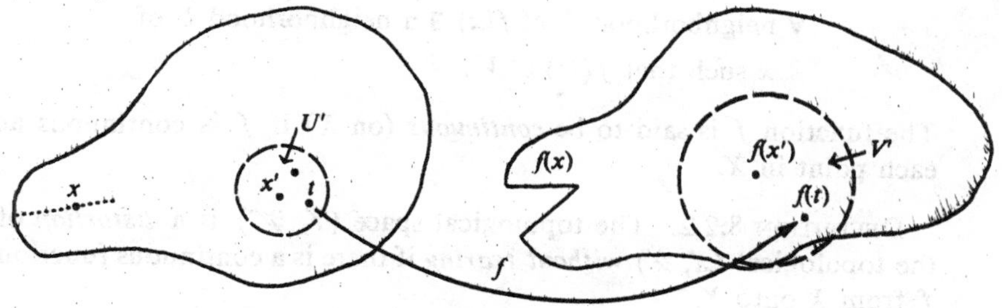

FIG. 8.2.

Suppose a neighborhood V' of x' is given. If things are as nice as the picture indicates, it seems that there must be a neighborhood U' of x' such that $f(U') \subset V'$; i.e., for any $t \in U', f(t) \in V'$. This is illustrated in Figure 8.2 and the following condition seems to hold:

(1) $\forall$ neighborhood V' of $f(x')$ $\exists$ a neighborhood U' of x' such that $\forall t \in U', f(t) \in V'$.

Take a look now at the situation near $f(x)$. Again, if things are as the picture indicates, it seems that the following is true: there is a neighborhood V of $f(x)$ such that, for any neighborhood U of x, $f(U) \not\subset V$; i.e., there is a $t \in U$ such that $f(t) \notin V$. This is indicated by Figure 8.3. Hence where there is a tear in the distortion the following condition holds:

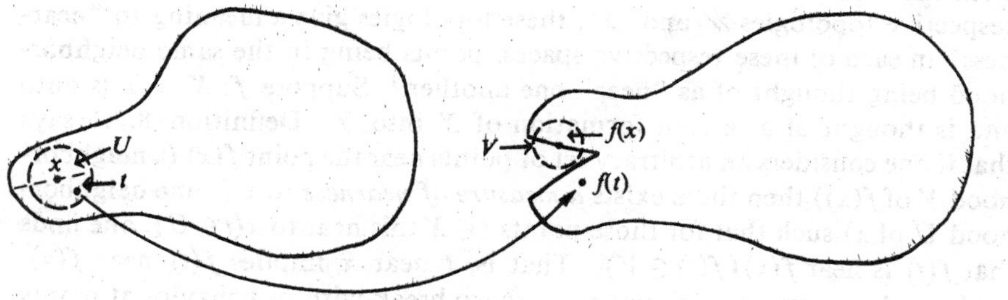

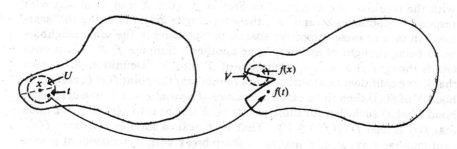

FIG. 8.3.

(2) $\exists$ a neighborhood V of $f(x)$ such that $\forall$ neighborhood U of x $\exists t \in U$ such that $f(t) \notin V$.

Notice that (1) and (2) are negations of one another. Hence the logic of one's senses (Figs. 8.1–8.3) meets well with the logic of one's mind ((1) and (2)). For this reason we will take (1) as an abstract definition of continuity, which we can, if we wish, think of as "non-tearing."

DEFINITION 8.2.1. Let $(X, \mathscr{U})$ and $(Y, \mathscr{V})$ be topological spaces and let $f: X \to Y$ (not necessarily onto). The function f is said to be *continuous at the point* $x \in X$ iff the following condition holds:

$\forall$ neighborhood V of $f(x)$ $\exists$ a neighborhood U of x such that $f(U) \subset V$.

The function f is said to be ***continuous*** (on X) if f is continuous at each point in X.

DEFINITION 8.2.2. The topological space $(Y, \mathscr{V})$ is a ***distortion*** of the topological $(X, \mathscr{U})$ ***without tearing*** if there is a continuous function f from X onto Y.

With these definitions we can now precisely state our conjectured answer to Question II.

If $(Y, \mathscr{V})$ is a distortion of $(X, \mathscr{U})$ without tearing and if $(X, \mathscr{U})$ is connected, then $(Y, \mathscr{V})$ connected.

It is often the case that stating a problem well is at least half and sometimes more of the difficulty of solving it. Question II is now well stated, every technical term has an exact meaning, and as we shall soon see, the question can be almost trivially settled using the theory of connectedness and topological spaces previously developed, along with a single result on continuity.

Before some examples and problems, a word or two will be given relating the technical concept of continuity expressed in Definition 8.2.1 with the intuitive one expressed in Section 1. Let X and Y be sets with respective topologies $\mathscr{U}$ and $\mathscr{V}$; these topologies give a meaning to "nearness" in each of these respective spaces, points being in the same neighborhood being thought of as "near" one another.[1] Suppose $f: X \to Y$ is onto and is thought of as a transformation of X into Y. Definition 8.2.1 says that if one considers an arbitrary set of points near the point $f(x)$ (a neighborhood V of $f(x)$) then there exists ***a measure of nearness*** to x (some neighborhood U of x) such that for those points $t \in X$ this near to $x (t \in U)$, one finds that $f(t)$ is near $f(x) (f(t) \in V)$. That is, t near x implies $f(t)$ near $f(x)$. Said another way, at x, f makes no sharp break with its behavior at points t near x. Hence the remarks of the introductory paragraph are also well met by Definition 8.2.1.

EXAMPLES. (a) Let $X = (0, 1)$ and $Y = (0, 2)$ with the usual topologies of all open subintervals of X and Y respectively. Let $f(x) = 2x$. Then f is continuous, for if $V = (c, d)$ is a neighborhood of some $f(x)$—that is, $f(x) \in (c, d) \subset (0, 2)$—then $2x \in (c, d)$ and hence $x \in (c/2, d/2)$. Since $(c, d) \subset (0, 2)$, then $(c/2, d/2) \subset (0, 1)$ and by definition of the topology on $(0, 1)$, $U = (c/2, d/2)$ is a neighborhood of x, and it is clear from the formula defining f that $f(U) \subset V$.

[1] We claim no precision here in the use of the word "near," the aim is to generate an intuitive view of the technical concept of continuity.

(b) Let $X = \mathscr{R} \times \mathscr{R}$ with the usual topology and let $Y = \mathscr{R}$ with the usual topology. Define $f\colon X \to Y$ be

$$f(x, y) = 2x + y.$$

We will show that f is continuous at $(x, y) = (0, 1)$, noting that $f(0, 1) = 1$.

Let

$$V = (1 - \varepsilon, 1 + \varepsilon)$$

be an arbitrary neighborhood of $1 = f(0, 1)$, where $\varepsilon > 0$. (Not quite completely arbitrary: a typical neighborhood of 1 is an interval (a, b) with $a < 1 < b$, but $(a, b) \supset (1 - \varepsilon, 1 + \varepsilon)$ when ε is taken to be the smaller of $1 - a$ and $b - 1$.) We must find a neighborhood U of $(0, 1)$ such that $t \in U$ implies $f(t) \in V$; that is

$$(x, y) \in U \to f(x, y) = 2x + y \in (1 - \varepsilon, 1 + \varepsilon)$$

or, equivalently,

$$1 - \varepsilon < 2x + y < 1 + \varepsilon.$$

Hence one must place nearness conditions (a neighborhood U) on points (x, y) relative to $(0, 1)$ such that

$$1 - \varepsilon < 2x + y < 1 + \varepsilon$$

or

$$-\varepsilon < 2x + y - 1 < \varepsilon.$$

Well, looking at this last inequality one sees that if y is within $\varepsilon/2$ of 1 and $2x$ is also with $\varepsilon/2$ of 0, we would have

$$2x + (y - 1) < \frac{\varepsilon}{2} + \frac{\varepsilon}{2} < \varepsilon.$$

This observation motivates the following tentative choice of U. Recall that a neighborhood of $(0, 1)$ in $\mathscr{R} \times \mathscr{R}$ is the interior of some circle centered at $(0, 1)$ of some radius. Let

$$U = \left\{(x, y)\colon\ (x - 0)^2 + (y - 1)^2 < \left(\frac{\varepsilon}{4}\right)^2\right\}$$

be the interior of the circle of radius $\varepsilon/4$ centered at $(0, 1)$. (We choose $\varepsilon/4$ rather than $\varepsilon/2$ because of the factor 2 in $2x + y$, as will shortly be clear.)

Suppose that $(x, y) \in U$. We claim $f(x, y) \in V$. Since $(x, y) \in U$ then

$$(x - 0)^2 \leq (x - 0)^2 + (y - 1)^2 < \left(\frac{\varepsilon}{4}\right)^2$$

and

$$(y-1)^2 \leq (x-0)^2 + (y-1)^2 < \left(\frac{\varepsilon}{4}\right)^2.$$

Hence, $|x - 0| < \varepsilon/4$ and $|y - 1| < \varepsilon/4$. Therefore, $|2x - 0| < \varepsilon/2$ and $|y - 1| < \varepsilon/4 < \varepsilon/2$. Thus,

$$\begin{aligned} |2x + y - 1| &= |(2x - 0) + (y - 1)| \\ &\leq |2x - 0| + |y - 1| \\ &< \frac{\varepsilon}{2} + \frac{\varepsilon}{2} = \varepsilon. \end{aligned}$$

Hence

$$-\varepsilon < 2x + y - 1 < \varepsilon$$

or,

$$1 - \varepsilon < 2x + y < 1 + \varepsilon$$

or, finally,

$$1 - \varepsilon < f(x, y) < 1 + \varepsilon.$$

Thus $f(x, y) \in V$.

The reader will notice in this example the use of several properties of inequalities and absolute values; these properties are summarized in the first problem set in Section 2.7.

(c) If one takes a string, cuts it in half and pulls the pieces apart, one would no doubt call this a distortion with tearing. Let us see how continuity in the formal sense relates to this operation. Suppose the string is represented by the unit interval [0, 1] (with the usual (relative) topology) and the distortion by the function $f: [0, 1] \to \mathscr{R}$ given by

$$f(x) = \begin{cases} x & \text{if } 0 \leq x < \frac{1}{2} \\ x + 1 & \text{if } \frac{1}{2} \leq x \leq 1. \end{cases}$$

We can picture f as indicated in Figure 8.4 or as a subset of $\mathscr{R} \times \mathscr{R}$ in the usual way, indicated in Figure 8.5.

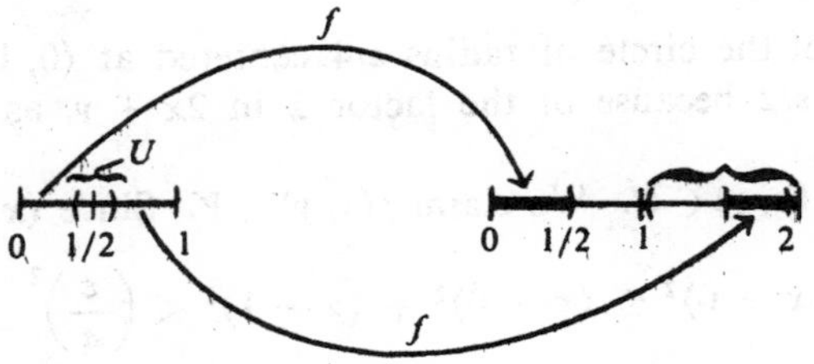

FIG. 8.4.

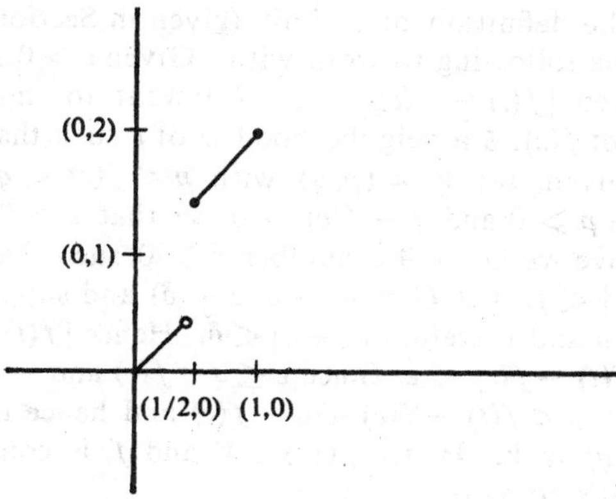

FIG. 8.5.

Evidently, f is not continuous at $1/2$. Let us tie this belief *precisely* to Definition 8.2.1. Accordingly, f is not continuous at $x = 1/2$ if $\exists$ a neighborhood V of $f(1/2)$ such that $\forall$ neighborhood U of $1/2, f(U) \not\subset V$.

Let $V = (1, 2)$; V is a neighborhood of $f(1/2) = 3/2$. Since *existence* is called for, any one neighborhood V will do. Let U be any neighborhood of $1/2$. We claim $f(U) \not\subset V$. Since $1/2 \in U$ and U is an open interval, say (c, d), then $c < 1/2 < d$. Let $t \in (c, 1/2)$. Then $t \in U$ and by definition of $f, f(t) = t < 1/2$. Hence $f(t) \notin V = (1, 2)$. Thus $f(U) \not\subset V$.

Hence, f is not continuous at $x = 1/2$. Notice however, that f is continuous at every *other* point, x, as one would expect, since no tearing occurs. For let $W = (a, b)$ be an neighborhood of $f(x) = x + 1$ where (say) $x > 1/2$. Then $a < x + 1 < b$ and $a - 1 < x < b - 1$. If

$$t \in U = (a - 1, b - 1) \cap [0, 1] \cap (1/2, \infty)$$

(this is a neighborhood of x in the relative topology), then $t \in (1/2, 1], f(t) = t + 1$ and since we also have $t \in (a - 1, b - 1)$, then

$$f(t) \in (a, b) = V.$$

Similarly, f is continuous at points $x < 1/2$.

(d) In calculus, one has the powful concept of *limit* to handle questions of continuity. Specifically, if $f: (a, b) \to \mathscr{R}$ and $c \in (a, b)$ then (in calculus) f is said to be continuous at c if $\lim_{x \to c} f(x)$ exists and equals $f(c)$; that is, $\lim_{x \to c} f(x) = f(c)$. The following argument shows that a function continuous by the limit definition is continuous by Definition 8.2.1. Hence one can use the calculus, at least for functions defined on intervals (a, b), to determine questions of continuity according to Definition 8.2.1.

To set things up, let $X = (a, b)$ with the usual topology of open subintervals and let Y be the set of real numbers with the usual topology of open intervals. We suppose that $f: X \to Y$ and that $\lim_{x \to c} f(c) = f(c)$ and will show that f is continuous at c in the sense of Definition 8.2.1.

According to the definition of a limit (given in Section 1.14, Problem III(5)) we have the following to work with. Given $\varepsilon > 0$ $\exists \delta > 0$ such that if $|t - c| < \delta$ then $|f(t) - f(c)| < \varepsilon$. We want to show that given a neighborhood V of $f(c)$, $\exists$ a neighborhood U of c such that $f(U) \subset V$.

Suppose V is given, say $V = (p, q)$ with $p < f(c) < q$. Let ε be the smaller of $f(c) - p > 0$ and $q - f(c) > 0$, so that $\varepsilon > 0$. According to the statement above we know $\exists$ a number $\delta > 0$ such that if $|t - c| < \delta$ then $|f(t) - f(c)| < \varepsilon$. Let $U = (c - \delta, c + \delta)$ and suppose $t \in U$. Then $c - \delta < t < c + \delta$ and therefore $|t - c| < \delta$. Hence $|f(t) - f(c)| < \varepsilon$ and therefore $-\varepsilon < f(t) - f(c) < \varepsilon$. Since $\varepsilon \leq q - f(c)$ and $-\varepsilon \geq -(f(c) - p)$ we have $-f(c) + p < f(t) - f(c) < q - f(c)$ and hence that $p < f(t) < q$ or that $f(t) \in (p, q) = V$. Hence $f(U) \subset V$ and f is continuous. Figure 8.6 indicates what's going on.

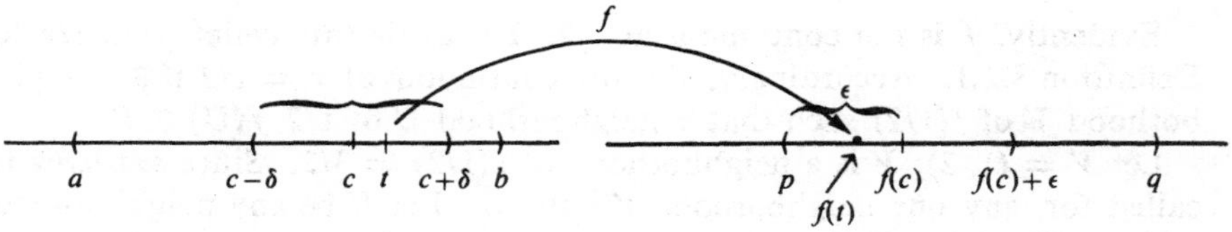

FIG. 8.6.

This last example was somewhat involved, but that's what can happen when one applies a neat abstract idea to a particular instance. While abstraction does allow one to think more clearly and to avoid harsh and often confusing details, one should recognize that abstraction only provides a special kind of overveiw, and in a specific instance of application one may have to get down into the nitty-gritty details which uniquely pertain to that instance in order to use the abstract thinking one might wish to use. Here are some problems on continuity, a few of which may be reminiscent of Problem 8, Section 5.4.

Problems.

(1) Let $X = (0, 1)$, $Y = \mathscr{R}$ and $f(x) = 2x + 1$. Show that f is continuous at $x = 1/2$ using Definition 8.2.1.

(2) Let $X = (-1, 1)$, Y the set of real numbers and $f(x) = x^2$. Show that f is continuous at 0 using Definition 8.2.1.

(3) For X and Y as above, let $g(x) = \sin x$. Given that $|\sin x| \leq |x|$ for all x show that g is continuous at 0 using Definition 8.2.1.

(4) Let $f: \mathscr{R} \to \mathscr{R}$ such that $\exists$ a number $M \ni \forall x, y \in \mathscr{R}$ $|f(x) - f(y)| < M|x - y|$. Show that f is continuous by Definition 8.2.1.

(5) Define $f: \mathscr{R} \times \mathscr{R} \to \mathscr{R}$ by $f(x, y) = x - y + 1$. Show that f is continuous at $(1, 1)$.

(6) Let $X = (0, 1)$ with the usual topology, and let $f(x)$ be 0 when x is rational, 1 when x is irrational. Show that f is not continuous

at any $x \in (0, 1)$.

(7) Let

$$X = \{(x, y): (x - 0)^2 + (y - 1)^2 = 1 \quad \text{and} \quad (x, y) \neq (0, 2)\}$$

with the relative topology in the plane. For each $(x, y) \in X$ let $f(x, y)$ be the intersection with the x-axis of a line drawn through $(0, 2)$ and (x, y). Using pictures, determine whether or not f is continuous. If f^{-1} exists, is f^{-1} continuous?

(8) Define $g: \mathscr{R} \times \mathscr{R} \to \mathscr{R}$ by $g(x, y) = x^2 + y^2$. Show that g is continuous at $(0, 1)$. A picture may help.

(9) Suppose X is a topological space and $X = A \cup B$ with A separated from B. Define $f: X \to \mathscr{R}$ by

$$f(x) = \begin{cases} 1 & \text{if } x \in A \\ 0 & \text{if } x \in B. \end{cases}$$

Show that f *is continuous* on X.

Thus any disconnected space admits a continuous function taking on only two distinct values. Now show that this property characterizes disconnected spaces. That is, show that if X is a topological space and $f: X \to \{0, 1\}$ which is continuous and onto, then X is not connected. *Hint*: show that $C = f^{-1}(1)$ is both open and closed.

(10) Let X, A, B and f be as in Problem 9, but this time suppose $A \cap \bar{B} \neq \square$. Let $x \in A \cap \bar{B}$ and show that f is *not* continuous at x.

(11) Let X be a set and $\mathscr{U}$ the discrete topology on X. Let Y be any

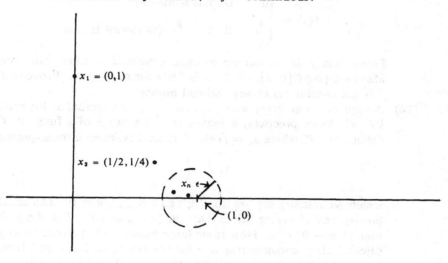

topological space. Show that any function $f: X \to Y$ is continuous.

(12) Recall an earlier note that the functions on a set mirror properties of the set. Suppose $(X, \mathscr{U})$ is a topological space and every function $f: X \to \mathscr{R}$ is continuous. Prove that $\mathscr{U}$ is the discrete topology on X. To do this it suffices to show that $\forall x \in X, \{x\} \in \mathscr{U}$. Consider the function $f(x) = 1, f(t) = 0$ for all $t \neq x$.

(13) Define $f: [0, 1] \to \mathscr{R}$ by

$$f(x) = \begin{cases} 0 & \text{if } x \text{ is irrational} \\ \dfrac{1}{q} & \text{if } x = \dfrac{p}{q} \text{ (in lowest terms)}. \end{cases}$$

Prove that f is continuous at each irrational number using the idea that $\{p/q \in [0, 1]: 1/q \geq \varepsilon\}$ is finite for any $\varepsilon > 0$. Prove that f is not continuous at any rational number.

(14) A sequence is an array $a_1, a_2, \ldots, a_n, \ldots$, one symbol a_k for each $k \in \mathscr{N}$. More precisely, a sequence is the range of a function f defined on $\mathscr{N}$, where $a_n = f(n)$. A typical example is the sequence

$$1, \frac{1}{2}, \frac{1}{3}, \ldots$$

which we casually say has limit 0; i.e., $\lim_{n\to\infty} 1/n = 0$. The usual precise way of saying this is that: $\forall \varepsilon > 0 \; \exists$ an $N \in \mathscr{N} \ni$ if $n \geq N$ then $|1/n - 0| < \varepsilon$. Now some large number of students in any calculus class, encountering this for the first time, insist on letting "n equal infinity" and concluding that $1/n = 0$, and hence reducing the whole business to an apparent triviality.

This attitude avoids the only point of interest! One has no ***number*** a_n for "n equal infinity," only a number a_n for each ***natural number*** n. Whatever may be one's thoughts as to the meaning of "infinity" here, it is no doubt far from being a natural number. In this kind of problem, given a sequence $a_1, a_2, \ldots, a_n, \ldots$ of (say) numbers, the ***whole point*** of interest is in deciding just what number a ***should*** correspond to "infinity," if one wishes to look at the problem in the "infinity" context. Thus to assume a value corresponding to "infinity" is to sidestep the only interesting point involved.

That said, topology does allow one to assign a value to be associated with infinity in a formal sense and to reduce the whole matter to a question of continuity. For the topological space X of Problem 8, Section 4.1.

(a) Let $f(n) = 1/n, f(\omega) = 0$. Show that f is continuous at ω.

(b) Let $g(n) = 1/n, g(\omega) = 1$. Show that g is not continuous at ω.

(c) Let $a_1, a_2, \ldots$ be any sequence of real numbers. Define

$\lim_{n\to\infty} a_n = L$ as usual to mean that: $\forall \varepsilon > 0 \ \exists$ an $N \in \mathcal{N} \ni n \geq N \rightarrow |a_n - L| < \varepsilon$. Let L be a real number and define $f: X \rightarrow \mathcal{R}$ by $f(n) = a_n$ and $f(\omega) = L$. Show that $\lim_{n\to\infty} a_n = L$ iff f is continuous at the point $\omega \in X$. One can think of ω as a point "at infinity," but to speak of infinity as a number would be somewhat naive in the light of Sections 2.9 and 6.1. Part (c) above should lend this clarification to the idea of the limit—$\lim_{n\to\infty} a_n$—of a sequence $a_1, a_2, \ldots$. The number $\lim_{n\to\infty} a_n$ is the one and only number which makes the function defined by $f(n) = a_n$ and $f(\omega) = \lim_{n\to\infty} a_n$, continuous at ω. The general notion of continuity given in Definition 8.2.1 appears to (and does) furnish a single framework within which one can realize, in the sense of precisely describing, all questions of convergence and approximation. One need not ever concern himself with any another version of continuity or approximation, such can always be realized in the context of Definition 8.2.1. In particular, the essential notion of Section 2.9, that one decides questions at infinity through knowledge in a finite context to an arbitrary good degree of approximation can now be seen for what it really is: the assumption (or desire) that behavior *at* infinity should be continuous, and not break sharply with the trend "up to," but not including, infinity.

(15) If anybody ever asks you, tell them topology is the study of those properties preserved by continuous functions. (They may not understand your answer but you will be free to talk on other things.) We will shortly see that connectedness is such a topological property. Thus continuity is the big thing in topology. In Example (d) above we saw that continuity as defined in calculus implies our definition of continuity. Now show the converse. Show that if f is a real-valued function defined on an interval (a, b) with the usual topology on both domain and range and which is continuous at a point $c \in (a, b)$ according to Definition 8.2.1, then $\lim_{x\to c} f(x) = f(c)$ according to the definition of *limit* given in Example (d) above. In other words, prove the converse of (d).

Hence, continuity for *any* function between *any* pair of topological spaces, coincides in the special case of a real-valued function defined on a set of real numbers, with continuity as defined in calculus.

(16) This is not really a problem but a point of clarification. Perhaps the reader has wondered: does topology furnish a unified setting for the statement of the limit concept? It does, and here is how.

Let $f: X \to Y$ be a function and $\mathscr{U}$ and $\mathscr{V}$ be topologies on X and Y, respectively. Let $L \in Y$ and let $a \in X$. One says that "the limit of $f(x)$ as x approaches a is L"—written $\lim_{x \to a} f(x) = L$—iff the following is true:

$\forall$ neighborhood V of L $\exists$ a neighborhood U of a such that $f(U) \subset V$.

Said in less precise language: if V is a set of points near L, then $\exists$ a set of points U near a such that for x near to a $(x \in U)$, $f(x)$ is near L $(f(x) \in V)$.

(17) We have heretofore avided any general discussion of sequences in a topological space. The concept of *convergence of sequences* originated long before the concept of a topological space, and although a natural idea, seems to cause a certain difficulty for the neophyte.

Let $(X, \mathscr{U})$ be a topological space. A sequence in X is a collection of points $x_1, x_2, x_3, \ldots, x_n, \ldots$ in X, exactly one for each $n \in \mathscr{N}$. That is, a sequence $\{x_n\}$ is the range of a function $f: \mathscr{N} \to X$ where $x_n = f(n)$. The natural question that goes with such an unending list of points $x_1, x_2, \ldots, x_n, \ldots$ is: is there a point $x \in X$ such that the points $x_1, x_2, \ldots, x_n, \ldots$ are "eventually close to" x? Stated precisely: is there an $x \in X$ such that $\forall$ neighborhood U (think of U as a measure of *closeness* to x) of x $\exists$ an $N \in \mathscr{N}$ such that for $n \geq N$, (*eventually*: if n is larger than some N), one has $x_n \in U$? If such a point x exists in X we say that $\{x_n\}$ converges to x and write $\lim_{n \to \infty} x_n = x$, or simply, $\{x_n\} \to x$. For purposes of intuition one thinks of the points x_n as "approaching" the point x.

For example, if

$$X = \mathscr{R} \times \mathscr{R}$$

and

$$x_n = \left(1 - \frac{1}{n}, \frac{1}{n^2}\right) \in \mathscr{R} \times \mathscr{R},$$

then $\{x_n\} \to (1, 0)$. For, given a neighborhood

$$U = \{(a, b)\colon\ [(a - 1)^2 + (b - 0)^2]^{1/2} < \varepsilon\},$$

if we choose N such that $1/N < \varepsilon/2$, then for $n \geq N$ one has

$$\left(1 - \left(1 - \frac{1}{n}\right)\right)^2 = \left(\frac{1}{n}\right)^2 < \left(\frac{1}{N}\right)^2 < \frac{\varepsilon^2}{4}$$

and

$$\left(\frac{1}{n^2} - 0\right)^2 = \left(\frac{1}{n^2}\right)^2 < \left(\frac{1}{N^2}\right)^2 < \frac{\varepsilon^2}{4}$$

and hence

$$\left[\left(\left(1-\frac{1}{n}\right)-1\right)^2+\left(\frac{1}{n^2}-0\right)^2\right]^{1/2}<\frac{\varepsilon}{\sqrt{2}}<\varepsilon .$$

Thus $x_n \in U$ for $n \geq N$. This argument is illustrated below.

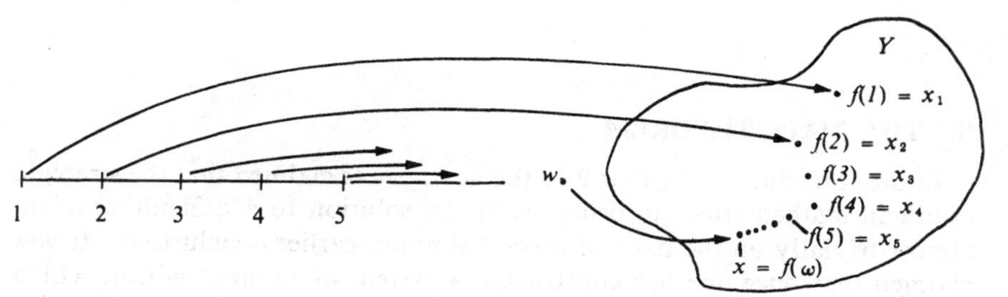

(a) Using the axiom of choice prove that if $r \in \mathscr{R}$ then $\exists$ a sequence r_n of rational numbers such that $\{r_n\} \to r$. *Hint*: Let r_n be a rational number in the interval $(r-1/n, r+1/n)$. Hence any real number is the limit of a sequence of rational numbers.

(b) Prove that sequences determine continuity on $\mathscr{R}$. That is, suppose $(X, \mathscr{U})$ is a topological space and $f: \mathscr{R} \to (X, \mathscr{U})$. Let $x \in \mathscr{R}$, $y = f(x)$. Suppose then $\forall$ sequence $\{x_n\} \subset \mathscr{R}$ such that $\{x_n\} \to x$ the corresponding sequence $\{y_n\} \to y$ where $y_n = f(x_n)$. Show that f is continuous at x. That is, if $\{x_n\}$ "approaching" x implies $\{f(x_n)\}$ "approaches" $f(x)$, then f is continuous at x.

(c) Prove the converse of (b): If f is continuous at x and $\{x_n\} \to X$ in $\mathscr{R}$, then $\{f(x_n)\} \to f(x)$ in X.

If the reader needs some feeling of intuition for (b) and (c), consider an example such as $f(x) = (x, x^2 - 1)$. Here $f: \mathscr{R} \to \mathscr{R} \times \mathscr{R}$. Notice that $\{1/n\} \to 0$ and $f(1/n) = (1/n, 1/n^2 - 1) \to (0, -1) = f(0)$.

(d) Let $(Y, \mathscr{U})$ be a topological space, $\{x_n\}$ a sequence in Y and $\{x_n\} \to x \in Y$. Define $f(n) = x_n$, $f(\omega) = x$ as in Problem 13 above. Show that this function $f: X \to Y$ is continuous at ω, where, as before, X is the space of Problem 8, Section 4.1.

3. THE MAIN THEOREM

In the introduction to Part 2 of this text, it was claimed that the strategic attack in mathematics can bring about the solution to a difficult problem almost trivially on the basis of several simpler, earlier conclusions. It was claimed that once one has constructed a system of thought within which one or more concepts and questions can be enunciated, one may find that others can be raised and/or settled as well. In answering Question II we will see that this is exactly the course our work has taken. In Chapter 4 we defined and characterized connected topological spaces, Theorem 4.6.7 being the result we will most make use of now. In Chapter 5 we introduced the function concept, enabling us to define the word *distortion*. We have just defined a distortion without tearing. All of this will now be put together to yield the main theorem quite easily.

Suppose $(Y, \mathscr{V})$ is a distortion without tearing of a topological space $(X, \mathscr{U})$ defined by some continuous function $f\colon X \to Y$. In Definition 8.2.2 we agreed that f is onto. We wish to conclude that Y is connected, given that X is connected. We have a criterion (Theorem 4.6.7) that Y be connected: any open and closed subset E in Y is either empty or the entire set Y. All the given information resides in X, and using f we can relate E to X through the set $f^{-1}(E) \subset X$. If we can say enough about $f^{-1}(E)$, given that f is continuous and E is open and closed, we may be able to conclude that $E = \square$ or $E = Y$. The best general result is given in

THEOREM 8.3.1. Let $(X, \mathscr{U})$ and $(Y, \mathscr{V})$ be topological spaces and let $f\colon X \to Y$. The following are logically equivalent:

(a) f is continuous.

(b) If O is open in Y, then $f^{-1}(O)$ is open in X.

(c) If C is closed in Y, then $f^{-1}(C)$ is closed in X.

We leave this to the reader to prove with the reminder that Section 5.4 may be of some help. Here is an answer to Question II.

THEOREM 8.3.2. If $(Y, \mathscr{V})$ is a distortion of $(X, \mathscr{U})$ without tearing, and $(X, \mathscr{U})$ is connected, then $(Y, \mathscr{V})$ is connected.

Again this is left to the reader. It is just about a trivial consequence of Theorem 8.3.1 and the remarks preceding it. Note in proving it that the question really wasn't so difficult, provided one picks out just those properties that really pertain: Theorem 4.6.7 and continuity. We will now use this result to obtain other conclusions. From Theorem 8.3.2 we obtain a corollary which will make such applications easier. Recall that in Theorem 8.3.2 we suppose the existence of an *onto* function. More generally,

THEOREM 8.3.3. Let $(X, \mathscr{U})$ be a connected topological space, $(Y, \mathscr{V})$ a topological space and $f: X \to Y$ a continuous function. Then $f(X)$ is a connected subspace of Y.

Proof. Let $W = f(X) \subset Y$ and let $\mathscr{W}$ denote the relative topology on W induced by $\mathscr{V}$. Thus, $\mathscr{W} = \{W \cap V: V \in \mathscr{V}\}$. Then f is *onto* W and to conclude that W is connected we need only show that f is continuous from $(X, \mathscr{U})$ to $(W, \mathscr{W})$ for by Theorem 8.3.2, $(W, \mathscr{W})$ will then be connected.

Let $x \in X$ and let $W \cap V$ be a neighborhood of $f(x)$ in W. Then, because f is continuous into $(Y, \mathscr{V})$, $\exists$ a neighborhood U of x such that $f(U) \subset V$. Since $f(U) \subset f(X) = W$, then $f(U) \subset W \cap V$ and f is continuous into W by Definition 8.2.1. Hence the proof.

These two theorems are the principle results concerning the preservation of the property of connectedness by a continuous transformation. We will now use these to derive some fundamental and well known conclusions. The first of these is an improved version of the so called "*intermediate value theorem*" of calculus.

Imagine a continuous function $f: [a, b] \to \mathscr{R}$. Suppose that $f(a) < f(b)$ and consider a number c between $f(a)$ and $f(b)$. With the assumption of continuity one believes that the graph of f must cross every horizontal line in the plane lying between $f(a)$ and f(b). In particular the graph must cross the line $y = c$, as indicated in Figure 8.7. Hence, it appears that there must exist an $x \in [a, b]$ such that $f(x) = c$. That is, any *intermediate* number c between $f(a)$ and $f(b)$ must be some value, $f(x)$, of f. One can easily prove the more general.

THEOREM 8.3.4. Let $(X, \mathscr{U})$ be any connected topological space, and let $f: X \to \mathscr{R}$ be continuous. If $x, y \in X$ and c is a number in $\mathscr{R}$ between $f(x)$ and $f(y)$, then $\exists z \in X$ such that $f(z) = c$.

Proof. By Theorem 8.3.3, $f(X)$ is a connected subset of $\mathscr{R}$. By Theorem 4.5.4, $f(X)$ must be an interval. Since $f(x)$ and $f(y)$ are numbers in the interval $f(X)$ and since c is between $f(x)$ and $f(y)$ then c must also be in the interval $f(X)$ by definition of an interval. But $c \in f(X)$ means $c = f(z)$ for some $z \in X$. Hence the proof.

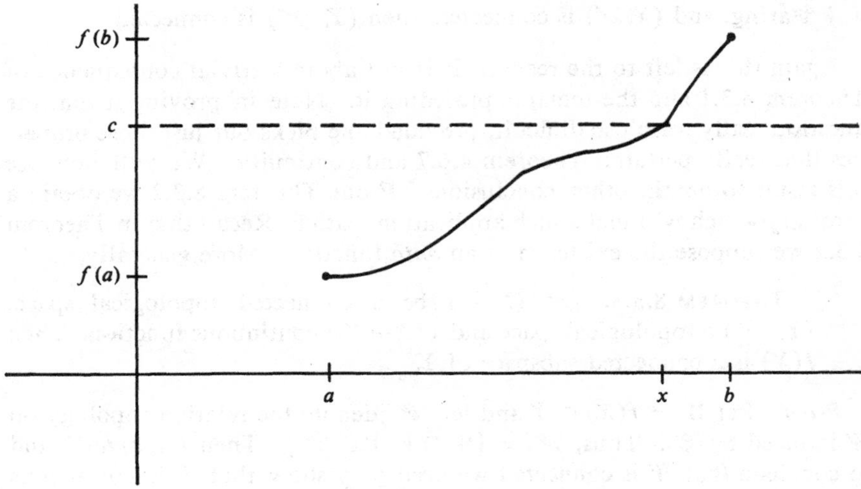

FIG. 8.7

With this result we can easily prove what is commonly called a fixed-point theorem. Fixed-point theorems are some of the more interesting and useful in mathematics, particularly in applications concerning the existence of solutions of differential and integral equations. The theorem we prove is not exactly in this category, but is interesting nonetheless. It can be interpreted as saying that if one takes a rubber band and stretches and/or contracts it within its original position, then at least one point on the band does not move. More precisely we have,

THEOREM 8.3.5. Let $g: [0, 1] \to [0, 1]$ be a continuous function. Then $\exists x \in [0, 1]$ such that $g(x) = x$.

Proof. To prove this we define an auxiliary function f that measures how far and in what direction each point $t \in [0, 1]$ is displaced by the correspondence g, where we think of $g(t)$ as the point t is displaced to. Precisely, let $f(t) = g(t) - t$. (For example, if g moves t to the left, then $g(t) \leq t$ and $f(t) = g(t) - t$ is a measure of how far t has traveled.) We claim there is a point x that does not move. For such a point we must have $f(x) = g(x) - x = 0$.

Here is the formal proof. Because g is continuous it is easy to prove that f is continuous. Note, too, that $f: [0, 1] \to \mathscr{R}$. Since $g(1) \in [0, 1]$ we have

$$f(1) = g(1) - 1 \leq 0;$$

similarly, since $g(0) \in [0, 1]$, we have

$$f(0) = g(0) - 0 \geq 0.$$

Hence $c = 0$ is between $f(0)$ and $f(1)$. By Theorem 8.3.4, $\exists x \in [0, 1]$ such that $f(x) = 0$. Hence $g(x) - x = 0$ or $g(x) = x$ and the proof is complete.

There is an amusing corollary to our answer to Question II which we will only outline, skipping over the somewhat tedious details. This result says that it is possible to slice any ham sandwhich consisting of a single (connected) piece of ham on a single (connected) piece of bread in such a way that both ham and bread are equally divided—with a *single* stroke of the knife! Unfortuately the result only says that one *can* do this, and provides no method as to how to do it.

The idea of the proof is as follows. One first proves that it is possible to slice bread and ham in half, but separately. One then proves that of the possible halving-cuts, at least one cut slices both pieces simultaneously in half. As it turns out the bread and ham need not even be overlapping. The general situation is indicated in Figure 8.8.

Here B and H are simply open connected sets in $\mathscr{R} \times \mathscr{R}$. For each x on the diameter pp' indicated in Figure 8.8, let $f(x) = R_x - L_x$, where R_x is the area of that part of H to the right of the cutting line perpendicular to pp' at x, and L_x is the area of the remaining region of H to the left of the cutting line. One wishes to show that there is an x for which $R_x = L_x$ or equivalently,

$$f(x) = R_x - L_x = 0\,,$$

for the line at this x then cuts H in half. It is apparent that

$$f(p) = 0 - L_p < 0$$

and

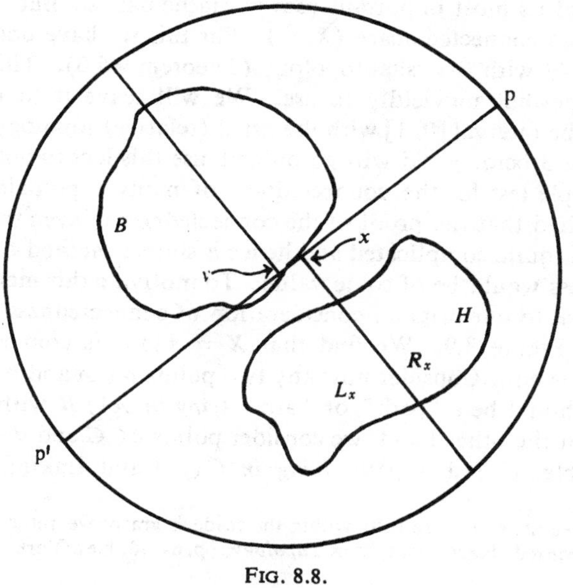

FIG. 8.8.

$$f(p') = R_{p'} - 0 > 0$$

so that $f(p) < 0 < f(p')$ and, hence—if it were known that f is continuous—there exists by Theorem 8.3.4 an x such that $f(x) = 0$, or $R_x = L_x$. The fact that f is indeed continuous follows from the rather believable inequality

$$|f(x) - f(y)| \leq 2r\, d(x, y)$$

where r is the radius of the indicated circle and $d(x, y)$ is the distance from x to y along the line pp'.

Similarly, one can show that along the line pp' one can find a point y from which the perpendicular at y cuts B in half. Moreover, as is intuitively acceptable, x and y are both unique; that is, along any diameter pp', there is only one point $x = x(p)$ and one point $y = y(p)$ for which the perpendiculars through $x(p)$ and $y(p)$ cut H and B in half, respectively.

The remainder of the argument is as follows. Let p vary around the circle C and for each p, let $g(p) = x(p) - y(p)$. The claim is that for some diameter with end point p, $g(p) = 0$. That is $x(p) = y(p)$ are the same point and, hence, the perpendicular through $x(p) = y(p)$ cuts *both* B and H in helf. Again, one shows that g is continuous and assumes at least one positive and at least one negative value and applies Definition 5.3.4 to yield a zero value. We remind the reader that this argument is far from a proof, being merely a suggested method of proof.[2]

We return now to some abstract mathematical results. The answer to Question II yields a method for showing that a topological space $(Y, \mathscr{V})$, no matter how wierd, is connected: Show that $(Y, \mathscr{V})$ is the range of some continuous function defined on a connected topological space $(X, \mathscr{U})$. This is one of its most important uses in mathematics. But this approach first requires a connected space $(X, \mathscr{U})$. For this we have only one certain candidate: $\mathscr{R}$ with the usual topology (Theorem 4.5.3). This topological space is somewhat unwieldly to use. We will leave it to the reader to prove that the interval $[0, 1]$ with the usual (relative) topology is connected (see Problem 3 below) and will go on and use this fact to obtain an operationally simple test for the connectedness of many topological spaces. It will be recalled that the proof of the connectedness of even the well known space $\mathscr{R}$ was quite complicated and hence a simple method of determining connectedness would be of some value. To motivate this method we go all the way back to our original consideration of connectedness itself.

Consider Figure 8.9. We find that $X = A \cup B$ is connected and that $Y = C \cup D$ is not. Consider now any two points $a \in A$ and $b \in B$. It seems that there should be a "path" or "arc," *lying in* $A \cup B$ with end points a and b. If on the other hand, we consider points $c \in C$ and $d \in D$ it does not seem possible to find a path, *lying in* $C \cup D$ and linking c to d. We

[2] For the details, which are fully within the reader's grasp, we refer to W. G. Chinn and N. E. Steenrod: *First Concepts in Topology*, pp. 64–70, New York: Random House, 1966.

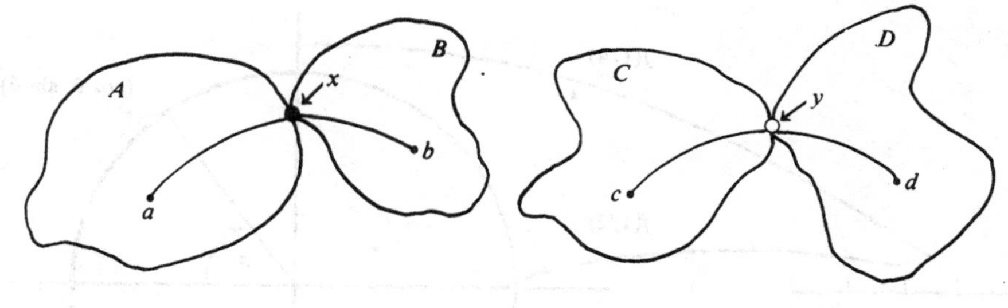

FIG. 8.9.

remarked in our earlier discussion (Section 4.5) of the definition of connectedness that this viewpoint might be a way of distinguishing connected from disconnected spaces, but that it involved a difficulty: that of defining, in a useful way, an arc or path joining two points. With the use of main theorem we can now make such a definition.

Consider the interval [0, 1] with the usual relative topology obtained from open intervals. An "arc" or "path," as we intuitively visualize it, is nothing more than a unit interval stretched, contracted and bent, *without tearing.* According to Theorem 8.3.3 the continuous image of the topological space [0, 1] in *any* topological space $(Y, \mathscr{V})$ is a connected subset of $(Y, \mathscr{V})$, given that [0, 1] is connected. This fact makes the following definition a useful one.

DEFINITION 8.3.6. Let $(X, \mathscr{U})$ be a topological space. If $a, b \in X$, and $f: [0, 1] \to X$ is a continuous function such that $f(0) = a$ and $f(1) = b$, then the range R_f of f is called an *arc* in X from a to b. If $a = f(0) = f(1) = b$, the arc is called a *closed curve.* If f is 1–1, the arc is called a *simple arc.*

Some examples of arcs in the plane are given in Figure 8.10, where (a) is an arc, (b) is a closed curve, (c) is a simple arc and one usually refers to (d) as a simple closed curve. That is to say, one believes that each of these figures can be shown to be the range of [0, 1] under a continuous

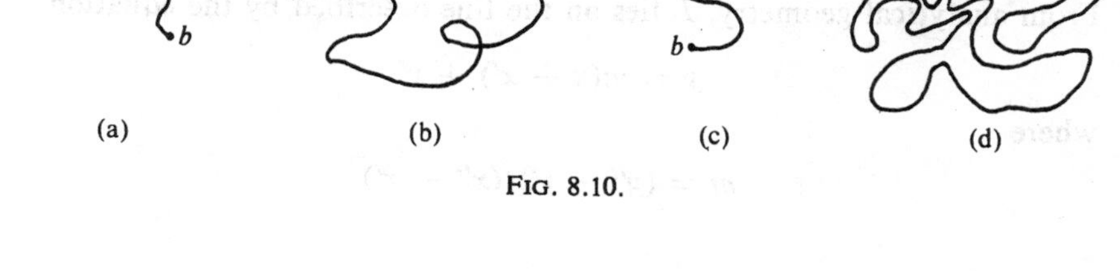

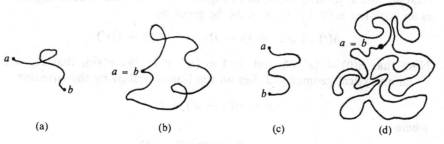

FIG. 8.10.

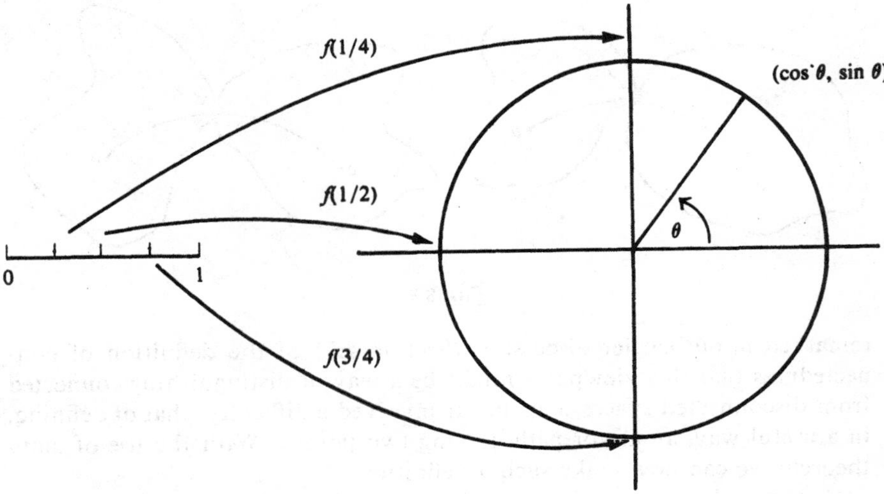

FIG. 8.11.

transformation. These arcs are however rather fancy, and a function that describes any one of them would be rather involved. Let us consider some examples of arcs given analytically.

EXAMPLES. (a) Consider a circle C of radius 1 centered at (0, 0) in $\mathscr{R} \times \mathscr{R}$, as in Figure 8.11. The well known trigonometric formulas $x = \cos\theta$, $y = \sin\theta$ imply that the formula $f(t) = (\cos 2\pi t, \sin 2\pi t)$ defines a function $f: [0, 1] \to C$. Morever, $f(0) = f(1) = (0, 1)$, $f(1/2) = (-1, 0)$, $f(1/4) = (0, 1)$, $f(3/4) = (0, -1)$, and in general f traces out C as t varies over $[0, 1]$. Given that the sine and cosine functions are continuous it follows that f is continuous. Hence C is an arc by Definition 8.3.6, and as in Figure 8.11(d) above, C would be called a simple closed curve.

(b) Similarly, the upper half C' of C lying above the x-axis is an arc from (1, 0) to (−1, 0) and is defined by the function $g: [0, 1] \to C'$ given by $g(t) = (\cos \pi t, \sin \pi t)$. Note that $g(0) = (1, 0)$ and $g(1) = (-1, 0)$.

(c) Consider now two points (x', y') and (x'', y'') in $\mathscr{R} \times \mathscr{R}$ and the straight line L joining them, as in Figure 8.12. As one should expect, L is an arc. Let $h: [0, 1] \to \mathscr{R} \times \mathscr{R}$ be given by

$$h(t) = (tx'' + (1 - t)x, ty'' + (1 - t)y').$$

Note that $h(0) = (x', y')$ and $h(1) = (x'', y'')$. We claim that $R_h = L$. From analytical geometry, L lies on the line described by the equation

$$y = m(x - x') + y'$$

where

$$m = (y'' - y')/(x'' - x')$$

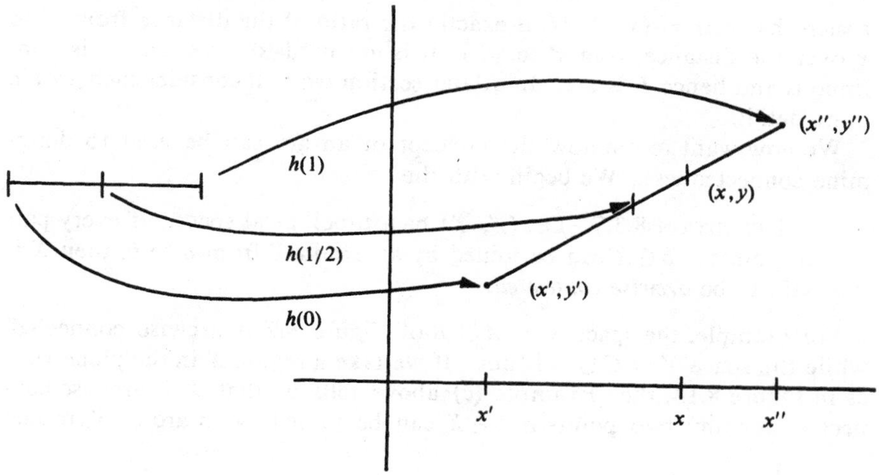

FIG. 8.12.

is the slope of the line L. If $(x, y) \in L$, then letting

$$t = \frac{(x' - x)}{(x' - x'')}$$

we have $t \in [0, 1]$. Morever,

$$x = tx'' + (1 - t)x'$$

and since

$$y = m(x - x') + y'$$

and

$$m = \frac{y - y''}{x - x''}$$

we also have

$$\begin{aligned} y &= \frac{y' - y''}{x' - x''}(x - x') + y \\ &= -t(y' - y'') + y' \\ &= ty'' + (1 - t)y' . \end{aligned}$$

Hence, $(x, y) = h(t)$ where

$$t = \frac{(x' - x)}{(x' - x'')} .$$

Thus $L \subset R_h$ and, moreover, it follows from ordinary algebraic manipulation that $R_h \subset L$ so that $R_h = L$. Hence, h describes the line L, $h(0) = (x', y')$ and $h(1) = (x'', y'')$. (Notice in fact that h is 1–1 and the number

t such that $h(t) = (x, y) \in L$ is exactly the ratio of the distance from x' to x over the distance from x' to x''.) It is not hard to show that h is continuous and hence L is arc. In a later section we will consider such arcs in more detail.

We now want to see how the concept of an arc can be used to determine connectedness. We begin with the

DEFINITION 8.3.7. Let $(X, \mathscr{U})$ be a topological space. If every pair of points $a, b \in X$ can be joined by an arc in X from a to b, then X is said to be *arcwise connected.*

For example, the space $X = A \cup B$ of Figure 8.9 is arcwise connected, while the space $Y = C \cup D$ is not. If we take a region X in the plane such as in Figure 8.13, then Example (c) above tells us that X is arcwise connected, for any two points $a, b \in X$ can be joined by an arc in X, in fact

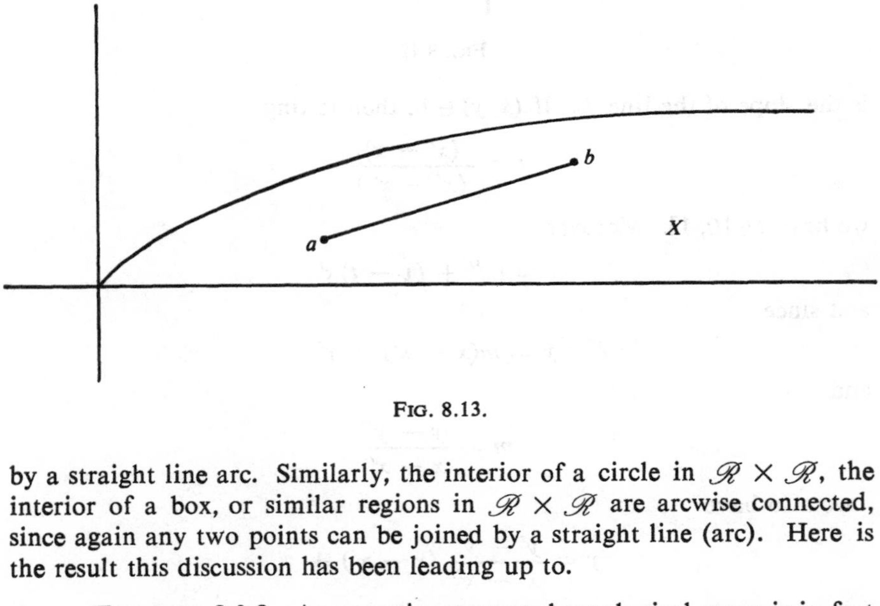

FIG. 8.13.

by a straight line arc. Similarly, the interior of a circle in $\mathscr{R} \times \mathscr{R}$, the interior of a box, or similar regions in $\mathscr{R} \times \mathscr{R}$ are arcwise connected, since again any two points can be joined by a straight line (arc). Here is the result this discussion has been leading up to.

THEOREM 8.3.8. Any arcwise connected topological space is in fact connected.

Proof. Suppose $(X, \mathscr{U})$ is a topological space which is arcwise connected but not connected. Then $X = A \cup B$, with $A \neq \Box$, $B \neq \Box$ and $A \cap \bar{B} = \bar{A} \cap B = \Box$. Let $a \in A$ and $b \in B$. Since X is arcwise connected, $\exists$ a continuous function $f: [0, 1] \to X$ such that $f(0) = a$ and $f(1) = b$. By Theorem 8.3.3, the range $R_f = f[0, 1]$ is a connected subspace of X. But by Theorem 4.6.4. we must have $f(X) \subset A$ or $f(X) \subset B$. Since $f(0) \in A$ and $f(1) \in B$ neither of these is possible and a contradiction results. Hence the proof.

Theorem 8.3.8, along with the (still yet to be proven) fact that [0, 1] is connected, gives a relatively easy way to prove that many common spaces are connected, helping to make the subject of connectedness a useful one. Before giving some problems and applications of the material of this section, we want to prove two general theorems which will make the problem of determining continuity of functions a whole lot easier. Just as we have just tied the property of connectedness to the connectedness of [0, 1], we would like to tie the property of continuity to continuity in $\mathscr{R}$. Recall that by Example (d), Section 8.2, continuity in $\mathscr{R}$ can be determined with the use of the limit theorems of the calculus, which we suppose the reader to be at least mildly familiar with.

Notice that each of the functions in the examples above were of the form $k(x) = (l(x), m(x))$ where $k\colon [0, 1] \to \mathscr{R} \times \mathscr{R}$ and $l, m\colon [0, 1] \to \mathscr{R}$. The use of limit concepts can be used to determine the continuity of l and m. But the reader has no limit concept to use to determine continuity for k, since $R_k \subset \mathscr{R} \times \mathscr{R}$. The following theorem is a help in this regard.

THEOREM 8.3.9. Let $(X, \mathscr{U})$ be any topological space, and let $h\colon X \to \mathscr{R}$ and $g\colon X \to \mathscr{R}$. Define $f\colon X \to \mathscr{R} \times \mathscr{R}$ by $f(x) = (h(x), g(x))$. If h and g are both continuous then so if f.

Proof. Let

$$f(x_0) = (h(x_0), g(x_0)) \in \mathscr{R} \times \mathscr{R}$$

and suppose

$$V = \{(u, v)\colon \sqrt{(u - h(x_0)^2) + (v - g(x_0)^2)} < \varepsilon\}$$

is a typical neighborhood of $(h(x_0), g(x_0))$. We claim $\exists$ a neighborhood U of x_0 such that $f(U) \subset V$.

Since h is continuous at x_0 and

$$W' = \left(h(x_0) - \frac{\varepsilon}{2}, h(x_0) + \frac{\varepsilon}{2}\right)$$

is an (open interval) neighborhood of $h(x_0) \in \mathscr{R}$, $\exists$ a neighborhood U' of x_0 such that $h(U') \subset W'$. Similarly, since g is continuous at x_0, $\exists$ a neighborhood U'' of x_0 such that

$$g(U'') \subset W'' = \left(g(x_0) - \frac{\varepsilon}{2}, g(x_0) + \frac{\varepsilon}{2}\right).$$

By the second axiom for a topological space $\exists$ a neighborhood U of x_0 such that $U \subset U' \cap U''$. We claim $f(U) \subset V$.

Let $x \in U$. Then $x \in U'$ and hence $h(x) \in W'$. Similarly, $x \in U''$ and $g(x) \in W''$. Hence

$$(h(x) - h(x_0))^2 + (g(x) - g(x_0))^2 < \left(\frac{\varepsilon}{2}\right)^2 + \left(\frac{\varepsilon}{2}\right)^2 = \frac{\varepsilon^2}{2} < \varepsilon^2 .$$

Thus

$$f(x) = (h(x), g(x)) \in V.$$

Hence the proof.

The next result we wish to prove concerns continuity and the composition $f \circ g$ of two functions f and g. Recall Example (a), where

$$f(t) = (\cos 2\pi t, \sin 2\pi t).$$

We can write

$$\cos 2\pi t = h(k(t)) = (h \circ k)(t)$$

where $h(x) = \cos x$ and $k(t) = 2\pi t$. One would like to be able to say that given that h is continuous and that k is continuous (both less complicated functions than $h \circ k$) then $h \circ k$ is continuous. This is the conclusion of

THEOREM 8.3.10. Let $(X, \mathscr{U})$, $(Y, \mathscr{V})$ and $(Z, \mathscr{W})$ be topological spaces. If $f: X \to Y$ and $g: Y \to Z$ are both continuous functions, then the function $g \circ f: X \to Z$ is also continuous.

Proof. Let W be an open set in Z. By Theorem 8.3.1 it suffices to show that $(g \circ f)^{-1}(W)$ is open in X.

It is easy to show that $(g \circ f)^{-1}(W) = f^{-1}(g^{-1}(W))$. Since g is continuous, $g^{-1}(W)$ is open in Y by Theorem 8.3.1. For the same reason, since f is continuous, $f^{-1}(g^{-1}(W))$ is open in X. Hence $(g \circ f)^{-1}(W)$ is open in X and hence $g \circ f$ is continuous.

Problems.

(1) Prove Theorem 8.3.1.

(2) Prove Theorem 8.3.2.

(3) Prove that the interval [0, 1] is a connected topological space in the relative topology. One way to do this is to first prove that (0, 1) is connected and then apply Theorem 4.6.6. To prove that (0, 1) is connected use the fact that $\mathscr{R}$ is connected (Theorem 4.5.2) and define a continuous function f from $\mathscr{R}$ onto (0, 1). Then use Theorem 8.3.2.

(4) Prove that any interval (a, b), $[a, b)$ and $[a, b]$ is connected using Problem 3.

(5) Find the fixed points of the mapping $f: [0, 1] \to [0, 1]$ given by $f(x) = (1 - x^4)^{1/2}$.

(6) Define a continuous function $f: (0, 1) \to (0, 1)$ which is onto but has no fixed points.

(7) Define a function $g: [0, 1) \to [0, 1)$ which is onto and continuous. and having a fixed point.

(8) Define a function $f: [0, 1] \to [0, 1]$ which is onto but has no fixed point.

(9) Knowing that a given function such as $f(x) = e^x$ takes on all

natural number values $n = 1, 2, \ldots$ (since $f(\log n) = e^{\log n} = n$) it is casually concluded that f takes on all values $y \geq n = 1$. Suppose $f: X \rightarrow [1, \infty)$ is a continuous function such that $\mathcal{N} \subset f(X)$. Give additional conditions which imply that $f(X) = [1, \infty)$; that is, conditions that imply that f is onto. Verify the validity of your choice of conditions and show by counter example that no weaker conditions on X will do.

(10) Given that the sine and cosine functions are continuous, Theorem 8.39, Problem 3 and Theorem 8.3.3 tell us that the circle C of radius 1 in the plane $\mathcal{R} \times \mathcal{R}$ is a connected topological space.

Let $g: C \rightarrow \mathcal{R}$ be any continuous function. Show that $\exists$ diametrically opposed points $x, y \in C$ such that $g(x) = g(y)$. *Hint*: for $x \in C$, let $x + \pi$ denote the point on C diametrically opposed (antipodal) to x. Let $f(x) = g(x + \pi) - g(x)$ measure the difference between g at x and g at the diametrically opposed point $x + \pi$. Then apply Theorem 8.3.4, paying due regard to questions of continuity.

(11) In Problem 10 suppose that C represents a circumference on the surface of this earth and that for $x \in C$, $g(x)$ is the temperature at x. What follows from Problem 10 concerning the temperature along a circumference of the earth.

(12) Show that $\mathcal{R} \times \mathcal{R}$ is connected by the following distinct methods.

(a) Let $C_x = \{(x, y): y \in \mathcal{R}\}$. Note that $\mathcal{R} \times \mathcal{R} = \bigcup_{x \in \mathcal{R}} C_x \cup D$ where $D = \{(x, 0): x \in \mathcal{R}\}$. Define $\phi: \mathcal{R} \rightarrow C_x$ by $\phi(y) = (x, y)$. Use Theorems 8.3.3 and 8.3.9 to see that C_x is connected and then use Theorem 4.6.5.

(b) Use Theorem 8.3.8 and Example (c) above.

(13) At the start of the discussion of connectedness we assumed for the sake of having examples that certain sets in $\mathcal{R} \times \mathcal{R}$ where connected. There is one set in particular that deserves more detailed study. This is the set

$$B = \left\{\left(x, \sin\frac{1}{x}\right): \ 0 < x \leq \frac{1}{\pi}\right\} \cup \left\{(0, y): \ -1 \leq y \leq 1\right\}$$

illustrated in Problem 8, Section 4.3. Show, in full detail, that B is a connected subset of $\mathcal{R} \times \mathcal{R}$ given that $h(x) = \sin 1/x$ is continuous on $(0, 1/\pi]$, as is $g(x) = x$.

(14) In $\mathcal{R} \times \mathcal{R}$ show that the following sets are connected.

(a) $A = \{(x, y): x^2 + y^2 < 1\}$.

(b) $B = \{(x, y): 0 \leq x \leq 1, 0 \leq y \leq 1\}$.

(c) $C = \{(x, y): y = ax + b\}$ where $a, b \in \mathcal{R}$ are fixed.

(15) Let $a, b, c \in X$, where X is a topological space with topology $\mathcal{U}$. Suppose $\exists$ arcs from a to b and from b to c described by continuous functions $f: [0, 1] \rightarrow X$ and $g: [0, 1] \rightarrow X$ such that

$$f(0) = a, \quad f(1) = b = g(0) \text{ and } g(1) = c.$$

Is there an arc in X from a to c? If so, define a function $h: [0, 1] \to X$ describing such an arc from a to c and show that it is continuous.

(16) Let X be a topological space. Define a relation R in X by: aRb iff $\exists$ an arc in X from a to b. Is R an equivalence relation?

(17) Which of the following subsets of $\mathscr{R} \times \mathscr{R}$ are connected and why?

$$D = \{(x, y): x \text{ or } y \text{ is a rational number}\}.$$

$$E = \{(x, y): x \text{ or } y \text{ is a rational number, but not both (are rational)}\}.$$

(18) The function $y = \sin 1/x$, $0 < x \leq 1/\pi$ is a great source of counter examples to what one would like to be true of continuity and connectedness. It is a continuous function and its graph $\{(x, \sin 1/x): 0 < x \leq 1/\pi\}$ is a connected subset of the plane according to Problem 13 above. According to Theorem 4.6.6, this graph along with the "limit line" $\{(0, y): -1 \leq y \leq 1\}$ is a connected set in $\mathscr{R} \times \mathscr{R}$. Now we will see a better reason for not using the concept of an arc in defining connectedness: interesting examples such as this one are left out of discussion. Recall that one always desires as inclusive a theory as is practical. Show that the set B in Problem 12, although connected, *cannot* be arcwise connected and hence that the converse of Theorem 8.3.8 does not hold. We will prove an amended converse to Theorem 8.3.8 in the sequel.

For a hint, suppose B is arcwise connected. Let $a = (1/\pi, 0)$ and $b = (0, 0)$. Then there would exist a continuous function $f: [0, 1] \to B$ such that $f(0) = a$ and $f(1) = b$. (Intuitively, the graph of f would have to follow the graph of $y = \sin 1/x$ in order to get continuously from a to b while remaining within the set B.) Let $L = \{(0, y): -1 \leq y \leq 1\} \subset B$. Let $C = B \backslash f^{-1}(b) = \{t: f(t) \notin L\}$. Since $a \notin L$, $0 \in C$ and hence $C \neq \square$. Show that C must be both open and closed, but that $1 \notin C$, a contradiction to the connectedness of $[0, 1]$.

(19) Consider the unit square $S = \{(x, y): 0 \leq x \leq 1, 0 \leq y \leq 1\}$ in $\mathscr{R} \times \mathscr{R}$ and suppose $f: S \to S$ is continuous and onto. Does f have a fixed point?

4. A SOURCE OF APPLICATION—COMPACTNESS

Connectedness is an important and interesting—and intuitive—topological property. From the viewpoint of applications within and without mathematics, there is another property of topological spaces. not at all related to connectedness, which proves to be of much more importance. It is the source of many classical applications of mathematics and arises from the following considerations. We can well handle it within the systems of thought thus far developed.

Suppose $(X, \mathscr{U})$ is any connected topological space and $f: X \to \mathscr{R}$ is a continuous function. We have see (Theorem 8.3.4 above) that the range $R_f = f(X)$ must be an interval in $\mathscr{R}$. Now there are several kinds of intervals in $\mathscr{R}$: (a, b), (a, ∞), $[a, b)$, $[a, b]$. Of these, the nicest is an interval of the form $[a, b]$. It is easy to find functions defined on connected topological spaces whose ranges are not closed intervals; for example if $X = [0, 1)$ and $f(x) = x$, then $R_f = f(x) = [0, 1)$ is not a closed interval, although f is continuous and X is connected. In this section we will introduce a fundamental condition on X such that $f(X)$ is a closed interval $[a, b]$ when X is connected.

The interest in this problem is not solely a matter of mathematical aesthetics, but concerns much of practical interest. Many applications of mathematics concern problems of maximization (or minimization). Mathematics is almost uniquely organized to measure extremes—the largest, the smallest; by nature of its generality, such concepts are indeed among the easiest to handle. Suppose one has a set X, a function $f: X \to \mathscr{R}$ and it is known that $f(X) = [a, b]$. Then, for any $x \in X$, $f(x) \in [a, b]$ and hence $a \leq f(x) \leq b$. Hence, b is the maximum of *all* the values of f and a is the minimum. Thus to know that $f(X) = [a, b]$ is to know that f has a maximum value b and minimum value a and assumes every value c in between a and b as well. This is sometimes a useful thing to know. We will now consider those conditions on X that have been found appropriate to the problem of the existence of maxima and minima.

Suppose $(X, \mathscr{U})$ is a topological space and $f: X \to \mathscr{R}$. It is possible that $f(X) = \mathscr{R}$, in which case f *no* maximum or minimum value. If f is to have a maximum *at all*, it must first be necessary that some number be larger than all the values of f; i.e., that $\exists$ a number M such that $f(x) < M$ for all $x \in X$. Otherwise, f would have a value larger than every given number and hence could have no maximum value.

Let us approach this matter from the opposite direction. If one did have a number M such that $f(x) < M$ for all $x \in X$, then one would have $X \subset f^{-1}(-\infty, M)$. Note that, if f were continuous, then since $(-\infty, M)$ is an open set, $f^{-1}(-\infty, M)$ would be open in X. In trying to decide whether or not such a number M exists at all, a reasonable approach would be to try various candidates for M. Try $M = 1$, then $M = 2$, then $M = 3$, and so on until one is found for which $f^{-1}(-\infty, M) \supset X$. In other words, consider the open subsets $f^{-1}(-\infty, n)$ of X where $n = 1, 2, 3, \ldots$. If, say, $X \subset f^{-1}(-\infty, 10)$, then $f(x) < 10$ for all $x \in X$ and hence there would be a number larger than all the values of f. But something a little weaker would do just as well as it is this weaker condition we wish to look at.

Consider the collection of sets $\mathscr{A} = \{f^{-1}(-\infty, n): n = 1, 2, 3, \ldots\}$. If $x \in X$, then $f(x) \in \mathscr{R}$ and by the Archimedian property (Section 2.7) of

$\mathscr{R}$, $\exists n \in \mathscr{N}$ such that $f(x) < n$. Hence $x \in f^{-1}(-\infty, n) \in \mathscr{A}$. This means that every $x \in X$ is in some set $A \in \mathscr{A}$—that is, $X \subset \bigcup_{A \in \mathscr{A}} \mathscr{A}$. The collection $\mathscr{A}$ is then called an (open) *cover* of X. Now $\mathscr{A}$ contains infinitely many sets, one for each $n \in \mathscr{N}$. Suppose that X were peculiar enough (or whatever adjective you might like) that only *finitely* many sets in $\mathscr{A}$ would "cover" X just as well. In other words, suppose there is some k number of sets $A_j = f^{-1}(-\infty, n_j) \in \mathscr{A}$, $j = 1, 2, \ldots, k$ such that $X \subset \bigcup_{i=1}^{k} A_j$. Consider then the natural numbers $n_1, n_2, \ldots, n_k$. Among these there is a largest, call it M; hence $n_i \leq M$ for $i = 1, 2, \ldots, k$. Now then, if $x \in X$, then $x \in A_i$ for some i. Hence $x \in f^{-1}(-\infty, n_i)$, thus $f(x) < n_i$ and hence $f(x) < M$ since $n_i \leq M$. Thus there is a single number M such that $f(x) < M$ for all $x \in X$.

The crux of this argument is that one only has to choose a maximum M from among *finitely* many numbers $n_1, n_2, \ldots, n_k$ such that for all $x \in X$, $f(x)$ is smaller than some n_i, and hence smaller than the single number M no matter what x is. If one had to choose from infinitely many numbers n_i, such a maximum would not necessarily be found, since, for example among the numbers

$$n_i = 2, n_2 = 2^2, \ldots, n_i = 2^i, \ldots,$$

there is no largest. We now give such "peculiar" spaces, X, the usual and universal title, and make the necessary formal definitions. It should be recalled that the sets in $\mathscr{A}$ were open sets.

DEFINITION 8.4.1. Let $(X, \mathscr{U})$ be topological space and let $Y \subset X$. A collection $\mathscr{A}$ of open subsets of X is called an *open cover* of Y if every $x \in Y$ belongs to at least one set $A \in \mathscr{A}$; equivalently, $Y \subset \bigcup_{A \in \mathscr{A}} A$. The subset Y of X is called *compact* if it has the (peculiar?) property that given *any* open cover $\mathscr{A}$ of Y $\exists$ finitely many sets $A_1, A_2, \ldots, A_n \in \mathscr{A}$ that themselves cover Y; that is, that $Y \subset \bigcup_{i=1}^{n} A_i$. If X itself is compact, then X is called a *compact topological space*.

It is much easier to find examples of non-compact topological spaces than to prove that a particular set is compact. The set $\mathscr{R}$ is not compact, for $\mathscr{A} = \{(-n, n): n \in \mathscr{N}\}$ is an open cover of $\mathscr{R}$ (since every $x \in (-n, n)$ for some choice of $n \in \mathscr{N}$), but no finite number of sets in $\mathscr{A}$ covers all of $\mathscr{R}$. Thus a connected set, such as $\mathscr{R}$, need not be compact. Any set X consisting of (say) two distinct points is trivially compact. Hence a compact set need not be connected and this is among our first important points: compactness is a distinctly different (and largely non-intuitive) topological concept from that of connectedness. For this reason we will now try to relate compactness to something familiar, but first a few problems.

Problems.

(1) Let $X = \mathscr{R}$, $Y = (0, 1)$ and let $\mathscr{A} = \{(0, 1 - 1/n): n \in \mathscr{N}\}$. Show

that $\mathscr{A}$ is an open cover of Y but no finite subcollection of $\mathscr{A}$ covers Y.

(2) All examples of open covers to this point have been countable. Here is an uncountable one. Let $X = (-1, 1)$ and let $\mathscr{A} = \{(-x, x): 0 < x < 1\}$. Show that $\mathscr{A}$ is an open cover of X for which no finite subcollection covers X.

(3) Can you find an open cover of $Y = [0, 1)$ in $\mathscr{R}$ for which no finite subcollection covers Y? What about $[0, 1]$?

(4) Let $Y = \{0, 1, 1/2, 1/3, 1/4, \ldots\}$. Show that Y is a compact subset of $\mathscr{R}$.

The remainder of this section is dedicated to proving a single result: if X is a compact and connected topological space and $f: X \to \mathscr{R}$ is continuous, then $f(X)$ is a closed interval $[a, b]$. We have seen that connectedness implies that $f(X)$ is an interval. We will now show that compactness implies that this interval must be closed and bounded. Our approach to this problem is to prove that

(a) There is a compact topological space, namely the interval $[0, 1]$. (Hence so is any closed interval by virtue of (b) below.)

(b) Any continuous function defined on any compact topological space has a compact range; hence compactness is a topological property.

(c) Any compact subset of $\mathscr{R}$ is closed and bounded and conversely.

We begin with (a), although (b) is much easier to prove. Although compactness is not related to connectedness we will use the connectedness of $[0, 1]$ to prove that $[0, 1]$ is compact, since we do know that $[0, 1]$ is connected.

THEOREM 8.4.3. $[0, 1]$ is a compact subset of $\mathscr{R}$.

Proof. Let $\mathscr{A}$ be an open cover of $[0, 1]$. We must show that there exists finitely many sets $A_1, \ldots, A_n \in \mathscr{A}$ such that $[0, 1] \subset \bigcup_{i=1}^{n} A_i$.

Let $T = \{x \in [0, 1]: \exists$ finitely many sets in $\mathscr{A}$ that cover $[0, x]\}$. We wish to show that $1 \in T$. (At this point it is not even evident that T is anything more than an empty set.)

We claim that, $0 \in T$. For, since $\mathscr{A}$ covers $[0, 1]$, $\exists$ a set $A \in \mathscr{A}$ such that $0 \in A$. Hence $[0, 0] = \{0\} \subset A$ and *one* set $A \in \mathscr{A}$ covers $[0, 0]$.

Thus $T \neq \square$. We claim $T = [0, 1]$. Since $[0, 1]$ is connected it suffices to show that T is both open and closed in the relative topology of $[0, 1]$. We first show that T is open.

Let $x \in T$. Then $\exists$ finitely many sets $A_1, A_2, \ldots, A_p$ such that $[0, x] \subset \bigcup_{i=1}^{p} A_i$. One of these, say A_j, contains x. Since $x \in A_j$ and A_j is open, $\exists$ a neighborhood (a, b) of x such that $x \in (a, b) \subset A_j$. Since $(a, b) \subset A_j$ and $[0, x] \subset \bigcup_{i=1}^{p} A_i$, then $[0, b) \subset \bigcup_{i=1}^{p} A_i$. If $t \in (a, b) \cap [0, 1]$, then $[0, t] \subset [0, b) \subset \bigcup_{i=1}^{p} A_i$ and hence the finite number of sets $A_1, \ldots, A_p$ covers $[0, t]$. Hence $t \in T$. Since $t \in (a, b) \cap [0, 1]$ was arbitrary, we have $x \in (a, b) \cap$

$[0, 1] \subset T$ and hence that T contains a relative neighborhood of any one of its points x. Hence T is open.

We now show that T is closed. Let $x \in \bar{T}$. We claim that $x \in T$. Hence we must show that $\exists$ finitely many sets in $\mathscr{A}$ which cover $[0, x]$. Since $\mathscr{A}$ covers all of $[0, 1]$ and $x \in [0\ 1]$, there exists an $A \in \mathscr{A}$ suct that $x \in A$. Since A is open $\exists$ a neighborhood (a, b) of x such that $x \in (a, b) \subset A$. Since $x \in \bar{T}$, $\exists y \in T \cap (a, b)$. Since $y \in T$, $\exists$ finitely many sets (say) $A_1, \ldots, A_n \in \mathscr{A}$ such that $[0, y] \subset \bigcup_{i=1}^{n} A_i$. Let $A_{n+1} = A$. Then $x, y \in (a, b) \subset A$ and if $s \in [0, x]$ then either $s \in [0, y] \subset \bigcup_{i=1}^{n} A_i$ or $s \in (y, x] \subset (a, b) \subset A_{n+1}$. In either case, $s \in [0, x]$ implies $s \in \bigcup_{i=1}^{n+1} A_i$. Hence $A_1, \ldots, A_n, A_{n+1} = A$ is a finite cover of $[0, x]$. Thus $x \in T$. Hence T is closed.

Since $[0, 1]$ is connected and $T \neq \square$ is open and closed, then $T = [0, 1]$. Hence $1 \in T$ and the proof is complete,

The argument above is a somewhat indirect one. Compactness is difficult to handle and unlike connectedness may require some ingenuity even for dealing with familiar sets. But *like connectedness*, compactness is preserved under countinuous transformation—under non-tearing distortion. With Theorem 8.4.3 and the next result we will be able to assert that any closed interval $[a, b]$ is compact.

THEOREM 8.4.4. Let $(X, \mathscr{U})$ be a compact topological space and suppose $(Z, \mathscr{W})$ is a topological space and $f: X \to Z$ is continuous. Then $f(X)$ is a compact subset of Z.

Proof. Let $\mathscr{A}$ be any open cover of $f(X)$ in Z. Let $\mathscr{B} = \{f^{-1}(A): A \in \mathscr{A}\}$; $\mathscr{B}$ is a collection of subsets of X. Each set in $\mathscr{B}$ is open by Theorem 8.3.1, because each set $A \in \mathscr{A}$ is open and f is continuous. If $x \in X$, then $f(x) \in f(X)$ and because $\mathscr{A}$ is an open cover of $f(X)$ $\exists A \in \mathscr{A} \ni f(x) \in A$. Hence $x \in f^{-1}(A) \in \mathscr{B}$. Thus $X \subset \bigcup_{B \in \mathscr{B}} B$. Since X is assumed to be compact, $\exists$ finitely many sets $B_1 = f^{-1}(A_1), \ldots, B_n = f^{-1}(A_n)$ in $\mathscr{B}$ such that $X \subset \bigcup_{i=1}^{n} B_i$. Hence, if $f(x) \in f(X)$, then $x \in B_i$ for some i, whence $f(x) \in A_i$. Thus $f(X) \subset \bigcup_{i=1}^{n} A_i$. By Definition 8.4.1, $f(X)$ is compact.

COROLLARY 8.4.5. The interval $[a, b]$ is a compact subset of $\mathscr{R}$ for any $a, b \in \mathscr{R}$ with $a < b$.

Proof. By Theorem 8.4.3, $[0, 1]$ is compact and it is easy to see that $f(x) = (b - a)x + a$ is continuous. Since $f: [0, 1] \to [a, b]$ and f is onto, the conclusion follows from Theorem 8.4.4.

We will now show that any compact set in $\mathscr{R}$ must be closed and bounded. This will substantially complete the theoretical work of this section.

THEOREM 8.4.6. Let C be a compact subset of $\mathscr{R}$. Then
(a) C is closed

(b) There exists $M \in \mathscr{R}$ such that $C \subset [-M, M]$.

Proof. Since (b) is easiest we will prove it first. Let

$$\mathscr{A} = \{(-n, n) \colon n \in \mathscr{N}\}.$$

Since $x \in C$ is a real number, and so is $|x|$, there is an $n \in \mathscr{N}$ such that $|x| < n$. Hence $x \in (-n, n) \in \mathscr{A}$. Thus $\mathscr{A}$ is an open cover of C. Since C is a compact set $\exists$ finitely many sets

$$(-n_1, n_1), (-n_2, n_2), \ldots, (-n_k, n_k) \in \mathscr{A}$$

such that

$$C \subset \bigcup_{i=1}^{k} (-n_i, n_i).$$

If M is the largest of the numbers $n_1, n_2, \ldots, n_k$, then $C \subset (-M, M)$ and (b) is proven.

For (a), suppose that $x \notin C$. We will find a neighborhood of x which lies in $\mathscr{R} \backslash C$ proving that $\mathscr{R} \backslash C$ is open and hence that C itself is closed. Let

$$\mathscr{A} = \left\{\left(-\infty, x - \frac{1}{n}\right) \cup \left(x + \frac{1}{n}, \infty\right) \colon n \in \mathscr{N}\right\}.$$

Each set in $\mathscr{A}$ is open. We claim $\mathscr{A}$ covers C. If $y \in C$, then since $x \notin C$, $x \neq y$. Hence $|x - y| > 0$ and hence $\exists n \in \mathscr{N}$ such that $1/n < |x - y|$. Thus

$$y \notin \left[x - \frac{1}{2n}, x + \frac{1}{2n}\right]$$

and hence

$$y \in \left(-\infty, x - \frac{1}{2n}\right) \cup \left(x + \frac{1}{2n}, \infty\right) \in \mathscr{A}.$$

Thus any $y \in C$ is in some set in $\mathscr{A}$.

Since C is compact $\exists$ finitely many sets $A_1, \ldots, A_k \in \mathscr{A}$ such that

$$C \subset \bigcup_{i=1}^{k} A_i.$$

Let

$$A_i = \left(-\infty, x - \frac{1}{n_i}\right) \cup \left(x + \frac{1}{n_i}, \infty\right)$$

and let m be the largest of the numbers n_i. We claim $(x - 1/m, x + 1/m)$ is a neighborhood of x contained in $\mathscr{R} \backslash C$.

If $t \in (x - 1/m, x + 1/m)$, then

$$|x - t| < \frac{1}{m} \leq \frac{1}{n_i}$$

for each $i = 1, 2, \ldots, k$. Hence

$$t \in \left[x - \frac{1}{n_i}, x + \frac{1}{n_i}\right] = \mathscr{R} \backslash A_i$$

for each $i = 1, 2, \ldots, k$ and thus

$$t \notin \bigcup_{i=1}^{k} A_i \,.$$

Since

$$C \subset \bigcup_{i=1}^{k} A_i \,,$$

then $t \notin C$. Hence C is closed.

We can now prove the principal result of this section.

THEOREM 8.4.7. Let $(X, \mathscr{U})$ be any compact, connected topological space and suppose that $f: X \to \mathscr{R}$ is continuous. Then $\exists$ numbers $a, b \in \mathscr{R}$ such that $f(X) = [a, b]$. That is

(a) f is a bounded function, all of its values bounded above by b and below by a.

(b) f has a maximum value b and a minimum value a and every value between a and b is in the range R_f of f.

Proof. Since X is connected, $f(X)$ is an interval in $\mathscr{R}$, according to Theorem 8.3.4. Since X is compact, $f(X)$ is a compact subset of $\mathscr{R}$ according to Theorem 8.4.4. From Theorem 8,4.6, $\exists M \in \mathscr{R}$ such that

$$f(X) \subset [-M, M] \,.$$

Let $Y = f(X)$. Since $Y \subset [-M, M]$; Y is a bounded set of real numbers and by the completeness axiom has a supremum $b \in \mathscr{R}$. Hence $y \in Y$ implies $y \leq b$. The set $Z = \{-y \colon y \in Y\} \subset [-M, M]$ and hence also has a supremum, call it a' in $\mathscr{R}$. If $a = -a'$ it follows that $a \leq y \; \forall y \in Y$. Hence, $Y \subset [a, b]$. Moreover, Y is closed by Theorem 8.4.6(a) and, by Theorem 4.4.3, $b \in Y$. For the same reason, $a' \in Z$ and hence $a \in Y$. Since $Y \subset [a, b]$ and $a, b \in Y$ and Y is an interval, $Y = [a, b]$ and the proof is complete.

We close this section with the converse of Theorem 8.4.6, hence obtaining a useful characterization of the compact subsets of $\mathscr{R}$ in terms of the (relatively simple) topological property of being closed and the arithmetical property of being bounded. That is,

THEOREM 8.4.8. Suppose that $C \subset \mathscr{R}$ such that (a) C is closed and (b) $\exists M \in \mathscr{R}$ such that $C \subset [-M, M]$. Then C is compact.

Proof. Let $\mathscr{A}$ be an open cover of C. We must show there is a finite subcollection $A_1, \ldots, A_n \in \mathscr{A}$ such that $C \subset \bigcup_{i=1}^n A_i$.

To do this we extend $\mathscr{A}$ to a cover of the compact set $[-M, M]$ which contains C by virtue of (b). Let $\mathscr{B} = \mathscr{A} \cup \{\mathscr{R} \backslash C\}$. Since C is closed, $\mathscr{R} \backslash C$ is open and hence $\mathscr{B}$ is a collection of open sets. We claim that $\mathscr{B}$ is a cover of $[-M, M]$.

Let $x \in [-M, M]$. Either $x \in C$ or $x \notin C$. If $x \notin C$, then $x \in \mathscr{R} \backslash C \in \mathscr{B}$. If $x \in C$, then since $\mathscr{A}$ is a cover of C $\exists A \in \mathscr{A} \subset \mathscr{B}$ such that $x \in A$. Hence $[-M, M] \subset \bigcup_{B \in \mathscr{B}} B$.

By Corollary 8.4.5, $[-M, M]$ is compact. Hence, since $\mathscr{B}$ is an open cover of $[-M, M]$, $\exists$ finitely many sets $B_1, \ldots, B_k \in \mathscr{B}$ such that

$$[-M, M] \subset \bigcup_{k=1}^{k} B_k .$$

Now $B_i \in \mathscr{B}$ implies $B_i \in \mathscr{A}$ or $B_i = \mathscr{R} \backslash C$. If $\mathscr{R} \backslash C$ is one of the sets B_i, remove it and let $A_1, A_2, \ldots, A_{k-1}$ denote the remaining sets B_j, each of which is then in the original collection $\mathscr{A}$. Then $C \subset \bigcup_{j=1}^{k-1} A_j$ for $x \in C$ implies $x \in [-M, M]$ which in turn implies $x \in B_j$ for some j. But since $x \in C$ then $x \notin \mathscr{R} \backslash C$ and hence x is in one of the remaining sets B_j. That is, $x \in A_i$ for some $i = 1, 2, \ldots, k-1$ and hence $C \subset \bigcup_{j=1}^n A_j$.

In the other case, where $\mathscr{R} \backslash C$ is not one of the sets B_i, one then has that each set $B_i \in \mathscr{A}$ and $C \subset [-M, M] \subset \bigcup_{j=1}^k B_j$ and hence that $B_1, \ldots, B_k$ is a finite subcollection of sets in $\mathscr{A}$ which covers C.

Thus C is compact and the proof is complete.

This completes the study of compactness in $\mathscr{R}$. In the next section we shall see some further examples of compact sets. The notion of a compact set is not easy to assimilate, but by virtue of Theorems 8.4.4 and 8.4.7 it is a vital one in advanced topology and analysis. The problems that follow provide further examples of compact sets.

Problems.

(1) By virtue of Theorem 8.4.6 the set $C = [0, 1)$ is not compact. Define a function $f: C \to \mathscr{R}$ having no maximum value. Thus the hypothesis of compactness in Theorem 8.4.7 is essential to the conclusion.

(2) Let $C \subset \mathscr{N} \subset \mathscr{R}$. Show that C is compact iff C is finite. This is why we have not concerned ourselves with subsets of $\mathscr{N}$ in discussing compactness, there just aren't any interesting compact subsets of $\mathscr{N}$. On the other hand, show that the space X of Problem 8, Section 4.1 is Compact.

(3) Which of the following subsets of $\mathscr{R}$ is compact?

 (a) $A = \{x \in [0, 1]: x \text{ rational}\}$.

 (b) $B = \{\sin 1/x: x \in (0, 1/\pi]\}$.

 (c) $C = \mathscr{R}$.

(d) $D = \{0\} \cup \{(-1)^n/n\colon n \in \mathscr{N}\}$.

(4) Prove that the circle $C = \{(x, y)\colon x^2 + y^2 = 1\}$ is compact in $\mathscr{R} \times \mathscr{R}$.

(5) Prove that an arc in any topological space is compact.

(6) Can you define a continuous function f from $[0, 1]$ onto $\mathscr{R}$? What about from $(0, 1)$ onto $\mathscr{R}$?

(7) Can you define a continuous function from $(0, 1)$ onto $[0, 1]$? What about a continuous function from $[0, 1]$ onto $(0, 1)$?

(8) In a previous problem it was claimed the sequence $1, 1/2, 1/3, \ldots, 1/n, \ldots$ along with the limit point 0 is compact in $\mathscr{R}$. This also follows quickly from Theorem 8.4.8 since the set $C = \{0, 1, 1/2, 1/3, \ldots\}$ is closed and $C \subset [-1, 1]$. Let $a_1, a_2, \ldots, a_n, \ldots$, be any sequence of real numbers and suppose there is a real number a such that $\lim_{k \to \infty} a_k = a$ (Problem 17, Section 8.3). Let $C = \{a, a_1, a_2, \ldots\}$. Show that C is compact using Theorem 8.4.8. Then show C is compact by proving that any open cover of C has a finite subcover.

(9) Consider the collection of sets $B_n = [1 - 1/n, 1) \subset [0, 1)$. These sets are closed in the relative topology on $(0, 1)$ and $\bigcap_{n=1}^{\infty} B_n = \square$. At times one would like to know whether or not an intersection of this kind is empty or not, and here compactness comes in. Suppose C is a compact subset of $\mathscr{R}$ and $B_n \subset C$ is a closed set in C for each $n \in \mathscr{N}$. Suppose further that $B_1 \supset B_2 \supset \ldots \supset B_n \supset \ldots$ and there each $B_n \neq \square$. Show that $\bigcap_{n=1}^{\infty} B_n \neq \square$ by considering the compactness of C and the collection of open sets $\mathscr{A} = \{\mathscr{R} \backslash B_n\colon n = 1, 2, \ldots\}$ and recalling De Morgan's laws (Theorem 3.6.3).

(10) Let C be a denumerable subset of $\mathscr{R}$ and suppose that $C \subset [-M, M]$ for some $M \in \mathscr{R}$. Since C is denumerable $\exists$ a 1-1 onto function $f\colon \mathscr{N} \to C$. Let $B_n = \overline{\{f(k)\colon k \geq n\}}$. Then $B_n \subset [-M, M]$ and according to Problem 9, $B = \bigcap_{n=1}^{\infty} B_n \neq \square$. Let $b \in B$. Prove that any neighborhood U of b contains denumerably many points of C. This result is known as the ***Bolzano-Weierstrass theorem***: any bounded infinite subset C of $\mathscr{R}$ has a limit point such that any neighborhood of that limit point contains infinitely many members of C.

(11) Show that a closed subset Y of a compact space X is itself compact. If X has the Hausdorff property: $\forall x, y \in X$ with $x \neq y$ $\exists$ neighborhoods U and $V \ni x \in U, y \in V$ and $U \cap V = \square$, show that any compact subset of X is closed. For a hint, consider the proof of Theorem 8.4.6(a) and assuming that $C \subset X$ is compact but not closed, invent an open cover of C for which no finite subcollection covers C.

(12) Let X be a compact topological space, Y a topological space and $f\colon X \to Y$ a continuous function which is 1-1 and onto. If Y has the Hausdorff property (see Problem (11) show that f^{-1} is con-

tinuous from Y to X using Theorem 8.3.1.

(13) Given that $f(x) = e^x$ is continuous and 1–1, prove that $f^{-1}(y) = \log y$, $y > 0$ is continuous. *Hint*: given $y_0 = e^{x_0}$, consider the function f restricted to the compact set $[x_0 - 1, x_0 + 1]$ and use Problem 12.

(14) Refer to Problem 10 again. Here is another method of proving the same conclusion that $\exists$ a number $b \in [-M, M]$ such that any neighborhood of b contains denumberably many points of C. This statement is equivalent to proving that if U is a neighborhood of b and $m \in \mathcal{N}$ then $\exists n \geq m$ such that $f(n) \in U$. Hence suppose this is not true for any number $b \in [-M, M]$. Obtain from this supposition an open cover of $[-M, M]$ which can have no finite subcover as follows: if no such b exists then $\forall b \in [-M, M]$, $\exists$ a neighborhood U_b of b and a number m_b such that $f(n) \notin U_b \forall n \geq m_b$. Obtain from this a contradiction.

(15) Show that the Cantor set C of Problem 8, Section 4.4, is compact. Hence compact sets need not be "nice" sets in the sense that connected sets are "nice" sets.

5. EUCLIDIAN *n*-DIMENSIONAL SPACE

You think of yourself as residing in a three-dimensional world. The reason no doubt is that you have found no reason to consider it otherwise. Scientists have found reason, from inside the atom all the way into the outer reaches of the universe, to consider events as points in a space of dimension greater than three. The source of this viewpoint is simple enough. If more than three measurements are required to precisely fix an event, thought of as a single entity, then that event can be most usefully viewed in the context of a space of dimension greater than three.

For example, consider the problem faced by a small city whose economy is supported by five manufacturing plants $P_1, P_2, \ldots, P_5$, but finds its air polluted by all of these and its local recreational waters slowly putrefying from their industrial waste. Its citizens, the quantity in their lives having long been on the increase, begin to wonder about its quality, as fathers can no longer take their sons fishing, mothers find their shrubs and flowers slowly fading and to attempt any outdoor activity on a warm and humid night is to inhale heavy quantities of noxious gases. So after 30 years of the same old thing, the citizens elect a really new local government pledged to do something about the matter. But people do have to eat, and there lies the rub, the city needs the employment provided by the plants. The new establishment decides that the way to proceed is to tax each plant according to the volume of harmful pollutants each belches forth and to use the revenue to improve recreational and educational facilities and thus provide some other employment for the townspeople as a result. It is found that to tax plant P_i by a rate t_i is to obtain a tax revenue of $a_i t_i$ but also a decrease in employment at P_i (as the plant managers cut production and raise prices to maximize net return) and loss in city income of $-b_i t_i$. It is further found that if the tax rate t_i is greater than a number u_i the plant will cease production entirely, but if smaller than a number v_i no appreciable lessening of pollution will occur. In other words, a reasonable tax range on P_i is defined by an interval $[v_i, u_i]$ from which the tax rate t_i can be chosen.

The progressive new government enlists thet aid of a like-mined mathematics student who has just completed this section of this book (!) who tells them that (from his standpoint) the problem is simple. Simply *maximize* the function $f: S \to \mathscr{R}$ given by

$$f(t_1, t_2, t_3, t_4, t_5) = (a_1 - b_2)t_1 + (a_2 - b_2)t_2 + (a_3 - b_3)t_3 + (a_4 - b_4)t_4 + (a_5 - b_5)t_5$$

where $(t_1, t_2, t_3, t_4, t_5)$ ranges over the set

$$S = [v_1, u_1] \times [v_2, u_2] \times [v_3, u_3] \times [v_4, u_4] \times [v_5, u_5]$$

for this function measures the *total* city income from a given choice of tax rates t_1, t_2, t_3, t_4, t_5. To this student the domain S of this function is a subset of a five-dimensional space, and whether or not he is right in believing that maximization is the desired result, we will point the development of these concluding sections to the mathematical solution of mathematical problems of this type.[3] In particular, we will try to completely describe the range of values of such a function on sets, such as S, which can best be thought of as subsets of a space of dimension greater than three.

Thus, in this section we wish to consider what is commonly called Euclidian n-dimensional space, a straight generalization of the plane $\mathscr{R} \times \mathscr{R}$ or of three-space $\mathscr{R} \times \mathscr{R} \times \mathscr{R}$. A point in three space is exactly specified by three numbers, commonly called its coordinates. This is presumed to be a familiar notion. Without, as in earlier sections, going into the question of mathematical existence, we begin with the following

DEFINITION 8.5.1. We denote by E^n the set of all n-tuples $x = (x_1, x_2, \ldots, x_n)$, where each $x_i \in \mathscr{R}$ and call the set E_n *n-dimensional space*, or simply *n-space*. The number x_i is called the i-th coordinate of the point $x = (x_1, x_2, \ldots, x_n) \in E^n$. The points $x = (x_1, x_2, \ldots, x_n)$ and $y = (y_1, y_2, \ldots, {}_n y)$ in E^n are equal iff $x_k = y_k$ for every $k = 1, 2 \ldots, n$. If $x = (x_1, \ldots, x_n)$ and $y = (y_1, y_2, \ldots, y_n)$ are in E^n, we define the distance $d(x, y)$ between x and y to be the number

[3] We do not want to mislead the reader here. The problem of finding the specific values a_i, b_i used to define such a function in such a problem is of tremendous difficulty. We will show only how to use such a function, how to discover its maximum once it is known.

$$d(x, y) = [\sum_{i=1}^{n} (x_i - y_i)^2]^{1/2}$$
$$= [(x_1 - y_1)^2 + (x_2 - y_2)^2 + \ldots + (x_n - y_n)^2]^{1/2}.$$

The set E^n with distance d is called Euclidian n-space.

EXAMPLES. (a) When $n = 1$, $E^1 = \{(x_1): x_1 \in \mathscr{R}\}$, which we will commonly think of as $\mathscr{R}$ itself. In other words, we identify $x = (x_1)$ with the number x_1 and write $E^1 = \mathscr{R}$. Here, if $x = (x_1)$ and $y = (y_1)$ are in E^1 then $d(x, y) = [(x_1 - y_1)^2]^{1/2} = |x_1 - y_1| = |x - y|$. Hence the distance between the two points (real numbers) x and y is the ordinary distance $|x - y|$ in $\mathscr{R}$.

(b) When $n = 2$,

$$E^2 = \{(x_1, x_2): x_i \in \mathscr{R}\} = \mathscr{R} \times \mathscr{R}$$

and

$$d(x, y) = \sqrt{(x_1 - y_1)^2 + (x_2 - y_2)^2}$$

is the ordinary Euclidian distance between (x_1, y_1) and (y_2, y_2). One commonly thinks of E^1 as the line and E^2 as the plane.

(c) For $n = 4$, $E^4 = \{(x_1, x_2, x_3, x_4): x_i \in \mathscr{R}\}$ can be thought of as $\mathscr{R} \times \mathscr{R} \times \mathscr{R} \times \mathscr{R}$. This is the space of relativity theory and can no longer be visualized (by anyone known to this writer). In use, the 4th coordinate may be thought of as the time-coordinate of a point, while the first three coordinates are thought of as the space coordinate of a point—its position in ordinary three-dimensional space. This interpretation leads to the common term "space-time" and E^4 is nowadays viewed as the most appropriate space for describing certain phenomena in this universe.

(d) If you believe yourself to be a person whose being is shaped by certain forces that you can attach a measure to, say cultural background, acquaintances, age, educational level, behavioral standards within your environment and (say) social attitudes, then perhaps you would like to think of yourself as a point floating (or grinding, bumping, bumbling (?)) through 6-dimensional space. If you're more complex than that, maybe you would prefer to see yourself as a path in an infinite-dimensional space, each point specified by countably many coordinates $(x_1, x_2, x_3, \ldots)$. There really is no reason to stop at that either (or even any reason to bring the matter up at all?).

The next stage in this game is to introduce a topology on E^n which in the case of $n = 1$, $n = 2$ and $n = 3$ is nothing more than the usual topology of open intervals, open circles and open spheres. Recall that a typical neighborhood of a point $x = (x_1, x_2) \in E^2 = \mathscr{R} \times \mathscr{R}$ is a set

$$U(x, r) = \{y = (y_1, y_2): (x_1 - y_1)^2 + (x_2 - y_2)^2 < r^2\}$$
$$= \{y: d(x, y) < r\}$$

—that is, a typical neighborhood is the interior of a circle of radius r

centered at $x = (x_1, x_2)$. We extend this to E^n in

DEFINITION 8.5.2. Let $x \in E^n$, $r \in \mathscr{R}$, $r > 0$. The set $U(x, r) = \{y \in E^n\colon d(x, y) < r\}$ is called a *neighborhood of x of radius r.*

For example if $\theta = (0, 0, 0) \in E^3$ and $r = 1/2$ then $U(\theta, 1/2)$ is the open sphere in E^3 centered at the origin θ of radius $1/2$.

Simply because we refer to the sets $U(x, r)$ as *neighborhoods* does not mean that they constitute neighborhoods for a topology on E^n. We have seen that these sets do yield a topology for $E^1 = \mathscr{R}$ and $E^2 = \mathscr{R} \times \mathscr{R}$. The result which is needed to prove that the sets $U(x, r)$ yield a topology for E^n is a purely algebraic one known as the triangle inequality, illustrated in Figure 8.14 for 2-space.

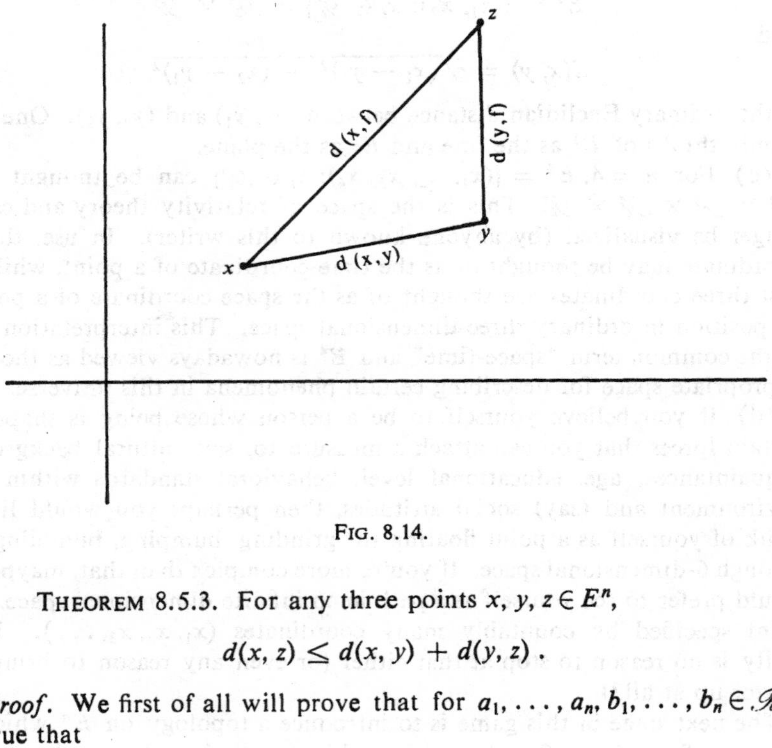

FIG. 8.14.

THEOREM 8.5.3. For any three points $x, y, z \in E^n$,

$$d(x, z) \leq d(x, y) + d(y, z) .$$

Proof. We first of all will prove that for $a_1, \ldots, a_n, b_1, \ldots, b_n \in \mathscr{R}$, it is true that

$$\Big(\sum_{i=1}^{n} a_i b_i\Big)^2 \leq \Big(\sum_{i=1}^{n} a_i^2\Big)\Big(\sum_{i=1}^{n} b_i^2\Big) .$$

For any $x \in \mathscr{R}$,

$$\sum_{i=1}^{n} (a_i x - b_i)^2 \geq 0 .$$

Hence

$$(\sum a_i^2)x^2 - 2(\sum_{i=1}^{n} a_ib_i)x + \sum_{i=1}^{n} b_i^2 \geq 0 .$$

Let

$$x = \frac{(\sum_{i=1}^{n} a_ib_i)}{(\sum_{i=1}^{n} a_i^2)}$$

provided $\sum_{i=1}^{n} a_i^2 \neq 0$. Substituting in the above inequality yields

$$(\sum_{i=1}^{n} a_ib_i)^2 \leq (\sum_{i=1}^{n} a_i^2)(\sum_{i=1}^{n} b_i^2),$$

the desired result.

If $\sum_{i=1}^{n} a_i^2 = 0$, then $a_2 = a_2 = \ldots = a_n = 0$ and the inequality is trivially satisfied for then $\sum_{i=1}^{n} a_ib_i = 0$.

We leave it to the reader to use this result to prove that

$$(\sum_{i=1}^{n} a_i^2)^{1/2} + (\sum_{i=1}^{n} b_i^2)^{1/2} \geq [\sum_{i=1}^{n} (a_i + b_i)^2]^{1/2}$$

using common algebra.

Now let $x = (x_1, x_2, \ldots, x_n)$, $y = (y_1, y_2, \ldots, y_n)$, and $z = (z_1, z_2, \ldots, z_n)$. Letting $a_i = x_i - y_i$ and $b_i = y_i - z_i$, we have $a_i + b_i = x_i - z_i$, and substitution in the last inequality yields $d(x, y) + d(y, z) \geq d(x, z)$, the desired result. As a consequence we have

THEOREM 8.5.4. For a given n, let $\mathcal{T}_n = \{U(x, r): x \in E^n, r > 0\}$ be the collection of all neighborhoods of all points in E^n. Then $\mathcal{T}_n$ satisfies Axioms 1 and 2 (Section 4.1) for topological space and hence $(E^n, \mathcal{T}_n)$ is a topological space.

Proof. For Axiom 1, if $x \in E^n$ is given, then for $r > 0$, $x \in U(x, r) \in \mathcal{T}_n$. Hence every point has least one neighborhood $U(x, r)$. (In fact, any point has infinitely many, one for each $r > 0$.)

For Axiom 2, suppose $U(x, p)$, $U(y, r) \in \mathcal{T}_n$ and $z \in U(x, p) \cap U(y, r)$. We must find a set $U(z, q) \in \mathcal{T}_n$ such that $U(z, q) \subset U(x, p) \cap U(y, r)$. Think in terms of Figure 8.15, and let q be the smaller of the two positive numbers $p - d(x, z)$ and $r - d(y, z)$. (Since $z \in U(x, p)$, we know $d(x, z) < p$ and hence $p - d(x, z) > 0$. Similarly, $r - d(y, z) > 0$.) We claim $U(z, q) \subset U(x, p) \cap U(y, r)$.

Let $u \in U(z, q)$. Then,

$$d(z, u) < q \leq p - d(x, z) .$$

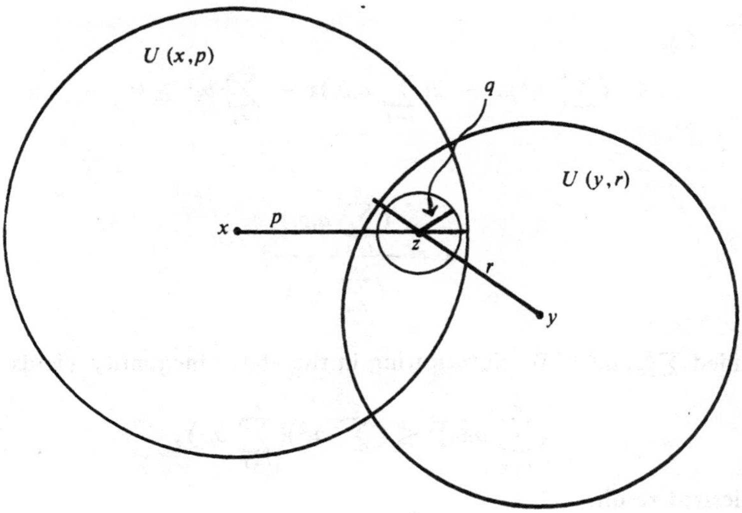

FIG. 8.15.

Hence from Theorem 8.5.4

$$d(x, u) \leq d(x, z) + d(z, u) < d(x, z) + [p - d(x, z)] = p .$$

Hence $d(x, u) < q$ and, therefore, $u \in U(x, p)$. Similarly, $d(y, u) < r$ and $u \in U(y, r)$. Thus $u \in U(z, q)$ implies $u \in U(x, p) \cap U(y, r)$ and the proof is complete.

Since we now know that $(E^n, \mathscr{T}_n)$ is a topological space, all previous topological results apply. Connectedness, continuity, compactness all have meaning. We will spend the remainder of this section investigating these three topological concepts in the space E^n with topology $\mathscr{T}_n$. The reader is advised to relate all statements and conclusions to what he already knows about $E^1 = \mathscr{R}$ and $E^2 = \mathscr{R} \times \mathscr{R}$.

We take up first the matter of continuity. Since we have theorems (8.3.3 and 8.4.4) concerning the range of a continuous function on a compact or connected set into a topological space, and since we wish to identify sets in E^n as compact and/or connected, we will be initially cancerned with the continuity of functions whose range is a subset of E^n. In particular we wish to prove an analogue of Theorem 8.3.9. To do this we begin (paradoxically) with a study of the so-called *projection functions* of E^n into E^1. These allow us to relate many properties $E^1 = \mathscr{R}$ to E^n.

DEFINITION 8.5.5. The function $\pi_k\, E^n \rightarrow E^1$ defined by

$$\pi_k(x_1, x_2, \ldots, x_n) = x_k$$

is called the k-th coordinate projection of E^n into E^1.

THEOREM 8.5.6. For each $k = 1, 2, \ldots, n$, the projection function π_k is continuous.

Proof. Let $x_k = \pi_k(x)$ where $x = (x_1, x_2, \ldots, x_n)$ and consider a neighborhood $V = (x_k - \varepsilon, x_k + \varepsilon)$ of the real number $x_k \in \mathscr{R}$. We must find a neighborhood $U = U(x, r)$ such that $\pi_k(U) \subset V$.

Let $r = \varepsilon$ and suppose $y \in U(x, \varepsilon)$. We claim $\pi_k(y) \in V$. If

$$y = (y_1, y_2, \ldots, y_n)\,,$$

then, because $y \in U(x, \varepsilon)$, we have

$$(x_k - y_k)^2 \leq (x_1 - y_1)^2 + (x_2 - y_2)^2 + \ldots + (x_n - y_n)^2 < \varepsilon^2$$

and hence that

$$|y_k - x_k| < \varepsilon\,.$$

Therefore

$$-\varepsilon < y_k - x_k < \varepsilon$$

or

$$x_k - \varepsilon < \pi_k(y) < \varepsilon + x_k\,.$$

Hence $\pi_k(y) \in V$ and the proof is complete.

Here now is the main theorem on functions and continuity into E^n.

THEOREM 8.5.7. Let $(X, \mathscr{U})$ be any topological space and let $f: X \to E^n$. Let $f_k = \pi_k \circ f: X \to E^1$. Then,

(a) $f(t) = (f_1(t), f_2(t), \ldots, f_n(t))$ for each $t \in X$.

(b) If each function f_k is continuous, then f is continuous.

Before a proof, a word of interpretation. The functions f_k, of which there are n, are simpler functions than f itself. The function f_k has domain X and range a subset of $E^1 = \mathscr{R}$—not E^n as f does. Furthermore, for a given $t \in X$, $f_k(t) = \pi_k(f(t))$ is simply the k-th coordinate of the point $f(t) \in E^n$. Finally, part (b) reduces the matter of continuity into E^n to one of continuity into the familiar space $\mathscr{R}$. Now for a proof.

Proof of Theorem 8.5.7. (a) If $t \in X$, let $f(t) = (x_1, x_2, \ldots, x_n)$. Then, $f_k(t) = [\pi_k \circ f](t) = \pi_k(f(t)) = x_k$. Hence $x_k = f_k(t)$ for $k = 1, 2, \ldots, n$ and, hence, $f(t) = (f_1(t), f_2(t), \ldots, f_n(t))$.

(b) Let $x = f(t) = (x_1, x_2, \ldots, x_n) \in E^n$ and let $V = U(x, r)$ be a neighborhood of $x = f(t)$ in E^n. We must find a neighborhood U of t such that $f(U) \subset V$.

Since each function f_k is continuous, given the neighborhood

$$V_k = \left(f_k(t) - \frac{r}{n}, f_k(t) + \frac{r}{n}\right)$$

of $f_k(t)$ in $\mathscr{R}$, $\exists$ a neighborhood U_k of t in X such that $f_k(U_k) \subset V_k$. Since $t \in \bigcap_{k=1}^n U_k$ and by $\bigcap_{k=1}^n U_k$ is open in X, $\exists$ a neighborhood U of t such that $t \in U \subset \bigcap_{k=1}^n U_k$. We claim that $f(U) \subset V$.

Let $s \in U$ and let $y = f(s) = (y_1, y_2, \ldots, y_n) = (f_1(s), f_2(s), \ldots, f_n(s))$. We must show $y \in V = U(x, r)$, or, equivalently, that $d(x, y) < r$. Since $y_k = f_k(s)$ and since $s \in U \subset U_k$, then $y_k = f_k(s) \in V_k$. That is,

$$|f_k(s) - f_k(t)| < \frac{r}{n}.$$

Since $x_k = f_k(t)$ and $y_k = f_k(s)$ we rewrite this is

$$|x_k - y_k| < \frac{r}{n}$$

and have

$$d(x, y) = [\sum_{i=1}^{n} (x_i - y_i)^2]^{1/2} < \left[\sum_{i=1}^{n} \frac{r^2}{n^2}\right]^{1/2}$$
$$< \left[\frac{r^2}{n}\right]^{1/2} = \frac{r}{\sqrt{n}} < r.$$

Hence $y = f(s) \in V$ when $s \in U$. Thus the proof.

EXAMPLES. (a) Define $f: \mathscr{R} \to E^4$ by $f(x) = (x, \sin x, x^2, x-1)$. Then $f_1(x) = x$, $f_2(x) = \sin x$, $f_3(x) = x^2$ and $f_4(x) = x - 1$. Each of these functions is continuous by the standard arguments (involving only real numbers) and hence according Theorem 8.5.7, f itself is continuous.

(b) Define $g: [0, 1] \to E^3$ by $g(t) = (t, -t, t)$. Then $g_1(t) = t$, $g_2(t) = -t$ and $g_3(t) = t$. Each function g_i is continuous, hence g is continuous. By Definition 8.3.6 the range $R_g = \{(t, -t, t): 0 \leq t \leq 1\}$ is an arc in E^3. By Theorem 8.4.4, R_g is compact in E^3. The range of g is a straight line in E^3 from $(0, 0, 0)$ to $(1, -1, 1)$.

Problems.

(1) Let $S = \{(x, 0, -1, x^2, 1 - x): x \in [0, 1]\} \subset E^5$. Describe $\pi_k(S)$ for $k = 1, 2, 3, 4, 5$.

(2) Let $T = \{(x, y, z): x^2 + y^2 = 1, z = 3\}$. Describe $\pi_k(T)$ for $k = 1, 2, 3$.

(3) Show that the set $A = \{(x_1, x_2, x_3, x_4): 0 \leq x_i \leq 1\}$ is arcwise connected and hence that A is connected in E^4. *Hint*: If $x = (x_1, x_2, x_3, x_4) \in A$ and $y = (y_1, y_2, y_3, y_4) \in A$, let $f(t) = (tx_1 + (1-t)y_1, tx_2 + (1-t)y_2, tx_3 + (1-t)y_4, tx_4 + (1-t)y_4)$ for $t \in [0, 1]$. Show that f is continuous and $f(0) = y, f(1) = x$.

(4) Show that the set $B = \{(x_1, x_2, x_3, x_4): x_1 + x_2 + x_3 + x_4 = 1\}$ is arcwise connected in E^4. How would you interpret $\{(x, y, z): x + y + z = 1\}$ geometrically in E^3?

(5) Is $C = \{(x_1, 0, x_3, 0, x_5): x_i \in \mathscr{R}\}$ a connected subset of E^5. *Hint*:

use the method of Problem 3.

(6) Is E^n a connected topological space?

(7) Prove the converse of Theorem 8.5.7(b): if f is continuous, then each function $f_k: X \to \mathscr{R}$ is continuous.

(8) Let $f: [0, 1] \to E^n$ be a continuous function. Hence, by Definition 8.3.6, f describes an arc from $x = f(0)$ to $y = f(1)$ in E^n. Show that f_k describes an arc (closed interval) in $\mathscr{R}$ from x_k to y_k where $x = (x_1, \ldots, x_n)$, $y = (y_1, \ldots, y_n)$.

(9) Prove that

$$(\sum_{n=1}^{n} a_i^2)^{1/2} + (\sum_{i=1}^{n} b_i^2)^{1/2} \geq [\sum_{i=1}^{n} (a_i + b_i)^2]^{1/2}.$$

6. CONNECTEDNESS IN E^n

We now turn to connectedness in E^n. We have a good criterion Theorem 8.3.8, for determining the connectedness of a subset of any topological space X—provided that one has a way to manufacture arcs in the space X. Because of certain algebraic properties of $X = E^n$, one can easily describe certain arcs in E^n which can well be thought of as analogues of straight lines in E^2 and E^3. We begin with

DEFINITION 8.6.1. Let $x = (x_1, x_2, \ldots, x_n)$ and $y = (y_1, y_2, \ldots, y_n)$ be points in E^n. Define

(a) $x + y$ to be the point $(x_1 + y_1, x_2 + y_2, \ldots, x_n + y_n)$ in E^n.

(b) If $a \in \mathscr{R}$ define ax to the point $(ax_1, ax_2, \ldots, ax_n)$.

DEFINITION 8.6.2. If $x, y \in E^n$ define the *line segment from x to y* to be the set

$$L(x, y) = \{ty + (1 - t)x: 0 \leq t \leq 1\} \subset E^n.$$

EXAMPLES. (a) If $x = (1, 2)$, $y = (0, -3)$ are in E^2 then $x + y = (1, -1)$ and $\sqrt{2}\,x = (\sqrt{2}, 2\sqrt{2})$. We can interpret $x + y$ and $\sqrt{2}\,x$ geometrically as in Figure 8.16. Thus addition in E^2 obeys the parallelogram law and multiplication is no more than a stretching (or contraction) of the line from the origin $(0, 0)$ to the point $x = (1, 2)$.

(b) If $x = (1, 2)$ and $y = (0, -3)$ then

$$\begin{aligned} L(x, y) &= \{(1 - t, -3t + 2(1 - t)):\ 0 \leq t \leq 1\} \\ &= \{(1 - t, -5t + 2):\ 0 \leq t \leq 1\} \\ &= \{(x, 5x - 3):\ x = 1 - t, 0 \leq x \leq 1\}. \end{aligned}$$

Hence $L(x, y)$ is that portion of the graph of $y = 5x - 3$ lying between $x = 0$ and $x = 1$ and hence is a straight line joining $(0, -3)$ to $(1, 2)$ as indicated in Figure 8.17.

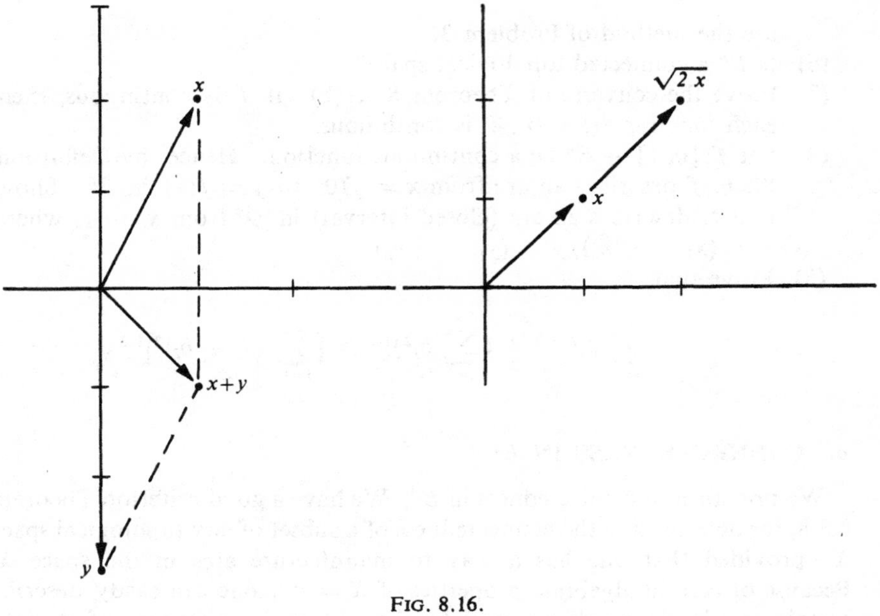

FIG. 8.16.

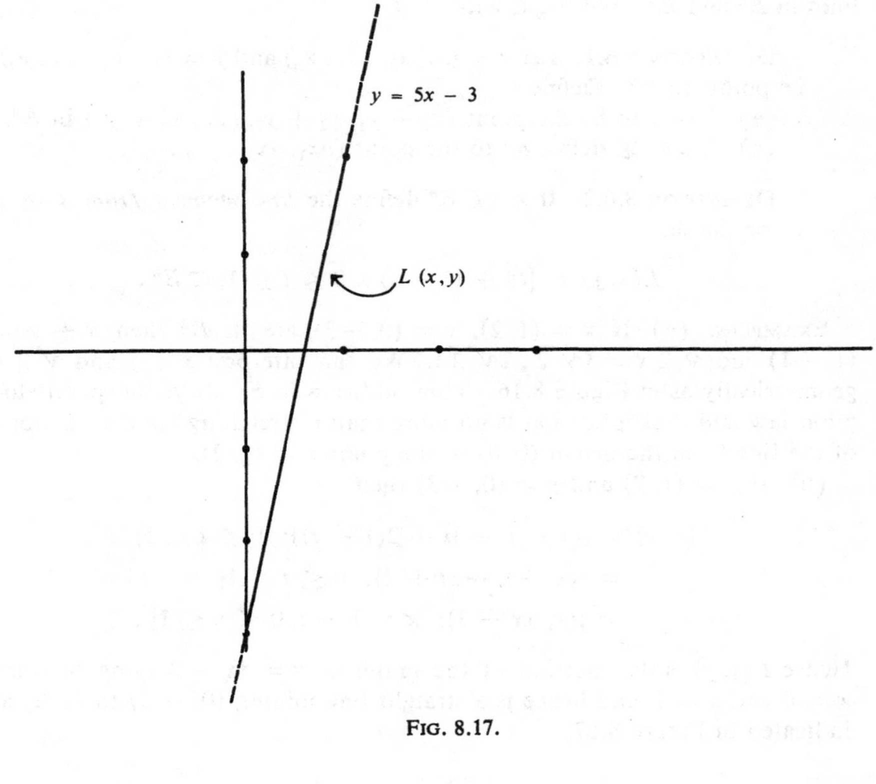

FIG. 8.17.

We have introduced the notion of a line segment in E^n because it is the simplest example of an arc in E^n.

THEOREM 8.6.3. If $x, y \in E^n$, then $L(x, y)$ is an arc in E^n. If $d(x, y) < r$, then $L(x, y) \subset U(x, r)$.

Proof. Define $f:[0, 1] \to L(x, y)$ by $f(t) = ty + (1-t)x$. Clearly f is onto $L(x, y)$ and $f(0) = x$, $f(1) = y$. We have only to prove that f is continuous.

Let $x = (x_1, x_2, \cdots, x_n)$, $y = (y_1, y_2, \cdots, y_n)$. By Theorem 8.5.7(a) and (b),

$$f_k(t) = ty_k + (1-t)x_k$$

where $f_k(t)$ is (as in Theorem 8.5.7) the k-th coordinate of $f(t)$. Since

$$f_k(t) = t(y_k - x_k) + x_k$$

is a simple linear function defined on $[0, 1]$ with range a subset of $\mathscr{R}$, it is easy to prove that f_k is continuous. By Theorem 8.5.7, f is continuous and $L(x, y)$ is an arc from x to y.

Suppose now that $d(x, y) < r$. Let $z \in L(x, y)$. We claim $z \in U(x, r)$. Since $z \in L(x, y)$, then $z = ty + (1-t)x$ for some t with $0 \leq t \leq 1$. If $x = (x_1, x_2, \cdots, x_n)$, $y = (y_1, y_2, \cdots, y_n)$ and $z = (z_1, z_2, \cdots, z_n)$, then $z_k = ty_k + (1-t)x_k$ and hence

$$(z_k - x_k)^2 = (ty_k - tx_k)^2 = t^2(y_k - x_k)^2 \leq (y_k - x_k)^2$$

since $t^2 \leq 1$. Hence

$$d(x, z) = [\sum_{k=1}^{n} (x_k - z_k)^2]^{1/2} \leq [\sum_{k=1}^{n} (y_k - x_k)^2]^{1/2} < r.$$

By definition of $U(x, r)$, we have $z \in U(x, r)$.

With Theorems 3.6.3 and 8.3.8 we have a simple criteria for connectedness in E^n. To introduce the usual terminology we make the

DEFINITION 8.6.4. A set $C \subset E^n$ is called *convex* if for every pair of points $x, y \in C$ the line segment $L(x, y)$ lies entirely in C.

THEOREM 8.6.5. Any convex set in E^n is connected.

Proof. Apply Theorems 8.6.3 and 8.3.8

With this we turn to what is truly a beautiful example of abstract mathematical thinking. We will prove that any open and connected set in E^n must be arcwise connected. Think about this a minute. We know that any arcwise connected set in E^n is connected. We have an example (Problem 18, Section 8.3) of a connected but not arcwise connected set in E^2. Note that this set is not open. We wish to prove that any open connected set is arcwise connected. But it is difficult (in fact, to the author, plainly unimaginable) to construct an arc in such an arbitrary kind of set in E^n. In the proof that follows we use a method of reasoning that avoids this considerable difficulty.

THEOREM 8.6.6. Any open connected subset A of E^n is arcwise connected.

Proof. Let $x \in A$. We must show that $\forall\, y \in A\ \exists$ an arc in A from x to y. Let $C = \{y \in A\colon \exists$ an arc in A from x to $y\}$. We wish to show that $C = A$.

First of all, $C \neq \square$ because $x \in C$. To see this define $f\colon [0, 1] \to A$ by $f(t) = x$. Then f is trivially continuous (being constant) and $f(0) = x$ and $f(1) = x$. Hence R_f is an arc in A from x to x and hence $x \in C$.

We can prove that $C = A$ by showing that C is both open and closed in the (relative) topology on A, according to Theorem 4.6.7(1) We first show that C is open.

Suppose $y \in C$. We claim $\exists$ a neighborhood $U(y, r) \subset C$. Since $y \in C \subset A$ and A is open, $\exists$ a neighborhood $U(y, r) \subset A$. We claim $U(y, r) \subset C$. To do this it must be shown that if $z \in U(y, r)$ then $\exists$ an arc in A from x to z. Since $y \in C$ we know $\exists$ an arc, defined by some continuous function $g\colon [0, 1] \to A$ such that $g(0) = x$, $g(1) = y$, from x to y. We will use g to construct an arc from y to z.

Since $z \in U(y, r)$ then by Theorem 8.6.3 $L(y, z) \subset U(y, r) \subset A$ and $L(y, z)$ is an arc described by the continuous function $h(t) = tz + (1 - t)y$, $t \in [0, 1]$.

$$f(t) = \begin{cases} g(2t) & 0 \leq t \leq \dfrac{1}{2} \\ h(2t - 1) & \dfrac{1}{2} < t \leq 1 \end{cases}$$

(At this point the reader might refer to Figure 8.18.) Hence, $f(0) = g(0) = x$ and $f(1) = h(1) = z$. Also, $f(1/2) = h(0) = g(1) = y$. From Definition 8.6.1 f is continuous at all points $t < 1/2$ and $t > 1/2$ because g and h are continuous. Because $h(0) = y = g(1) = f(1/2)$ it also follows that f is continuous at $1/2$ and this detail is left to the reader. Hence f describes an arc in A from x to z. Thus $z \in C$. Since $z \in U(y, r)$ was arbitrary, $U(y, r) \subset C$ and hence C is an open set.

We now show that C is closed in A. Suppose $z \in A$ is a limit point of C. Since A is open, there is a neighborhood $U(z, r) \subset A$. Since z is a limit point of C, $U(z, r) \cap C \neq \square$. Let $y \in U(z, r) \cap C$. Then, because $y \in C$, $\exists$ an arc described by some continuous function $g\colon [0, 1] \to A$ such that $g(0) = x$, $g(1) = y$. But then, $L(y, z) \subset U(z, r)$ and as above we can obtain an arc in A from x to z by piecing together the arc from x to y described by g to the arc $L(y, z)$ from y to z. Hence there is an arc in A from x to z. Hence $z \in C$ and thus C contains all its limit points and is, therefore, closed.

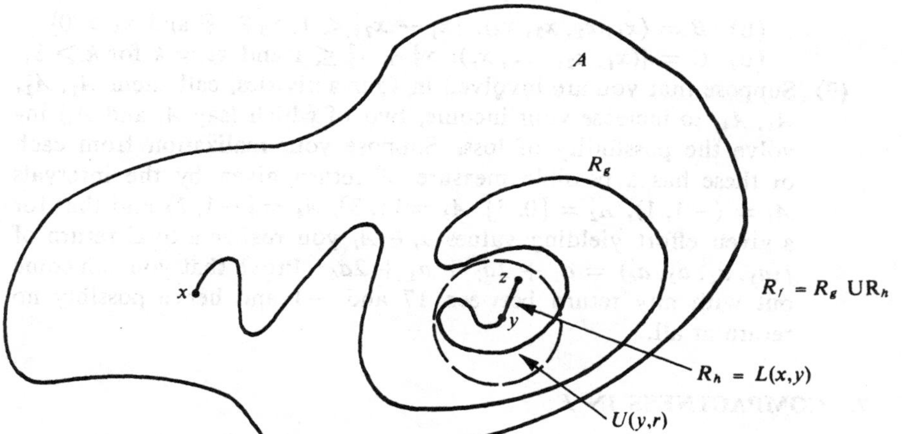

FIG. 8.18.

Since C is a non-empty open and closed subset of A, and A is connected, then $C = A$ and the proof is complete.

This concludes our brief study of connectedness in E^n. From it we know that connectedness and arcwise connectedness are equivalent properties on open subsets of E^n and that any arcwise connected set is connected.

Problems.

(1) Prove that any neighborhood in E^n is convex and hence is connected.

(2) Interpret Theorem 8.6.3 geometrically in E^2 and E^3 and give examples of convex sets in E^2 having and not having various topological properties.

(3) Let M be a convex set in E^n. Is $\bar{M}$ connected? Is $\bar{M}$ convex?

(4) Is the closure of an arcwise connected set arcwise connected?

(5) Prove that *any* connected set in $E^1 = \mathscr{R}$ is arcwise connected.

(6) Let A, B be connected subsets of $\mathscr{R}$. Prove that $A \times B$ is connected in $E^2 = \mathscr{R} \times \mathscr{R}$. *Hint*: for $a \in A$, define $f_a : B \to A \times B$ by $f_a(b) = (a, b)$. Prove that $A \times B = \bigcup_{a \in A} R_{f_a}$ and hence that A is a union of connected sets R_{f_a}, each of which intersects the connected set $C = \{(a, B) : a \in A\}$ where b is a fixed element of C.

(7) Use Theorems 8.3.8 and 8.5.7 to prove that if $A_1, \cdots, A_n$ are connected sets in $\mathscr{R}$, then

$$A_1 \times A_2 \times \cdots \times A_n = \{(a_1, a_2, \cdots, a_n) : a_i \in A\}$$

is an arcwise connected set in E^n and hence connected.

(8) Which of the following are connected? Which are arcwise connected?

(a) $A = \{(x_1, 2, 0, x_4) \colon x_1 \in [0, 1], x_4 \in (0, 2)\}$

(b) $B = (x_1, x_2, x_3, x_4)$: $|x_1 - x_2| < 1$, $x_3 \in \mathscr{R}$ and $x_4 > 0\}$
(c) $C = \{(x_1, x_2, \ldots, x_n)$: $x_1^2 + x_2^2 \leq 1$ and $x_k = k$ for $k \geq 3\}$.

(9) Suppose that you are involved in four activities, call them A_1, A_2, A_3, A_4, to increase your income, two of which (say A_1 and A_4) involve the possibility of loss. Suppose your realization from each of these has a possible measure of return given by the intervals $A_1 = (-1, 1)$, $A_2 = [0, 5]$, $A_3 = [1, 3]$, $A_4 = [-1, 2)$ and that for a given effort yielding values $a_i \in A_i$ you realize a total return of $f(a_1, a_2, a_3, a_4) = a_1 + 2a_2 + a_3 + 2a_4$. Prove that you can come out with any return between 17 and -1 and hence possibly no return at all.

7. COMPACTNESS IN E^n

In E^n all previous difficulties with compactness in $\mathscr{R}$ are literally compounded n-times, and a simple characterization of compact sets is extremely useful. We will show that the criterial is the same as that for $\mathscr{R}$ itself—closed (a topological property) and bounded (an arithmetical one).

DEFINITION 8.7.1. A set $B \subset E^n$ is bounded if $\exists$ a number $M > 0$ such that $B \subset U(\theta, M)$ where $\theta = (0, 0, \ldots, 0)$ is the origin E^n.

In the special case of E^2, a set is bounded if it is contained inside some circle of some radius M centered at (0, 0). A set in E^3 is bounded if it is contained within some sphere in E^3 of some radius M centered at (0, 0, 0). We wish to prove

THEOREM 8.7.2. If $C \subset E^n$ and C is closed and bounded, then C is compact.

The proof of this result is quite complicated and we develop it in a series of lemmae. The underlying reason for its validity is again the completeness axiom for $\mathscr{R}$ itself. The underlying idea in making the proof is that given a countably infinite subset S of C there must exist a point in C such that every neighborhood of this point contains infinitely many members of the set S. That is, infinitely many elements of S must be *near* a single point in C. This is what we will first prove, and prove it first for a closed and bounded set in $\mathscr{R}$.

LEMMA 8.7.3. If $B \subset \mathscr{R}$ is closed and bounded and $f: \mathscr{N} \to B$ then $\exists x \in B$ such that if U is a neighborhood of x and $k \in \mathscr{N}$, then

$$\exists\, m > k \ni f(m) \in U.$$

The latter part of this statement of course implies that U must contain the images $f(n)$ of infinitely many n. The set $f(\mathscr{N})$ could be in fact finite.

Proof. Suppose not. Then $\forall x \in B \exists$ a neighborhood U_x of x such that $\exists k_x \in \mathscr{N}$ such that for all $n > k_x$, $f(n) \notin U_x$. (This is the negation of the conclusion of the lemma. The subscripted U_x and k_x are to help us recall that U_x and k_x depend on what $x \in B$ we are talking about and may differ from point to point.)

Let $\mathscr{A} = \{U_x: x \in B\}$. Since $x \in U_x$, $\mathscr{A}$ is an open cover of B. Since B is closed and bounded in $\mathscr{R}$, then by Theorem 8.4.8 B is compact and hence there are finitely many sets $U_{x_1}, \ldots, U_{x_n} \in \mathscr{A}$ such that $B \subset \bigcup_{i=1}^n U_{x_i}$.

Consider the n numbers $k_{x_1}, \ldots, k_{x_n}$. Let $M = k_{x_1} + \cdots + k_{x_n}$. Since $M > k_{x_i}$ for each i, then $f(M) \notin U_{x_i}$. Thus, $f(M) \notin \bigcup_{i=1}^n U_{k_x} \supset B$. But $f(M) \in B$ by hypothesis on f. This is a contradiction. Hence the conclusion must hold.

We now extend this to any closed and bounded set $C \subset E^n$.

LEMMA 8.7.4. If $C \subset E^n$ is closed and bounded and $f: \mathscr{N} \to C$, then $\exists x \in C$ such that if U is a neighborhood of x and $k \in \mathscr{N}$, $\exists\, m > k$ such that $f(m) \in U$.

Proof. By Lemma 8.7.3, the conclusion holds for $n=1$. Suppose Lemma 8.7.4 is true for $n = p$. We claim it holds for $n = p + 1$. Thus suppose that $C \subset E^{p+1}$ is closed and bounded and that $f: \mathscr{N} \to C$.

Since C is bounded, $\exists\, M \in \mathscr{R}$ such that $C \subset U(\theta, M)$. Define $\pi: E^{p+1} \to E^p$ by

$$\pi(x_1, \ldots, x_p, x_{p+1}) = (x_1, x_2, \ldots, x_p)\,.$$

Consider the closed and bounded set

$$C_0 = \Big\{(y_1, \ldots, y_p): \Big[\sum_{i=1}^{p} y_i^2\Big]^{1/2} \leq M\Big\} \subset E^p$$

and the function $g = \pi \circ f$. Since $(x_1, \ldots, x_p, x_{p+1}) \in C$ implies

$$\Big[\sum_{i=1}^{p} x_i^2\Big]^{1/2} \leq \Big[\sum_{i=1}^{p1+} x_i^2\Big]^{1/2} < M,$$

then $\pi(C) \subset C_0$ and hence

$$g(\mathscr{N}) = \pi(f(\mathscr{N})) \subset \pi(C) \subset C_0\,.$$

Thus $g: \mathscr{N} \to C_0$.

Since $C_0 \subset E^p$ and C_0 is closed and bounded, then by the induction hypothesis there is a $\bar{y} = (y_1, \ldots, y_p) \in C_0$ such that given the neighborhood $U(\bar{y}, 1/n)$ of $\bar{y}$ $\exists$ a $k > n$ such that $g(k) \in U(\bar{y}, 1/n)$. Define $h: \mathscr{N} \to \mathscr{N}$ as follows: for each $n \in \mathscr{N}$ let $h(n)$ be the smallest natural number $k > n$ such that $g(k) \in U(\bar{y}, 1/n)$. Thus, $g(h(n)) \in U(\bar{y}, 1/n)$.

Consider now the function

$$\pi_{p+1} \circ f \circ h: \mathscr{N} \to \mathscr{R}.$$

$$h(n) \in \mathcal{N}, \quad f(h(n)) \in C \quad \text{and} \quad \pi_{p+1}(f(h(n)) \in \mathcal{R}.$$

Since $f(h(n)) \in C$, then if

$$f(h(n)) = (x_1, \ldots, x_{p+1})$$

then

$$|x_{p+1}| \leq [\sum_{i=1}^{p+1} x_i^2]^{1/2} < M$$

and hence

$$\pi_{p+1}(f(h(n)) = x_{p+1} \in [-M, M].$$

In other words,

$$\pi_{p+1} \circ f \circ h \colon \mathcal{N} \to [-M, M].$$

Since $[-M, M]$ is closed and bounded there is a $y_0 \in [-M, M]$ such that, if U is a neighborhood of y_0 and $k \in \mathcal{N}$ is given, $\exists\, m > k$ such that

$$\pi_{p+1}(f(h(m))) \in U.$$

Let $y = (y_1, \ldots, y_p, y_0)$ where $\bar{y} = (y_1, \ldots, y_p)$. We claim $y \in C$ and that if V is a neighborhood of y and $k \in \mathcal{N}$ is given, then $\exists\, m > k$ such that $f(m) \in V$. This of course is what we wished to prove.

Suppose that

$$V = \{x \in E^{p+1} \colon d(x, y) < \varepsilon\}$$

and that $k \in \mathcal{N}$ is given. By the Archimedian property of $\mathcal{R}$, $\exists$ a $j \in \mathcal{N}$ such that $j > k$ and $1/j < \varepsilon/2$. Let

$$U = \left(y_0 - \frac{\varepsilon}{2}, \quad y_0 + \frac{\varepsilon}{2}\right)$$

be a neighborhood of y_0. We have just proven that $\exists\, i > j$ such that $\pi_{p+1}(f(h(i)) \in U$. Let

$$(x_1, \ldots, x_p, x_{p+1}) = f(h(i)).$$

We then have $x_{p+1} \in U$. Moreover,

$$g(h(i)) = \pi \circ f(h(i)) = (x_1, \ldots, x_p) \in N\left(\bar{y}, \frac{1}{i}\right).$$

Hence,

$$d((x_1, \ldots, x_p), (y_1, \ldots, y_p)) = [\sum_{q=1}^{p} (x_q - y_q)^2]^{1/2} < \frac{1}{i} < \frac{1}{j} < \frac{\varepsilon}{2}.$$

Since $x_{p+1} \in U$, then

$$|x_{p+1} - y_0| < \frac{\varepsilon}{2}$$

and hence

$$\begin{aligned} d(x, f(h(i)))^2 &= [\sum_{q=1}^{p+1} (y_q - x_q)^2] \quad (\text{where} \quad y_{p+1} = y_0) \\ &\leq \sum_{q=1}^{p} (y_q - x_q)^2 + |x_{p+1} - y|^2 \\ &\leq \left(\frac{\varepsilon}{2}\right)^2 + \left(\frac{\varepsilon}{2}\right)^2 = \frac{\varepsilon^2}{2}. \end{aligned}$$

Thus $d(x, f(h(i)) < \varepsilon$ and hence $f(h(i)) \in V$. Since $h(i) > i > j > k$ this means that we have shown that given a neighborhood V of y, a number $k \in \mathscr{N}$, there then exists a number $m = h(i) > k$ such that $f(m) \in V$.

To complete the proof we only need show that $y \in C$. But C is closed and we have just shown that every neighborhood V of y contains a point $f(m) \in C$. Thus $y \in C$. This completes the proof.

The remainder of the proof of Theorem 8.7.2 is relatively easy.

LEMMA 8.7.5. Let $\mathscr{A}$ be an open cover of C. Then $\exists$ a number $r > 0$ such that $\forall x \in C \exists A \in \mathscr{A}$ such that $U(x, r) \subset A$.

A literal interpretation of this lemma is that an open cover of C can consist of open sets that can only be so small (at least one no smaller than $U(x, r)$) about each point.

Proof. Suppose not. Then $\forall$ number $r > 0 \; \exists \; x \in C$ such that $\forall \; A \in \mathscr{A}$, $U(x, r) \not\subset A$. In particular, for each number $1/n > 0 \; \exists \; x_n \in C$ such that $\forall \; A \in \mathscr{A}$, $U(x_n, 1/n) \not\subset A$. Define f: $\mathscr{N} \to C$ by $f(n) = x_n$. (This makes use of the axiom of choice—choosing a point x_n in C.)

By Lemma 8.7.4, $\exists \; x \in C$, satisfying the conclusion of Lemma 8.7.4. Since $\mathscr{A}$ is an open cover of $C \; \exists$ a set $A \in \mathscr{A}$ such that $x \in A$. Since A is open $\exists$ a neighborhood U of x contained in A. Now

$$U = \{y \in E^n \colon \; d(x, y) < \varepsilon\}$$

for some $\varepsilon < 0$. Choose $k \in \mathscr{N}$ such that $1/k < \varepsilon/2$. Then $\exists \; n > k$ such that $x_n = f(n) \in \{y \colon d(x, y) < \varepsilon/2\}$. We claim $U(x_n, 1/n) \subset A$ which will be a contradiction.

If $y \in U(x_n, 1/2)$, then $d(y, x_n) < 1/n < 1/k < \varepsilon/2$. Since $d(x, x_n) < \varepsilon/2$, then $d(x, y) < d(x, x_n) + d(x_n, y) < \varepsilon$. Thus $y \in U \subset A$. Hence $U(x_n, 1/n) \subset A$ to complete the proof.

Finally,

Proof of Theorem 8.7.2. Let $\mathscr{A}$ be an open cover of C. By Lemma 8.7.5

$\exists$ an $r > 0$ such that $x \in C$ implies $\exists$ $A \in \mathscr{A}$ such that $U(x, r) \subset A$.

Consider the open cover $\mathscr{B} = \{U(x, r): x \in C\}$ of C. We claim that finitely many members of $\mathscr{B}$ cover C. Suppose not. Pick $x_1 \in C$. Then, $C \subset U(x_1, r)$. Hence $\exists$ $x_2 \in C \backslash U(x_1, r)$ and $C \not\subset \bigcup_{i=1}^{2} U(x_i, r)$. Suppose that $x_1, \ldots, x_n$ have been chosen such that

$$x_k \notin \bigcup_{i=1}^{k-1} U(x_i, r) \quad \text{for} \quad k = 1, 2, \ldots, n.$$

Then, since no finite number of sets $U(x_i, r) \in \mathscr{B}$ are supposed to cover C, then $C \not\subset \bigcup_{i=1}^{n} U(x_i, r)$ and hence $\exists$ $x_{n+1} \in C \backslash \bigcup_{i=1}^{n} U(x_i, r)$. Thus, $\forall n \in \mathscr{N}$ $\exists$ a point $x_n \in C$ such that $x_n \in C \backslash \bigcup_{i=1}^{n-1} U(x_i, r)$. Define $f: \mathscr{N} \to C$ by $f(n) = x_n$.

Let x be the point in C guaranteed by Lemma 8.7.4. Let $U = U(x, r/2)$. Then if $1/k < r/2$ $\exists$ $n > k$ such that $x_n = f(n) \in U$. Hence, $d(x_n, x) < r/2$. Applying Lemma 8.7.4 again, $\exists$ $m > n$ such that $x_m = f(n) \in U$. Hence $d(x_m, x) < r/2$. Thus, by Theorem 8.5.3

$$d(x_n, x_m) \leq d(x_n, x) + d(x, x_m) < \frac{r}{2} + \frac{r}{2} = r.$$

Hence, $x_m \in U(x_n, r)$. But also, $x_m \in C \backslash \bigcup_{i=1}^{m-1} U(x_i, r)$ since $n \leq m - 1$. This is contradictory. Hence the proof.

Problems.

(1) Which of the following sets are compact?
- (a) $A = \{(x, y, z): x, y, z \in [0, 1]\} \subset E^3$.
- (b) $B = \{(x, y, z): x^2 + y^2 + z^2 = 1\} \subset E^3$.
- (c) $C = \{(x_1, x_2, x_3, x_4): |x_1| + |x_2| + |x_3| + |x_4| \leq 1\} \subset E^4$.
- (d) $D = \{(x_1, x_2, \ldots, x_n): x_i = 0$ iff i is even, and $|x_i| \leq 1$ if i is odd$\} \subset E^n$.
- (e) $E = \{(x, y): x + y = 1\} \subset E^2$.
- (f) $F = \{(x_1, x_2, x_3, x_4): |x_1| + |x_2| \leq 1$ and $x_3 + x_4 = 0\}$.
- (g) $G = \{(x_1, 0, x_3, 0, x_5): x_1^2 + x_3^2 + x_5^2 \leq 1\}$.
- (h) $H = \{(1, 2 - 1/n, 3 + 2/n, 4 - 3/n^2): n \in \mathscr{N}\} \cup \{(1, 2, 3, 4)\}$.

(2) According to Theorems 8.4.4 and 8.4.6, and continuous function defined on a compact set must be bounded. On each non-compact set in Problem 1 define a continuous function with range in $\mathscr{R}$ that is not bounded.

(3) We can more clearly indicate the link between compactness and the boundedness of continuous functions through the use of Lemma 8.7.4. Suppose $C \subset E^n$ is closed and bounded (hence compact) and that $f: C \to [0, \infty)$ is a continuous function. If f were not bounded then $\forall$ $n \in \mathscr{N}$ there would exist a point $x_n \in C$ such that $f(x_n) \geq n$. Use Lemma 8.7.4 to conclude that this is nonsense.

(4) Let C be a bounded set in E^n which is not closed. Since C is not

closed $\exists\ y \in E^n \ni y \in \bar{C} \setminus C$. For each $x \in C$, $d(x, y) \neq 0$ since $y \notin C$. Define $f: C \to \mathscr{R}$ by $f(x) = 1/d(x, y)$. Show that f is continuous. but not bounded, on C.

(5) According to Theorem 8.4.7, if $C \subset E^n$ is compact and connected and $f: C \to \mathscr{R}$ is continuous then $f(C) = [a, b]$ and b is the maximum value of f. Hence $b = f(x)$ *for* some $x \in C$. This result only asserts that f has a maximum value, but does not say show to find it. Suppose $C = \{x, y, z): 0 \leq x, y, z \leq 1\} = [0, 1] \times [0, 1] \times [0, 1]$ is the cube of side one in E^3 and that $f: C \to \mathscr{R}$ is defined by $f(x, y, z) = x - y + 2z$. Show that the maximum value of f occurs at one of the eight corners of the cube C. What about the minimum value of f on C?

(6) In general, one uses the technique of differentiation, as in calculus, to locate the maximum value of a function on a set. Many functions that occur in practice are what are called linear. The general linear function $f: E^n \to \mathscr{R}$ is of the form

$$f(x_1, x_2, \ldots, x_n) = a_1x_1 + a_2x_2 + \cdots + a_nx_n$$

where $a_1, \ldots, r_n$ are n fixed real numbers. For example, the function $f(x, y, z) = 2x + y - z$ is a linear function.

Such functions occur in mathematical economics (as well as physics) quite easily. If a manufacturing concern has n production units $P_1, P_2, \ldots, P_n$ and for an input of x_i raw materials into plant P_i the company realizes a profit a_ix_i from P_i, then the total profit of the n-plants, with inputs $x_1, \ldots, x_n$ is

$$f(x_1, x_2, \ldots, x_n) = a_1x_1 + a_2x_2 + \cdots + a_nx_n\,.$$

(At some future date, one may have

$$f(a_1, \ldots, x_n) = (a_1b_1)x + \cdots + (a_n - b_n)x_n$$

where $-b_ix_i$ measures the damage done to the environment, for which the plant is taxed. As usual mathematics is far ahead of its times and uses!) Wishing to maximize production, one needs to maximize f. This problem concerns the finding of maximum values for such functions such as f above. Such maximization problems are in general realizable without the techniques of calculus and are quite adaptable to computerized procedures (see the term "linear programming" for additional details) in large part because of the conclusion of the following problem.

Suppose that $C \subset E^n$ is a compact convex set and that $f: C \to \mathscr{R}$ is a linear function, say $f(x_1, \ldots, x_n) = a_1x_2 + \cdots + a_nx_n$. It follows that f is continuous and hence that some value $f(x_0)$ is the maximum of all its values—i.e., $f(x) \leq f(x_0)\ \forall\ x \in C$. Prove first that $f(x + y) = f(x) + f(y)$ and $f(ax) = af(x)$ for all $a \in \mathscr{R}$,

$x, y \in C$. Then prove that x_0 cannot lie on the inside of any line segment $L(x, y) \subset C$. That is, show that if $x_0 \in L(x, y) \subset C$, then $x_0 = x$ or $x_0 = y$. Notice that the corners of the cube in Problem 5 have exactly this property: they can only be end points of lines lying in the cube, and moreover, no other points of the cube but the corners have this property. Such points are commonly called *extreme points* and the conclusion of this problem is that the maximum value of a linear function on a compact, convex set can be found by checking only its values at the extreme points of the set. Note that among the infinitely many points of the cube in Problem 5 there are only eight extreme points, considerably fewer values to consider in finding the maximum.

(7) We have proven that a closed and bounded set in E^n is compact. Now prove the converse. Note that a compact set in E^n is closed by Problem 11, Section 8.4, and that the function $f(x) = d(\theta, x)$ is continuous on E^n.

(8) Let $C \subset E^n$ and suppose that *every* continuous function $f: C \rightarrow \mathscr{R}$ is bounded. Prove that C is compact. *Hint*: Show that C is closed using the idea in Problem 4 above. To prove that C is bounded consider the function $f: C \rightarrow \mathscr{R}$ given by $f(x) = d(\theta, x)$. The meaning of this is that only on compact sets can one always expect to have a maximum and minimum value for *every* continuous function.

(9) Suppose that each of the following functions is continuous. Describe their range as a subset of $\mathscr{R}$ and find there maximum and minimum values.

(a) Let $f: C \rightarrow \mathscr{R}$ be defined

$$f(x_1, x_2, x_3, x_4) = 2x_1 - x_2 - 3x_3 + 4x_4$$

where

$$C = \{(x_1, x_2, x_3, x_4): x_i \in [-1, 1]\}$$

(b) Let $g: C \rightarrow \mathscr{R}$ be defined

$$f(x_1, x_2, x_3, x_4) = x_1^2 + x_2^2 + x_3^2 + x_4^2$$

where

$$C = \left\{(x_1, x_2, x_3, x_4): \sum_{i-1}^{4} x_i^2 = 1\right\} \cup \left\{(x_1, x_2, x_3, x_4): |x_1| \leq \frac{1}{2},\ x_1 - x_2 = \frac{1}{2},\ x_3 = x_4 = 0\right\}.$$

Related Readinds

1. Apostol, T. M.: *Mathematical Analysis*, Reading, Mass.: Addison-Wesley Publishing Co., Inc., 1957. A good text with discussions of continuity, connectedness, compact-

ness, convexity and arcwise connectivity.

2. Baum, J. D.: *Elements of Point Set Topology*, Englewood Cliffs, N. J.: Prentice-Hall, 1964. A general reference.
3. Chinn, W.. G,, and Steenrod, N. E.: *First Concepts in Topology*, New York: Random House, Inc., 1966. Part I of this book is concerned, in detail, with compactness and connectedness in $\mathscr{R}$ and $\mathscr{R} \times \mathscr{R}$.
4. Mansfield, M. J.: *Introduction to Topology*. Princeton: Van Nostrand, 1963. A general reference.
5. Simmons, G. F.: *Introduction to Topology and Modern Analysis*, New York: McGraw Hill Book Co., Inc., 1963. A wonderfully written book, from which this author did learn much pleasing mathematics. Here you will find discussions of compactness, continuity and connectedness.

APPENDIX A

Mathematics in its most meaningful form is the art of establishing that, if *A* is given, then *B* must follow. It is never a question of whether *A* is true of anything. If, indeed, one is aware of a particular instance in which *A* is true, it is essential that this instance never be mentioned in establishing *B*. For if this is done, then *B* could not be certain except in that one instance of *A*, and something less than mathematics would have been done. If the mathematician has established that *B* follows from *A* without regard to any instance of *A*, then he or anyone else is certain that on *all* actual instances of *A*, *B* persists. This doesn't make mathematics easy, but it does make the task *der mathematiker* clear-cut. These two facets seem to be its attraction for many mathematicians.

This essay concerns the art of establishing mathematical proof—and the intent is to bring to the point of consciousness things you may know already. The art and mode of expression of a proof in mathematics is a separate and distinct thing from the uses of language or art that we are accustomed to. Indeed it is possible to do mathematics without the aid of any written language, in a form understandable to any mathematician of any national origin. The *Principia Mathematica*, in which Sir Bertrand Russell and Sir Alfred North Whitehead attempted to show that all of mathematics could be seen as consequences of a few primitive tautologies expressed in a symbolic manner, can be read by anyone, if it can be read at all!

The written-spoken language is an amazing thing. Within large limits it can communicate a tremendous variety of feelings, impressions and beliefs. It can also be used as a barrier to communication, as when a word or rule is used as a substitute for thought or understanding, or when a catch-phrase is waved around and planted is one's line of thought to be followed by clouds of enfoiling smoke with the apparent hope that what is really going on may never be discovered, a seemingly natural propensity of that professional class which dominates the six o'clock news.[1] The relevance of this barb is that this mode of expression is the very antithesis of that used by the mathematician, whose vices are at least not professional (if you will excuse all the notation, which although abstract, is not obscure). In consequence, much of the facility of language is denied the mathematician, who cannot use shades of meaning but who must have precise meaning because of the ever-present abstraction. Thus, although it is of little use to carry the matter of language to extremes, all technical terms must be given a precise meaning, and the only other language used in mathematical proofs is found in a non-technical language basis. Words such as *it suffices, for,*

[1] The author does not mean the newsmen!

from, therefore, hence, and other variants of the connectives of logic are used in the usual way. A word like ***smallest*** or ***least*** must be precisely defined. Suffice it to say that, in contrast to political discussions, a vague word or phrase is ***never*** used to arrive at a precise or particular conclusion!

Mathematical proof has been referred to as the art of establishing that B follows from A. There is an art to it, and a beautiful proof is what one ultimately seeks. A beautiful proof is clear and to the point, and it especially strives to make the crux of the implication of B by A as clear as possible so that its revelation is not a thing of drudgery; he who understands it might even enjoy it as much as he who reveals it. Unfortunately, such elegance is not always obtained, and, even more unfortunately, before one can get to the level where much art is possible, a good deal of not very lively, straightforward verification is often needed.

The form that art may take is rather free and let us be grateful. Not so with mathematics. In this, it is more like music, and in fact music is governed by certain relationships between the various notes that are subject to precise mathematical description involving the ratios of the frequencies of each note. Although some might say you couldn't tell it to hear it, if there's sound and rhythm, then there is a certain order to it and once you've picked it up you would sense when it's gone. There is an analogy here with the nature of a mathematical proof. Order is there and it is dictated by the logic used for mathematics. Once you've gotten onto the train of thought, the flow or rhythm of a good proof, you would sense when things have gone astray, if they do. Unfortunately, there are valid proofs that might not be called good in this sense, and they are often replaced by better ones as understanding increases.

What about the specifics of mathematical proof? How does one come to weave the tapestry of a mathematical proof? Most often, by sheer hard work, even for the very best, as you would learn if you were to read the biographies of the more prominent mathematicians.[2] But that is not unique to mathematics or much of anything else that we bother to pay attention to. And, if the hard work leads to a good proof, there is usually a good feeling to go with it.[3]

Given the will to work at it, what then? There is no formula for a proof and for that reason this discussion will have a quite nebulous character. The only real rules are those dictated by logic. If one wishes to prove that A implies B, one could try to show that not B implies not A or that A and

[2] E. T. Bell: *Men of Mathematics*, New York: Simon and Schuster, 1937.

[3] Sometimes the feeling is accompanied by a certain frustration because you can't find anyone to appreciate the great thing you've done. This is the particular plight of the mathematician who is the only man in his department who works in a particular area of mathematics. He can sometimes be seen prowling the halls hoping to con someone in too near to his blackboard to escape when he begins to explain his proof, whereupon the poor fellow politely nods his head at each conclusion while having only the faintest idea of what's going on. If the poor unfortunate is a graduate student, he must also endure the dread that he will be asked a question and have his ignorance exposed.

not *B* is absurd, or another method dictated by logic. But, whatever method is tried, there still is the necessity to begin with something and end with something else and with something definitive in between. The discovery of what goes in between is called finding a proof.

How does one find a proof? We consider this in a general sense at first, as much to indicate directions as to warn of difficulties and disappointment. To begin a proof that *A* implies *B*, one can just sit and think about *A* and *B* or fiddle with the two a bit to see if an idea comes or perhaps begin by making one small deduction after another from *A* in what is hoped is the general direction of *B*. Often one works backward from *B*, trying to find something, say *C*, which implies *B* and then showing that *A* implies *C*. This is common in limit proofs in calculus. The main idea is to try to accomplish at least a little, and then a little more. Two steps forward, perhaps one step back, as the saying goes. The process is not really logical or reasonable, but after working with what is there for a while, understanding increases, the fog lifts a bit, and sometimes, in what can only be called a flash, one sees how to link *A* to *B*.[4]

This is not always the case. Some proofs are what are called in the trade "trival" or "obvious"; they are sometimes wrong. Those that are not wrong are the verification type. One begins with *A*, reinterprets it or uses some previous property about whatever *A* concerns, writes it down and notices that (perhaps after a few more steps) *B* is a consequence of this property.

The main thing in getting a proof is to try something. Deduce anything, no matter how simple it seems, from *A*. Then extract something from this new thought. You should expect not to see how *B* follows from *A* at the start of trying to prove that *A* implies *B*, save in rare instances. You should hope only to get a start and to carry it a bit further. If you finally get even a tenuous train of thought from *A* to *B*, it usually becomes clear how to simplify and certify it beyond all doubt. This stage is often accompanied by the feeling that you were a fool not to have seen this simple line of reasoning in the first place.

There is something else to the art of mathematical proof that is difficult to convey though it seems to present real difficulty to some who try for the first time to construct a proof in mathematics. There is a kind of mindlessness that one must have about the matter at hand. That is to say, you are given *A* and somehow, even though it seems contrary to what you first thought about *A*, *A* does imply (say) *C* by some (usually) very simple deduction. So you accept *C* as a consequence of *A* even though it doesn't seem quite "reasonable" (intuitively reasonable) but because it follows logically from this little manipulation, or whatever, that you came up with. The thing is not to let this bother you but to go on from there. Use *C* and

[4] The essay by Poincaré in J. Hadamard: *The Psychology of Invention in the Mathematical Field* (Princeton: Princeton University Press, 1945) is of interest regarding inspirational proofs.

deduce something from it. The chances are that, if you eventually conclude *B*, *C* will, in retrospect, seem immensely reasonable.

This essay is at a point where all meaning may be lost on the reader! Perhaps it is the nature of the subject. The whole point of the last few paragraphs is this: in trying to show *A* implies *B*, dredge up anything you can think of that seems to form a chain of thought from *A* to *B*, even if this chain is rusty and obscure, with odd-size links twisted and half broken, perhaps with extra pieces hanging on but going nowhere. This piece of mental debris is all that you can sometimes get at first: it is an *idea* for a proof.

This brings on phase two. One must clean and polish the links, straighten and join the broken ones or replace them if necessary, getting rid of hangers-on. This is difficult work. It can be as demanding as the original dredging operation. It is no small matter to polish up a proof, for this is the stage in which the proof is put into communicable form understandable and believable by anyone with the relevant technical knowledge at his command. It is also at this stage that the whole proof is most likely to break down, where as a link is being polished it instead crumbles into useless pieces, the cohesive bond of logic having been broken by some small essential flaw unnoticed in the rough. This is why mathematicians are so careful and precise in the way a proof is given, such flaws are not easy to spot. This is why the demand is made that each successive step in a proof follows as an immediate—immediate—consequence of some previous deduction.

The above very general remarks are intended to give you some idea of what processes a mathematician goes through in finding a proof that *A* implies *B* and some idea of the attendant difficulties. These remarks should also tell you that this is not a well defined process, that the finished proof that appears is only the proverbial tip of an iceberg of thought. It is the usual thing that a research mathematician works from several weeks to several months to construct the proof of an essential theorem; the finished proof may only involve a few pages or paragraphs of material. In the initial stages of such work, matters are usually a blur of hazy notions and intuitive feelings. Be assured that it is extremely rare to begin with *A* and, with the first attempted sequence of deductions from *A*, to arrive at *B*. But you might never tell it from the appearance of a finished proof.

Here are some specific directions to follow in attempting to find a proof of a theorem. The first thing is to determine exactly what all the technical words used in the statement of the theorem "*A* implies *B*" mean. For some reason the beginner in the art of mathematical proof often seems unwilling to go through this essential, initial step. This hang-up appears to involve an unwareness of the status of definitions in mathematics. A definition of a technical term is ***not*** a statement intended to give you some idea or sense of what the term is! (*Note*: not.) The statements, if any, which are intended to give you some intuitive feeling for the term being defined usually precede or follow the actual definition and are usually accompanied by examples of things which have the properties stated in the definition. If there is a directive that can always be followed in proving theorems it is this: the *definition* of a term is a *rule for identifying* the thing defined no matter what its abstract appearance.

By way of explanation of this important though thoroughly reasonable point, suppose that a *confusion* space is defined to be one which is both recondite and impalpable and an *amphigoric* space is one which is defined to be either fatuous or sheer drivel (please do not attach any meaning to these words, this is an example). Suppose further that one wishes to show that if X is a confusion space then X is amphigoric. The only way to do this is to argue that *any* X which is both recondite and impalpable is either fatuous or sheer drivel (or, of course, some logical equivalent of this statement). Moreover, the use of any other property of X other than that it is recondite and impalpable could never lead to a valid proof. Finally, and this is really the key point, a rule for showing that X is an amphigoric space (i.e., a directive for proving the theorem) is given by the definition of an amphigoric space—X must be shown to be fatuous or sheer drivel. If you have found yourself wanting to look up the meaning of these words in order to understand this paragraph, then you have misunderstood the point. You must find meaning in the paragraph without looking up these words!

Supposing then that this point is accepted and that the technical terms are understood, their definition precisely known. At this stage it is often helpful to state the matter of proving that A implies B as a problem. That is, one should complete the sentence: "It must be shown . . . given that . . . ," using the definitions of the terms involved. This statement is then the problem posed by the theorem. It is this that must be solved if the theorem is to be proven. But it should also be noted that it can sometimes be difficult to decide upon the real problem posed by a theorem in a form which has meaning to one who would solve it; usually once this is done, one begins to believe that it can be solved.

Once one has involved his mind with the substance of the problem posed by the theorem in a form which has meaning to him, the next reasonable step is to cast around for the crux of the difficulty in solving the problem. This is not necessarily a rational process and, as indicated earlier, one is *free to think about this in any way that seems appropriate*; intuition, a feeling for what is going on, is often a powerful tool at this stage. Once the crux of the problem has been settled on and solved, then a rusty chain of reasoning has been substantially exposed and one is at the point touched on earlier. And as noted earlier, it is here that matters become demanding again, but in a different sense. One has a chain of reasons that A implies B. This chain must be expressed in a communicable form before it will be admitted to be a proof by anyone else. Learning how to do this is not especially straightforward. The only general rule is that the proof be a precise explanation of why A implies B and that one take the most special care *not to be misunderstood.* Thus you should write up your proof as though you were *explaining why A* implies B. It is not a bad idea to imagine that you are writing this explanation for a veritable moron who is capable of reading only the definitions of the terms involved and who somehow knows the rules of logic. Do not write it up as though your instructor is the only soul to read it. This attitude makes things initially tedious, but in time it will pay off, as you will learn how to express mathematical implications more concisely and come to know just what is appropriate for a good proof.

Intimately involved in all this is the need for, and the use of, notation. Again, this is no small hang-up. Although the use of notation is learned only over a period of time, through experience and by imitation of how others use notation, a few general comments are possible. The first thing to keep in mind is that notation does not exist only in or of itself; notation is a substitute for a concept, an idea and, more usually, a complex of ideas. That is why it is useful, for good notation which can be altered and manipulated by precise rules allows one to mindlessly transform one complex of ideas, expressed through notation, into another complex of ideas expressed by the resulting notation. The thing about this is that one can do this without having to involve one's mind with the whole complex of ideas involved. Thus with good notation one can be rather simple minded about difficult concepts. An exceedingly good example of this is the limit concept and its notation. This concept has been around in rough form since the time of Aristotle but it was not precisely formulated until the early 19th century. Not until this was done and a good notation was developed was it possible to mindlessly manipulate this very difficult concept and obtain consistently accurate results.

Althought you must admit that it is a nice thing to be able to avoid difficult conceptual thinking, there is a service charge that goes with it and it has two parts. The first is that one always faces the difficulty of choosing appropriate notation that is subject to precise formulation and manipulation and which, in itself, incorporates the relevant complex of ideas. To learn to do this takes some time and, from time-to-time in the text more specific examples of this problem are considered. The second cost is in the loss of intuition and feeling for the complex of ideas that has been condensed into a concise bit of notation. Perhaps, it is more appropriate to say that this cost is the loss to the study of mathematics of those persons whose minds cannot think through the device of abstract notation—it is only a device, though a fantastically useful one. The individual who can surmount this loss of intuition and substitute for it an ability to reason with notation is then likely to enjoy mathematics as well as make use of it. Abstract notation is a most useful device for avoiding difficult conceptual thought.

There are a couple of specific practical points regarding notation that can be mentioned in this general discussion. Suppose that you are trying to prove some theorem about (say) a given set of numbers. Suppose you get the idea that a method of proof can be found by utilizing what you feel is likely to be true about what you are thinking of as the least number in this set. At some stage, in expressing the proof you hope to obtain, it will be convenient to say: consider the least number in this set and denote it by L. Now there are two things that you have done here: (1) you have introduced a symbol, L, to take the place of the idea "the least number in the set" and (2) you have assumed that there is a least number in the set. Before (1) would be accepted for use in your proof, your audience should demand that you convince them that (2) is valid—that such a number exists. Once that is done, the notation L will be accepted as an expression for a particular thing that does exist. The point is that notation is only a representation of something that is already present and is not a device for sneaking into existence things that are not there, only a tool for discussion of what is there.

The second specific point regarding notation is its subsequent use in a discussion (proof) after it has been introduced. The key here is that the notation has only the properties given it in its definition. Thus the properties of the number L mentioned above are that: (a) L is a number in the set (and in consequence has whatever properties given about numbers in this set) and (b) L is the least number in the set. These are the only properties that L possesses. Only these properties of L, or properties deducted from these, can be used in using L to prove whatever you intend to prove. To give L any other properties would be to make an unwarranted assumption and in effect introduce an additional hypothesis.

There is one final point to be covered in this discussion. What if you have tried to prove a theorem and have begun to suspect that it isn't even true? How do you determine if that expectation is indeed the correct one? Here abstraction can be cast away. To show that A does not imply B, it suffices to find one particular instance in which A holds but B does not. Such an instance is referred to as a *counter example*. It can, and should, be as specific as you can make it. Thus the conjecture that: if $a^2 = b^2$, then $a = b$, can be shown to be false by the single example: $a = 1, b = -1$.

It was once remarked by a well known mathematician that a proof is not understood until it is seen as a single idea. What is meant by this is that once a proof is obtained and understood, he who has found it sees it as a kernel of thought which he may not even be able to express as such in a verbal way, but which lies at the back of his mind and guides him through his explanation, his proof, of why the theorem is valid. This is a good criterion to apply to your own attempts at mathematical proof. It may happen that you have a valid proof without seeding it as a single idea, but if you can see it as such, you will undoubtedly feel more certain of its validity.

Does this mean that after all this discussion about precision, about precise logic, it is yet possible to have a proof and not be absolutely certain of it, or worse to believe one has a proof that in fact has a hole in it? Yes. Good mathematicians have had mistakes and errors pointed out in their work as it went through the process of review and publication, and they in turn have found errors and invalid theorems in the published work of others, and this is not a unique experience. Mathematics is a demanding art; it is not certain, but it can be made more certain, than many other things.

In the following, the assumption is made that you are familiar with the standard definitions of continuity and differentiability as studied in calculus. The idea is to indicate by example that there is a lot more to a proof than its final appearance. Consider proving the

> THEOREM. If a function is differentiable on an interval (a, b) then it is continuous on (a, b).

which is of the form A implies B.

The first thing is to decide what must be shown. It is helpful, even at this early stage, to have some notation which is not present in the statement of the theorem. Let f be a function defined on the interval (a, b).

We are to show that if f is differentiable on (a, b), then f is continuous on (a, b). What does this mean? To say that f is continuous on (a, b), means by the definition of continuity, that if $a < c < b$, then $\lim_{x \to c} f(x)$ exists and equals $f(c)$. To say that f is differentiable at c means that

$$\lim_{x \to c} \frac{f(x) - f(c)}{x - c}$$

exists. This limit is usually denoted by $f'(c)$. Thus it must be shown that: If

$$\lim_{x \to c} \frac{f(x) - f(c)}{x - c}$$

exists, then, $\lim_{x \to c} f(x)$ exists and equals $f(c)$—i.e., $\lim_{x \to c} f(x) = f(c)$.

This is the point noted earlier; this is the problem posed by the theorem stated in terminology that can be used or manipulated. It is a significant step forward.

At this point one need not be logical about the matter. One is given:

$$\lim_{x \to c} \frac{f(x) - f(c)}{x - c} = f'(c)$$

and one needs $\lim_{x \to c} f(x) = f(c)$.

The crux of the matter seems to be in finding a relationship between these two expressions.

After awhile perhaps, one notices that

$$\lim_{x \to c} f(x) = f(c)$$

is the same thing as

$$\lim_{x \to c} [f(x) - f(c)] = 0\,,$$

because of previously verified theorems on limits.

Another small step has been made. The question then is how to relate

$$\lim_{x \to c} [f(x) - f(c)] = 0$$

to

$$\lim_{x \to c} \frac{f(x) - f(c)}{x - c} = f'(c)\,.$$

Since it usually helps to try something, notice that for $x \neq c$,

$$f(x) - f(c) = \frac{f(x) - f(c)}{x - c} \cdot (x - c)\,.$$

There, at least, is a relationship. Try it for what it's worth. Applying $\lim_{x \to c}$ to both sides,

$$(1)\quad \lim_{x \to c} f(x) - f(c) = \lim_{x \to c} \frac{f(x) - f(c)}{x - c} \cdot (x - c)$$

$$(2)\qquad = \lim_{x \to c} \frac{f(x) - f(c)}{x - c} \cdot \lim_{x \to c} (x - c)$$

$$(3)\qquad = f'(c) \cdot 0 = 0$$

It works! That is, it works if all the steps are okay. Here is where the polishing up comes in. After all, the transformation from (1) to (2) is questionable, since, for example, if one is to imprecise about things of this sort one can get the following;

$$0 = \lim_{x \to 0+} x^2\left(\frac{1}{x}\right) = \lim_{x \to 0+} x^2 \lim_{x \to 0+} \frac{1}{x} = 0 \cdot \infty$$

while

$$\infty = \lim_{x \to 0+} x\left(\frac{1}{x^2}\right) = \lim_{x \to 0+} x \lim_{x \to 0+} \frac{1}{x^2} = 0 \cdot \infty$$

which indicates that one can't mindlessly manipulate formulas unless certain conditions are present.

A polished version of the above tenuous chain of reasoning—a complete proof—might appear as follows.

Proof. Let c be a number such that $a < c < b$. Suppose f is differentiable at c. We will show f is continuous at c. Hence we must show $\lim_{x \to c} f(x)$ exists and equals $f(c)$. For $x \neq c$, define the function h by the formula

$$h(x) = \frac{f(x) - f(c)}{x - c}$$

Define the function g by the formula

$$g(x) = f(x) - f(c) .$$

Then for all $x \neq c$,

$$h(x)(x - c) = g(x) .$$

Hence, by a previous theorem,

$$\lim_{x \to c} h(x)(x - c) = \lim_{x \to c} g(x)$$

if either of these limits exist. Since f is differentiable,

$$\lim_{x \to c} \frac{f(x) - f(c)}{x - c} = \lim_{x \to c} h(x)$$

exists and equals the number $f'(c)$. Since $\lim_{x \to c} (x - c)$ exists and is 0, then, by a previous limit theorem, $\lim_{x \to c} h(x) \cdot (x - c)$ exists and

$$\begin{aligned} \lim_{x \to c} h(x)(x - c) &= [\lim_{x \to c} h(x)][\lim_{x \to c} (x - c)] \\ &= f'(c) \cdot 0 = 0 . \end{aligned}$$

Thus, by our earlier equation and the definition of g,

$$0 = \lim_{x \to c} g(x) = \lim_{x \to c} (f(x) - f(c)) .$$

Since $f(c)$ is a constant, then by a previous theorem this means

$$0 = [\lim_{x \to c} f(x)] - f(c)$$

or

$$\lim_{x \to c} f(x) = f(c) .$$

This is what was to be shown. Q.E.D.

This example indicates the considerable difference between the rough idea for a proof and the completed proof. Note the introduction of the auxiliary notation, $h(x)$ and $g(x)$. These are like catalysts in a chemical reaction; they help things along by making the explanation of the reasoning easier to state. Notice also that at the beginning the problem posed by the theorem was stated exactly: it must be shown that $\lim_{x \to c} f(x)$ exists and equals $f(c)$. Thus, one has a clear idea at least of what one is looking for.

This text gives you many chances to try your thinking in the field of abstract mathematical thought. It is no easy thing for most at first, but it is like most things. Success and confidence breed interest and enjoyment. It is for you to make as much or as little as you wish of it. The hope is that this essay will be of some help, the expectation is that its aid will actually be minimal. If you will return to read parts of it from time-to-time as you progress through this text it may become more and more meaningful to you.

NOTATIONAL INDEX

SUBJECT INDEX

A CATALOG OF SELECTED

DOVER BOOKS

IN SCIENCE AND MATHEMATICS

A SURVEY OF INDUSTRIAL MATHEMATICS, Charles R. MacCluer. Students learn how to solve problems they'll encounter in their professional lives with this concise single-volume treatment. It employs MATLAB and other strategies to explore typical industrial problems. 2000 edition. 384pp. 5 3/8 x 8 1/2. 0-486-47702-9

NUMBER SYSTEMS AND THE FOUNDATIONS OF ANALYSIS, Elliott Mendelson. Geared toward undergraduate and beginning graduate students, this study explores natural numbers, integers, rational numbers, real numbers, and complex numbers. Numerous exercises and appendixes supplement the text. 1973 edition. 368pp. 5 3/8 x 8 1/2. 0-486-45792-3

A FIRST LOOK AT NUMERICAL FUNCTIONAL ANALYSIS, W. W. Sawyer. Text by renowned educator shows how problems in numerical analysis lead to concepts of functional analysis. Topics include Banach and Hilbert spaces, contraction mappings, convergence, differentiation and integration, and Euclidean space. 1978 edition. 208pp. 5 3/8 x 8 1/2. 0-486-47882-3

FRACTALS, CHAOS, POWER LAWS: Minutes from an Infinite Paradise, Manfred Schroeder. A fascinating exploration of the connections between chaos theory, physics, biology, and mathematics, this book abounds in award-winning computer graphics, optical illusions, and games that clarify memorable insights into self-similarity. 1992 edition. 448pp. 6 1/8 x 9 1/4. 0-486-47204-3

SET THEORY AND THE CONTINUUM PROBLEM, Raymond M. Smullyan and Melvin Fitting. A lucid, elegant, and complete survey of set theory, this three-part treatment explores axiomatic set theory, the consistency of the continuum hypothesis, and forcing and independence results. 1996 edition. 336pp. 6 x 9. 0-486-47484-4

DYNAMICAL SYSTEMS, Shlomo Sternberg. A pioneer in the field of dynamical systems discusses one-dimensional dynamics, differential equations, random walks, iterated function systems, symbolic dynamics, and Markov chains. Supplementary materials include PowerPoint slides and MATLAB exercises. 2010 edition. 272pp. 6 1/8 x 9 1/4. 0-486-47705-3

ORDINARY DIFFERENTIAL EQUATIONS, Morris Tenenbaum and Harry Pollard. Skillfully organized introductory text examines origin of differential equations, then defines basic terms and outlines general solution of a differential equation. Explores integrating factors; dilution and accretion problems; Laplace Transforms; Newton's Interpolation Formulas, more. 818pp. 5 3/8 x 8 1/2. 0-486-64940-7

MATROID THEORY, D. J. A. Welsh. Text by a noted expert describes standard examples and investigation results, using elementary proofs to develop basic matroid properties before advancing to a more sophisticated treatment. Includes numerous exercises. 1976 edition. 448pp. 5 3/8 x 8 1/2. 0-486-47439-9

THE CONCEPT OF A RIEMANN SURFACE, Hermann Weyl. This classic on the general history of functions combines function theory and geometry, forming the basis of the modern approach to analysis, geometry, and topology. 1955 edition. 208pp. 5 3/8 x 8 1/2. 0-486-47004-0

THE LAPLACE TRANSFORM, David Vernon Widder. This volume focuses on the Laplace and Stieltjes transforms, offering a highly theoretical treatment. Topics include fundamental formulas, the moment problem, monotonic functions, and Tauberian theorems. 1941 edition. 416pp. 5 3/8 x 8 1/2. 0-486-47755-X

Mathematics–Logic and Problem Solving

PERPLEXING PUZZLES AND TANTALIZING TEASERS, Martin Gardner. Ninety-three riddles, mazes, illusions, tricky questions, word and picture puzzles, and other challenges offer hours of entertainment for youngsters. Filled with rib-tickling drawings. Solutions. 224pp. 5 3/8 x 8 1/2. 0-486-25637-5

MY BEST MATHEMATICAL AND LOGIC PUZZLES, Martin Gardner. The noted expert selects 70 of his favorite "short" puzzles. Includes The Returning Explorer, The Mutilated Chessboard, Scrambled Box Tops, and dozens more. Complete solutions included. 96pp. 5 3/8 x 8 1/2. 0-486-28152-3

THE LADY OR THE TIGER?: and Other Logic Puzzles, Raymond M. Smullyan. Created by a renowned puzzle master, these whimsically themed challenges involve paradoxes about probability, time, and change; metapuzzles; and self-referentiality. Nineteen chapters advance in difficulty from relatively simple to highly complex. 1982 edition. 240pp. 5 3/8 x 8 1/2. 0-486-47027-X

SATAN, CANTOR AND INFINITY: Mind-Boggling Puzzles, Raymond M. Smullyan. A renowned mathematician tells stories of knights and knaves in an entertaining look at the logical precepts behind infinity, probability, time, and change. Requires a strong background in mathematics. Complete solutions. 288pp. 5 3/8 x 8 1/2. 0-486-47036-9

THE RED BOOK OF MATHEMATICAL PROBLEMS, Kenneth S. Williams and Kenneth Hardy. Handy compilation of 100 practice problems, hints and solutions indispensable for students preparing for the William Lowell Putnam and other mathematical competitions. Preface to the First Edition. Sources. 1988 edition. 192pp. 5 3/8 x 8 1/2. 0-486-69415-1

KING ARTHUR IN SEARCH OF HIS DOG AND OTHER CURIOUS PUZZLES, Raymond M. Smullyan. This fanciful, original collection for readers of all ages features arithmetic puzzles, logic problems related to crime detection, and logic and arithmetic puzzles involving King Arthur and his Dogs of the Round Table. 160pp. 5 3/8 x 8 1/2. 0-486-47435-6

UNDECIDABLE THEORIES: Studies in Logic and the Foundation of Mathematics, Alfred Tarski in collaboration with Andrzej Mostowski and Raphael M. Robinson. This well-known book by the famed logician consists of three treatises: "A General Method in Proofs of Undecidability," "Undecidability and Essential Undecidability in Mathematics," and "Undecidability of the Elementary Theory of Groups." 1953 edition. 112pp. 5 3/8 x 8 1/2. 0-486-47703-7

LOGIC FOR MATHEMATICIANS, J. Barkley Rosser. Examination of essential topics and theorems assumes no background in logic. "Undoubtedly a major addition to the literature of mathematical logic." – *Bulletin of the American Mathematical Society.* 1978 edition. 592pp. 6 1/8 x 9 1/4. 0-486-46898-4

INTRODUCTION TO PROOF IN ABSTRACT MATHEMATICS, Andrew Wohlgemuth. This undergraduate text teaches students what constitutes an acceptable proof, and it develops their ability to do proofs of routine problems as well as those requiring creative insights. 1990 edition. 384pp. 6 1/2 x 9 1/4. 0-486-47854-8

FIRST COURSE IN MATHEMATICAL LOGIC, Patrick Suppes and Shirley Hill. Rigorous introduction is simple enough in presentation and context for wide range of students. Symbolizing sentences; logical inference; truth and validity; truth tables; terms, predicates, universal quantifiers; universal specification and laws of identity; more. 288pp. 5 3/8 x 8 1/2. 0-486-42259-3

Mathematics–Algebra and Calculus

VECTOR CALCULUS, Peter Baxandall and Hans Liebeck. This introductory text offers a rigorous, comprehensive treatment. Classical theorems of vector calculus are amply illustrated with figures, worked examples, physical applications, and exercises with hints and answers. 1986 edition. 560pp. 5 3/8 x 8 1/2. 0-486-46620-5

ADVANCED CALCULUS: An Introduction to Classical Analysis, Louis Brand. A course in analysis that focuses on the functions of a real variable, this text introduces the basic concepts in their simplest setting and illustrates its teachings with numerous examples, theorems, and proofs. 1955 edition. 592pp. 5 3/8 x 8 1/2. 0-486-44548-8

ADVANCED CALCULUS, Avner Friedman. Intended for students who have already completed a one-year course in elementary calculus, this two-part treatment advances from functions of one variable to those of several variables. Solutions. 1971 edition. 432pp. 5 3/8 x 8 1/2. 0-486-45795-8

METHODS OF MATHEMATICS APPLIED TO CALCULUS, PROBABILITY, AND STATISTICS, Richard W. Hamming. This 4-part treatment begins with algebra and analytic geometry and proceeds to an exploration of the calculus of algebraic functions and transcendental functions and applications. 1985 edition. Includes 310 figures and 18 tables. 880pp. 6 1/2 x 9 1/4. 0-486-43945-3

BASIC ALGEBRA I: Second Edition, Nathan Jacobson. A classic text and standard reference for a generation, this volume covers all undergraduate algebra topics, including groups, rings, modules, Galois theory, polynomials, linear algebra, and associative algebra. 1985 edition. 528pp. 6 1/8 x 9 1/4. 0-486-47189-6

BASIC ALGEBRA II: Second Edition, Nathan Jacobson. This classic text and standard reference comprises all subjects of a first-year graduate-level course, including in-depth coverage of groups and polynomials and extensive use of categories and functors. 1989 edition. 704pp. 6 1/8 x 9 1/4. 0-486-47187-X

CALCULUS: An Intuitive and Physical Approach (Second Edition), Morris Kline. Application-oriented introduction relates the subject as closely as possible to science with explorations of the derivative; differentiation and integration of the powers of x; theorems on differentiation, antidifferentiation; the chain rule; trigonometric functions; more. Examples. 1967 edition. 960pp. 6 1/2 x 9 1/4. 0-486-40453-6

ABSTRACT ALGEBRA AND SOLUTION BY RADICALS, John E. Maxfield and Margaret W. Maxfield. Accessible advanced undergraduate-level text starts with groups, rings, fields, and polynomials and advances to Galois theory, radicals and roots of unity, and solution by radicals. Numerous examples, illustrations, exercises, appendixes. 1971 edition. 224pp. 6 1/8 x 9 1/4. 0-486-47723-1

AN INTRODUCTION TO THE THEORY OF LINEAR SPACES, Georgi E. Shilov. Translated by Richard A. Silverman. Introductory treatment offers a clear exposition of algebra, geometry, and analysis as parts of an integrated whole rather than separate subjects. Numerous examples illustrate many different fields, and problems include hints or answers. 1961 edition. 320pp. 5 3/8 x 8 1/2. 0-486-63070-6

LINEAR ALGEBRA, Georgi E. Shilov. Covers determinants, linear spaces, systems of linear equations, linear functions of a vector argument, coordinate transformations, the canonical form of the matrix of a linear operator, bilinear and quadratic forms, and more. 387pp. 5 3/8 x 8 1/2. 0-486-63518-X

Mathematics-Probability and Statistics

BASIC PROBABILITY THEORY, Robert B. Ash. This text emphasizes the probabilistic way of thinking, rather than measure-theoretic concepts. Geared toward advanced undergraduates and graduate students, it features solutions to some of the problems. 1970 edition. 352pp. 5 3/8 x 8 1/2. 0-486-46628-0

PRINCIPLES OF STATISTICS, M. G. Bulmer. Concise description of classical statistics, from basic dice probabilities to modern regression analysis. Equal stress on theory and applications. Moderate difficulty; only basic calculus required. Includes problems with answers. 252pp. 5 5/8 x 8 1/4. 0-486-63760-3

OUTLINE OF BASIC STATISTICS: Dictionary and Formulas, John E. Freund and Frank J. Williams. Handy guide includes a 70-page outline of essential statistical formulas covering grouped and ungrouped data, finite populations, probability, and more, plus over 1,000 clear, concise definitions of statistical terms. 1966 edition. 208pp. 5 3/8 x 8 1/2. 0-486-47769-X

GOOD THINKING: The Foundations of Probability and Its Applications, Irving J. Good. This in-depth treatment of probability theory by a famous British statistician explores Keynesian principles and surveys such topics as Bayesian rationality, corroboration, hypothesis testing, and mathematical tools for induction and simplicity. 1983 edition. 352pp. 5 3/8 x 8 1/2. 0-486-47438-0

INTRODUCTION TO PROBABILITY THEORY WITH CONTEMPORARY APPLICATIONS, Lester L. Helms. Extensive discussions and clear examples, written in plain language, expose students to the rules and methods of probability. Exercises foster problem-solving skills, and all problems feature step-by-step solutions. 1997 edition. 368pp. 6 1/2 x 9 1/4. 0-486-47418-6

CHANCE, LUCK, AND STATISTICS, Horace C. Levinson. In simple, non-technical language, this volume explores the fundamentals governing chance and applies them to sports, government, and business. "Clear and lively ... remarkably accurate." – *Scientific Monthly.* 384pp. 5 3/8 x 8 1/2. 0-486-41997-5

FIFTY CHALLENGING PROBLEMS IN PROBABILITY WITH SOLUTIONS, Frederick Mosteller. Remarkable puzzlers, graded in difficulty, illustrate elementary and advanced aspects of probability. These problems were selected for originality, general interest, or because they demonstrate valuable techniques. Also includes detailed solutions. 88pp. 5 3/8 x 8 1/2. 0-486-65355-2

EXPERIMENTAL STATISTICS, Mary Gibbons Natrella. A handbook for those seeking engineering information and quantitative data for designing, developing, constructing, and testing equipment. Covers the planning of experiments, the analyzing of extreme-value data; and more. 1966 edition. Index. Includes 52 figures and 76 tables. 560pp. 8 3/8 x 11. 0-486-43937-2

STOCHASTIC MODELING: Analysis and Simulation, Barry L. Nelson. Coherent introduction to techniques also offers a guide to the mathematical, numerical, and simulation tools of systems analysis. Includes formulation of models, analysis, and interpretation of results. 1995 edition. 336pp. 6 1/8 x 9 1/4. 0-486-47770-3

INTRODUCTION TO BIOSTATISTICS: Second Edition, Robert R. Sokal and F. James Rohlf. Suitable for undergraduates with a minimal background in mathematics, this introduction ranges from descriptive statistics to fundamental distributions and the testing of hypotheses. Includes numerous worked-out problems and examples. 1987 edition. 384pp. 6 1/8 x 9 1/4. 0-486-46961-1

Mathematics–Geometry and Topology

PROBLEMS AND SOLUTIONS IN EUCLIDEAN GEOMETRY, M. N. Aref and William Wernick. Based on classical principles, this book is intended for a second course in Euclidean geometry and can be used as a refresher. More than 200 problems include hints and solutions. 1968 edition. 272pp. 5 3/8 x 8 1/2. 0-486-47720-7

TOPOLOGY OF 3-MANIFOLDS AND RELATED TOPICS, Edited by M. K. Fort, Jr. With a New Introduction by Daniel Silver. Summaries and full reports from a 1961 conference discuss decompositions and subsets of 3-space; n-manifolds; knot theory; the Poincaré conjecture; and periodic maps and isotopies. Familiarity with algebraic topology required. 1962 edition. 272pp. 6 1/8 x 9 1/4. 0-486-47753-3

POINT SET TOPOLOGY, Steven A. Gaal. Suitable for a complete course in topology, this text also functions as a self-contained treatment for independent study. Additional enrichment materials make it equally valuable as a reference. 1964 edition. 336pp. 5 3/8 x 8 1/2. 0-486-47222-1

INVITATION TO GEOMETRY, Z. A. Melzak. Intended for students of many different backgrounds with only a modest knowledge of mathematics, this text features self-contained chapters that can be adapted to several types of geometry courses. 1983 edition. 240pp. 5 3/8 x 8 1/2. 0-486-46626-4

TOPOLOGY AND GEOMETRY FOR PHYSICISTS, Charles Nash and Siddhartha Sen. Written by physicists for physics students, this text assumes no detailed background in topology or geometry. Topics include differential forms, homotopy, homology, cohomology, fiber bundles, connection and covariant derivatives, and Morse theory. 1983 edition. 320pp. 5 3/8 x 8 1/2. 0-486-47852-1

BEYOND GEOMETRY: Classic Papers from Riemann to Einstein, Edited with an Introduction and Notes by Peter Pesic. This is the only English-language collection of these 8 accessible essays. They trace seminal ideas about the foundations of geometry that led to Einstein's general theory of relativity. 224pp. 6 1/8 x 9 1/4. 0-486-45350-2

GEOMETRY FROM EUCLID TO KNOTS, Saul Stahl. This text provides a historical perspective on plane geometry and covers non-neutral Euclidean geometry, circles and regular polygons, projective geometry, symmetries, inversions, informal topology, and more. Includes 1,000 practice problems. Solutions available. 2003 edition. 480pp. 6 1/8 x 9 1/4. 0-486-47459-3

TOPOLOGICAL VECTOR SPACES, DISTRIBUTIONS AND KERNELS, François Trèves. Extending beyond the boundaries of Hilbert and Banach space theory, this text focuses on key aspects of functional analysis, particularly in regard to solving partial differential equations. 1967 edition. 592pp. 5 3/8 x 8 1/2. 0-486-45352-9

INTRODUCTION TO PROJECTIVE GEOMETRY, C. R. Wylie, Jr. This introductory volume offers strong reinforcement for its teachings, with detailed examples and numerous theorems, proofs, and exercises, plus complete answers to all odd-numbered end-of-chapter problems. 1970 edition. 576pp. 6 1/8 x 9 1/4. 0-486-46895-X

FOUNDATIONS OF GEOMETRY, C. R. Wylie, Jr. Geared toward students preparing to teach high school mathematics, this text explores the principles of Euclidean and non-Euclidean geometry and covers both generalities and specifics of the axiomatic method. 1964 edition. 352pp. 6 x 9. 0-486-47214-0

Mathematics-History

THE WORKS OF ARCHIMEDES, Archimedes. Translated by Sir Thomas Heath. Complete works of ancient geometer feature such topics as the famous problems of the ratio of the areas of a cylinder and an inscribed sphere; the properties of conoids, spheroids, and spirals; more. 326pp. 5 3/8 x 8 1/2. 0-486-42084-1

THE HISTORICAL ROOTS OF ELEMENTARY MATHEMATICS, Lucas N. H. Bunt, Phillip S. Jones, and Jack D. Bedient. Exciting, hands-on approach to understanding fundamental underpinnings of modern arithmetic, algebra, geometry and number systems examines their origins in early Egyptian, Babylonian, and Greek sources. 336pp. 5 3/8 x 8 1/2. 0-486-25563-8

THE THIRTEEN BOOKS OF EUCLID'S ELEMENTS, Euclid. Contains complete English text of all 13 books of the Elements plus critical apparatus analyzing each definition, postulate, and proposition in great detail. Covers textual and linguistic matters; mathematical analyses of Euclid's ideas; classical, medieval, Renaissance and modern commentators; refutations, supports, extrapolations, reinterpretations and historical notes. 995 figures. Total of 1,425pp. All books 5 3/8 x 8 1/2.
Vol. I: 443pp. 0-486-60088-2
Vol. II: 464pp. 0-486-60089-0
Vol. III: 546pp. 0-486-60090-4

A HISTORY OF GREEK MATHEMATICS, Sir Thomas Heath. This authoritative two-volume set that covers the essentials of mathematics and features every landmark innovation and every important figure, including Euclid, Apollonius, and others. 5 3/8 x 8 1/2.
Vol. I: 461pp. 0-486-24073-8
Vol. II: 597pp. 0-486-24074-6

A MANUAL OF GREEK MATHEMATICS, Sir Thomas L. Heath. This concise but thorough history encompasses the enduring contributions of the ancient Greek mathematicians whose works form the basis of most modern mathematics. Discusses Pythagorean arithmetic, Plato, Euclid, more. 1931 edition. 576pp. 5 3/8 x 8 1/2. 0-486-43231-9

CHINESE MATHEMATICS IN THE THIRTEENTH CENTURY, Ulrich Libbrecht. An exploration of the 13th-century mathematician Ch'in, this fascinating book combines what is known of the mathematician's life with a history of his only extant work, the Shu-shu chiu-chang. 1973 edition. 592pp. 5 3/8 x 8 1/2. 0-486-44619-0

PHILOSOPHY OF MATHEMATICS AND DEDUCTIVE STRUCTURE IN EUCLID'S ELEMENTS, Ian Mueller. This text provides an understanding of the classical Greek conception of mathematics as expressed in Euclid's Elements. It focuses on philosophical, foundational, and logical questions and features helpful appendixes. 400pp. 6 1/2 x 9 1/4. 0-486-45300-6

BEYOND GEOMETRY: Classic Papers from Riemann to Einstein, Edited with an Introduction and Notes by Peter Pesic. This is the only English-language collection of these 8 accessible essays. They trace seminal ideas about the foundations of geometry that led to Einstein's general theory of relativity. 224pp. 6 1/8 x 9 1/4. 0-486-45350-2

HISTORY OF MATHEMATICS, David E. Smith. Two-volume history – from Egyptian papyri and medieval maps to modern graphs and diagrams. Non-technical chronological survey with thousands of biographical notes, critical evaluations, and contemporary opinions on over 1,100 mathematicians. 5 3/8 x 8 1/2.
Vol. I: 618pp. 0-486-20429-4
Vol. II: 736pp. 0-486-20430-8

Engineering

FUNDAMENTALS OF ASTRODYNAMICS, Roger R. Bate, Donald D. Mueller, and Jerry E. White. Teaching text developed by U.S. Air Force Academy develops the basic two-body and n-body equations of motion; orbit determination; classical orbital elements, coordinate transformations; differential correction; more. 1971 edition. 455pp. 5 3/8 x 8 1/2. 0-486-60061-0

INTRODUCTION TO CONTINUUM MECHANICS FOR ENGINEERS: Revised Edition, Ray M. Bowen. This self-contained text introduces classical continuum models within a modern framework. Its numerous exercises illustrate the governing principles, linearizations, and other approximations that constitute classical continuum models. 2007 edition. 320pp. 6 1/8 x 9 1/4. 0-486-47460-7

ENGINEERING MECHANICS FOR STRUCTURES, Louis L. Bucciarelli. This text explores the mechanics of solids and statics as well as the strength of materials and elasticity theory. Its many design exercises encourage creative initiative and systems thinking. 2009 edition. 320pp. 6 1/8 x 9 1/4. 0-486-46855-0

FEEDBACK CONTROL THEORY, John C. Doyle, Bruce A. Francis and Allen R. Tannenbaum. This excellent introduction to feedback control system design offers a theoretical approach that captures the essential issues and can be applied to a wide range of practical problems. 1992 edition. 224pp. 6 1/2 x 9 1/4. 0-486-46933-6

THE FORCES OF MATTER, Michael Faraday. These lectures by a famous inventor offer an easy-to-understand introduction to the interactions of the universe's physical forces. Six essays explore gravitation, cohesion, chemical affinity, heat, magnetism, and electricity. 1993 edition. 96pp. 5 3/8 x 8 1/2. 0-486-47482-8

DYNAMICS, Lawrence E. Goodman and William H. Warner. Beginning engineering text introduces calculus of vectors, particle motion, dynamics of particle systems and plane rigid bodies, technical applications in plane motions, and more. Exercises and answers in every chapter. 619pp. 5 3/8 x 8 1/2. 0-486-42006-X

ADAPTIVE FILTERING PREDICTION AND CONTROL, Graham C. Goodwin and Kwai Sang Sin. This unified survey focuses on linear discrete-time systems and explores natural extensions to nonlinear systems. It emphasizes discrete-time systems, summarizing theoretical and practical aspects of a large class of adaptive algorithms. 1984 edition. 560pp. 6 1/2 x 9 1/4. 0-486-46932-8

INDUCTANCE CALCULATIONS, Frederick W. Grover. This authoritative reference enables the design of virtually every type of inductor. It features a single simple formula for each type of inductor, together with tables containing essential numerical factors. 1946 edition. 304pp. 5 3/8 x 8 1/2. 0-486-47440-2

THERMODYNAMICS: Foundations and Applications, Elias P. Gyftopoulos and Gian Paolo Beretta. Designed by two MIT professors, this authoritative text discusses basic concepts and applications in detail, emphasizing generality, definitions, and logical consistency. More than 300 solved problems cover realistic energy systems and processes. 800pp. 6 1/8 x 9 1/4. 0-486-43932-1

THE FINITE ELEMENT METHOD: Linear Static and Dynamic Finite Element Analysis, Thomas J. R. Hughes. Text for students without in-depth mathematical training, this text includes a comprehensive presentation and analysis of algorithms of time-dependent phenomena plus beam, plate, and shell theories. Solution guide available upon request. 672pp. 6 1/2 x 9 1/4. 0-486-41181-8

Physics

THEORETICAL NUCLEAR PHYSICS, John M. Blatt and Victor F. Weisskopf. An uncommonly clear and cogent investigation and correlation of key aspects of theoretical nuclear physics by leading experts: the nucleus, nuclear forces, nuclear spectroscopy, two-, three- and four-body problems, nuclear reactions, beta-decay and nuclear shell structure. 896pp. 5 3/8 x 8 1/2. 0-486-66827-4

QUANTUM THEORY, David Bohm. This advanced undergraduate-level text presents the quantum theory in terms of qualitative and imaginative concepts, followed by specific applications worked out in mathematical detail. 655pp. 5 3/8 x 8 1/2. 0-486-65969-0

ATOMIC PHYSICS AND HUMAN KNOWLEDGE, Niels Bohr. Articles and speeches by the Nobel Prize–winning physicist, dating from 1934 to 1958, offer philosophical explorations of the relevance of atomic physics to many areas of human endeavor. 1961 edition. 112pp. 5 3/8 x 8 1/2. 0-486-47928-5

COSMOLOGY, Hermann Bondi. A co-developer of the steady-state theory explores his conception of the expanding universe. This historic book was among the first to present cosmology as a separate branch of physics. 1961 edition. 192pp. 5 3/8 x 8 1/2. 0-486-47483-6

LECTURES ON QUANTUM MECHANICS, Paul A. M. Dirac. Four concise, brilliant lectures on mathematical methods in quantum mechanics from Nobel Prize-winning quantum pioneer build on idea of visualizing quantum theory through the use of classical mechanics. 96pp. 5 3/8 x 8 1/2. 0-486-41713-1

THE PRINCIPLE OF RELATIVITY, Albert Einstein and Frances A. Davis. Eleven papers that forged the general and special theories of relativity include seven papers by Einstein, two by Lorentz, and one each by Minkowski and Weyl. 1923 edition. 240pp. 5 3/8 x 8 1/2. 0-486-60081-5

PHYSICS OF WAVES, William C. Elmore and Mark A. Heald. Ideal as a classroom text or for individual study, this unique one-volume overview of classical wave theory covers wave phenomena of acoustics, optics, electromagnetic radiations, and more. 477pp. 5 3/8 x 8 1/2. 0-486-64926-1

THERMODYNAMICS, Enrico Fermi. In this classic of modern science, the Nobel Laureate presents a clear treatment of systems, the First and Second Laws of Thermodynamics, entropy, thermodynamic potentials, and much more. Calculus required. 160pp. 5 3/8 x 8 1/2. 0-486-60361-X

QUANTUM THEORY OF MANY-PARTICLE SYSTEMS, Alexander L. Fetter and John Dirk Walecka. Self-contained treatment of nonrelativistic many-particle systems discusses both formalism and applications in terms of ground-state (zero-temperature) formalism, finite-temperature formalism, canonical transformations, and applications to physical systems. 1971 edition. 640pp. 5 3/8 x 8 1/2. 0-486-42827-3

QUANTUM MECHANICS AND PATH INTEGRALS: Emended Edition, Richard P. Feynman and Albert R. Hibbs. Emended by Daniel F. Styer. The Nobel Prize–winning physicist presents unique insights into his theory and its applications. Feynman starts with fundamentals and advances to the perturbation method, quantum electrodynamics, and statistical mechanics. 1965 edition, emended in 2005. 384pp. 6 1/8 x 9 1/4. 0-486-47722-3